Renewable Energy

Renewable Energy

POWER FOR A SUSTAINABLE FUTURE
FOURTH EDITION

Edited by Stephen Peake

OXFORD
UNIVERSITY PRESS

Published by Oxford University Press, Great Clarendon Street, Oxford OX2 6DP in association with The Open University, Walton Hall, Milton Keynes MK7 6AA.

Oxford University Press is a department of the University of Oxford. It furthers the University's objective of excellence in research, scholarship, and education by publishing worldwide. Oxford is a registered trade mark of Oxford University Press in the UK and in certain other countries.

Published in the United States of America by Oxford University Press
198 Madison Avenue, New York, NY 10016, United States of America

Edited and designed by The Open University.

Typeset by The Open University.

British Library Cataloguing in Publication Data available on request.

Library of Congress Cataloguing in Publication Data available on request.

Printed and bound in the United Kingdom by Bell & Bain Ltd, Glasgow.

This book forms part of Open University teaching materials. Details of these and other Open University modules can be obtained from the Student Recruitment, The Open University, PO Box 197, Milton Keynes MK7 6BJ, United Kingdom (tel. +44 (0)300 303 5303, email general-enquiries@open.ac.uk).

www.open.ac.uk

ISBN 978-0-19-875975-1

4.1

Preface

The transition away from fossil fuels and towards renewable energy technologies is now fully underway. This fourth edition of *Renewable Energy* reflects the remarkable progress that has been made in the field since the publication of the first edition in 1996.

The recent rapid global growth in renewable energy production is no accident. It is the result of three decades of policy-driven research, development, investment, and policy support for multiple renewable energy technologies. It has resulted in crashing prices for large-scale wind and solar PV. So much so in fact, that in this next period, renewable energy will continue to grow robustly and become *the dominant* source of our global energy supply.

A great many challenges still need to be overcome. Helping greater numbers of people access more sustainable supplies of energy is now seen as a critical part of solving some of the world's key global health, climate change and energy security issues. Globally, around 1 billion people still do not have access to reliable, affordable mains electricity. An astonishing 3 billion people (over 40% of the world's population or more than the size of China and India combined) do not have access to modern cooking fuel. Air pollution is a serious and growing problem around the world and remains one of the key drivers of policies to reduce coal consumption in China. Growing populations and incomes will continue to put more pressure on our evolving energy systems.

The renewable energy sources described in this book are essentially carbon-free or low-carbon and appear to be generally more sustainable than fossil or nuclear fuels. However, many technologies are still under development and, at the time of writing, the costs of some remain high.

What happens next in our energy systems is a balancing act performed at two main levels. Firstly, there is the overall balance of investment (and market sentiment) between fossil fuel systems and renewables and other alternatives. Secondly, markets must then decide where and how within the 'alternatives to fossil fuels' market, renewable energy investments should be directed.

If the potential of renewables is to be realized, the world will need many more professional people with a thorough knowledge of renewable energy systems, their underlying physical and technological principles, their economics, their environmental impact and how they can be integrated into the world's energy systems. *This book* is intended to provide a foundation for this knowledge.

Renewable Energy is aimed at students and staff in universities, and at professionals, policymakers and members of the public interested in creating a sustainable energy future.

We hope that *Renewable Energy* will contribute to an improved understanding of renewable energy as a key potential solution to the sustainability problems of our present energy systems. We also hope that it conveys something of the enthusiasm we feel for this complex, fascinating and increasingly important subject.

Stephen Peake
Senior Lecturer in Environmental Technologies, The Open University

New to this edition

■ New editor Stephen Peake, Fellow of the Judge Business School and Senior Lecturer at The Open University, brings a fresh perspective to the current state of renewable energy

■ A new chapter on 'Thermodynamics' explores the physical basis of energy

■ A new chapter on 'Renewable energy futures' looks ahead to potential developments in the field

■ An updated chapter on the status, challenges and opportunities for the integration of renewable energy technologies in rapidly evolving electric grid systems

■ Enhanced pedagogy includes expanded chapter summaries.

About the authors

Godfrey Boyle is Emeritus Professor of Renewable Energy at The Open University

Caspar Donnison is a Postgraduate Researcher in the School of Biological Sciences at the University of Southampton

Les Duckers is Principal Lecturer in Energy and Environment at the School of Energy, Construction and Environment, Coventry University

David Elliott is Emeritus Professor of Technology Policy at The Open University

Bob Everett is a Lecturer in Renewable Technology at The Open University

Astley Hastings is a Senior Research Fellow in the Institute of Biological and Environmental Science at the University of Aberdeen

Mark Knös is Senior Research Engineer Hydrodynamics at the Offshore Renewable Energy Catapult, UK

Kevin Lindegaard is a Director of Crops for Energy Ltd

Ned Minns is the Offshore Group Manager at the environmental consultancy ITPEnergised

Stephen Peake is Senior Lecturer in Environmental Technologies at The Open University

Janet Ramage is a visiting lecturer at The Open University

Jonathan Scurlock is Chief Adviser, Renewable Energy and Climate Change at National Farmers' Union and a visiting research fellow at The Open University

Hazel Smith is a Postdoctoral Researcher in Biological Sciences at the University of Southampton

Derek Taylor is an architect and visiting lecturer at The Open University

James Warren is Senior Lecturer at The Open University.

Contents

Chapter 1

Introducing renewable energy

by Stephen Peake and Bob Everett

1.1 Introduction

Renewable energy sources, derived principally from the enormous power of the Sun's radiation, are at once the most ancient and the most modern forms of energy used by humanity.

Solar power, both in the form of direct solar radiation and in indirect forms such as bioenergy, water or wind power, was the energy source upon which early human societies were based. When our ancestors first used fire, they were harnessing the power of photosynthesis, the solar-driven process by which plants are created from water and atmospheric carbon dioxide. Societies went on to develop ways of harnessing the movements of water and wind, both caused by solar heating of the oceans and atmosphere, to grind corn, irrigate crops and propel ships. As civilizations became more sophisticated, architects began to design buildings to take advantage of the Sun's energy by enhancing their natural use of its heat and light, so reducing the need for artificial sources of warmth and illumination.

Technologies for harnessing the power of Sun, firewood, water and wind continued to improve right up to the early years of the industrial revolution. However, by then the advantages of coal, the first of the fossil fuels to be exploited on a large scale, had become apparent. These highly concentrated energy sources soon displaced wood, wind and water in the homes, industries and transport systems of the industrial nations. Today the fossil fuel trio of coal, oil and natural gas provides just over 80% of the world's energy supply.

Concerns about the adverse environmental and social consequences of fossil fuel use, such as air pollution or mining accidents, and about the finite nature of supplies, have been voiced intermittently for several centuries. But it was not until the 1970s, with the steep price rises of the 'oil crisis' and the advent of the environmental movement, that humanity began to take more seriously the prospect of fossil fuels 'running out'. Since the 1990s the possibility that their continued use could be destabilizing the planet's natural ecosystems and the global climate has become a major concern (see Section 1.3 below).

The development of nuclear energy following the Second World War raised hopes of a cheap, plentiful and clean alternative to fossil fuels. However, nuclear power development has stalled in many countries following a series of major incidents, the most recent being the 2011 Fukushima disaster in Japan, and there are continuing concerns about capital costs and radioactive waste disposal. Nuclear expansion is continuing in some countries, in particular in Russia and China.

Continuing concerns about the 'sustainability' of both fossil and nuclear fuel use have been a major catalyst for the enormous investment in renewable energy sources in recent decades. Ideally, a **sustainable energy source** is one that:

■ is not substantially depleted by continued use

■ does not entail significant pollutant emissions or other environmental problems

■ does not involve the perpetuation of substantial health hazards or social injustices.

In practice, only a few energy sources come close to this ideal, but as this and subsequent chapters will show, the 'renewables' (see Section 1.4 for an explicit definition) appear generally more sustainable than fossil or nuclear fuels: they are essentially inexhaustible and their use usually entails fewer health hazards and much lower emissions of greenhouse gases or other pollutants.

Before going on to introduce the renewables in more detail, it is first useful to review some basic energy concepts that may be unfamiliar to readers who do not have a scientific background. For a more detailed discussion of basic energy concepts see for example, *Energy Systems and Sustainability* (Everett et al., 2012).

Force, energy and power

The word 'energy' is derived from the Greek *en* (in) and *ergon* (work). The scientific concept of **energy** (broadly defined as 'the capacity to do work') serves to reveal the common features in processes as diverse as burning fuels, propelling machines or charging batteries. These and other processes can be described in terms of diverse **forms of energy**, such as *thermal* energy (heat), *chemical* energy (in fuels or batteries), *kinetic* energy (in moving masses), *electrical* energy, *gravitational* potential energy, and various others.

In the main, this book uses the international SI system of units. The conversion factors between these and other units commonly used in the field of energy can be found in Appendix A.

The scientific world agreed on a single set of units, the SI system (Système International d'Unités) in 1960. There are seven basic units, of which the three which are relevant here are the metre (m), the kilogram (kg) and the second (s). The units for many other quantities are derived from the basic units. For example, the unit for speed is metres per second, which can be written as m/s. In this book we have adopted the convention of using negative powers, so it is written as $m\ s^{-1}$.

Other units have been given specific names, such as the:

■ newton (N) for force

■ the joule (J) for energy

■ the watt (W) for power.

Large quantities are specified using multipliers (see Table 1.1 for some examples). Thus, a kilowatt (written as 1 kW) is a thousand watts.

Note that, unless otherwise stated, in this book the multiplier M and the terms billion and trillion are as defined in Table 1.1 (Appendix A has a fuller list of multiplier prefixes and more details about international variations in definition).

Table 1.1 Multiplier prefixes

Symbol	Prefix	Multiply by	... as a power of ten
k	kilo-	one thousand	10^3
M	mega-	one million	10^6
G	giga-	one billion (one thousand million)	10^9
T	tera-	one trillion (one million million)	10^{12}
P	peta-	one quadrillion (one million billion)	10^{15}
E	exa-	one quintillion (one billion billion)	10^{18}

In order to change the motion of any object, a force is needed, and the formal SI unit for force, the **newton (N)**, is defined as that force which will accelerate a mass of one kilogram (kg) at a rate of one metre per second per second (m s^{-2}). Expressed more generally:

force (N) = mass (kg) × acceleration (m s^{-2}).

Thus the derived unit, the newton, is equivalent to kg m s^{-2}.

In the real world, force is often needed to move an object even at a steady speed, but this is because there are opposing forces such as friction to be overcome.

Whenever a force is accelerating something or moving it against an opposing force, it must be providing energy. The unit of energy, the **joule (J)**, is defined as the energy supplied by a force of one newton in causing movement through a distance of 1 metre. In general:

energy (J) = force (N) × distance (m).

So a joule is dimensionally equivalent to one newton metre (N m).

The terms *energy* and *power* are often used informally as though they were synonymous (e.g. wind energy/wind power), but in scientific discussion it is important to distinguish between them. **Power** is the *rate* at which energy is being converted from one form to another, or transferred from one place to another. Its unit is the **watt (W)**, and one watt is defined as one joule per second (J s^{-1}). A 100 watt incandescent light bulb, for example, is converting one hundred joules of electrical energy into light (and 'waste' heat) each second. In popular speech, the terms 'power' and 'energy' are often taken to denote *electricity*, but scientifically they apply to any situation where energy is transferred or converted. Occasionally a power rating maybe specifically defined as MW_e or MW_t where the subscripts e and t refer to electrical and thermal energy respectively.

In practice, it is often convenient to measure energy in terms of the power used over a given time period. If the power of an electric heater is 1 kW, and it runs for an hour, we say that it has consumed one **kilowatt-hour (kWh)** of energy. As 1 kilowatt is 1000 watts, from the definition of the watt this is 1000 joules per second. There are 3600 seconds in an hour, so:

$$1 \text{ kWh} = 3600 \times 1000 = 3.6 \times 10^6 \text{ joules (3.6 MJ)}.$$

Energy is also often measured simply in terms of quantities of fuel, and national energy statistics often use the units 'tonnes of coal equivalent' (tce), 'tonnes of oil equivalent' (toe) or even 'barrels of oil equivalent' (boe). The most common units and their conversion factors are listed in Appendix A.

Energy conservation: the First Law of Thermodynamics

The renewable energy technologies described in this book transform one form of energy into another (the final form in many cases being electricity). In any such transformation of energy, the total quantity of energy remains unchanged. This principle, that energy is always conserved, is expressed by the **First Law of Thermodynamics**. So if the electrical energy *output* of a power station, for example, is less than the energy content of the fuel *input*, then some of the energy must have been converted to another form (usually waste heat).

If the total quantity of energy is always the same, how can we talk of *consuming* it? Strictly speaking, we don't: we just convert it from one form into other forms. We consume *fuels*, which are sources of readily available energy. We may burn fuel in a vehicle engine, converting its stored chemical energy into heat and then into the kinetic energy of the moving vehicle. By using a wind turbine we can extract kinetic energy from moving air and convert it into electrical energy, which can in turn be used to heat the filament of a halogen incandescent lamp, causing it to radiate light energy.

Forms of energy

At the most basic level, energy forms can be reduced to four types:
- kinetic
- gravitational
- electrical
- nuclear.

Kinetic energy

The **kinetic energy** possessed by any moving object is equal to half the mass (m) of the object times the square of its velocity (v), i.e.:

$$\text{kinetic energy} = \tfrac{1}{2} mv^2$$

where energy is in joules (J), mass in kilograms (kg) and velocity in metres per second (m s^{-1}).

Less obviously, the kinetic energy *within* a material determines its temperature. All matter consists of atoms, or combinations of atoms called molecules. In a gas, such as the air that surrounds us, these move freely. In a solid or a liquid, they form a more or less loosely linked network in which every particle is constantly vibrating. **Thermal energy**, or heat, is the name given to the energy associated with this rapid molecular random motion. The higher the temperature of a body, the faster its molecules are moving. In the temperature scale that is most natural to scientific theory, the Kelvin (K) scale (described a little more in Chapter 2), zero corresponds to zero molecular motion. In the more commonly used Celsius scale of temperature (written as °C), the size of one degree is the same as 1 kelvin, but zero corresponds to the freezing point of water and 100 °C to the boiling point of water at atmospheric pressure. The two scales are therefore related by a simple formula:

$$\text{temperature (K)} = \text{temperature (°C)} + 273.$$

Gravitational energy

A second fundamental form of energy is **gravitational energy**. On Earth, an input of energy is required to lift an object because the gravitational pull of the Earth opposes that movement. If an object, such as an apple, is lifted above your head, the input energy is stored in a form called **gravitational potential energy** (often just 'potential energy' or 'gravitational energy'). That this stored energy exists is obvious if you release the apple and observe the subsequent conversion to kinetic energy. The gravitational force pulling an object towards the Earth is called the *weight* of the object, and is equal to its *mass*, m, multiplied by the acceleration due to gravity, g (which is 9.81 m s^{-2}, although for rough calculations needing less than 2% precision a value of 10 m s^{-2} is often used). Note that although everyday language may treat mass and weight as the same, science does not. The potential energy (in joules) stored in raising an object of mass m (in kilograms) to a height H (in metres) is given by the following equation (see Figure 1.1):

$$\text{potential energy} = \text{force} \times \text{distance} = \text{weight} \times \text{height} = m \times g \times H.$$

Electrical energy

Gravity is not the only force influencing the objects around us. On a scale far too small for the eye to see, electrical forces hold together the atoms and molecules of all materials; gravity is an insignificant force at the molecular level. The **electrical energy** associated with these forces is the third of the basic forms. Every atom can be considered to consist of a cloud of electrically charged particles, called electrons, moving incessantly around a central nucleus. When atoms bond with other atoms to form molecules, the distribution of electrons is changed, often with dramatic effect. Thus **chemical energy**, viewed at the atomic level, can be considered to be a form of electrical energy. When a fuel is burned, the energy liberated (the chemical energy) is converted into heat energy. Essentially, the electrical energy released as the electrons are rearranged (that is, the net release of energy from the breaking and forming of bonds) is converted to the kinetic energy of the molecules of the combustion products.

A more familiar form of electrical energy is that carried by **electric currents** – organized flows of electrons in a material, usually a metal. In metals, one or two electrons from each atom can become detached and move freely

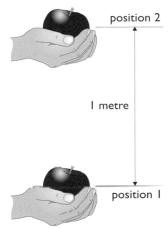

Figure 1.1 The amount of energy required to raise a 100 g apple vertically through 1 m is approximately one joule (1 J)

through the lattice structure of the material. These 'free electrons' allow metals to carry electrical currents.

To maintain a steady current of electrons requires a constant input of energy because the electrons continually lose energy in collisions with the metal lattice (which is why wires get warm when they carry electric currents). Voltage (in volts) is a measure of the electrical 'potential difference' between two points in an electrical circuit, analogous to height in the measurement of gravitational potential energy (see above). The power (in watts) delivered by an electrical supply, or used by an appliance, is given by multiplying the voltage (in volts) by the current (measured in amperes, or 'amps'):

$$\text{power (W)} = \text{voltage (V)} \times \text{current (I)}.$$

In a typical themal power station, the input fuel is burned and used to produce high-pressure steam, which drives a rotating turbine. This in turn drives an electrical generator, which operates on a principle discovered by Michael Faraday in 1832: a voltage is induced in a coil of wire that spins in a magnetic field. Connecting the coil to an electric circuit will then allow a current to flow. The electrical energy can in turn be transformed into heat, light, motion or whatever, depending upon what is connected to the circuit. Electricity is often used in this way, as an *intermediary* form of energy: it allows energy released from one source to be converted to another quite different form, usually at some distance from the source.

Another form of electrical energy is that carried by electromagnetic radiation. More properly called **electromagnetic energy**, this is the form in which, for example, solar energy reaches the Earth. Electromagnetic energy is radiated in greater or lesser amounts by every object. It travels as a wave that can carry energy through empty space. The length of the wave (its wavelength) characterizes its form, which includes X-rays, ultraviolet and infrared radiation, visible light, radio waves and microwaves.

Nuclear energy

The fourth and final basic form of energy, bound up in the central nuclei of atoms, is called **nuclear energy**. The technology for releasing it was developed during the Second World War for military purposes, and subsequently in a more controlled version for the commercial production of electricity. Nuclear power stations operate on much the same principles as fossil fuel plants, except that the furnace in which the fuel burns is replaced by a nuclear reactor in which atoms of uranium are split apart in a 'fission' process that generates large amounts of heat.

The energy source of the Sun is also of nuclear origin. Here the process is not nuclear fission but nuclear *fusion*, in which hydrogen atoms fuse to form helium atoms – such enormous numbers of these reactions take place that massive amounts of solar radiation are generated in the process. Attempts to imitate the Sun by creating power-producing nuclear fusion reactors have been the subject of many decades of research and development effort but have yet to come to fruition.

Conversion, efficiencies and capacity factors

When energy is converted from one form to another, the useful output is never as much as the input. The ratio of the useful output to the required input (usually expressed as a percentage) is called the **efficiency** of the process:

percentage efficiency = (energy output/energy input) × 100.

This efficiency can be as high as 90% in a water turbine or well-run electric motor. Some inefficiencies can be avoided by good design, but others are inherent in the nature of the type of energy conversion. For example a coal fired power station may have an efficiency of 35–40% (if the 'waste' heat is not put to use), but the figure for a geothermal power station may be only 10%.

In the systems mentioned above, the difference between the high and low conversion efficiencies is because the latter are **heat engines** involving the conversion of heat into mechanical or electrical energy. Heat, as already indicated, is the kinetic energy of randomly moving molecules, an essentially chaotic form of energy. No machine can convert this chaos completely into the ordered state associated with mechanical or electrical energy. This is the essential message of the **Second Law of Thermodynamics**: that there is necessarily a limit to the efficiency of any heat engine. Some energy must always be lost to the external environment, usually as low temperature heat. (Section 2.3 of Chapter 2 looks at the efficiency of heat engines in more detail.)

When considering the economics of a power plant, rather than just its efficiency, it is useful to have a measure of its productivity in practice.

One measure of this is the plant's capacity factor (CF): its actual output over a given period of time divided by the maximum possible output. The units for the output quantities can be kWh, MWh, GWh, etc., and the result can be expressed as either a fraction or a percentage.

There are 8760 hours in a year (365 days × 24 hrs/day = 8760 hours), so a 1 MW plant running constantly at its full rated capacity for one year would generate 8760 MWh of output, and would have an annual capacity factor of 1, or 100%.

A 1 MW wind turbine might, in practice, typically produce 3000 MWh of electricity in a year (because the wind doesn't always blow at the full rated speed for which the turbine is designed) – in such a case its annual capacity factor would be:

(3000/8760) = 0.342 = 34.2%.

The period to which a capacity factor relates is not always a year – weekly or monthly capacity factors are often quoted.

The terms 'plant factor' and 'load factor' are also sometimes used as synonyms for 'capacity factor' in the context of power systems.

1.2 Present-day energy use

World energy supplies

The energy used by a final consumer is usually the end result of a series of energy conversions. For example, energy from burning coal may be converted in a power station to electricity, which is then distributed to households and used in immersion heaters to heat water in domestic hot water tanks. The energy released when the coal is burned is called the **primary energy** required for that use. The amount of electricity reaching the consumer, after conversion losses in the power station and transmission losses in the electricity grid, is the **delivered energy**. This is what the consumer actually receives (and pays for). After further losses in the water tank and pipes, a final quantity, called the **useful energy**, comes out of the hot tap.

World total annual consumption of all forms of primary energy increased more than tenfold during the twentieth century, and by the year 2014 had reached an estimated 575 EJ (exajoules), or about 13 700 million tonnes of oil equivalent (Mtoe) (see Figure 1.2). As the figure reveals, fossil fuels provided more than four fifths of the total. The world population in 2014 was some 7.2 billion, so the annual average energy consumption per person was about 80 GJ (gigajoules), equivalent to the energy content of approximately 6 litres of oil per day for every man, woman and child.

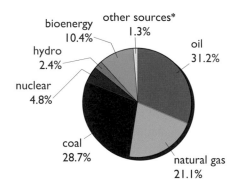

Total: about 575 EJ, equivalent
to 13.7 billion tonnes of oil

Figure 1.2 Percentage contributions to world primary energy consumption in 2014 (source IEA, 2016a) Notes: The bioenergy contribution includes traditional use of solid biomass and modern use of bioenergy. The nuclear contribution is the notional primary energy that would be needed to produce the actual output at an efficiency of 33%. The hydro contribution is the actual electrical output. *'Other sources' are solar and geothermal energy, and energy from wind, wave, tide and wastes

But these figures conceal major differences. The average North American consumes more than 250 GJ per year, most people in Europe use roughly half this amount, and many of those in the poorer countries of the world less than one fifth – much of it in the form of local 'biomass' (see Chapter 5).

How much do renewables contribute to world energy supplies? As Figure 1.2 shows, traditional biomass, hydro power and a range of other renewable sources contributed an estimated 14.2% of world primary energy in 2014.

Over a quarter of the world's fossil fuel consumption is used for electricity generation. World electricity consumption has been increasing at 3% per annum over the last 20 years and this growth seems likely to continue into the foreseeable future.

The generation fuels used vary from country to country, but coal (often of poor quality) is a major source. Table 1.2 shows the generation fuel mix for the world and some sample countries. Worldwide, renewables contributed 22% of world electricity demand.

Table 1.2 Electricity generation by fuel

	Coal	Oil	Natural gas	Nuclear	Hydro	Wind and other sources*
UK – 2015	21%	0%	29%	25%	2%	22%
USA – 2015	33%	1%	33%	20%	6%	7%
China – 2014	72%	0%	2%	2%	19%	4%
World – 2014	41%	4%	22%	11%	16%	6%

Note: *'Other sources' includes other electricity generating renewables and energy from wastes. Proportions may not add to 100% due to rounding
Sources: BEIS, 2016a; BEIS, 2016c; EIA, 2016; IEA, 2016a; IEA, 2016b; IEA, 2016c

In practice, many electricity-generating technologies produce large amounts of unused 'waste' heat. Quoting statistics about proportions of renewable energy based on *primary* energy may thus give a misleading picture. National (and global) statistics are now often quoted in terms of gross *final energy consumption* (see Box 1.1).

BOX 1.1 Primary energy, delivered energy and gross final energy

Figure 1.2 showed the estimated total global *primary energy* consumption for 2014. About 25–30% of this was turned into waste heat in power stations, most of which is dumped uselessly into seas, lakes or the sky via cooling towers. Only a small percentage of this waste heat is put to good use in district heating schemes. This wastage occurs in fossil fuelled power stations, nuclear power plants and renewable power plants fuelled by wood or landfill gas. For example, a landfill gas plant may consume 4 kWh of (primary) gas to produce 1 kWh of output electricity. However, other technologies such as wind, PV and hydro power can generate useful electricity directly with minimal losses.

When comparing technologies or compiling national statistics, those based on *delivered energy*, i.e. the fuel, useful heat or electricity actually received by an end user, probably give a better representation of the overall picture.

The 2009 European Union Renewable Energy Directive (CEC, 2009) sets out requirements for expressing future national renewable energy contributions in terms of the **gross final energy consumption**. This is basically defined as the *delivered energy* to the end users but with two additional small contributions: firstly, the losses in transmission of electricity and heat (in district heating schemes); and secondly, the electricity and heat consumed in energy industries (such as in oil refineries or within power stations). However, it does *not* include the very large waste heat losses in electricity generation that feature in primary energy figures. For most practical purposes the terms delivered energy and final energy can be taken to be equivalent within 10%.

The overall effect of using statistics based on gross final energy is to give more prominence to hydro, wind and PV technologies and less to low efficiency thermal electricity generation technologies where the waste heat is not put to good use. It also focuses attention on the need to improve electricity generation efficiencies.

How long will the world's fossil fuel reserves last? At current consumption rates, it is estimated that world coal reserves could last for over 150 years, and oil and natural gas reserves could last for approximately 50 years (BP, 2017).

The concern that fossil fuels will eventually 'run out' has long been a reason for promoting the use of renewable energy. The world's supply of oil has been a matter of concern ever since the 'oil price shocks' of 1973 and 1979 when the world price of oil rose by a factor of 10 in less than 10 years (see Figure 1.3). This was the result of a war in the Middle East followed by a revolution in Iran.

Figure 1.3 World oil price 1950–2015 in US$2015 dollars per barrel (source: BP, 2017)

New supplies of oil from fields in the North Sea and the Arctic led to another period of cheap oil during the 1990s, but these are now depleting. The UK North Sea oilfields reached their peak production in 1999 and production has more than halved since then.

World oil prices rose again after 2000, reaching over US$100 per barrel between 2011 and 2013. Existing oilfields have a limited life; once exhausted they have to be replaced with new ones. In order just to maintain the world's oil production at its current level, a large number of new oil fields have to be developed continuously. According to the International Energy Agency, the 'easy oil' has now been largely used up. What remains is likely to be more expensive and in inaccessible areas such as the Arctic or in deep offshore wells, or produced by unconventional techniques such as 'fracking', which involves fracturing the underground rock to allow the oil to flow.

Since 2014 there has been a global price war between major oil suppliers and the world price has fallen. At the time of writing (early 2017) it is around US$55 per barrel.

Although some worries about the short-term supply of oil remain, in the longer term concerns about climate change and the urgent need to reduce CO_2 emissions from fossil fuel combustion mean that most of the world's proven fossil fuel reserves will have to stay in the ground.

1.3 Fossil fuels and climate change

Society's current use of fossil and nuclear fuels has many adverse consequences. These include air pollution, acid rain, the depletion of natural resources and the dangers of nuclear radiation. This brief introduction concentrates on one of these problems: global climate change caused principally by emissions of greenhouse gases from fossil fuel combustion.

The surface temperature of the Earth establishes itself at an equilibrium level where the incoming energy from the Sun balances the outgoing infrared energy re-radiated from the surface back into space (see Chapter 3, Figure 3.5). If the Earth had no atmosphere its surface temperature would be −18 °C; but its atmosphere, which includes 'greenhouse gases' – principally, water vapour, carbon dioxide and methane – acts like the panes of a greenhouse, allowing solar radiation (which lies in the range from ultra-violet to short wave infrared) to enter but inhibiting the outflow of long-wave infrared radiation. The natural 'greenhouse effect' that these gases cause is essential in maintaining the Earth's average surface temperature at a level suitable for life, at around 15 °C.

Since the Industrial Revolution, however, human activities have been adding extra greenhouse gases to the atmosphere. The principal contributor to these increased emissions is carbon dioxide (CO_2) from the combustion of fossil fuels. Humanity's rate of emission of CO_2 from these fuels has increased enormously, particularly since 1950 (see Figure 1.4). A result of this has been that the concentration of CO_2 in the atmosphere has risen from 280 parts per million (ppm) to 400 ppm in 2016 and is currently rising at around 3 ppm per year.

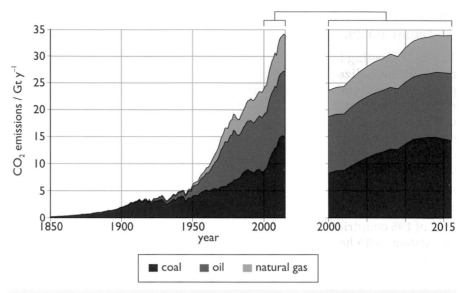

Figure 1.4 CO_2 emissions from the burning of fossil fuels 1850–2016 (sources: Boden et al., 2016; BP, 2017)

As will be explained in Chapter 2, different fuels emit different amounts of CO_2 when burned per unit of energy produced. Producing a GJ of heat by burning coal produces 25% more CO_2 than producing a GJ from oil and 60% more than from natural gas. Cutting global coal use is thus of prime importance in reducing total CO_2 emissions. Renewable electricity generation technologies that can substitute for electricity produced from coal fired power stations are being strongly promoted in countries such as China.

In addition to CO_2 emissions there have also been significant additional greenhouse gas contributions from emissions of methane, largely from agriculture, but also as leakage of natural gas, which mainly consists of methane.

According to NASA's Goddard Institute for Space Studies, the global surface temperature in 2016, the warmest year on record, was 0.99 °C above the 1951-1980 average (NASA, 2017).

International research on climate change is coordinated by the Intergovernmental Panel on Climate Change (IPCC). If emissions were to continue rising at the same rate as they did between 2000 and 2010, they estimate that the Earth's surface temperature could rise by around 4 °C by the end of the twenty-first century. Such rises would probably be associated with an increased frequency of climatic extremes, such as floods or droughts, and serious disruptions to agriculture and natural ecosystems. The thermal expansion of the world's oceans could mean that sea levels would rise by around 0.5 m by the end of the century, which could inundate some low-lying areas. Beyond 2100, or perhaps before, much greater sea level rises could occur if major Antarctic ice sheets were to rapidly collapse and melt.

The threat of global climate change has spurred urgent international action. Starting in 1995 the United Nations has convened a series of annual

Climate Change conferences, each normally now called a 'Conference of the Parties' (COP).

The 1997 meeting in Kyoto, Japan, agreed the Kyoto Protocol. Under this, many industrialized countries agreed to legally binding reductions in greenhouse gas emissions of an average of 6–8% below 1990 levels to be reached between the years 2008–2012. Many countries managed to meet or exceed these targets. For example, by 2008 the UK's greenhouse gas emissions had fallen by 20%. In that year the UK Parliament passed a Climate Change Act requiring that the country's greenhouse gas emissions should fall by 80% from their 1990 levels by 2050. By 2015, UK greenhouse gas emissions were 38% below its 1990 levels (BEIS, 2016a).

Following the 21st COP meeting (COP 21) held in Paris in late 2015, a total of 195 countries committed to curbing their greenhouse gas emissions 'consistent with holding the increase in the global average temperature to well below 2 °C above pre-industrial levels and pursuing efforts to limit the temperature increase to 1.5 °C above pre-industrial levels' (UNFCCC, 2015).

A 2 °C target will mean limiting the atmospheric concentration of CO_2 to 450 ppm. Global CO_2 emissions will need to peak immediately and fall to almost *zero* by 2100 (IEA, 2016a). A 1.5 °C target would require that CO_2 emissions will need to fall even faster, possibly to zero by 2050. Emission reductions on this scale will inevitably involve a switch to low- or zero-carbon energy sources such as renewables. It means scaling up investment in renewable energy (and energy efficiency) enormously and reducing investments in fossil fuel energy systems, particularly in the use of coal.

This may sound like an overwhelmingly difficult task, but there are already signs of change. Most of the increase in CO_2 emissions from coal since 2000 shown in Figure 1.4 has taken place in China, but, faced with the resulting severe ground level air pollution, the Chinese government has now reversed its policy of burning ever more coal (see Section 1.5). Global CO_2 emissions from coal burning may now have peaked and, given the large global investments in energy efficiency and the high growth rates in renewable energy, the world may now be approaching its peak of CO_2 emissions from *all* fossil fuels.

1.4 Renewable energy sources

Renewable energy can be defined as:

> *energy obtained from the continuous or repetitive currents of energy recurring in the natural environment*
>
> (Twidell and Weir, 1986)

or as

> *energy flows which are replenished at the same rate as they are 'used'*
>
> (Sorensen, 2000)

The main sources and magnitudes of the Earth's renewable energy are shown in Figure 1.5. Their principal source is solar radiation. Approximately 30% of the 5.4 million EJ per year of solar radiation arriving at the Earth is reflected back into space. The remaining 70% is, in principle, available for use on Earth, and amounts to approximately 3.8 million EJ, more than 6500 times the total world primary energy consumption of 575 EJ. Two non-solar, renewable energy sources are also shown on the figure: the motion of the ocean tides and geothermal heat from the Earth's interior, which manifests itself in convection in volcanoes and hot springs, and in conduction in rocks.

Solar energy: direct uses

Solar radiation can be converted into useful energy *directly*, using various technologies. Absorbed in solar 'collectors', it can provide hot water or space heating (i.e. heating the interior spaces of buildings). Buildings can also be designed with 'passive solar' features that enhance the contribution of solar energy to their space heating and lighting requirements.

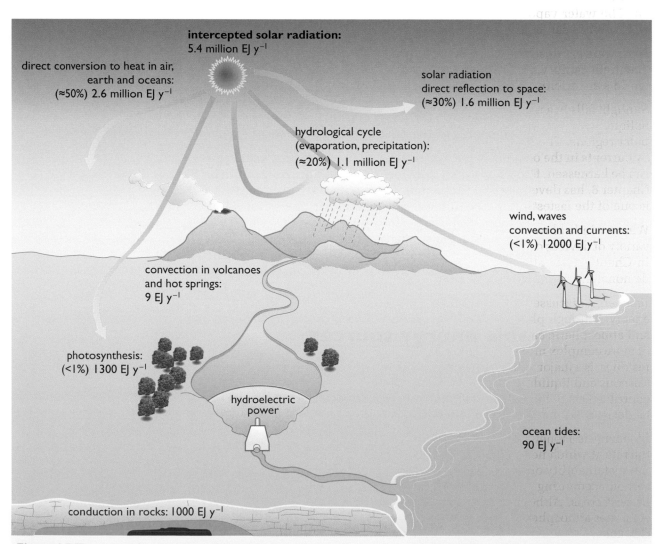

Figure 1.5 The various forms of renewable energy depend primarily on incoming solar radiation, which totals some 5.4 million EJ per year

Solar energy can also be concentrated by mirrors to provide high-temperature heat for generating electricity. Such concentrating solar power (CSP) stations are in commercial operation in countries like the USA and Spain. *Solar thermal energy* conversion is described in Chapter 3.

Solar radiation can also be converted directly into electricity using photovoltaic (PV) modules, mounted on the roofs or facades of buildings, or deployed in large ground-mounted arrays called 'solar farms'. At the time of writing, the costs of electricity from photovoltaics have been falling fast and in many countries it is becoming competitive with that generated from fossil fuels. *Solar photovoltaics* is described in Chapter 4.

Solar energy: indirect uses

Solar radiation can be converted to useful energy *indirectly*, via other energy forms. A large fraction of the radiation reaching the Earth's surface is absorbed by the oceans, warming them and adding water vapour to the air. The water vapour condenses as rain to feed rivers, into which dams and turbines can be located to extract the energy of the flowing water.

Hydropower, described in Chapter 6, has steadily grown during the twentieth century, and at the time of writing provides about a sixth of the world's electricity.

Sunlight falls in a more perpendicular direction in tropical regions and more obliquely at high latitudes, heating the tropics to a greater degree than the polar regions. The result is a massive heat flow towards the poles, carried by currents in the oceans and the atmosphere. The energy in such currents can be harnessed, for example by wind turbines. *Wind power*, described in Chapter 8, has developed on a large scale only in the past few decades, but is one of the fastest-growing of the 'new' renewable sources of electricity.

Where winds blow over long stretches of ocean, they create waves, and a variety of devices can be used to extract that energy. *Wave power*, described in Chapter 9, is attracting new funding for research, development and demonstration in several countries.

Bioenergy, discussed in Chapter 5, is another indirect manifestation of solar energy. Through photosynthesis in plants, solar radiation converts water and atmospheric carbon dioxide into carbohydrates, which form the basis of more complex molecules. Biomass, in the form of wood and agricultural residues, is a major world energy source, especially in the developing world. Gaseous and liquid fuels derived from biological sources make significant contributions to the energy supplies of some countries. Bioenergy can also be derived from wastes, many of which are biological in origin.

Bioenergy is a renewable resource if the rate it is consumed is no greater than the rate at which new plants are re-grown. This is not always the case, and the sustainability of some biomass systems has been challenged - however, carbon accounting should be conducted over appropriate timescales and harvest areas. Although the production and combustion of biomass fuels generates atmospheric CO_2 emissions, these should be largely offset by the CO_2 absorbed when the plants were growing.

Non-solar renewables

Three other sources of renewable energy do not depend on solar radiation: *tidal* and *deep geothermal* energy and the heat gains from *heat pumps*.

Tidal energy, discussed in Chapter 7, is often confused with wave energy, but its origins are quite different. Ocean tides are caused by the gravitational pull of the Moon (with a small contribution from the Sun) on the world's oceans, causing a regular rise and fall in water levels as the Earth rotates. The power of the tides can be harnessed by building a low dam or 'barrage' behind which the rising waters are captured and then allowed to flow back through electricity-generating turbines.

It is also possible to harness the power of strong underwater currents, which are mainly tidal in origin. Various devices for exploiting this energy source, such as marine current turbines (rather like underwater wind turbines) are now beginning to enter commercial service.

Heat from within the Earth is the source of geothermal energy, discussed in Chapter 10. The high temperature of the interior was originally caused by gravitational contraction of the planet as it was formed, but has since been enhanced by the heat from the decay of radioactive materials deep within the Earth.

In some places where hot rocks are very near to the surface, water is heated in underground aquifers. These have been used for centuries to provide hot water or steam. In some countries, geothermal steam is used to produce electricity and, in others, hot water from geothermal wells is used for heating.

If steam or hot water is extracted at a greater rate than heat is replenished from surrounding rocks, a geothermal site will cool down and new holes will have to be drilled nearby. When operated in this way, geothermal energy is not strictly renewable. However, it is possible to operate in a renewable mode by keeping the rate of extraction below the rate of heat replenishment.

Another technology that may be classified as renewable energy is the use of heat pumps. These are devices that use electrical energy to drive a compressor to pump a flow of heat from a low temperature to a higher one. The domestic refrigerator is an example. Heat pumps are widely used to cool buildings. This is commonly known as *air-conditioning*. Their use for *heating* buildings and providing hot water has been increasing over the past 20 years. In principle the technology has the potential to turn a low carbon flow of electricity (from renewable energy or nuclear power) into a low carbon flow of heat, allowing it to substitute for fossil fuel heating.

Although it can be seen as a form of *energy efficiency*, giving more heating value to a kilowatt-hour of electricity, under the 2009 EU Renewable Energy Directive, heat gains from heat pumps may be classified as renewable energy (CEC, 2009). Heat pumps that draw heat from the ground, ground-source heat pumps, may be called *geothermal heat pumps*, though this is not the same form of 'geothermal energy' as the 'deep geothermal' technology mentioned above. The principles of heat pumps and their use are described in Chapter 2.

1.5   Renewable energy use today

Renewable energy sources are already providing a significant proportion of the world's primary energy and some technologies are growing very rapidly.

World renewable energy use

As shown earlier in Figure 1.2 renewables contributed about 14% of world primary energy in 2014. The proportion of the world's *final energy* consumption was about 18%. Figure 1.6 shows a breakdown:

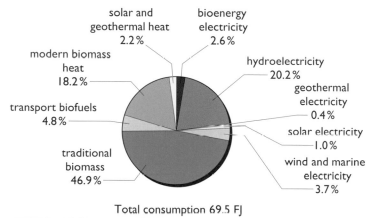

Total consumption 69.5 FJ

Figure 1.6 Percentage breakdown of renewable energy sources' contributions to world final energy supplies in 2014 (sources: IEA, 2016a; REN21, 2016)

The largest contribution is an estimated value of about 33 EJ from 'traditional biomass' (wood, straw, dung, etc.) mainly used in developing countries. Since most of this isn't traded, it often doesn't enter into national economic statistics and its true magnitude is only known approximately.

The various forms of 'electricity from renewables' make up a further 19 EJ or 5400 TWh, 22% of global electricity generation (as shown earlier in Table 1.2). About three quarters of this was supplied by hydropower. Electricity generated from bioenergy includes that made from landfill and sewage gas and from municipal waste incineration. Solar electricity includes that produced by photovoltaics (PV) and concentrating solar power (CSP). Wind and marine electricity comes mainly from wind, but also includes small amounts from wave power and tidal power projects.

Various forms of renewable heat supplied another 14 EJ. These include direct geothermal energy use, the output of solar thermal water heaters and heat produced by combustion of commercially traded biomass fuels such as wood and other crops specifically grown for energy use.

Finally, transport biofuels such as biogas, bioethanol and biodiesel supplied another 3.3 EJ. Perhaps what is most striking about the year by year changes in world renewable energy supply is the extraordinary growth rates of some of the technologies, particularly those generating electricity. Table 1.3 shows some values:

Table 1.3 Global annual growth in renewable energy technologies, end-2010 to end-2015

Renewable technology	Annual growth rate
Solar PV	42%
Concentrating solar power	9.7%
Wind	17%
Geothermal electricity	3.7%
Hydro	2.9%
Solar water heating	12%
Biodiesel	6.5%
Bioethanol	3.0%

Source: REN21, 2016

Over this period world wind power capacity has been doubling every four years and solar PV capacity has been doubling every two years! The various national policy pledges given following the COP 21 meeting would suggest that these high growth rates are likely to continue.

Energy use in the UK

For most of the nineteenth and twentieth centuries coal was the dominant fuel and the UK was a coal exporter. During the 1950s and 1960s it imported large amounts of cheap oil. In the 1970s the discovery of North Sea oil and gas made it once again a net energy exporter and natural gas became the main fuel of choice for heating and electricity generation. The North Sea oil and gas reserves are now running down and the UK is presently an importer of coal, oil and gas. There are now serious questions about the country's future energy supplies.

Figure 1.7 shows the primary and delivered energy use of the UK. The main sources of primary energy in 2015, shown in the top bar, were:

- natural gas, used for electricity generation and many forms of heating
- oil, almost exclusively used for transport
- coal, almost all used for electricity generation
- nuclear heat, used for electricity generation
- a variety of forms of renewable energy, used for electricity generation, heating and transport fuels.

Although coal made up 13% of the primary energy supply in 2015, this proportion has been falling and is likely to fall further in the future as the UK government has promised to phase out all coal fired power stations by 2025. In April 2017 the UK experienced its first whole day without needing electricity generated from coal since 1882.

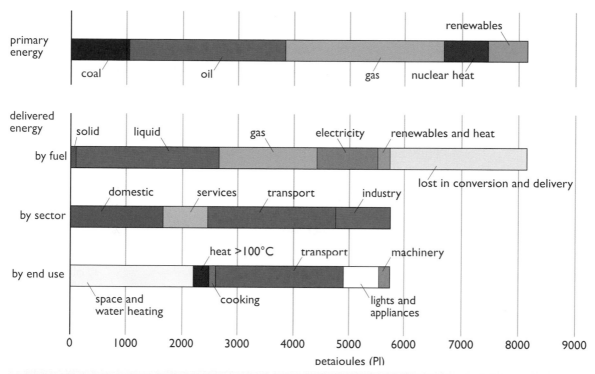

Figure 1.7 UK primary and delivered energy use, 2015 (sources: BEIS, 2016a; BEIS, 2016b)
Note: in the second bar, 'electricity' includes renewable electricity; 'renewables and heat' includes biofuels for transport and heat from CHP plants

The other three bars show UK *delivered energy* broken down in three different ways.

The first of these bars shows the delivered fuels: solid fuel (mainly coal and coke), liquid fuels (petrol and diesel fuels and aviation kerosene), natural gas, electricity and some renewable energy (mainly wood used for heating and transport biofuels) plus heat supplied by district heating schemes.

It is obvious that much of the energy itemized in the top, primary energy, bar has disappeared from the second: most of the coal, a large amount of natural gas, all of the nuclear heat and some of the renewable energy. These are fuels used for electricity generation. The second bar contains a large 'electricity' component. It also contains an enormous component, which accounts for 'losses in conversion and delivery', 30% of UK primary energy. This is mostly waste heat produced at power stations but it also includes losses in refining crude oil into liquid fuels.

The second delivered energy bar shows end use by sector. As in most countries, energy demand is categorized in official statistics into four of these: domestic or residential, services (or commercial and institutional), industrial and transport.

The third delivered energy bar shows the end uses. Nearly 40% of the energy demand is for space and water heating and drying. These are uses that could be met with low temperature heat produced by technologies such as solar heating, bioenergy, heat pumps or by using waste heat from power stations.

Even when energy has been delivered to customers in the various sectors, it is often used very wastefully so using *energy efficiency* to reduce demand is every bit as important as new renewable energy supply.

The total contribution of renewables to UK *primary* energy supply in 2015 was only 8.6%. Over 70% of this came from various forms of biomass and wastes (see Figure 1.8a). The percentage contribution of renewables to *electricity supplies* was much larger. In 2015 nearly a quarter of UK electricity came from renewable sources, mainly in the form of wind power and bioenergy, with smaller contributions from solar and hydro power (see Figure 1.8b).

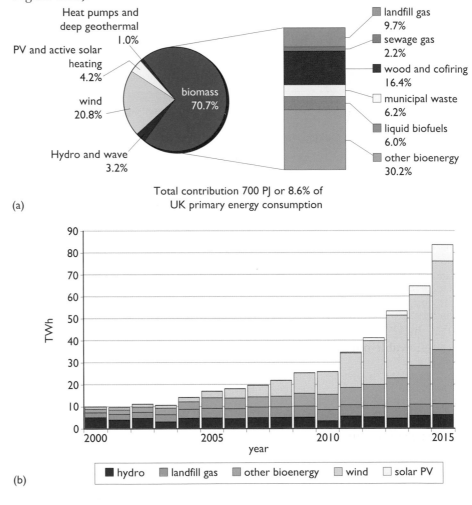

(a)

(b)

Figure 1.8 (a) Primary energy contributions from renewable energy in the UK, 2015 Note: 'Other bioenergy' includes straw and energy crops used for electricity generation. (b) Growth in electricity generation from renewable sources in the UK 2000–2015. In 2015 renewables contributed nearly a quarter of UK electricity (sources: BEIS 2016a; BEIS, 2016c)

The overall renewable energy contribution to UK *final energy* in 2015 was 8.3% but this has been increasing rapidly. In order to comply with the EU's Renewable Energy Directive this needs to reach 15% by 2020, although it is presently expected to fall short of the target.

Energy in the USA

The USA is a country with large reserves of coal but only limited reserves of oil and natural gas. Today the main primary energy fuels (see Figure 1.9) are:

- oil, used for transport
- natural gas, used for electricity generation
- coal, used for electricity generation
- nuclear heat, used for electricity generation
- a range of renewable energy sources which made up nearly 7% of the USA's primary energy.

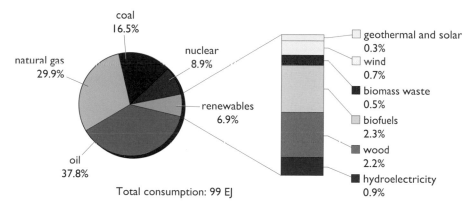

Figure 1.9 Primary energy use in the USA 2015 (source: EIA, 2016)
Notes: The value for nuclear is the heat used to generate electricity. The values for hydro, wind, geothermal and solar electricity are for the electricity actually produced

The amount of renewable energy has increased by over 50% since 2000. In 2015, the USA was second only to China in renewable energy investment and renewable electricity generating capacity (REN21, 2016).

Although the USA is largely self-sufficient in coal and natural gas, it imports large amounts of oil as a transport fuel. This has long posed an 'energy security' problem ever since the oil price shocks of the 1970s (shown earlier in Figure 1.3). One response to this has been a continuing growth in the use of transport biofuels such as bioethanol and biodiesel. The USA is currently the world's largest consumer of transport biofuels, accounting for 6% of its road transport energy use.

Other growth areas have been solar and wind power for electricity generation. Concentrating solar power (CSP) has been in use in California since the 1970s (see Chapter 3) but since 2007 there has been a rapid growth in both CSP and solar PV electricity generation. In 2015 the State of California accounted for two thirds of all of the USA's solar electricity.

By autumn 2016 the USA had a total installed wind power capacity of 76 GW (a gigawatt of wind power is equivalent to 1000 large 1 MW turbines). A quarter of these were in the state of Texas.

Energy in China

China's energy use has had a dramatic effect on global CO_2 emissions, particularly since 2000. Until the 1980s it was a largely agricultural country, but has since embarked on policies of industrial expansion and supplying electricity to all of its population. Its total primary energy consumption has risen five-fold since 1980 and its electricity generation has been growing at a rate of 10% per year. China does not have large oil or gas reserves, so the main increase has been in the use of coal, dramatically so since 2000 (see Figure 1.10).

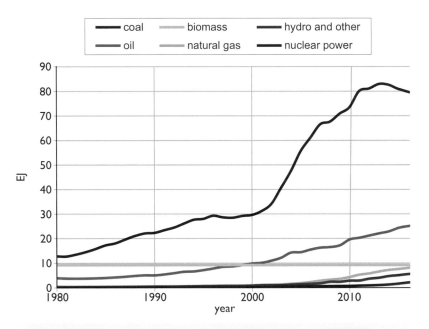

Figure 1.10 Primary energy use in China 1980–2016 (source: BP, 2017).
Notes: The amount of biomass is only approximately known and is taken to be 9.2 EJ per year based on estimates in IEA, 2016a. 'Hydro and other' includes wind and PV electricity generation

In 2016 China's coal consumption made up a half of the world's total. One result of this has been serious air pollution in most of the major Chinese cities. Another is a serious depletion of the country's coal reserves.

When the problems of global climate change are added (China contributed 28% of the world's CO_2 emissions in 2016), it is clear that the country has many reasons to switch from coal to renewable energy, particularly for electricity generation.

China has invested heavily in hydropower, which in 2016 made up 19% of its electricity supply with enormous projects such as the Three Gorges Dam (described in Chapter 6).

It is also investing heavily in wind turbines. In 2016, China had an estimated 149 GW of installed wind power. This supplied about 4% of China's electricity supply and this proportion is likely to continue to increase.

China is also the world's largest installer of solar PV electricity generation. In 2016 it had approximately 78 GW of PV capacity. The country also accounts for 70% of the world's sales of solar water heaters.

Overall, in 2016 China was the world leader in both renewable electricity generating capacity and in investment in renewable energy technologies (REN21, 2017).

1.6 Renewable energy policies for the future

A global transition away from fossil fuels and towards 'modern' renewable energy sources is underway.

The 2015 Paris COP 21 agreement has resulted in calls to promote renewable energy in many countries and at all levels. For example:

- the United Nations General Assembly Sustainable Development Goals which include a dedicated goal on 'sustainable energy for all'
- the 2015 Paris City Hall Declaration, signed by over 1000 city mayors and local leaders, which calls for 100% renewables or 80% reductions in greenhouse gases by 2050
- Catholic, Islamic, Hindu and Buddhist religious declarations urging billions of people of faith to work towards zero- or low-carbon futures through renewable energy.

Individual countries have made COP 21 'pledges' of **Intended Nationally Determined Contributions (INDCs)**. Some of these are statements of intended cuts in greenhouse gas emissions, others include specific renewable energy targets. For example:

- the UK is bound by its Climate Change Act to reduce its greenhouse gas emissions by 80% from 1990 levels by 2050 and has promised to phase out coal-fired electricity generation by 2025
- the EU is committed to a 40% reduction in greenhouse gas emissions from a 1990 baseline and a minimum contribution of 27% renewable energy in its final energy consumption by 2030
- Brazil has set a target to meet 45% of its primary energy needs from renewables by 2030
- China is committed to increasing its share of non-fossil fuel energy to 20% by 2030. More immediately, its installed wind power will be increased to 200 GW and its PV capacity to 150 GW by 2020
- in the USA, some states have adopted strong renewable energy targets. Both California and New York now have targets of obtaining 50% of their electricity from renewable sources by 2030.

Limiting the global temperature rise to 2 °C will involve an enormous financial investment in renewable energy. The International Energy Agency (IEA) have estimated that globally, over the period 2010–2015, an average of US$280 billion per year was spent on renewable energy (while four times that amount was spent on fossil fuels!). However, in 2015 renewables accounted for 70% of total electricity generation investment. The IEA have modelled the effects of changes in future patterns of energy use in various 'scenarios' (descriptions of possible actions in the future). In their '450 Scenario' (limiting the atmospheric concentration of CO_2 to 450 ppm) they see a major shift in investment away from fossil fuels and into energy efficiency and renewables. They estimate that over the period 2016–2040 investment in renewables will almost double to US$500 billion per year (IEA, 2016a). Other scenarios for the possible future expansion of renewable energy use are described in Chapter 12.

1.7　Summary

This first chapter has looked at many basic aspects of renewable energy.

Renewable energy is potentially a far more **sustainable** source than energy sourced from burning fossil fuels or from nuclear power.

Energy is basically the '**capacity to do work**'. It may be specified in units of kilowatt-hours (kWh) in electricity bills, joules (J) in scientific literature, or tonnes of oil equivalent (toe) in national energy statistics.

Power is the *rate* of use of energy and commonly has units of the watt (W) or kilowatt (kW). In common speech the terms 'power' and 'energy' may be used just to refer to *electricity*, but it is always worth checking to see what is actually being discussed.

The **First Law of Thermodynamics** says that (unfortunately) energy cannot be created out of nothing. Most of our energy supply comes from the combustion of **fuels** (coal, oil or natural gas) which are readily available sources.

There are four main forms of energy:

- **kinetic**, that of a mass in motion. **Heat energy** can be considered as a form of this
- **gravitational** or **potential**, that acquired by a mass moving against gravity
- **electrical**, which may involve currents flowing in wires but also includes **electromagnetic energy**, waves travelling through space such as light, and **chemical energy**, that of the chemical bonds within fuels
- **nuclear** energy, energy derived from the fission of atoms.

The **efficiency** of an energy process is the ratio of its output energy to its input.

The **Second Law of Thermodynamics** says that it is not possible to make a perfect heat engine (such as a power station). There will always be some waste heat.

The **capacity factor** of an electricity generation plant is the ratio of its actual output to what it might produce if it could run at full power all of the time.

Primary energy is the energy content of basic fuels, and **delivered energy** is what the consumer actually receives (and pays for). The difference between these two is made up of **losses in conversion and delivery**: waste heat at power stations and energy losses in oil refineries. The term **final energy** can be taken to be similar to delivered energy.

In 2014 **world primary energy demand** was 575 EJ equivalent to 13.7 billion tonnes of oil.

Are fossil fuels running out? At present rates of use, oil and gas reserves could last for 50 years and coal for 150 years. In practice, concerns about climate change will mean that most of these fossil fuel reserves will have to remain in the ground.

The Earth sits in a radiation balance between the high temperature of the Sun and the cold of deep space. Its surface temperature is 'just right' for human existence. Increasing emissions of **greenhouse gases**, such as **carbon dioxide (CO_2)** produced by the combustion of fossil fuels, have upset the radiation balance and have been warming the planet. Per unit of heat produced coal emits 25% more CO_2 than oil and 60% more than natural gas.

A succession of **Conferences of the Parties (COPs)** organised by the United Nations have been held since 1995. At the 21st meeting, COP21, an international agreement was reached that countries should reduce their greenhouse gas emissions to keep the future warming of the Earth's surface '**well below 2 °C**'. Fortunately, CO_2 emissions from fossil fuels may now be reaching a peak.

Renewable energy sources are essentially those that are continuously replenished. Most are derived from energy supplied by the Sun. In principle it provides over 6500 times the total world primary energy consumption.

Direct solar energy uses include **solar thermal water heating** using solar collectors, **passive solar heating** of buildings, and electricity generation using **solar photovoltaics (PV)** or **concentrating solar power (CSP)**.

Indirect solar energy uses include **hydropower**, derived from rainfall, **wind power** and **wave power**, both derived from the world's air currents, and **bioenergy**, derived from plant materials created by photosynthesis.

Non-solar renewables include **tidal energy** derived from the gravitational effect of the Moon and Sun on the world's oceans, **deep geothermal energy** using the energy of hot rocks beneath the Earth's surface, and **heat gains from heat pumps**.

In 2014, renewables provided 14% of the world's primary energy and 18% of the world's final energy supply, nearly a half from 'traditional biomass', mainly used in developing countries.

Electricity from renewables in 2015 produced 22% of global electricity generation, with **hydroelectricity** providing about three quarters of this. **Solar electricity** only produced 1% of the world's renewables supply in 2014, but the installed capacity has recently been doubling every two years. Similarly **wind power** supplied under 4% of the world's renewables but the installed capacity has been doubling every four years.

In 2015, the main **primary energy sources** in the UK were natural gas, oil, coal, nuclear heat and renewable energy. The coal was almost entirely used for electricity generation and its use is to be phased out by 2025. The oil is almost entirely used for transport. Losses in conversion and delivery, mostly in the form of waste heat at power stations, made up nearly 30% of the total.

The **main distributed fuels in the UK** are natural gas for heating, oil for transport, and electricity. The four sectors of the energy economy are **domestic**, **services**, **transport** and **industry**. Low temperature final uses such as space heating and water heating make up over 40% of the energy demand.

In 2015, **renewable energy in the UK** made up 8.6% of primary energy demand, over 70% of this coming from various forms of biomass, mostly used for generating electricity. A further 21% came from wind power. A quarter of UK electricity was generated from renewables and there has been a large growth in electrical generation from biomass, wind power and solar PV.

The **main fuels in use in the USA** are oil, natural gas, coal and nuclear heat. Although largely self-sufficient in coal and gas, it has to import large amounts of oil. In 2015 renewables supplied nearly 7% of US primary energy with biomass as the main source. The amount of renewable energy has doubled since 2000.

The USA is the world's largest consumer of transport biofuels and there has been considerable growth in solar electricity (both concentrating solar and PV) particularly in the state of California. Wind power use has been increasing right across the USA. A quarter of the installed wind turbines are in the state of Texas.

China has large reserves of coal but little oil or natural gas. Its recent industrial expansion has been mainly fuelled by coal. In 2016, China's coal consumption made up half the world's total and **it emitted 28% of the world's CO_2 emissions**. Most Chinese cities now suffer from serious air pollution. Since 2012 its policies of burning ever more coal have been reversed in favour of more renewable energy and its coal consumption may now have peaked. The country has considerable hydroelectricity resources, which in 2016 supplied nearly a fifth of its electricity. **It is currently the world's largest investor in renewable energy**.

Following the 2015 United Nations COP21 Paris agreement, over 100 countries have formally committed to reduce their greenhouse gas emissions to keep the global temperature rise 'well below 2 °C'. This will require a major switch in investment away from fossil fuels towards energy efficiency and renewable energy use.

In order to meet this target, it is estimated that the US$280 billion spent globally each year on renewable energy over the period 2010–2015 will need to rise to US$500 billion per year over the next 25 years. This suggests that the prospects for renewable energy in the coming decades look bright.

Finally, in the chapters that follow, we take a tour of each of the principal renewable energy sources in turn: in each case their physical principles, the main technologies involved, their costs and environmental impact, the size of the potential resource and their future prospects are discussed. But to start, in the next chapter we look at some of the important thermodynamic principles underpinning the production of work and the generation of electricity from heat and at the technology of heat pumps.

References

BEIS (2016a) *UK Energy in Brief 2016*, Department for Business, Energy and Industrial Strategy [Online]. Available at https://www.gov.uk/government/collections/uk-energy-in-brief#2016 (Accessed 4 November 2016).

BEIS (2016b) *Energy Consumption in the UK: 2016 data tables*, Department for Business, Energy and Industrial Strategy [Online]. Available at https://www.gov.uk/government/statistics/energy-consumption-in-the-uk (Accessed 8 November 2016).

BEIS (2016c) *Digest of UK Energy Statistics (DUKES): long term trends*, Department for Business, Energy and Industrial Strategy [Online]. Available at https://www.gov.uk/government/collections/digest-of-uk-energy-statistics-dukes (Accessed 12 November 2016).

Boden, T. A., Marland, G. and Andres, R. J. (2016) *Global, Regional, and National Fossil-Fuel CO2 Emissions*, Carbon Dioxide Information Analysis Center, Oak Ridge National Laboratory, U.S. Department of Energy, Oak Ridge, TN [Online]. Available at http://cdiac.ornl.gov/trends/emis/meth_reg.html (Accessed 5 November 2016).

BP (2017) *BP Statistical Review of World Energy 2017*, London, The British Petroleum Company [Online]. Available at https://www.bp.com/content/dam/bp/en/corporate/pdf/energy-economics/statistical-review-2017/bp-statistical-review-of-world-energy-2017-full-report.pdf (Accessed 17 August 2017).

CEC (2009) *Directive 2009/28/EC on the promotion of energy from renewable sources*, Commission of the European Communities [Online]. Available at http://eur-lex.europa.eu (Accessed 16 October 2016).

EIA (2016) *Monthly Energy Review (MER) October 2016*, Energy Information Administration [Online]. Available at http://www.eia.gov/totalenergy/data/monthly/ (Accessed 9 November 2016).

European Union (2011) *Renewable Energy targets by 2020* [Online]. Available at: http://ec.europa.eu/energy/renewables/targets_en.htm (Accessed 1 November 2011).

Everett, B., Boyle, G. A., Peake, S. and Ramage, J. (eds) (2012) *Energy Systems and Sustainability: Power for a Sustainable Future* (2nd edn), Oxford, Oxford University Press/Milton Keynes, The Open University.

IEA (2016a) *World Energy Outlook 2016*, Paris, International Energy Agency.

IEA (2016b) *Key World Energy Statistics 2016*, Paris, International Energy Agency [Online]. Available at https://www.iea.org/publications/freepublications/publication/key-world-energy-statistics.html (Accessed 13 November 2016).

IEA (2016c) *China, People's Republic of: Electricity and Heat for 2014*, Paris, International Energy Agency [Online]. Available at http://www.iea.org/statistics/statisticssearch/report/?country=China&product=electricity and heat (Accessed 13 November 2016).

NASA (2017) *Global temperature* [online] available at: https://climate.nasa.gov/vital-signs/global-temperature/ (Accessed 11 May 2017).

REN21 (2016) Renewables 2016 Global Status Report [online] available at http://www.ren21.net/resources/publications (accessed 10 September 2017)

REN21 (2017) Renewables 2017 Global Status Report [online] available at http://www.ren21.net/resources/publications (accessed 10 September 2017)

Sorensen, B. (2000) *Renewable Energy* (2nd edn), London, Academic Press.

Twidell, J. and Weir, A. (1986) *Renewable Energy Resources*, London, E. and F. N. Spon.

UNFCCC (2015) *Adoption of the Paris Agreement*, United Nations Framework Convention on Climate Change, Bonn [Online]. Available at https://unfccc.int/resource/docs/2015/cop21/eng/l09r01.pdf (Accessed 19 November 2016).

Acknowledgements

The authors would like to thank Godfrey Boyle, Gary Alexander and Janet Ramage for their contributions to this chapter in previous editions.

Chapter 2

Thermodynamics, heat engines and heat pumps

by Bob Everett

2.1 Introduction

Chapter 1 has introduced some basic energy concepts and units. It has described how there are many forms of renewable energy. Some, such as PV, wind, wave and tidal power, can be used to generate electricity directly. Others produce heat. This chapter is about the problems of turning heat (renewable or otherwise) into useful work and vice versa.

Heat engines are used to turn heat into work to drive pumps or vehicles, or more commonly to generate electricity.

Heat for heat engines and heat pumps can come from:

- burning fuels – fossil, biomass or waste
- nuclear fission in reactors
- 'solar energy' – high temperatures from concentrating solar power using mirrors, or lower temperatures using systems such as ocean thermal energy conversion (OTEC)
- 'deep geothermal energy' – hot water from deep below the Earth's surface.

Heat pumps carry out the reverse process. They are used to 'pump' low grade heat, usually drawn from the air or the ground, from a lower to a higher temperature. This requires an input of mechanical work (usually from an electric motor) in the process.

The performance of both heat engines and heat pumps are limited by the laws of thermodynamics.

The laws of thermodynamics

The word *thermodynamics* (from *thermo*: concerning heat, and *dynamic*: concerning force, or power) describes a field of scientific study that started in the 1820s and its three laws had been established by the end of the nineteenth century.

- The **First Law of Thermodynamics** is essentially the law of *conservation of energy*, already mentioned in Chapter 1. It rules out perpetual motion machines, and can be summarized as 'you can't get something for nothing in the world of energy'.

- The **Second Law of Thermodynamics** imposes more severe constraints on any heat engine. It says that when *heat* is the input you can't break even. There must inevitably be some wasted heat energy. The second law and its consequences impose limitations to the efficiency of both heat engines and heat pumps.

- The **Third Law of Thermodynamics** states that there is a lowest possible temperature, an *absolute zero* below which nothing can be cooled further. This temperature is –273 °C. It gives rise to the Kelvin temperature scale which starts from absolute zero. It is named after William Thompson, Lord Kelvin, one of the early pioneers of thermodynamics. As mentioned in Chapter 1, the size of a degree in this scale, a kelvin (K), is the same as one in the more familiar Celsius scale (°C) the only difference being that 0 °C = 273 K. Note that there is no degree symbol used with temperatures in the Kelvin scale.

2.2 Fuels and combustion

The burning (or combustion) of fuels is currently the most common source of useful heat. In 2015 this source made up over 90% of the UK's primary energy and supplied two thirds of the UK's electricity.

What are fuels?

Fuels are materials from which useful energy can be extracted. Common examples are coal, oil, natural gas and wood. The release of energy usually involves combustion. Energy *can* be extracted from some fuels without combustion as, for example, in the digestion of food in the human body, or the use of hydrogen as a fuel in fuel cells (to be discussed in Chapter 11); but here we focus on what happens when combustion occurs.

Some essential features of combustion are:

- it needs air – or to be more precise, oxygen
- the fuel undergoes a major change of chemical composition
- heat is produced, i.e. energy is released.

Consider, as an example, methane. This is the principal component of the fossil fuel natural gas and of biofuels such as landfill gas (see Chapter 5). Each methane molecule consists of one carbon and four hydrogen atoms: CH_4. Atmospheric oxygen has molecules consisting of two atoms (O_2), so in full combustion each methane molecule reacts with two oxygen molecules:

$$CH_4 + 2O_2 \rightarrow CO_2 + 2H_2O + energy$$

The heat energy released in this process is the *difference* between the chemical energy of the original fuel plus oxygen, and the chemical energy of the resulting carbon dioxide plus water. In practice, it is usual to refer to it as the **energy content** (or **calorific value** or **heat value**) of the fuel – the methane in this case.

The reaction shown above shows the essential features of the combustion of many common fuels which contain carbon and hydrogen. The fuel interacts with oxygen from the air to produce carbon dioxide and water, the latter usually as water vapour in the form of steam.

If we know the composition of the fuel and the relative masses of the chemical elements making up its molecules, we can predict how much carbon dioxide will be produced in burning a given amount of fuel (see Box 2.1).

BOX 2.1 CO_2 from fuel combustion

As an example, consider the combustion of methane (CH_4). The masses of the atoms of carbon and oxygen are respectively 12 times and 16 times the mass of a hydrogen atom, so we can associate masses with the items in the combustion equation:

$$CH_4 \quad + \quad 2O_2 \quad \rightarrow \quad CO_2 \quad + \quad 2H_2O$$

$$12 + (4 \times 1) + 2 \times (2 \times 16) \quad \rightarrow \quad 12 + (2 \times 16) + 2 \times (2 \times 1 + 16)$$

We can see, therefore, that burning 16 tonnes of CH_4 releases 44 tonnes of CO_2.

So burning one tonne of methane releases 44/16 = 2.75 tonnes (2750 kg) of CO_2.

Higher and lower heating values

Other fuels are chemically more complex than methane. Many are **hydrocarbons** like methane that consist almost entirely of hydrogen and carbon. Petrol and diesel fuel are common examples. Their combustion follows a similar process. An essential feature is the production of water vapour (steam). In most combustion processes this is just released to the atmosphere, but some useful energy can be obtained by cooling and condensing it, as in a domestic condensing gas boiler.

Fuels are often quoted with two different calorific values giving their potential heat output when burned. Values for biomass, industrial waste, coal and oil are normally quoted as the lower heat value (LHV) or net calorific value (NCV) of the fuel. This assumes that any water vapour produced is not condensed. However, if it is condensed, then a higher heat value (HHV), also called the gross calorific value (GCV), is appropriate. For fuels such as coal and oil the difference between LHV and HHV is small enough to be ignored in rough calculations, but this is not the case for methane or natural gas where water is a major combustion product. The LHV for methane is 50 MJ kg^{-1} but the HHV is 11% higher at 55.5 MJ kg^{-1}. For hydrogen (to be discussed in Chapter 11) the difference between its LHV and HHV is 17%.

Biomass fuels such as wood may also contain considerable amounts of moisture prior to combustion which makes it difficult to specify a precise calorific value. Box 2.2 gives details of the basic properties of a range of common fuels.

> **BOX 2.2 Properties of fuels**
>
> Table 2.1 gives the energy densities and *direct* CO_2 emission factors of different fuels, i.e. the emissions at the point of use. There are other *indirect* CO_2 emissions involved in the supply of fuels (such as the refining of petrol). These are discussed in Chapter 11.
>
> Although all fuels have higher and lower heating values, in practice the higher heat values are only quoted for fuels where condensing of the flue gases is likely to take place, for example in condensing boilers.
>
> The direct CO_2 emissions produced by the combustion of a biofuel such as wood depends on how it has been produced (as will be discussed in Chapter 5). If the wood is grown sustainably, so that trees are planted to replace those harvested and combustion is complete, its life cycle CO_2 emissions should be close to zero. However, there are likely to be CO_2 emissions involved in associated fertilizer production, harvesting and transport.

Table 2.1 Heating values and direct CO_2 emissions of some fuels

Fuel	Lower heating value / MJ kg^{-1}	Higher heating value / MJ kg^{-1}	CO_2 emissions LHV / gCO$_2$ MJ^{-1}
Coal (electricity generation)	24.9	–	90
Road diesel/light heating oil*	42.6 (9.9 kWh litre^{-1})	45.3	74 (2.7 kg litre^{-1})
Petrol (gasoline)*	44.8 (9.1 kWh litre^{-1})	–	70 (2.3 kg litre^{-1})
Natural gas	47.8	53.1	57
Hydrogen	120	142	zero
Wood (air-dry – 20% moisture)	~15	–	~ zero over life cycle

Source: BEIS, 2016a; BEIS, 2016b; AFDC, 2014
Note:*100% mineral fuel - not blended with biofuels

2.3 Heat engines

The steam turbine power station

Electricity generation consumes about a third of the world's fossil fuel supply. Power stations use the heat from burning fuel in a **boiler** to produce hot, high-pressure steam for the **steam turbines** that drive the generators. Such turbine systems are often referred to as **Rankine cycles**, after another of the pioneers of thermodynamics, William Rankine. Figure 2.1 shows such a system.

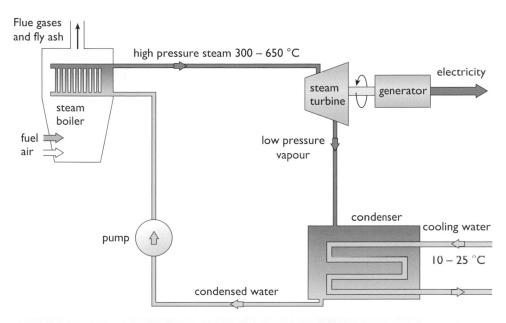

Figure 2.1 Basic power station system with turbine and generator

Steam is raised in a boiler fed by a mixture of fuel and air. The fuel may be coal (often of poor quality and containing sulfur), wood, a 'difficult fuel' such as municipal solid waste, or a mixture. It is important to ensure complete combustion and to minimize the emissions of pollutants such as particulates, oxides of nitrogen (NO_x) and sulfur dioxide (SO_2). Two common types of steam boiler are the **pulverized fuel boiler** and the **fluidized bed boiler**.

In the pulverized fuel boiler the fuel is ground down to a fine powder that burns almost instantly. The energy from this burning powder is transferred to water circulating through pipes in the boiler. The combustion can be carefully controlled by adjusting the ratio of air to the fuel. The combustion temperature must be high enough to produce steam, but not high enough to oxidize the nitrogen in the combustion air to NO_x. The flue gases will contain a large quantity of fine ash (fly ash) and, depending on the sulfur content of the fuel, some sulfur dioxide. The fly ash can be separated out using techniques such as cyclones and electrostatic precipitators. The SO_2 can be separated by reacting the flue gases with wet limestone, i.e. calcium carbonate ($CaCO_3$), turning it into gypsum ($CaSO_4$).

The fluidized bed boiler contains a large bed of inert material such as sand or gravel. This is fed with high pressure air so that it starts to move freely like a liquid. The bed contains heat exchanger pipes that the water is circulated through to be heated.

The fuel is ground down into small particles (but not a powder) and fed into the moving bed. There it burns in close contact with the particles of the bed and the heat exchanger pipes. Limestone can be added to the fuel to absorb any sulfur content. The ash is continually extracted from the bed, rather than from the flue gases as in the pulverized fuel boiler.

In both types the heat exchanger in the boiler is fed with water and produces steam at a high temperature and pressure. Steam is very corrosive (it can slowly 'rust' the pipework). Even with the best modern steel alloys, steam temperatures are limited to about 650 °C.

The steam then travels from the boiler to a steam turbine. Here it is 'expanded', so a small volume of high pressure, high temperature steam is converted to a larger volume of steam at a lower temperature and pressure. This process produces work: the turbine drives a generator to produce electricity.

From the turbine the low pressure steam travels on to a condenser where it is cooled and condensed back to water. The condenser is fed by a large amount of cooling water. This may come directly from a river or the sea, from a large cooling tower inside which river water is evaporated, or from a so-called 'dry cooling' heat exchanger, rather like a very large car radiator through which large volumes of air are circulated. Finally the condensed water is pumped back to the boiler to be reheated.

This basic high temperature steam cycle is used in the combined cycle gas turbine (CCGT) power plant (see page 37), concentrating solar power (CSP) plants (described in Chapter 3), biomass power plants (see Chapter 5) and in 'dry steam' geothermal plants (see Chapter 10).

Organic Rankine Cycle (ORC) engines

The normal working fluid for steam engines is water, which boils at 100 °C at atmospheric pressure. The use of water thus poses difficulties if the supply of heat is below this temperature. The **Organic Rankine Cycle** (ORC) uses a working fluid with a lower boiling point. This can be a common hydrocarbon such as butane, which boils at only 0 °C at atmospheric pressure. Although cheaply available this has the disadvantage of being inflammable (it is sold as fuel for camping stoves!). Alternatively an ORC may use one of a range of hydrofluorocarbons (HFCs) used as refrigerants in refrigerators and heat pumps (to be discussed later in Box 2.4). The **Kalina cycle** is an ORC variant that uses a mixture of ammonia and water as a working fluid.

An ORC power plant uses low temperature heat to boil the working fluid and drive a turbine with the high pressure vapour in exactly the same manner as a steam turbine power plant. It can be used in an Ocean Thermal Energy Conversion (OTEC) system making use of the (small) temperature difference between the surface and bottom of deep sea water (see Chapter 3) or in a 'binary cycle' geothermal plant (see Chapter 10).

Internal combustion engines

Another common heat engine is the **internal combustion engine**, i.e. the fuel is actually burned inside the 'expander' part of the engine rather than in a separate 'boiler'. In its spark ignition form it is used in petrol cars (which may be fuelled with petrol/bioethanol mixtures) and in landfill or sewage gas engines running on methane (see Chapter 5). Its diesel form is used in road vehicles which may run on biodiesel (also described in Chapter 5).

In their most common form both petrol and diesel engines use a piston which is driven up and down inside a cylinder and connected to the drive section by a rotating crankshaft. Figure 2.2 shows the basic operating cycle of a four-stroke petrol or gas engine. At the top of the engine there is a cylinder head containing a number of valves controlling the flow of gas in and out. The four 'strokes' are: induction, compression, power and exhaust.

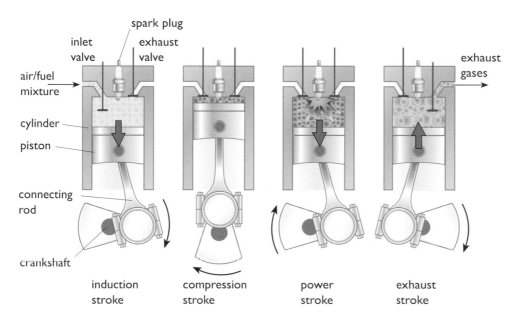

Figure 2.2 The petrol or gas-fuelled internal combustion engine

(1) On the induction stroke a small amount of fuel and air are drawn into a cylinder through the open inlet valve, which then closes.

(2) On the compression stroke this air–fuel mixture is then compressed into typically one-tenth of its original volume, creating a highly inflammable mixture.

(3) This mixture is then ignited using an electric spark on a sparking plug. The gases then burn very rapidly reaching a high temperature (750 °C or more) and expand, pushing down the piston on the power stroke.

(4) Finally, on the exhaust stroke, the burnt gases are pushed out into the exhaust system through the open exhaust valve. The whole cycle then repeats.

The reduction in volume during the second stroke is a rather critical factor called the **compression ratio**. If the volume of the cylinder is 300 cc when the piston is at the bottom of its stroke and the mixture is compressed down to only 30 cc when the piston is right at the top, then the compression ratio is 300:30 or 10:1. This is a typical figure for a modern petrol car engine.

The diesel engine is similar, except that the compression ratio is higher, around 20:1, and the diesel fuel spontaneously ignites (without spark plugs) at the high temperature and pressure produced by the compression. Once ignited, the fuel rapidly burns, reaching a higher temperature than that likely in a petrol engine. Diesel engines are generally more thermally efficient than petrol engines and can use a lower grade of fuel. However they produce more pollution in the form of NO_x and particulates than petrol engines.

Stirling engine

The steam engine is not the only type of 'external combustion engine'. The Stirling engine was quite popular during the nineteenth century for a range of uses. In its modern form it uses reciprocating pistons in cylinders to expand a working fluid such as high pressure helium. It has been used in concentrating solar power (CSP) applications (see Chapter 3).

Thermodynamic limits to heat engine efficiency

In the early years of the nineteenth century, steam engines were being developed and their efficiencies were increasing slowly. It took careful mathematical and logical reasoning by a young French engineer, Sadi Carnot, to show that *it is impossible to have a perfect heat engine*. This effectively is the *Second Law of Thermodynamics*. Carnot took, as his basis for discussion, an extremely idealized version of a steam engine (Figure 2.3).

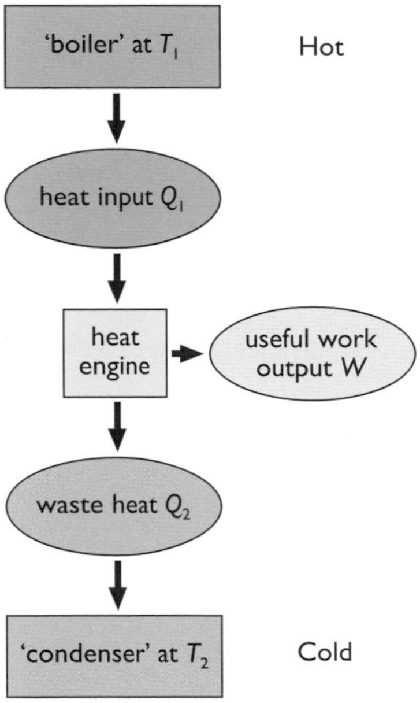

The **working fluid** of his heat engine goes through a continuous cyclic process, taking in heat from the 'boiler' at an **input temperature**, T_1, producing a mechanical output (W), and rejecting waste heat to a 'condenser' at a lower **exhaust temperature**, T_2. Everything is ideal: no friction, perfect insulation, no sudden changes of any sort.

Suppose the quantity of heat taken in from the boiler by each kilogram of working fluid is Q_1 and the quantity of 'waste heat' rejected to the condenser is Q_2. Because energy is conserved (as in the First Law of Thermodynamics), the mechanical output – the work done – must be equal to the difference:

$$W = Q_1 - Q_2$$

The efficiency of the engine, the output divided by the input, is then

$$\text{efficiency} = \frac{W}{Q_1} = \frac{Q_1 - Q_2}{Q_1} = 1 - \frac{Q_2}{Q_1}$$

There are other details in the full specification of the idealized **Carnot engine**, but for our purposes we need only consider the results, the most important of which was Carnot's proof that, for his engine:

$$\frac{Q_2}{Q_1} = \frac{T_2}{T_1}$$

That is, the waste heat divided by the input heat is exactly equal to the condenser temperature divided by the boiler temperature, where these are specified on the Kelvin temperature scale.

This leads to a new simple formula for the efficiency of the Carnot engine:

$$\textbf{efficiency} = 1 - \frac{T_2}{T_1}$$

Or an alternative that is often useful in practice:

$$\textbf{efficiency} = \frac{T_1 - T_2}{T_1}$$

(This can be multiplied by 100 to give the percentage efficiency.)

Carnot then proved the important point that any real heat engine operating between the same two temperatures must be less efficient than his idealized one. In other words, the formula above allows us to calculate the *maximum possible efficiency* for any type of heat engine if we know the boiler and condenser temperatures.

'boiler' at T_1 Hot

heat input Q_1

heat engine → useful work output W

waste heat Q_2

'condenser' at T_2 Cold

Figure 2.3 Carnot diagram for a heat engine

For example, a modern coal fired power station can produce steam at 620 °C (T_1). This would be fed to a steam turbine and low temperature heat might be rejected in river water at 25 °C (T_2). The theoretical efficiency of the system would therefore be:

$$\frac{T_1 - T_2}{T_1} = (620 - 25) / (620 + 273) = 67\%$$

Its practical efficiency is more likely to be about 39%, due to various losses. Low temperature systems using Organic Rankine Cycles are likely to have poor efficiencies. For example, the theoretical Carnot efficiency of a heat engine that was fed with relatively low-temperature vapour at 85 °C, such as that from a geothermal well, and that exhausted heat at 25 °C would be only 17%. In practice its efficiency is likely to be well under 10%.

A highly efficient system – the combined cycle gas turbine (CCGT)

The key message from Carnot's theory is that a high inlet temperature, T_1, is necessary for a high thermal efficiency. Steam temperatures are limited by the corrosive effects of steam on steel, yet combustion temperatures in *gas* turbines used in aircraft can exceed 1500 °C. This temperature is limited only by the properties of the latest metal alloys and ceramics. Industrial gas turbines tend to use lower temperatures to give a longer life.

The Carnot efficiency of a heat engine operating between 1500 °C and 25 °C would obviously be very good (over 80%), but designing a single heat engine that can operate across this wide temperature range is best done by using a combined cycle, with two separate stages.

Such a power station system, as used at Shoreham Power Station in Sussex in the UK, opened in 2002, is shown in Figure 2.4.

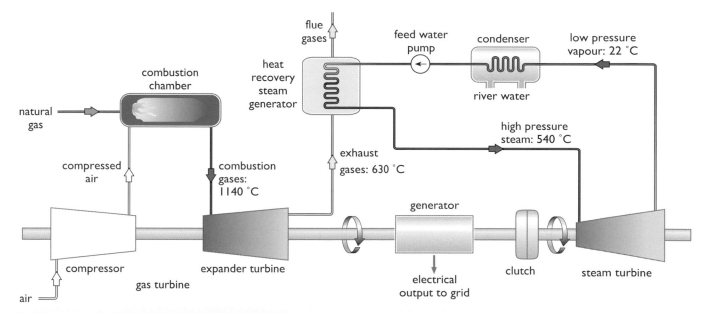

Figure 2.4 Combined cycle gas turbine generation plant

The power plant has an electrical generator mounted on a single drive shaft, which spins at 3000 rpm. This is connected both to an industrial gas turbine and, through a clutch, to a steam turbine. At the far left end of the drive shaft a compressor is driven producing compressed air. Natural gas is then burned in a combustion chamber feeding an expander turbine mounted on the drive shaft with hot gases at 1140 °C. This extracts energy and provides 240 MW of power to the generator.

The exhaust gases from the gas turbine, which leave at about 630 °C are fed to a steam boiler, more formally known as a **heat recovery steam generator (HRSG)**, and used to raise steam at 540 °C. This is then fed to a conventional steam turbine connected to the opposite end of the generator via a clutch, producing a further 140 MW of electricity. The station draws cooling water from a river estuary giving an average working temperature in the condenser of 22 °C .

When starting from cold, the steam turbine is initially left disconnected from the generator. The gas turbine is run up to 3000 rpm using the generator as a motor and the gas burners are lit. It takes about 30 minutes to get the gas turbine stage fully operational and running at low power. Over the next hour steam starts to be produced in the boiler. This is initially used to clean out condensed water in the various stages of the steam turbine. Then the high pressure steam is used to run this turbine up to 3000 rpm and the clutch connects it to the generator. It takes about three hours for the whole system to reach full power of 380 MW. The overall electricity generation efficiency is then over 50% and it will require a natural gas fuel input of around 750 MW (over 50 tonnes per hour). The most recent CCGT designs use combustion temperatures as high as 1600 °C giving an efficiency approaching 60%.

The ability of a system such as this to run up to full power in a short period of time (its so-called **ramp rate**) is important if it is to be used on an electricity system where there is a large proportion of variable renewable electricity from sources such as wind or PV, a topic to be discussed further in Chapter 11. New CCGT designs can reach full power in under an hour.

BOX 2.3 CO_2 emission factors for fossil fuel generated electricity

The overall amount of CO_2 emitted in generating a kilowatt hour of electricity by a conventional fossil fuelled power station depends on its thermal efficiency and the lower heating value of the fuel used.

For example, let us consider a typical UK coal fired power station with an efficiency of 36%.

Efficiency = electricity output / heat input

Rearranging the equation:

Heat input = electricity output / efficiency

Thus to generate 1 kWh of electricity will require 1 / 36% = 2.78 kWh of heat.

The CO_2 emission factor for coal given earlier in Table 2.1 is 90 g CO_2 MJ^{-1}. As noted in Chapter 1, 1 kWh = 3.6 MJ, so the amount of CO_2 emitted in producing 1 kWh of *heat* = 90 × 3.6 = 324 grams.

The amount of CO_2 emitted in producing 2.78 kWh of heat (and thus 1 kWh of electricity) = 2.78 x 324 ≈ 900 grams or 0.90 kg.

Table 2.2 gives average UK emission factor values for the two main fossil fuel technologies in competition with renewable electricity generation, coal and gas. Other technologies, such as nuclear power or the various forms of renewable electricity generation, are likely to have low values of under 0.1 kg CO_2 kWh^{-1}.

In 2015 UK electricity was produced by a mix of generation technologies, giving an average emission factor of 0.33 kg CO_2 kWh^{-1}. This figure has been decreasing since the 1980s as gas and nuclear power have replaced coal. It will decrease further as coal fired generation is totally phased out in the UK and the proportion of low carbon electricity generation (renewables and nuclear power) increases.

Table 2.2 CO_2 emission factors for UK electricity generation, 2015 averages

Generation technology	CO_2 emission factor/kg CO_2 kWh^{-1}
Coal	0.92
CCGT natural gas fired	0.38
UK overall mix	0.33

Source: BEIS, 2016b

2.4 Heat pumps

The need for alternative heating systems

As noted in Chapter 1, approximately a third of UK delivered energy is used for low temperature heating; *space heating* to warm the internal spaces of buildings, *domestic water heating*, and *low temperature processing and drying in industry*. In principle, these tasks only require heat at a temperature of 60 °C or less. 90% of the heat for this currently comes from the combustion of natural gas or heating oil, and this combustion produces about a quarter of the UK's CO_2 emissions (BEIS, 2016c; BEIS, 2016d).

As will be described in Chapter 3, the space heating demand of buildings can be dramatically reduced by the use of insulation. Also in urban areas there are opportunities for supplying low temperature heat to buildings from conventional power stations, tapping off some of the waste heat. This is known as **combined heat and power generation (CHP)**, a topic to be discussed further in Chapter 11. Also **active solar heating** (to be discussed in Chapter 3) has a future role to play in supplying domestic water heating in both urban and rural areas.

There is a particular need to develop non-fossil fuel heating for buildings in rural areas, particularly those currently using heating oil or direct electric resistance heating. In principle, heat pumps can turn a supply of *low carbon electricity* into a supply of *low carbon heat*. They have long been used to provide *cooling* (or *air conditioning*) in buildings but in the UK it is only in the past 20 years that there has been interest in using them for heating.

Basic principles

A heat engine converts a flow of heat (from a high temperature to a lower one) into work. A heat pump does the reverse: it is a device for pumping heat from a lower temperature to a higher one using an input of work, usually supplied by an electric motor. The most common example is the domestic refrigerator. In this, heat is absorbed in an evaporator inside the refrigerated compartment, thus lowering its temperature, and pumped to a condenser on the back of the refrigerator, where the heat is released.

Figure 2.5 shows the key elements of the most basic type of heat pump.

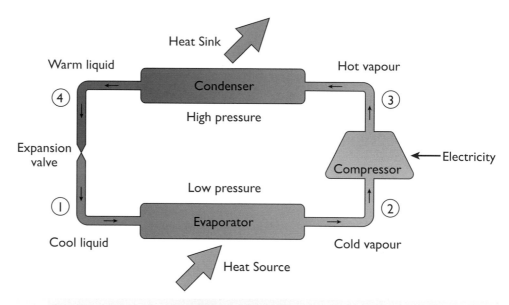

Figure 2.5 Schematic diagram of a heat pump

The heat pumping process is made possible by the use of a special refrigerant liquid that boils at a low temperature at atmospheric pressure.

- At point (1) in Figure 2.5 the refrigerant starts as a cool liquid. In order to convert a liquid to a vapour, it must be given energy – the so-called latent heat of evaporation. The refrigerant absorbs heat in a heat exchanger called the **evaporator** and vaporizes.

- At point (2) the vapour enters an electrically driven compressor that raises its pressure and temperature.
- At point (3) the hot vapour then enters another heat exchanger, the **condenser**, where it condenses to a warm liquid and gives up its latent heat of vaporization.
- At point (4) it is forced through a fine expansion valve or throttle where it loses pressure and drops in temperature.

It then repeats the cycle. Overall, heat is 'pumped' from a low temperature in the evaporator to a higher one in the condenser.

The choice of refrigerant fluid has a slightly dark history since one refrigerant in particular has contributed to the 'ozone hole' over the Antarctic (see Box 2.4).

BOX 2.4 Choosing an environmentally friendly refrigerant

Ozone depletion

In order to be effective a refrigerant fluid should boil at just the right temperature (about −30 °C at atmospheric pressure), be non-flammable and non-corrosive. Such a chemical is Freon-12, a chlorofluorocarbon (CFC), a hydrocarbon where some of the hydrogen atoms have been substituted by atoms of chlorine and fluorine. This was developed in the 1930s and became the standard working fluid for domestic and industrial refrigerators. Then in 1974 it was discovered that CFCs leaking from fridges (and other appliances) were making their way high into the stratosphere and breaking down the ozone layer which shields the Earth's surface from ultra violet radiation from the Sun. A large 'ozone hole' had been developing over Antarctica that was eventually reported in 1985.

The result was an urgent series of meetings during the 1980s and 1990s coordinated by the United Nations Environment Programme eventually leading to a global ban on the use of Freon-12 and a range of other refrigerants containing chlorine. This agreement is usually referred to as the 1987 Montreal Protocol. Other readily available refrigerant chemicals which do not damage the ozone layer such as ammonia, butane and propane have their own problems of toxicity or inflammability restricting their widespread use (see Table 2.3 below). New commercial refrigerants containing fluorine but no chlorine (**hydrofluorocarbons or HFCs**) were developed to fill the gap.

Following the ban on chlorinated refrigerants the ozone hole over Antarctica has been healing slowly and may disappear by about 2070.

Global warming

In the 1990s it was pointed out that as well as damaging the ozone layer Freon-12 was a very powerful greenhouse gas with a very high direct global warming potential (DGWP) of 10 900. This is the ratio of its potency as a greenhouse gas to that of CO_2. Thus the leakage of 1 g of this refrigerant would have a global warming effect equivalent to 10.9 kg of CO_2.

While the use of Freon-12 is now banned, its HFC replacements are also strong greenhouse gases with DGWPs of around 1500. In 2016 international agreement was reached to phase out their use.

Table 2.3 Refrigerants and their environmental properties

Refrigerant	DGWP	Environmental problems
Freon-12	10 900	A chlorofluorocarbon (CFC). Damaging to the ozone layer and a strong greenhouse gas. Now banned.
HFC 134a	1430	A hydrofluorocarbon (HFC). Developed as a replacement for Freon-12. Commonly used in heat pumps. Now being phased out because of its high DGWP
HFO-1234yf	<1	Hydrofluoroolefins (HFO). Developed as a
HFO-1234ze	<1	replacement for HFC 134a
Ammonia	0	In use as a refrigerant since the 1850s. Often used in large heat pumps. Highly toxic with a strong smell.
Isobutane	3	Sometimes used in ORCs. Highly inflammable.
Propane	3.3	Highly inflammable

Source: BEIS, 2016a

A whole set of new refrigerants, **hydrofluoroolefins (HFOs)** have now been developed that do not damage the ozone layer and have low direct global warming potentials. Like HFCs these consist of hydrogen, fluorine and carbon, but the term HFO has been used to distinguish them from their high DGWP relatives.

Domestic heat pumps

In warm countries domestic heat pumps are commonly used for **air conditioning**, *cooling* a building (or even a car) in summer. In this case the heat pump has a fan coil heat exchanger located inside as the evaporator and another one in the external environment as the condenser. Heat is pumped from the inside to the outside cooling the interior space. Electrical energy is used to drive the heat pump compressor. Many such systems used in southern Europe or the southern states of the USA are *reversible* and can be used to provide *heating* in winter. In this case the external heat exchanger becomes the evaporator and the internal one the condenser.

The main focus of domestic heat pump systems for the UK has been on those that only provide heating. Figure 2.6(a) shows the basic layout of such a heat pump with three different possibilities for the heat source.

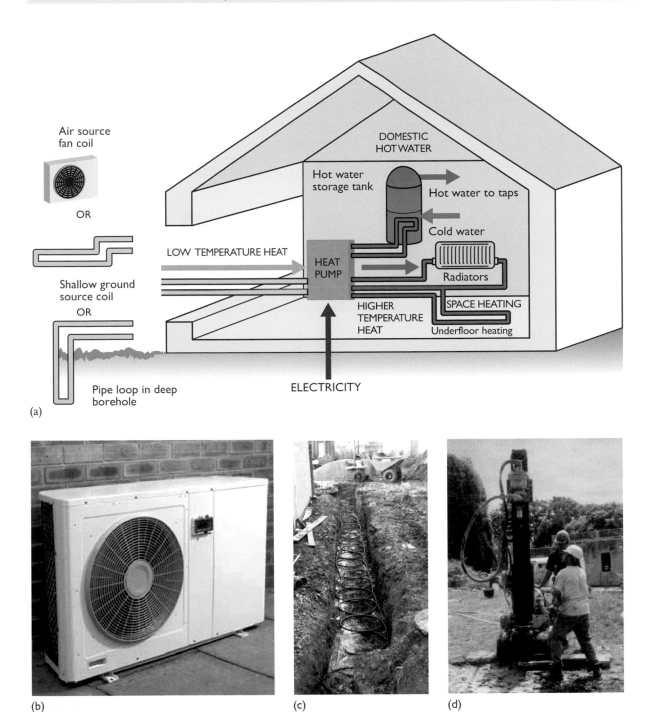

Figure 2.6 (a) Layout of a heat pump system used for heating a building and supplying domestic hot water (b) A fan-coil evaporator unit for an air source heat pump (c) Evaporator ground source heat exchanger. Slinky coils in a trench about to be buried (d) Drilling a borehole for a ground source evaporator

The evaporator is located outside in the external environment. There are three possible alternatives. An **air source heat pump** is likely to have a fan coil unit such as that shown in Figure 2.6(b). A **ground source heat pump** uses pipes buried in the soil. These may be laid in a shallow trench in a large garden, as in Figure 2.6(c) or in a deep vertical borehole that may be 30 metres or more deep, as seen being drilled in Figure 2.6(d).

Low temperature heat is then pumped from the outside environment to a condenser inside the building, normally connected to a central heating system. This is likely to have a storage tank for the domestic hot water. The space heating may use conventional radiators or, preferably, underfloor heating which can use hot water with a lower temperature. A good practical system might use underfloor heating on the ground floor and radiators in upstairs rooms. The temperature of the heat from the condenser must be sufficient to be useful for heating purposes, in the range of 40–60 °C. Electrical energy is, of course, required to operate the compressor.

A key performance parameter is the **coefficient of performance** (COP):

COP = Heat output from condenser / Electrical work input

The maximum value of this is limited by Carnot's theorem (see Box 2.5).

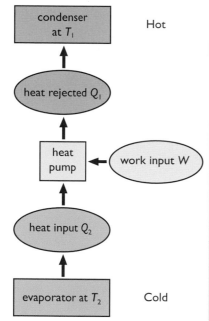

Figure 2.7 Carnot diagram for a heat pump

BOX 2.5 The Carnot efficiency of heat pumps

The theoretical Carnot heat engine described earlier is completely reversible. It is possible to drive it backwards supplying work and extracting heat.

The work input from an electric motor is W.

The useful heat output from the condenser at temperature T_1 is Q_1.

From consideration of the conservation of energy $W = Q_1 - Q_2$.

The coefficient of performance (COP) =

$$Q_1 / W = Q_1 / (Q_1 - Q_2)$$

Carnot's theorem says that this ratio will be the same as that of the associated temperatures (in the Kelvin scale, of course):

$$COP = T_1 / (T_1 - T_2)$$

This can obviously be greater than 1.

For example, if the evaporator was at 0 °C and heat was required at 60 °C, the maximum theoretical COP would be

$$COP = (60 + 273) / (60 - 0) = 5.55$$

A practical value for these temperatures might be less than 4.

It is possible to have a high COP if the temperature difference through which the heat is raised is small, but the value may drop significantly if the temperature difference is large. This may be particularly true where domestic hot water has to be heated to 60 °C in mid-winter conditions.

In order to function in mid-winter, the evaporator has to be able to absorb heat from the external environment even though the external temperature may be very low (below −5 °C in the UK). The fluid circulated between the heat exchanger and the heat pump must not freeze so is usually a water/antifreeze mixture using ethylene glycol or propylene glycol, which is less toxic.

Burying the evaporator coil in the ground (or in a lake or river) provides a more stable temperature environment in extreme winter conditions and can result in higher COPs. For design purposes the undisturbed ground temperature below about 10 metres depth can be assumed to be almost constant at the annual mean air temperature for that location. For central Europe this temperature is about 10 °C but at only two metres depth it will vary by about plus and minus 3 °C over the year (BSI, 2007). More precise values will depend on the actual site location.

In an air source heat pump, the heat that is drawn from the external environment is taken immediately from the outside air, cooling it in the process. The evaporator may build up an external layer of ice and require an occasional 'defrost cycle' in which hot water is circulated from the heat pump through the evaporator, wasting some energy in the process.

In a ground source heat pump, heat is extracted from the external environment by cooling the ground (but only by a degree or two) so that heat flows down into it from the air above over a large area and a long time period.

It is important that the evaporator is made large enough to supply the heat pump. This is not particularly a problem with air source heat pumps, but installing ground source loops or boreholes is expensive and they must not be undersized. Given this expense, it is important that proper consideration is given to insulating a house to minimize the heat demand before installing a heat pump.

The precise ground source evaporator size depends on the soil conditions. Dry soil with a poor thermal conductivity will require larger ground loops or boreholes than ground with a high thermal conductivity such as moist or water-saturated soil, or solid rock.

Sizing a shallow ground source evaporator

A shallow ground source evaporator is likely to take the form of a set of trenches one to two metres deep into which a long pipe is laid backwards and forwards, or as a set of loops. These are so-called Slinky coils, shown earlier in Figure 2.6(c).

In simple cases (i.e. domestic buildings), the average specific values for the heat extraction rate can be obtained by determining an extraction rate *per square metre* of ground collector area in W m^{-2}. The extraction rate depends upon the thermal conductivity of the ground (the higher the conductivity the better) and the overall duration of heat extraction by the heat pump in hours per year. This in turn will depend on the length of the heating season and the level of insulation of the house (a topic to be discussed in Chapter 3).

Table 2.4 gives some sample figures for a shallow ground source evaporator assuming an operation period of 2400 hours per year.

Table 2.4 Heat extraction rates for shallow ground source evaporators - operation period 2400 hours per year

Ground quality	Specific heat extraction flow rate / W m^{-2}
Dry, non – cohesive soil	8
Moist cohesive soil	16 to 24
Water saturated sand or gravel	32

Source BSI, 2007 (see MCS, 2011 for more detailed UK figures)

Thus, for example, over a year a square metre of evaporator in moist cohesive soil with a heat extraction flow rate of 20 W m^{-2} should be capable of supplying 2400 × 20 watt hours = 48 kWh of heat.

A small, very well-insulated house might use a heat pump drawing 3 kW (3000 W) of heat from the ground. With a soil heat extraction flow rate of 20 W m^{-2} this will require 3000 / 20 = 150 m^2 of evaporator area (i.e. an area 10 metres by 15 metres). Finding this amount of space to dig up in a domestic garden may prove difficult, so it may be preferable to drill one or more deep boreholes.

Sizing a ground source borehole evaporator

For small domestic applications a borehole can be sized by calculating its *overall length*. Table 2.5 shows some sample heat extraction rates for boreholes in an assumed location in central Europe.

Table 2.5 Heat extraction rates for borehole ground source evaporators - operation period 2400 hours per year

Ground quality	Specific heat extraction flow rate / W m^{-1}
Dry gravel or sand	< 20
Moist clay	30 – 40
Gravel or sand saturated with water	55 – 65
Gravel or sand and strong ground water flow	80 – 100

Source BSI, 2007 (see MCS, 2011 for more detailed UK figures)

In this case for a heat pump requiring 3 kW of heat, a borehole in moist clay might be conservatively assumed to be capable of supplying 30 watts per metre length of borehole. The total length of borehole would thus have to be 3000 / 30 = 100 metres. This is most likely to take the form of two separate 50 metre deep boreholes.

Practical heating system performance

An actual *heating system* using a heat pump may include hot water storage tanks, radiators, fans and pumps. It may also require top-up direct electric heating of the circulated water in midwinter. In order to minimize this it is important to ensure that the heat pump is properly sized to meet the full design heat load of the building (MCS, 2013).

A more accurate measure of the overall performance than the COP is thus the **system performance factor (SPF)**. This is the ratio of the heat produced by the heating system to the electricity consumed. Measured SPF values for a properly installed system in the UK have ranged from 2.2 to 4.0, with ground source heat pumps performing better than air source ones (Energy Saving Trust, 2013). Put simply:

Useful heating energy produced = electricity consumed × SPF

Although heat pumps driven by gas engines or absorption heat cycles have been produced, this is essentially an electric heating technology. It can be thought of as reclaiming some of the primary energy losses at the power station.

As described earlier, a modern gas-fuelled CCGT power can have an efficiency of 50%. It thus takes 2 kWh of gas to produce 1 kWh of electricity. However, a heat pump with an SPF of 3.0 can turn this back into 3.0 kWh of heat (although only at a low temperature).

The overall fuel efficiency = 3.0 kWh / 2 kWh = 150%

This makes it an attractive alternative to fossil fuel heating and direct electric resistance heating.

Large heat pump projects

There are obviously economies of scale where a single large heat pump can be used in conjunction with an existing district heating network to supply a large number of buildings or flats within a single block.

Aquifer Thermal Energy Storage: London Wandsworth Riverside Quarter

Many cities are built on the flood plains of large rivers, London being a good example. Much of the underlying ground along the banks of the Thames in London is an **aquifer**, a layer of gravel saturated with water. As shown in Table 2.4, this type of ground has a high thermal conductivity and is ideal for heat pump use. The Aquifer Thermal Energy Storage (ATES) concept uses a heat pump to cool a large volume of underground water saturated soil in winter to provide space heating in buildings. Then in summer, when space cooling is required, the heat pump is reversed and used to warm the soil.

Such a system has been installed in south-west London at the London Wandsworth Riverside Quarter, a development of over 500 flats with commercial space. Three heat pumps are connected to a set of eight 120 m deep boreholes. These actually tap into the groundwater and circulate it through the heat pump.

The heat pumps supply a peak cooling capacity of 2.25 MW and a heating peak output of 1.2 MW. The aquifer is normally at a temperature of about 15 °C but warms slightly over the summer due to the injection of the waste heat from the cooling loads, leading to better heat pump performance in winter. In the winter the aquifer is cooled as heat for the space heating is drawn out, and this cooling of the aquifer leads to higher cooling COPs during summertime operation. Top-up heat is supplied from gas boilers and a small gas combined heat and power (CHP) system.

Helsinki district heating

The city of Helsinki in Finland has a well-established district heating network that supplies 90% of the city's heat demand. It also has a district cooling network supplying chilled water. The main fuel for these systems is natural gas. In 2006, 84 MW of heat pump capacity was integrated into the existing system, covering 4% of the network's total heat and 33% of total cooling load.

Five 16.8 MW heat pumps are connected to both the heating and cooling networks. In winter the heat pumps recover heat sourced from the city's warm outflowing sewage at 10 °C. The heat is raised in temperature and used to preheat the district heating hot water supply from 50 °C to 62 °C. This supply is then topped up to the final supply temperature using natural gas. Combining the two methods of heating allows the heat pump to work at its best thermal efficiency, raising a large amount of heat through a small temperature difference.

In summer the heat pumps remove heat from the water in the district cooling network, reducing the temperature to 4 °C. Some of this heat, again raised in temperature, is then pumped into the district heating network to provide the city's domestic hot water load. The rest is rejected through heat exchangers into the Baltic Sea (DECC, 2016).

Environmental benefits

Although it is possible to build heat pumps directly fuelled by gas, their use is primarily an electric heating technology competing with conventional heating fuels. Given a supply of low carbon electricity, it has potential to reduce the CO_2 emissions of heating systems. The amount of CO_2 emitted in producing a kilowatt hour of useful heat from natural gas or light heating oil is fixed by the heating value of the fuel and the thermal efficiency of the heating system.

As noted earlier in Table 2.2, the overall emissions produced by direct electric resistance heating are dependent on the fuels used to generate electricity. These will vary widely between different countries.

Table 2.6 gives the CO_2 emission factors for heat from common UK heating systems and for a heat pump with an assumed SPF of 3.0. Since the heat pump can produce three units of heat for every unit of electricity consumed, its CO_2 emission factor is only a third of that of direct electric heating.

Table 2.6 Direct CO_2 emissions for different fuels and heating methods

Heating fuel and method	CO_2 emissions /kg per kWh of heat
Natural gas, 90% efficient condensing boiler	0.20
Light heating oil, 90% efficient condensing boiler	0.28
Direct resistance heating, 100% efficiency – 2015 UK generation mix	0.33
Electric heat pump, SPF = 3.0 – 2015 UK generation mix	0.11

Source: BEIS, 2016a

In 2014 a typical UK household required about 13 000 kWh of heating energy per year (BEIS, 2016d). For such a household using light heating oil the annual heating emissions would be 13 000 × 0.28 = 3640 kg CO_2.

If a heat pump with an SPF of 3.0 was installed then the annual emissions would be 13 000 × 0.11 = 1430 kg.

The annual emissions saving =

3640 − 1430 = 2210 kg or 2.21 tonnes of CO_2 per year.

These potential annual savings are likely to increase as the carbon intensity of UK electricity generation continues to fall.

Electricity supply considerations

The energy demand for heating domestic hot water is roughly constant throughout the year. However, demand for space heating peaks in the winter months. Heating oil is an ideal fuel for supplying space heating since it is easy to store throughout the year. Natural gas can also be stored but less easily. However, electricity essentially has to be generated on demand.

The use of any form of electric heating thus requires the existence of some form of power station to supply it when required. Seen from the perspective of an electricity supply industry, the best electricity load is one that is constant throughout the year. This could be supplied from a 'base load' power station that ran flat out uniformly 365 days a year.

An electricity demand that only occurs in mid-winter would have to be supplied by a power station that sits idle for the rest of the year. Although it wouldn't be consuming any fuel, its capital cost still has to be paid for. Such winter peak loads are usually met by older power stations on the retirement list.

Any technology that *reduces* the mid-winter electricity demand peak is likely to be welcomed by the electricity supply industry. Using heat pumps to replace direct electric resistance heating is an example. However, replacing oil or gas heating by electric heat pumps will *increase* the mid-winter electricity demand and may create electricity supply problems. This is discussed further in Chapter 11.

Economics

Heat pumps have a high initial capital cost. For a domestic installation of a particular power rating an air source heat pump is likely to be cheaper than a ground source one. An installed ground source heat pump system is likely to cost over £10 000, about two thirds of which will be for the ground source evaporator. Given this expense it is essential that the house is properly insulated first.

When taken as one of a number of options for reducing national CO_2 emissions, heat pumps are considered by the UK Committee on Climate Change as being largely cost-effective in buildings not connected to the natural gas grid, as an alternative to oil or electric heating (CCC, 2016).

Heat pumps are supported in the UK under the government's Renewable Heat Incentive scheme. Under this, home owners who install heat pumps receive annual payments based on the heat load of the house for the first seven years of operation. There is also a similar non-domestic scheme.

Is this really renewable energy?

The energy savings of heat pumps can be thought of as a form of *energy efficiency*, getting more heating value for a kilowatt hour of electricity.

It is thus perhaps rather confusing to find that the *heat gains* from heat pumps, i.e. the difference between the heat output and the electricity input, have been classified as *renewable energy* but only if the SPF is greater than 2.5. Under the 2009 EU Renewable Energy Directive (CEC, 2009) gains from air source heat pumps are termed *aerothermal*; those from rivers or lakes are *hydrothermal* and those from 'energy stored in the form of heat beneath the surface of solid earth' are classified as *geothermal* energy.

Ground source heat pumps, particularly those using vertical borehole heat exchangers are likely to be called 'geothermal heat pumps'. Even though these boreholes may be 30 metres deep, they are essentially only dealing with energy flows near the surface of the ground. This differs from the 'deep geothermal energy', described in Chapter 10, using water from hot underground rocks which may be a kilometre below the surface.

For the UK energy statistics (as shown in Figure 1.8 in Chapter 1) the heat gains from *all* heat pumps (ground source and air source) are classified together with heat from deep geothermal wells. Together in 2015 they supplied an estimated 7.1 PJ of heat (BEIS, 2016c). This was roughly 1% of the UK's total renewable energy supply for that year.

Deployment and future prospects

In 2014 there were an estimated 120 000 heat pumps in the UK of which about 15% were ground source (EurObserv'ER, 2015). Given that there are 26 million households in the UK the low proportion reflects the strong competition from natural gas as a heating fuel.

In Europe heat pumps have been strongly promoted in countries such as France and Sweden. In France, which has almost no indigenous oil or gas resources, heat pumps are more widely used, largely supplied with electricity from its many nuclear power stations. In 2014, there were an estimated 4.4 million heat pumps in France, almost all air-sourced.

Sweden, likewise, is reliant on imported oil but has a plentiful supply of hydroelectricity. In 2014 there were an estimated 1.4 million heat pumps in Sweden, about a third of them ground source.

There is obviously a considerable potential for their use in the UK. The Committee on Climate Change has suggested that a total of 1.1 million could be retrofitted in UK homes by 2030 (CCC, 2016).

2.5 Summary

This chapter has described some basic physics that underpins many renewable technologies using heat. It has described the three Laws of Thermodynamics:

- The **First Law** is the **law of conservation of energy**.
- The **Second Law** (proposed by the French engineer **Carnot**) sets limits to the maximum efficiency of both *heat engines* and *heat pumps*.
- The **Third Law** defines an 'absolute zero' temperature (-273 °C) at which substances contain *no thermal energy*.
- The **Kelvin temperature scale** is used for Carnot efficiency calculations and starts at absolute zero. A degree kelvin (K) is the same size as a degree Celsius (°C). A kelvin temperature = Celsius temperature + 273.

The burning (or combustion) of fuels is the most common source of useful heat. Common fuels are coal, oil and natural gas. Their combustion involves a reaction with oxygen to produce heat and the emission of carbon dioxide (CO_2). The combustion of natural gas, which is mainly methane, or oil also produces water vapour (steam).

The **heating value of a fuel** is the amount of heat produced by the combustion of one kilogram. **The higher heating value** or **gross calorific value** is the amount of heat produced if the water vapour is condensed back to water. **The lower heating value** or lower calorific value is the amount of heat produced if the water vapour is simply released into the atmosphere.

The carbon dioxide emission factor, the amount of CO_2 emitted per unit of heat produced, varies between fuels. Coal has the highest emissions, followed by diesel fuel, petrol and natural gas as the lowest.

This chapter has also focussed on heat engines and heat pumps.

Heat engines turn energy into work, either generating electricity or providing motive power for transportation.

The **steam turbine power station cycle** (also called the **Rankine cycle**) is used to generate most of the world's electricity. A fuel such as coal, wood or municipal solid waste is burned in a **boiler** to produce high pressure steam to drive a **steam turbine**. Alternatively, steam can be raised using geothermal energy or concentrating solar power.

The **Organic Rankine Cycle (ORC)** engine uses a refrigerant fluid with a lower boiling point than water to drive a turbine enabling it to use sources of heat below 100 °C.

Internal combustion engines (or piston engines) burn fuels such as diesel, petrol or a gas to drive a rotating crankshaft. They are mainly used in road vehicles and ships or for electricity generation using landfill or sewage gas.

The **Stirling engine** is an external combustion piston engine with potential for electricity generation use in small Concentrating Solar Power (CSP) plants.

The **Second Law of Thermodynamics (Carnot's Theorem)** sets limits to the maximum thermal efficiency of a heat engine. It depends critically on the highest temperature that can be produced by burning a fuel. The maximum temperature in a steam turbine is limited by the properties of steel. This limits the practical overall efficiency of a conventional coal fired power station to about 39%.

The **Combined Cycle Gas Turbine (CCGT)** is a heat engine with a high thermodynamic efficiency. It normally burns natural gas at a temperature of over 1000 °C to drive a gas turbine. The waste heat from this is then used to raise steam to drive a conventional steam turbine. The gas and steam turbines together drive an electric generator. Practical CCGT power stations can achieve a thermal efficiency of 50% or more.

A kilowatt hour of electricity produced by burning natural gas in a CCGT power station only involves the emission of 40% of the CO_2 that would be emitted for a kWh of electricity from a coal fired power station.

Heat pumps use electrical energy to drive a compressor raising a flow of heat from a low temperature to a higher one.

The heat pump is a very common technology used in the domestic fridge. Electrical energy is used to drive a compressor cycling a refrigerant fluid between two heat exchangers: an **evaporator** on the 'cold' side and a **condenser** on the 'hot' side. A heat pump uses electrical work to raise a flow of heat from a low temperature to a higher one. In a fridge it pumps heat from the interior, lowering its temperature, and out into the room. When used for heating, a heat pump draws heat from the external environment, raises its temperature and pumps it into the interior of a building.

Refrigerants are fluids with low boiling points. They may be used in ORC heat engines or in heat pumps. The long-term use of **Chlorofluorocarbon (CFC)** refrigerants has **damaged the ozone layer** in the upper atmosphere, which protects the Earth's surface from ultra-violet rays from the Sun. There is now a global ban on their use. They have been replaced by **Hydrofluorocarbon (HFC) refrigerants**. Even these are now being phased out because they are strong greenhouse gases and contribute to global warming. They are being replaced by **Hydrofluoroolefin (HFO) refrigerants**.

Domestic heat pumps may be **air source**, taking heat from a evaporator fan coil unit in the external air or **ground source**, taking heat from the ground, either from a long heat exchanger pipe buried in a shallow trench (usually as a set of 'Slinky' loops) or pipes buried in deep vertical boreholes.

The **coefficient of performance (COP)** is a key performance parameter.

COP = Heat output from condenser / electrical work input.

Like heat engines, **the maximum possible performance is limited by the Second Law of Thermodynamics** and Carnot's theorem. It is possible to have a high COP if the heat is only raised through a small temperature difference, but a higher temperature difference will mean a lower COP.

Air source heat pumps are likely to require regular 'defrosting' in winter and will generally have a lower COP than a ground source heat pump.

Ground source evaporators should be adequately sized. The exact size is dependent on the thermal conductivity of the soil. Moist or water-saturated soils have a high thermal conductivity. A shallow pipe evaporator for a small house may require 150 m^2 or more of ground area.

The **System Performance Factor (SPF)** is a better guide to overall performance than the COP. It is the ratio of useful heat produced to the total electricity consumed, including that used for pumps and top-up electric heating. A reasonable value is about 3.0.

An aquifer is an underground layer of water-saturated soil. Aquifer Thermal Energy Storage uses a heat pump to heat a building in winter, cooling the underground soil and then to cool the building in summer, re-warming the ground beneath.

Large heat pumps can be used in conjunction with city heating schemes that provide both district heating and cooling. Heat can even be usefully drawn from a city's waste sewage water.

Heat pumps can turn a supply of low carbon electricity into low carbon heat. They are considered a cost-effective option to reduce the CO_2 emissions of UK homes that do not have a natural gas supply.

Replacing direct resistance heating by heat pumps will *reduce* the peak winter electricity demand and is likely to be welcomed by the electricity supply industry. Replacing fossil fuelled heating by heat pumps will *increase* the winter electricity demand and is not likely to be welcomed.

The use of heat pumps can be considered as a form of *energy efficiency*, but in Europe the energy gains from heat pumps have been classified as *renewable energy* for statistical reporting purposes. Ground source heat pumps are often referred to as *geothermal* heat pumps, and their energy gains as *geothermal energy*. But this is not the same as the 'deep geothermal energy' described in Chapter 10.

References

AFDC (2014) *Fuel Properties Comparison*, Washington, Alternative Fuels Data Center, AFDC [Online]. Available at http://www.afdc.energy.gov/fuels/fuel_comparison_chart.pdf (Accessed 16 October 2016).

BEIS (2016a) *Government emission conversion factors for greenhouse gas company reporting*, London, Department for Business, Energy and Industrial Strategy, BEIS [Online]. Available at https://www.gov.uk/government/collections/government-conversion-factors-for-company-reporting#conversion-factors-2016 (Accessed 16 October 2016).

BEIS (2016b) *Digest of United Kingdom energy statistics (DUKES)*, London, Department for Business, Energy and Industrial Strategy, BEIS [Online]. Available at https://www.gov.uk/government/collections/digest-of-uk-energy-statistics-dukes (Accessed 16 October 2016).

BEIS (2016c) *UK Energy in Brief 2016*, London, Department for Business, Energy and Industrial Strategy, BEIS [Online]. Available at https://www.gov.uk/government/collections/uk-energy-in-brief#2016 (Accessed 16 October 2016).

BEIS (2016d) *Energy Consumption in the UK (ECUK)*, London, Department for Business, Energy and Industrial Strategy, BEIS [Online]. Available at https://www.gov.uk/government/statistics/energy-consumption-in-the-uk (Accessed 16 October 2016).

BSI (2007) *Heating systems in buildings: Design of heat pump heating systems, BS EN 15450:2007* [Online]. Available at www.standardsuk.com (Accessed 16 October 2016).

CCC (2016) *Next Steps for UK Heat Policy*, London, Committee on Climate Change [Online]. Available at https://www.theccc.org.uk/wp-content/uploads/2016/10/Next-steps-for-UK-heat-policy-Committee-on-Climate-Change-October-2016.pdf (Accessed 18 October 2016).

CEC (2009) *Directive 2009/28/EC on the promotion of energy from renewable sources*, Commission of the European Communities, CEC [Online]. Available at http://eur-lex.europa.eu (Accessed 16 October 2016).

DECC (2016) *Heat Pumps in District Heating: Case Studies*, London, Department of Energy and Climate Change, DECC [Online]. Available at https://www.gov.uk/government/uploads/system/uploads/attachment_data/file/489605/DECC_Heat_Pumps_in_District_Heating_-_Case_studies.pdf (Accessed 27 October 2016).

Energy Saving Trust EST (2013) *The heat is on: heat pump field trials phase 2*, Energy Saving Trust [Online]. Available at http://www.energysavingtrust.org.uk (Accessed 11 May 2016).

EurObserv'ER (2015) *Heat Pumps Barometer 2016* [Online]. Available at http://www.eurobserv-er.org/category/all-heat-pumps-barometers/ (Accessed 28 October 2016).

MCS (2011) MCS 022: *Ground heat exchanger look-up tables*, Microgeneration Certification Scheme, MCS [Online]. Available at http://www.microgenerationcertification.org (Accessed 28 October 2016).

MCS (2013), *Microgeneration Installation Standard: MIS 3005, Issue 4.3*, Microgeneration Certification Scheme, MCS [Online]. Available at http://www.microgenerationcertification.org (Accessed 28 October 2016).

Further reading

GSHPA (2007) CE82. London, Ground Source Heat Pump Association [Online]. Available at http://www.gshp.org.uk/documents/CE82-DomesticGroundSourceHeatPumps.pdf (Accessed 17 October 2016).

Acknowledgements

The author would like to thank Janet Ramage for her contribution on some of these topics in previous editions of this book.

Chapter 3

Solar thermal energy

By Bob Everett

3.1 Introduction

As we saw in Chapter 1, the Sun is the ultimate source of most of our renewable energy supplies. Since there is a long history of the Sun being regarded as a deity, the direct use of solar radiation has a deep appeal to engineer and architect alike.

In this chapter, we look at some of the methods employed to gather solar thermal or heat energy. Solar photovoltaic (PV) energy, the direct conversion of the Sun's rays to electricity, is dealt with in Chapter 4. Solar thermal collection methods are many and varied, so we can only give the briefest introduction and supply pointers to further reading for those interested in studying the subject in greater depth.

What sorts of system can be used to collect solar thermal energy?

Most systems for low-temperature solar heating depend on the use of glazing, in particular its ability to transmit visible light but block infrared radiation. High-temperature solar collection is more likely to employ mirrors to concentrate the Sun's radiation. In practice, solar energy systems of both types can take a wide range of forms. These include:

Active solar heating. This always involves a discrete **solar collector**, usually mounted on the roof of a building, to gather solar radiation. Mostly, collectors are quite simple and the heat will be at low temperature (under 100 °C) and used for domestic hot water or swimming pool heating.

Passive solar heating. This term has two slightly different meanings.

- In the 'narrow' sense, it means the absorption of solar energy directly into a building to reduce the energy required for heating the habitable spaces (i.e. what is called **space heating**). Passive solar heating systems mostly use air to circulate the collected energy, usually without pumps or fans – indeed the 'collector' is often an integral part of the building.

- In the 'broad' sense, it means the whole process of integrated low-energy building design, effectively to reduce the heat demand to the point where small passive solar and other 'free heat' gains make a significant contribution in winter. A large solar contribution to a large heat load may look impressive, but what really counts is to minimize the total fossil fuel consumption and thus achieve the minimum cost. This is a key feature of *superinsulated* or *Passivhaus* building design.

Daylighting. This means making the best use of natural daylight, through both careful building design and the use of controls to switch off artificial lighting when there is sufficient natural light available.

Solar thermal engines (Concentrating Solar Power, or CSP). These are an extension of active solar heating, usually using more complex collectors to produce temperatures high enough to drive steam turbines to produce electric power. They can come in a number of different types, but most of the world's solar thermally generated electricity is produced in California and Spain in multi-megawatt plants using large parabolic mirrors.

It must be stressed at the outset that making the best use of solar energy requires a careful understanding of the climate of any particular location. Indeed, many of our present energy problems stem from attempts to produce buildings inappropriate to the local climate. This can mean that the economics of solar technologies commonly used in southern Europe may be disappointing when transferred, for example, to northern Scotland.

However, most of the methods described in this chapter have been well tried and tested over the past century. Even the most spectacular of modern solar thermal-electric power stations are just uprated versions of inventive systems built at the beginning of the 20th century. The skill of using solar thermal energy, in all its forms, perhaps lies in producing systems that are cheap enough to compete with 'conventional' systems based on fossil fuels at current prices.

3.2 The rooftop solar water heater

For most people, 'solar heating' means the rooftop solar water heater. In Europe by 2014 there were over 45 million square metres (45 square kilometres) of solar collectors installed, nearly 18 million m^2 in Germany but only about 780 000 m^2 in the UK (ESTIF, 2015). Most use simple flat plate collectors. There are two basic forms of system: pumped or thermosyphon.

The pumped solar water heater

This is the form most common in northern Europe, normally roof-mounted (Figure 3.1). A typical flat plate pumped system consists of three elements as shown in Figure 3.2.

Figure 3.1 Solar panels mounted on roof (photo courtesy of Arcon)

(1) A collector panel, typically of 3–5 square metres in area, tilted to face the Sun and mounted on the normal pitched roof of a house, as in Figure 3.1. This panel itself normally consists of three components (see Figure 3.3). The main absorber might be a steel plate bonded to copper or steel tubing through which water circulates. The plate is sprayed with a special black paint or coated with a selective surface to maximize the solar absorption. It is normally covered with a single sheet of glass or plastic and the whole assembly is insulated on the back to cut heat losses.

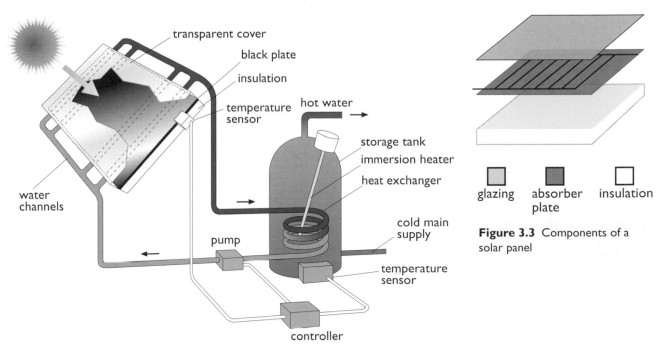

Figure 3.3 Components of a solar panel

Figure 3.2 Pumped active solar water heater

(2) A storage tank, typically of around 200 litres capacity, which often doubles as the normal domestic hot water cylinder. This often contains an electric immersion heater for winter use. The tank is insulated all round, typically with 50 mm of glass fibre or polyurethane foam. The hot water from the panel circulates through a heat exchanger at the bottom of the tank.

(3) A pumped circulation system to transfer the heat from the panel to the store. Sensors detect when the collector is becoming hot and switch on an electric circulating pump. Since in northern Europe the collector has to be able to survive freezing temperatures, the circulating water contains an antifreeze. Non-toxic propylene glycol is often used (instead of the poisonous ethylene glycol commonly used in car engines).

In the UK, field trial results suggest that such a system can typically provide over 1100 kWh y^{-1} or about 40% of a household's hot water (EST, 2011).

The thermosyphon solar water heater

In frost-free climates where it is safe to mount the storage tank outdoors, a simpler **thermosyphon** arrangement can be used, as shown in Figure 3.4.

This design dispenses with the circulation pump. It relies on the natural convection of hot water rising from the collector panel to carry heat up to the storage tank, which must be installed above the collector. There is no need for a heat exchanger as the required domestic hot water circulates directly through the panel.

Normally the storage tank also contains an electric immersion heater for top-up and use on cloudy days. Mediterranean systems are usually designed

Figure 3.4 A typical Mediterranean thermosyphon solar water heater – the insulated storage tank is at the top

to be free-standing for mounting on buildings with flat roofs. Given the higher levels of solar radiation in these countries, they are usually sold with only around 2 m² of collector area.

3.3 The nature and availability of solar radiation

The wavelengths of solar radiation

The Sun is an enormous nuclear fusion reactor, which converts hydrogen into helium at the rate of 4 million tonnes per second. It radiates energy by virtue of its high surface temperature, approximately 6000 °C. Of this radiation, approximately one-third of that incident on Earth is simply reflected back. The rest is absorbed and eventually retransmitted to deep space as long-wave infrared radiation. On average the Earth re-radiates just as much energy as it receives and sits in a stable energy balance at a temperature suitable for life.

We perceive solar radiation as white light. In fact it spreads over a wide spectrum of wavelengths, from 'short-wave' infrared (longer than red light) to ultraviolet (shorter than violet). The pattern of wavelength distribution is critically determined by the temperature of the surface of the Sun (see Figure 3.5).

The Earth, which has an average atmospheric temperature of –20 °C and a surface temperature of 15 °C, radiates energy as long-wave infrared to deep space, the temperature of which is only a few degrees above the absolute zero value, –273 °C. We tend to forget this outgoing radiation, but its effects can be observed on a clear night when a ground frost can occur as heat radiates first to the cold upper atmosphere and then out into space.

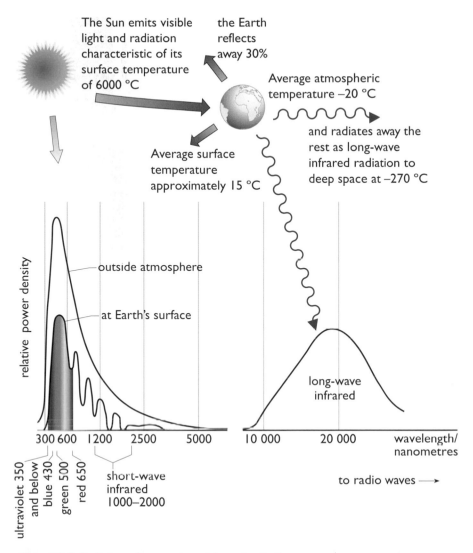

Figure 3.5 Radiation of energy to and from the Earth

As we shall see, most of the art of low-temperature solar energy collection depends on our ability to use glass and surfaces with selective properties that allow solar radiation to pass through but block the re-radiation of long-wave infrared. The gathering of solar energy for high-temperature applications, such as driving steam engines, mainly involves concentrating solar energy using complex mirrors.

Direct and diffuse radiation

When the Sun's rays hit the atmosphere, more or less of the light is scattered, depending on the cloud cover. A proportion of this scattered light comes to Earth as **diffuse radiation**. On the ground this appears to come from all over the sky. Some of it we see as the blue colour of a clear sky, but most is the 'white' light scattered from clouds.

What we normally call 'sunshine', that portion of light that appears to come straight from the Sun, is known as **direct radiation**. On a clear day, this can approach a power density of 1 kilowatt per square metre (1 kW m^{-2}), known as '1 sun' for solar collector testing purposes. Generally in northern Europe and in urban locations in southern Europe, practical peak power densities are around 900–1000 watts per square metre.

In northern Europe, on average over the year approximately 50% of the radiation is diffuse and 50% is direct. In southern Europe, where solar radiation levels are higher, most of the extra contribution is in direct radiation, especially in summer. Both diffuse and direct radiation are useful for most solar thermal applications, but only direct radiation can be focused to generate very high temperatures. On the other hand it is the diffuse radiation that provides most of the 'daylight' in buildings, particularly in north-facing rooms.

Availability of solar radiation

Figure 3.6 A pyranometer (also known as a solarimeter)

Interest in solar energy has prompted the accurate measurement and mapping of solar energy resources over the globe. This is normally done using **pyranometers** (see Figure 3.6). These contain carefully calibrated thermoelectric elements fitted under a glass cover, which is open to the whole vault of the sky. A voltage proportional to the total incident light energy is produced and then recorded electronically.

Most pyranometer measurements are recorded simply as **total energy incident on the horizontal surface**. More detailed measurements separate the direct and diffuse radiation. These can be mathematically recombined to calculate the radiation on tilted and vertical surfaces.

As we might expect, annual total solar radiation on a horizontal surface is highest near the equator, over 2000 kilowatt-hours per square metre per year (kWh m^{-2} y^{-1}), and especially high in sunny desert areas. These are more favoured than northern Europe, which typically only receives about 1000 kWh m^{-2} y^{-1}. Many experimental projects, such as solar thermal power stations, have been built in areas like southern France or Spain, where radiation levels are around 1500 kWh m^{-2} y^{-1}, or the southern USA, where levels can reach 2500 kWh m^{-2} y^{-1}.

It is obvious that in Europe summers are sunnier than winters, but what does that mean in energy terms?

On average in July, the solar radiation on a horizontal surface in northern Europe (e.g. Ireland, UK, Denmark and northern Germany) is between 4.5 and 5 kWh m^{-2} per day (see Figure 3.7). Five kilowatt-hours represents about three quarters of the daily average energy consumption for water heating for an average UK household. At 2016 UK domestic fuel prices, this amount of heat would cost approximately 23p if it was obtained from a high-efficiency gas boiler, or about 33p using off-peak electricity. In southern Europe (Spain, Italy and Greece), July solar radiation levels are higher, between 6 and 7.5 kWh m^{-2} per day.

In winter, however, the amount of solar radiation is far lower. In January on average in northern Europe it can be only one-tenth of its July value, around 0.5 kWh m^{-2} per day (see Figure 3.8), yet in southern Europe there may still be appreciable amounts, 1.5 to 2 kWh m^{-2} per day.

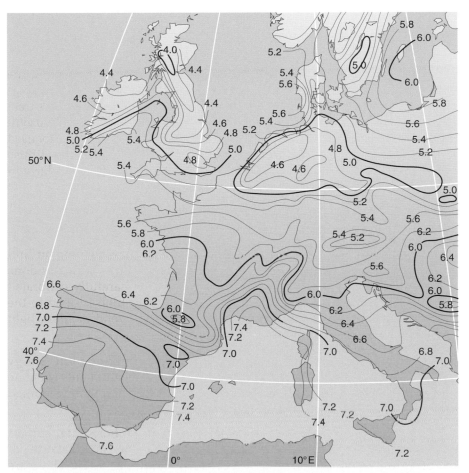

Figure 3.7 Solar radiation on horizontal surface (kWh per square metre per day), Europe, July (source: CEC, 1994)

The implications of this are that in northern Europe we need to look for applications that require energy mainly in the summer. In southern Europe, there may be enough radiation in the winter to consider year-round applications.

Tilt and orientation

So far, we have talked about solar radiation on the horizontal surface. To collect as much radiation as possible, a surface should face south (assuming it is in the northern hemisphere) and must be tilted towards the Sun. How much it should be tilted is dependent on the latitude and at what time of year most solar collection is required.

If the tilt angle between a surface and the horizontal is equal to the latitude, it will be perpendicular to the Sun's rays at midday in March and September (see Figure 3.9).

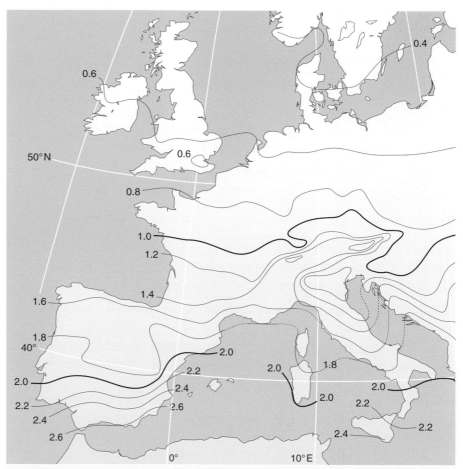

Figure 3.8 Solar radiation on horizontal surface (kWh per square metre per day), Europe, January (source: CEC, 1994)

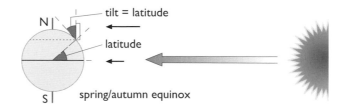

Figure 3.9 A surface tilted at the latitude angle will be perpendicular to the Sun's rays at mid-day on the spring or autumn equinox

There is also the difference between summer and winter to consider. Box 3.1 explains why countries at high latitudes (such as the UK) receive more solar energy in summer than winter.

To maximize solar collection in summer (when there is most radiation to be had), the tilt angle should be less than the latitude. To maximize solar collection in winter (when more solar radiation may be needed) the tilt angle should be greater than the latitude angle (see Figure 3.12).

BOX 3.1 Solar radiation and the seasons

If the energy output of the Sun is constant, why does the UK receive more radiation in summer than in winter?

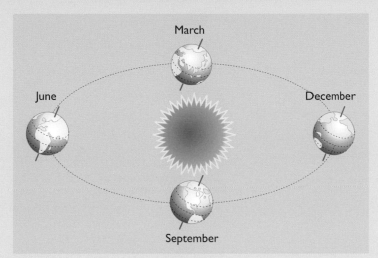

Figure 3.10 The Earth revolves around the Sun with its axis tilted at an angle of 23.5° to the plane of rotation

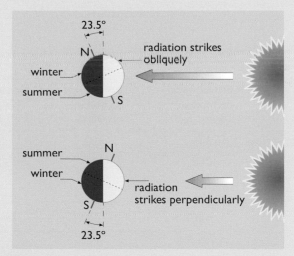

Figure 3.11 The tilt of the Earth's axis creates summer and winter

The Earth circles the Sun with its polar axis tilted towards the plane of rotation (Figure 3.10). In June, the North Pole is tilted towards the Sun. The Sun's rays thus strike the northern hemisphere more perpendicularly and the Sun appears higher in the sky (Figure 3.11). In December the North Pole is tilted away from the Sun and its rays strike more obliquely, giving a lower energy density on the ground (i.e. fewer kilowatt-hours reach each square metre of ground per day).

Another important factor is that the lower the Sun in the sky, the further its rays have to pass through the atmosphere, giving them more opportunity to be scattered back into space. When the Sun is at 60° to the vertical its peak energy density will have fallen to one-quarter of that when it is vertically overhead. This topic will be revisited in Chapter 4.

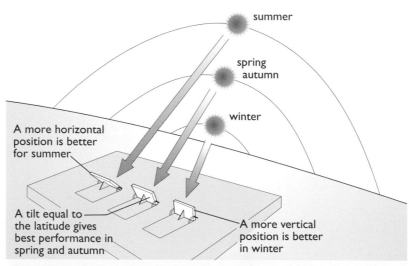

Figure 3.12 Optimizing the tilt for different seasons

Fortunately the effects of tilt and orientation are not particularly critical. Table 3.1 gives totals of energy incident on various tilted surfaces for Kew, near London.

Table 3.1 Effect of tilting a south-facing collection surface (data for Kew, near London, latitude 52° N)

Tilt /°	Annual total radiation /kWh m^{-2}	June total radiation /kWh m^{-2}	December total radiation /kWh m^{-2}
0 – Horizontal	944	153	16
30	1068	153	25
45	1053	143	29
60	990	126	30
90 – Vertical	745	82	29

Source: Achard and Gicquel, 1986

Similarly, the effects of orientation away from south are relatively small. For most solar heating applications, collectors can be faced anywhere from south-east to south-west. This relative flexibility means that a large proportion of existing buildings have roof orientations suitable for solar energy systems. This conclusion applies to both solar thermal and solar photovoltaic (PV) systems.

3.4 The magic of glass

Most low-temperature solar collection is dependent on the properties of one rather curious substance – glass. It is hard to imagine a world without glazed windows. They have been around since the time of the Romans, who invented a process for making plate glass, although the ability to make large sheets of glass was lost in the aptly named 'Dark Ages' and did not reappear until the 17th century.

Transparency

Glass is transparent to visible light and short-wave infrared radiation but has the added advantage of being opaque to long-wave infrared re-radiated from a solar collector or building behind it (see Figure 3.13).

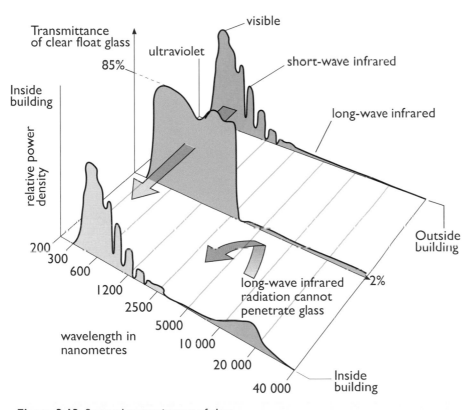

Figure 3.13 Spectral transmittance of glass

Over the past few decades, enormous effort has been put into improving the performance of glazing, both to increase its transparency to visible radiation, and to prevent heat escaping through it.

Manufacturers strive to make glass as transparent as possible, i.e. they try to maximize its **transmittance**, the fraction of incident light that passes through it. They usually do this by minimizing the iron content of the glass. Certain plastics that have optical properties similar to glass can be used instead, although normally they must be protected from the damaging effects of ultraviolet light.

Table 3.2 shows the optical properties of commonly used glazing materials. They share the property of a high solar transmittance (close to 1.0), but the long-wave infrared transmittance is very low by comparison.

Table 3.2 Optical properties of commonly used glazing materials

Material	Thickness /mm	Solar transmittance	Long-wave infrared transmittance
Float glass (normal window glass)	3.9	0.83	0.02
Low-iron glass	3.2	0.90	0.02
Perspex	3.1	0.82	0.02
Polyvinyl fluoride (tedlar)	0.1	0.92	0.22
Polyester (mylar)	0.1	0.87	0.18

Heat loss mechanisms

Much development work has also gone into reducing the heat loss through windows and solar collector glazing. Heat energy will flow through any substance where the temperature on the two sides is different.

The *rate* of this energy flow depends on:

■ the temperature difference between the two sides

■ the total area available for the flow

■ the insulating qualities of the material.

It is obvious that more heat is lost through a large window than a small one, and on a cold day than a warm one. In order to understand how this heat loss occurs, and how it can be minimized, we need to look at the three mechanisms that are involved in the transmission of heat: conduction, convection and radiation.

Conduction

In any material, heat energy will flow by **conduction** from hotter to colder regions. The rate of flow will depend on the first two factors listed above, and on the **thermal conductivity** of the material.

Generally, *metals* have very high thermal conductivities and can transmit large amounts of heat for small temperature differences. Where the frames of glazing systems are made of metal, they should include an insulated thermal break to minimize heat loss.

Insulators require a large temperature differential to conduct only a small amount of heat. Still air is a very good insulator. Most practical forms of insulation rely on very small pockets of air, trapped for example between the panes of glazing, as bubbles in a plastic medium, or between the fibres of mineral wool.

Convection

A warmed fluid, such as air, will expand as it warms, becoming less dense and rising as a result, creating a fluid flow known as **convection**. This is one of the principal modes of heat transfer through windows and out to the environment (see Figure 3.14). It occurs between the air and the glass on the inside and outside surfaces, and, in double glazing, in the air space

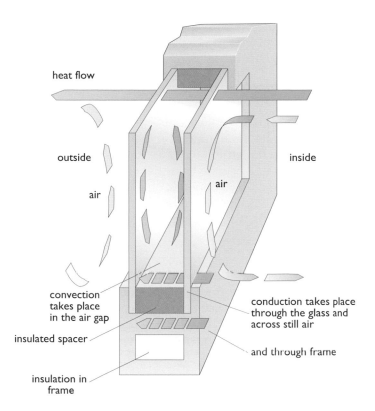

heat flow

outside   inside

air   air

convection
takes place
in the air gap   conduction takes place
                                           through the glass and
                                           across still air
insulated spacer

                                           and through frame

insulation in
frame

Figure 3.14 How heat escapes from a double-glazed window. Most use a sealed double-glazing unit with a space between the panes of about 16 mm and filled with air or argon gas. If the space is too narrow, convection will be difficult but conduction will be easy because there is only a small thickness of air to conduct across. If it is too wide, convection currents can easily circulate. In addition infra-red radiation across the air space can be reduced by using low emissivity coatings and conduction through the glazing spacer and frame blocked with insulation.

between the panes. The convection effects can be reduced by filling double glazing with heavier, less mobile gas molecules, most commonly argon, though krypton can be used.

Convection can also be reduced by limiting the space available for gas movement. This is the principle used in the insulation materials mentioned above.

Various forms of **transparent insulation** have been developed that use a transparent plastic medium containing bubbles of trapped insulating gas. These materials could eventually revolutionize the concept of windows (and walls), but at present the materials are expensive, not robust and need protection from the rigours of weather and ultraviolet light.

Alternatively double glazing can be evacuated. Convection currents cannot flow in a vacuum. However, a very high vacuum is required and it needs to last for the whole life of the window, 50 years or more. Also the window will need internal structural spacers to stop it collapsing inwards under the air pressure on the outside. These spacers conduct heat across the gap, slightly reducing the overall performance. The gap can be very small, typically 0.2 mm in commercial vacuum double glazing units.

A simpler way to reduce the convection effects is to insert extra panes of glass or of transparent plastic film between the other two, turning double glazing into triple or quadruple glazing.

Radiation

Heat energy can be **radiated**, in the same manner as it is radiated from the Sun to the Earth. The quantity of radiation is dependent on the temperature difference between the radiating body and its surroundings. The roof of a building, for example, will radiate heat (i.e. long-wave infrared radiation) away to the atmosphere. It also depends on a quality of the surface known as **emissivity**. Most materials used in buildings have high emissivities of approximately 0.9, that is, they radiate 90% of the theoretical maximum for a given temperature.

Other surfaces can be produced that have low emissivities. This means that although they may be hot, they will radiate little heat outwards. **'Low-E' coatings** are now normally used inside double glazing to cut radiated heat losses from the inner pane to the outer one across the air gap. There are two basic types. 'Hard coat' uses a thin layer of tin oxide, giving an emissivity of about 0.15. 'Soft coat' uses very thin layers of optically transparent silver sandwiched between layers of metal oxide and gives a lower emissivity of about 0.05.

Window *U*-value

Conduction, convection and radiation all contribute to the complex process of heat loss through a wall, window, roof, etc. In practice, the actual performance of any particular building element is usually specified by a **U-value**, defined so that:

heat flow rate per square metre = *U*-value × temperature difference.

The units in which *U*-values are expressed are thus watts per square metre per kelvin (W m^{-2} K^{-1}). As pointed out in Chapters 1 and 2 temperatures can be measured in degrees Celsius (°C) or kelvins (K). The size of a degree is the same on both scales, so temperature differences are identical in °C and K and *U*-values will often be seen written in units of W m^{-2} °C^{-1}.

The lower the *U*-value, the better the insulation performance. Table 3.3 gives typical *U*-values of various types of window glazing system (the precise values will depend on construction details, particularly the details of the frames). By way of comparison: 10 cm of opaque fibreglass insulation has a *U*-value of 0.35 W m^{-2} K^{-1}.

Table 3.3 Indicative *U*-values for windows with wood or PVC-U frames

Glazing type	W m^{-2} K^{-1}
Single glazing	4.8
Double glazing (normal glass, air filled)	2.7
Double glazing (hard coat low-e, emissivity = 0.15, air filled)	2.0
Double glazing (hard coat low-e, emissivity = 0.2, argon filled)	2.0
Double glazing (soft coat low-e, emissivity = 0.05, argon filled)	1.7
Triple glazing (soft coat low-e, emissivity = 0.05, argon filled)	1.3

Source: BRE, 2013

The figures in Table 3.3 are conservative. Currently commercially available double glazed windows with insulated frames may have U-values of 1.5 W m^{-2} K^{-1} or better. Box 3.2 gives a sample energy calculation.

BOX 3.2 *U-value and heat loss*

What is the rate of heat loss through a large single-glazed window with an area of 2 m^2, on a day when the outdoor and indoor temperatures are 5 °C and 20 °C respectively?

Table 2.3 shows that the U-value for this window is 4.8 W m^{-2} K^{-1}, so the loss rate is

$$2 \times 4.8 \times (20 - 5) = 144 \text{ W}$$

Note that, if the temperature difference remained the same throughout 24 hours, the total loss would be almost 3.5 kWh. If this window was replaced with the best of the glazing types shown in Table 3.3, this loss would be reduced to under 1 kWh. Although windows are an important route by which solar energy can be collected, their role in cutting unnecessary heat losses is just as important.

3.5 Low-temperature solar energy applications

We have seen how solar radiation can produce low-temperature heat. Just how useful is this?

As we saw in Chapter 1, Figure 1.7, in the UK about a third of all the end-use of fuel is for low-temperature space and water heating. Over 80% of delivered energy use in the domestic sector is in this form (see Figure 3.15).

Although simple solar systems are in principle ideal for supplying this heat, there are other potential competitors. These include:

- district heating fed by waste heat from existing conventional power stations or from industrial processes
- heat from small-scale combined heat and power generation plant
- heat pumps (described in Chapter 2).

All of these merit further development, and unlike solar heating, most have the advantage of being able to run all year round.

Swimming pools are another potential solar application. They do not use a significant proportion of Europe's total energy consumption, since there are not very many of them, but individually they can be enormous energy users. A large, indoor leisure pool in northern Europe can use 1 kW of power for every square metre of pool area continuously throughout the year. This kind of establishment is a prime candidate for the technologies listed above.

Outdoor pools, usually unheated, are rather different. Here the aim is to make the water a little more attractive when people come to use them, which is usually on sunny, warm days. This is ideal solar heating territory.

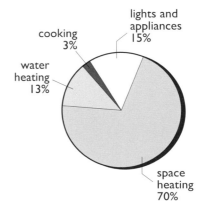

Figure 3.15 Breakdown of UK domestic sector energy use 2013 (BEIS, 2016a)

Domestic water heating

Domestic water heating is perhaps the best overall potential application for active solar heating in Europe. It is a demand that continues all year round and still needs to be satisfied in the summer when there is plenty of sunshine. In the UK in 2013 it accounted for approximately 4% of the total national delivered energy use. A typical UK household uses approximately 7 kWh per day of delivered energy for this purpose (BEIS, 2016a). In practice, much of this can be simply lost as waste heat. Uninsulated hot water cylinders and unlagged pipework are common causes of such losses and even solar water heaters can suffer from this failing.

Incoming mains water is usually at a temperature close to that of the ground at about 1 metre depth, approximately 12 °C in the UK, varying only slightly over the year, and it has to be heated up to 60 °C. In some books it is suggested that temperatures as low as 45 °C are adequate, but recent concerns over Legionnaires' Disease, caused by *Legionella pneumophila* bacteria multiplying in warm water, have highlighted the need for a higher temperature.

Domestic water heating is usually done in one of three ways:

- By electricity, with an immersion heater in a hot-water storage cylinder.
- Again using a storage cylinder, but with a heat exchanger coil inside connected to a central heating boiler (usually gas-fired) or possibly to a district heating supply system.
- By an 'instantaneous' heater, usually powered by gas or electricity.

Each of these will have different CO_2 emissions per kilowatt-hour of heat produced (see Table 3.4). Where solar heat can usefully substitute for these alternative heat sources these will be the relevant CO_2 savings.

Table 3.4 Direct CO_2 emissions of different alternative heating systems

Heating system and fuel	Emission factor /kg CO_2 kWh^{-1}
90% efficiency condensing gas boiler – natural gas	0.20
90% efficiency condensing gas boiler – light heating oil	0.28
Electric heating – 100% efficiency – 2015 UK fuel mix	0.33

Note: See Chapter 11 for more detail on 'direct' and 'indirect' CO_2 emissions.

In the UK, natural gas is the dominant fuel for domestic heating. At present (2016) the best CO_2 savings are likely to result from substituting solar energy for heat from heating oil or electricity use (both probably in rural areas beyond the gas grid). However the UK electricity emission factor is likely to fall in the future as more renewably-generated electricity is connected to the grid, limiting the overall possible CO_2 savings.

In many sunnier countries the majority of homes may use electric water heating throughout the year and the electricity generation fuel mix may have a higher proportion of fossil fuels. For example in Greece, the electricity generation mix currently has a high proportion of coal use and the emission savings may be closer to 0.65 kg CO_2 per kWh of heat produced (IEA, 2016, BP, 2016).

Also, in such countries, the national electricity demand is likely to peak in the summer, with ever-increasing demands for refrigeration and air-conditioning, rather than in the winter as in the UK. Thus every solar water heater installed saves not only on fuel, but, equally importantly, on building power plants to meet the peak summer electricity demand. Put another way, where solar heat can be used to *reduce* the peak summer electricity load, this can be considered as beneficial as building a PV or solar thermal-electric power plant to *supply* an equivalent amount of extra electricity.

Domestic space heating

Space heating involves warming the interior spaces of buildings to internal temperatures of approximately 20 °C. In the UK, it consumes about 30% of the country's delivered energy, yet with an appropriate heating system it can in principle be carried out with water at only 45 °C. It is an activity that only occurs over the **heating season**. For normal UK buildings, this extends from about mid-September to April, although, as we shall see later in the section on passive solar heating, this can vary considerably with location and level of insulation.

However, there is a fundamental problem that for this application in the UK, as in much of northern Europe, the availability of solar radiation is completely out of phase with the overall demand for heat (see Figure 3.16). Although the total amount of solar radiation over a whole year on a particular site may far exceed the total building heating needs, the amount available during the heating season may be quite small.

Even with south-facing vertical surfaces, the amount of radiation intercepted in the UK over the winter is relatively small. In London, for example, over a typical six-month winter period of October to March, 1 m² of south-facing vertical surface will only intercept 250 kWh of solar radiation.

It is important to emphasize that the suitability of solar energy for space heating is dependent on the local climate. Textbooks may show quite grandiose solar buildings, but these may only be appropriate in particular locations, often places that are both cold and sunny in winter.

In summer, the UK has similar temperatures, and receives a similar amount of solar radiation, to other European countries on the same latitude.

In winter, the picture is different. The UK has relatively mild winters. However, the winter solar radiation remains largely dependent on latitude alone. As can be seen in Figure 3.17, average January temperatures in London are virtually identical to those in the south of France (follow the 5 °C contour).

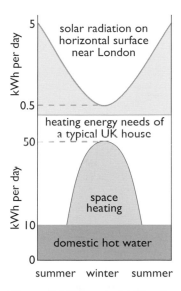

Figure 3.16 The availability of solar radiation is out of phase with space heating demand in the UK

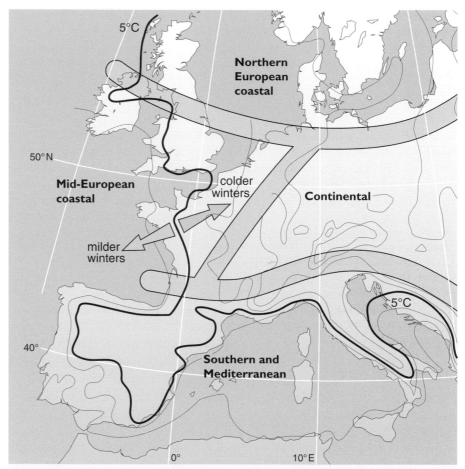

Figure 3.17 The different climatic zones of Europe. UK winters are mild compared with much of the rest of Europe. The 5 °C contour is for average January temperatures.

Why then do northern Europeans go south for the winter? Because it is sunnier. As we saw from Figure 2.8, the south of France receives three times as much solar radiation on the horizontal surface in mid-winter as does London.

Broadly speaking, the climate of western Europe can be split into four regions (Figure 3.17):

(1) Northern European coastal zone: cold winters with little solar radiation in mid-winter; mild summers.

(2) Mid-European coastal zone: cool winters with modest amounts of solar radiation; mild summers.

(3) Continental zone: very cold winters with modest amounts of solar radiation; hot summers.

(4) Southern and Mediterranean zone: mild winters with high solar radiation; hot sunny summers.

It is no coincidence that many solar experimental projects have been built on the boundaries of regions 3 and 4, in the Pyrenees and the area around the Alps. This kind of climate is also typical of Colorado in the central USA, another area where solar-heated houses abound.

The broad view of passive solar heating is really about the subtle influence of climate on building design. Without this appreciation, it is all too easy to design buildings that are inappropriate to their surroundings.

As the Roman architect Vitruvius said in the first century BC:

> We must begin by taking note of the countries and climates in which homes are to be built if our designs for them are to be correct. One type of house seems appropriate for Egypt, another for Spain … one still different for Rome, and so on with lands of varying characteristics. This is because one part of the Earth is directly under the Sun's course, another is far away from it … It is obvious that designs for homes ought to conform to the diversities of climate.
>
> (Cited in Butti and Perlin, 1980)

Varieties of solar heating system

In practice, the categories 'active' and 'passive' are not clear cut: they blend into each other, with a whole range of possibilities in between. The following examples illustrate the range of solar heating systems available in addition to the roof-mounted solar water heaters that we have already considered.

Swimming pool heating

For swimming pool heating, the solar system can be extremely simple. Pool water is pumped through a large area of collector, usually unglazed. Typically, the collector will be about half the area of the pool itself. The best results are achieved with pools that do not have other forms of heating and are consequently at relatively low temperatures (under 20 °C). The aim may not necessarily be to save energy as much as to make the pool temperature more acceptable to bathers.

Conservatory (or 'sunspace')

A conservatory or greenhouse on the south side of a building can be thought of as a kind of habitable solar collector (see Figure 3.18(a)). Air is the heat transfer fluid, carrying energy into the building behind. The energy store is the building itself, especially the wall at the back of the conservatory.

Trombe wall

With a Trombe wall (named after its French inventor, Félix Trombe), the conservatory is replaced by a thin air space in front of a storage wall (see Figure 3.18(b)). This is a solar collector with the storage immediately behind. Solar radiation warms the store and is radiated into the house in an even fashion from its inner side. In addition, on sunny days, air is circulated through the air space into the house behind. At night and on cold days, the air flow is cut off.

This concept can take many forms. Small collector panels can be built directly on to the existing walls of buildings. In the extreme, the air path can be omitted, and walls simply covered with 'transparent insulation'.

Direct gain

Direct gain is the simplest and most common of all passive solar heating systems (see Figure 3.18(c)). All glazed buildings make use of this to some degree. The Sun's rays simply penetrate the windows and are absorbed into the interior. If the building is 'thermally massive' enough, i.e. built of heavy materials such as concrete, and the heating system responsive, the gains are likely to be useful. If the building is too 'thermally lightweight', such as one of timber frame construction, it may overheat on sunny days and the occupants will perceive the effect as a nuisance.

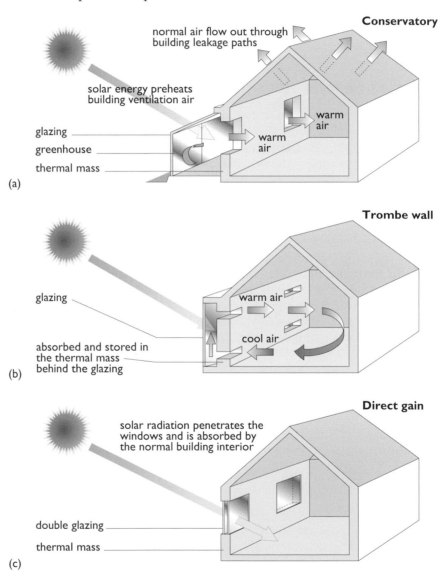

Figures 3.18 Different types of passive solar heating system: (a) conservatory; (b) Trombe wall; (c) direct gain

3.6 Active solar heating

History

A solar water heater could be made simply by placing a tank of water behind a normal window. Indeed, many of the first systems produced in the USA in the 1890s were little more than this.

The thermosyphon solar water heater as we know it was patented in 1909 by William J. Bailey in California. Since the system had an insulated tank, which could keep water hot overnight, Bailey called his business the 'Day and Night' Solar Water Heater Company (Figure 3.19). He sold approximately 4000 systems before the local discovery of cheap natural gas in the 1920s virtually closed his business.

In Florida, the solar water heating business flourished until the 1940s. Eighty per cent of new homes built in Miami between 1935 and 1941 had solar systems. Possibly as many as 60 000 were sold over this period in this area alone. Yet by 1950 the US solar heating industry had completely succumbed to cheap fossil fuel (Butti and Perlin, 1980).

It was not until the oil price rises of the 1970s that the commercial solar collector reappeared. Low fuel prices in the 1990s discouraged their use, but they since have been heavily promoted in countries such as Austria and Germany and, more recently, China. By the end of 2015 an estimated total of 622 million m² (i.e. 622 square kilometres) of solar collectors had been installed worldwide, the figure having quadrupled since 2005. Over 70% of them have been installed in China (REN21, 2016).

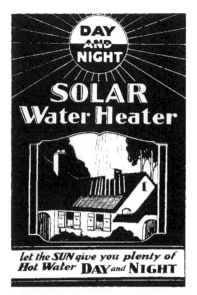

Figure 3.19 An advertisement for Bailey's thermosyphon solar water heaters, circa 1915

As will be explained in Chapter 4, solar photovoltaic (PV) panels are normally specified by their peak electrical output expressed in kW_p under full sunshine. For statistical comparability solar thermal collectors are now often quoted in terms of their peak *thermal* output, where 1 m² of collector is taken to have a peak thermal output of 0.7 kW_t. The estimated total world installed capacity, given the installed collector areas above, is thus 435 GW_t.

Solar collectors

We have already considered the basic form of a solar water heating system, but what about the choices involved in selecting the components?

Just as solar heating systems can have several variants, so can solar collectors. Figure 3.20 illustrates the most common types for low-temperature use.

Unglazed panels (Figure 3.20(a)) are most suitable for swimming pool heating, where it is only necessary for the water temperature to rise by a few degrees above ambient air temperature, so heat losses are relatively unimportant.

Flat plate air collectors (Figure 3.20(b)) are not so common as water collectors and are mainly used for applications such as crop drying.

Glazed flat plate water collectors (Figure 3.20(c)) are, outside of China, the mainstay of domestic solar water heating. Usually they are only single glazed but may have an additional second glazing layer, sometimes of plastic. The more elaborate the glazing system, the higher the temperature difference that can be sustained between the absorber and the external air.

The absorber plate usually has a very black surface that absorbs nearly all of the incident solar radiation, i.e. it has a high **absorptivity**. Most normal black paints still reflect approximately 10% of the incident radiation (a white surface, by way of comparison, might reflect back 70–80%). Some panels use a **selective surface** that has both high absorptivity in the visible region and a low emissivity in the long-wave infrared to cut heat losses.

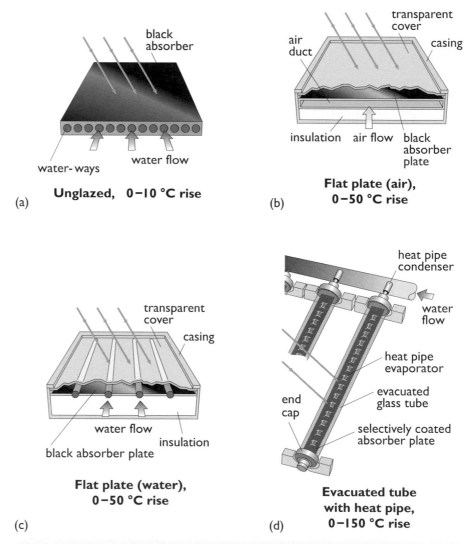

Figure 3.20 Solar collectors for low-temperature collection

Many designs of absorber plate have been tried with success, including specially made pressed aluminium panels and small-bore copper pipes soldered to thick copper or steel sheet. Generally, an absorber plate must have high thermal conductivity, to transfer the collected energy to the water with minimum temperature loss.

Evacuated tube collectors. The example shown in Figure 3.20(d) takes the form of a set of modular tubes superficially similar to fluorescent lamps. The absorber plate is a metal strip down the centre of each tube. Convective heat losses are suppressed by a vacuum in the tube. The absorber plate uses a special 'heat pipe' to carry the collected energy to the water, which circulates along a header pipe at the top of the array. A **heat pipe** is a device that takes advantage of the thermal properties of a boiling fluid to carry large amounts of heat. A hollow tube is filled with a liquid at a pressure chosen so that it can be made to boil at the 'hot' end, but the vapour will condense at the 'cold' end. The tube in effect has a thermal conductivity many times greater than if it had been made of solid metal, and is capable of transferring large amounts of heat for a small temperature rise.

An alternative design, commonly used in China, uses a narrow collector blade through which water circulates contained within a double-walled glass tube. Heat losses are suppressed by having a vacuum between the inner and outer walls of the tube.

Evacuated collectors are generally more expensive than flat plate ones, but they have a lower heat loss, allowing a better performance in winter.

Other *concentrating collector* designs used for high-temperature steam raising and electricity generation are described later, in Sections 3.9 and 3.10.

Robustness, mounting and orientation

Solar collectors are usually roof-mounted and once installed are often difficult to reach for maintenance and repairs. They must be firmly attached to the roof in a leak-proof manner and then must withstand everything that nature can throw at them – frost, wind, acid rain, sea spray and hailstones. They also have to be proof against internal corrosion and very large temperature swings. A double-glazed collector is potentially capable of producing boiling water in high summer if the heat is not carried away fast enough. It is quite an achievement to make something that can survive 20 or more years of this treatment.

Fortunately, as we have seen, panels do not have to be installed to a precise tilt or orientation for acceptable performance. This, in turn, means that a large portion of the current building stock, possibly 50% or more in the UK, could support a solar collector.

Active solar space heating and interseasonal storage

So far, we have looked in detail at domestic solar water heaters with only a few square metres of collector. It might be tempting to think that if a larger collector together with a much larger storage tank were fitted, solar energy could supply the annual low-temperature space heating needs of a building. However, as pointed out earlier, solar radiation is least available in mid-winter when it would be most needed for space heating.

One possibility is to increase the size of the storage tank so that summer sun could be saved right through to the winter. This is known as **interseasonal storage**. However, the difficulties of this should not be underestimated. The volume of hot water storage needed to supply the heating needs of a house may be almost the same size as the house itself. Also such a storage tank might need insulation half a metre thick to retain most of its heat from summer to winter. In order to reduce the ratio of surface area to volume, and hence heat loss, it pays to make the storage tank very large. This essentially limits the technology to large buildings or communal district heating schemes.

Although experimental active solar space heating systems for single buildings have been built, in practice it has generally proved simpler and more economical to *save* a kilowatt-hour of space heating energy through better insulation than to *supply* an extra one from active solar heating.

Solar district heating

There are considerable economies of scale for large projects where solar collectors can be purchased and erected in bulk. Since the 1980s there has been a steady stream of construction of large arrays of solar collectors for district heating systems in mainland Europe, mainly in Denmark, Sweden and Germany. The arrays can be very large. Originally started in 1994 and successively expanded over time, the system in the town of Marstal in Denmark has a total of over 33 000 m² of solar collectors used in conjunction with a 75 000 m³ heat store in a single large pit. This provides up to 50% of the heating needs for 1650 households, the remainder being supplied by burning wood chips. This represents a collector area of 20 m² and a storage volume of 45 m³ per household. (State of Green, n.d)

Figure 3.21 An 18 000 m² array of collectors feeding a district heating system at Marstal in Denmark. A further 15 000 m² of collectors and a 75 000 m³ pit heat store have been installed to the left of this picture.

Starting in 2012, an even larger system has been constructed for the Danish town of Vojens. This has 70 000 m² of collector area and a 200 000 m³ pit heat store aiming to supply about 50% of the heating needs of a population of about 7500 (Arcon-Sunmark, 2016).

3.7 Passive solar heating

History

All glazed buildings are already to some extent passively solar heated – effectively they are live-in solar collectors. The art of making the best use of this dates back to the Romans, who put glass to good use in their favourite communal meeting place, the bath house. Window openings 2 m wide and 3 m high have been found at Pompeii.

After the fall of the Roman Empire, the ability to make really large sheets of glass vanished for over a millennium. It was not until the end of the

17th century that the plate glass process reappeared in France, allowing sheets of 2 m² or more to be made.

Even so, cities of the 18th and 19th centuries were overcrowded and the houses ill-lit. It was not until the late 19th century that pioneering urban planners set out to design better conditions. They became obsessed with the medical benefits of sunlight after it was discovered that ultraviolet light killed bacteria. Sunshine and fresh air became the watchwords of 'new towns' in the UK like Port Sunlight near Liverpool, built to accommodate the workers of a soap factory.

The planners then did not realize that ultraviolet light does not penetrate windows, but the tradition of allowing access for plenty of sunlight continues, reinforced by findings that exposure to bright light in winter is essential to maintain human hormone balances. Without it, people are likely to develop 'seasonal affective disorder (SAD)' with symptoms of mid-winter depression.

Given the UK's past plentiful supply of coal, there was little interest in using solar energy to cut fuel bills. The construction of the Wallasey School building (the 'Solar Campus') in Cheshire in 1961, inspired by earlier US and French buildings, was thus something of a novelty (see Figures 3.22 and 3.23).

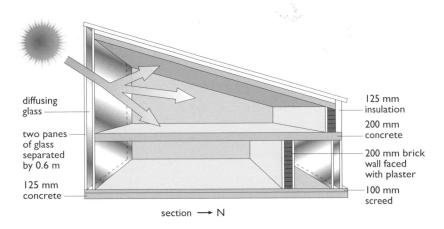

Figure 3.22 Wallasey 'Solar Campus', Cheshire, UK – built in 1961

diffusing glass

two panes of glass separated by 0.6 m

125 mm concrete

125 mm insulation

200 mm concrete

200 mm brick wall faced with plaster

100 mm screed

section ⟶ N

Figure 3.23 Solar Campus – section

Direct gain buildings as solar collectors

The Wallasey Solar Campus building is a classic direct gain design. It has the essential features required for passive solar heating:

(1) a large area of south-facing glazing to capture the sunlight

(2) thermally heavyweight construction (dense concrete or brickwork). This stores the thermal energy through the day and into the night

(3) thick insulation on the outside of the structure to retain the heat.

After its construction, the oil-fired heating system originally installed was found to be unnecessary and was for a time removed, leaving the building totally heated by a mixture of solar energy, heat from incandescent lights and the body heat of the students.

Its architectural design has now been regarded as sufficiently innovative for it to be awarded the status of a 'Grade II listed building'.

Passive solar heating versus superinsulation

Although the Wallasey School building is one style of low-energy building, there are others. The Wates house, built at Machynlleth in Wales in 1975 (Figure 3.24) was one of the first 'superinsulated' buildings in the UK. It features 450 mm of wall insulation and small quadruple-glazed windows. This was a radical design, given that at this time normal UK houses were built with single glazing, no wall insulation, and new building regulations requiring a mere 25 mm of loft insulation were only just being introduced.

Figure 3.24 The superinsulated Wates House at Machynlleth (photographed in 2003)

Situated low in a mountain valley, the Wates house is certainly not well-placed for passive solar heating. In fact, it was intended to be heated and lit by electricity from a wind turbine.

Which of the two design approaches – passive solar or superinsulation – is better? There are no easy answers to this question. The art of design for passive solar heating is to understand the energy flows in a building and make the most of them. There need to be sufficient solar gains to meet a substantial proportion of the winter heating needs. These can be reduced by good levels of insulation. Solar energy is also needed to provide adequate lighting, but not so much in summer that there is overheating.

Window energy balance

We can think of a south-facing window as a kind of passive solar heating element. Solar radiation enters during the day, and, if the building's internal temperature is higher than that outside, heat will be conducted, convected and radiated back out.

The question is whether more heat flows in than out, so that the window provides a net energy benefit. The answer depends on several things:

(1) the building's average internal temperature

(2) the average external temperature

(3) the available solar radiation

(4) the transmittance characteristics of the window, its orientation and shading

(5) the U-value of the window, which is, in turn, dependent on whether it is single or double glazed (or even better insulated).

Figure 3.25 shows the average monthly 'energy balance' of a south-facing window in the vicinity of London for a building with an average internal temperature of 18 °C. In the dull, cold months of December and January, a single-glazed window is a net energy loser and a double-glazed one only just breaks even. However, in the autumn and spring months, November and March, a double-glazed window becomes a positive contributor to space heating needs. Its performance can be further improved by insulating it at night.

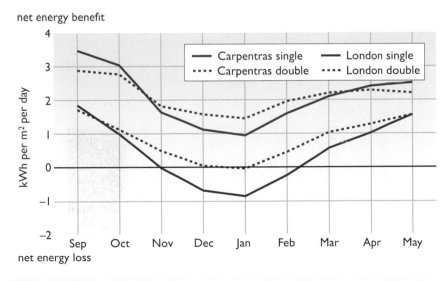

Figure 3.25 Window energy balance: London and Carpentras, in the south of France

We can compare this with a similar energy balance for Carpentras, near Avignon, in the south of France (also shown in Figure 3.25). Although the mid-winter months there are almost as cold as in London, they are far sunnier, being at a lower latitude. The incoming solar radiation is far greater than the heat flowing out, even in mid-winter, and the energy balance is markedly positive.

The heating season and free heat gains

To return to the UK, we need to consider how best use can be made of the solar energy available.

With a long heating season, a south-facing double-glazed window is a good thing. It can perhaps supply extra heat during October and November, March and April. On the other hand, with a very short heating season confined to the dullest months, say just December and January, it is not really much use at all.

How long is the heating season?

In order to answer this question, we must consider the rest of the building, its insulation standards and its so-called 'free' heat gains.

In a typical house, to keep the inside warmer than the outside air temperature, it is necessary to inject heat. The greater the temperature difference between the inside and the outside, the more heat needs to be supplied. In summer it may not be necessary to supply any heat at all, but in mid-winter large amounts will be needed. The total amount of heat that needs to be supplied over the year can be called the **gross heating demand**.

This will have to be supplied from three sources:

(1) 'free heat gains', which are those energy contributions to the space heating load of the building from the normal activities that take place in it: the body heat of people, and heat from cooking, washing, lighting and appliances. Taken individually, these are quite small. In total, they can make a significant contribution to the total heating needs. In a typical UK house, this can amount to 15 kWh per day

(2) passive solar gains, mainly through the windows

(3) fossil fuel energy, from the normal heating system.

Let us now consider, for example, the monthly average gross heat demand of a poorly insulated 1970s UK house (similar houses will be found right across northern and central Europe). As shown in Figure 3.26,

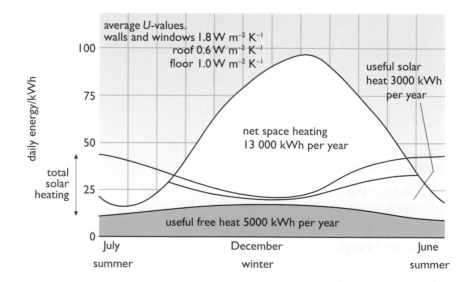

Figure 3.26 Contribution to the net space heating demand in a typical poorly insulated UK house of the 1970s

this will be higher in the cold mid-winter months than in the warmer spring and autumn. In summer, when the outside air temperature is high, this heating requirement almost drops to zero.

As shown in Figure 3.26, for this particular house, over the whole year, out of a total gross heating demand of 21 000 kWh, 5000 kWh come from free heat gains and 3000 kWh from solar gains (we have assumed to count free heat gains before solar gains).

Put another way, a perfectly ordinary house is already 14% passive solar heated. The **net heating demand**, to be supplied by the normal fossil fuel heating system, is simply the outstanding heat requirement, namely 13 000 kWh. This will have to be supplied from mid-September to the end of May.

It is possible to cut the house heat demand by putting in cavity wall and loft insulation and double instead of single glazing. This will reduce the gross heating demand and, as shown in Figure 3.27, allow the free heat gains and normal solar gains to maintain the internal temperature of the house for a longer period of the year. The insulation levels shown are approximately those of the 2002 UK Building Regulations for new housing. They have since been tightened further.

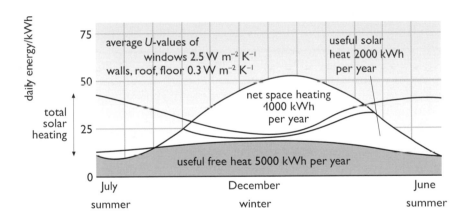

Figure 3.27 Contribution to net space heating demand in a house of normal design but reasonably well insulated

The heating season will then be reduced to between October and the end of April. Out of a total gross heat demand of 11 000 kWh, 5000 kWh still come from free heat gains, but as a result of the improved insulation, it will be possible to utilize only 2000 kWh of solar gains. Finally, 4000 kWh will remain to be supplied from the normal heating system.

By insulating the house, 9000 kWh per year will have been saved in fossil fuel heating, but the solar contribution (using our slightly arbitrary accounting system) will have fallen from 3000 to 2000 kWh.

It might be thought that improvements in the efficiency of domestic appliances and lighting would be leading to reduced levels of free heat gains from their use. UK statistics do show that in 2013 homes used less energy for cooking than they did in 1970, and that energy use for lighting and refrigerators has been falling since the late 1990s. However most of these reductions have been offset by increased electricity use for consumer electronics, including televisions and home computers (BEIS, 2016a). All of these unintentionally provide heating energy as well.

There are two ways in which the space heating demand could be cut further.

(1) By providing extra insulation. If the house was superinsulated, using insulation of 200 mm or greater thickness, the space heating load might disappear almost completely, leaving just a small need on the coldest, dullest days. Solar gains might not be essential.

(2) By providing appropriate glazing to ensure that the best use is made of the mid-winter sun.

Which of these methods is chosen will depend on the local climate and the relative expense of insulation materials and glazing. Per square metre, insulated wall tends to be a lot cheaper than good quality insulated glazing. The desired aesthetics of the building and the need for natural daylight inside will dictate whether or not it is easier to collect an extra 100 kWh of solar energy or to save 100 kWh with extra insulation.

Passivhaus design

In Germany the idea of superinsulation has been promoted in the form of the **PassivHaus standard**, developed during the 1990s. The house is 'passive' in the sense that it does not need a conventional large heating system (BRE, 2008). This form of design has been used in over 30 000 buildings across the world to date. It involves using thick insulation, good quality windows and airtight construction to reduce the space heating demand of a building to a low level. It can then be heated mainly by solar gains and heat from appliances and the occupants themselves.

Although the *Passivhaus* approach has mainly been applied to new buildings, even the thermal performance of existing buildings can be radically improved if they are adequately insulated. In the late 1990s an estate of apartment blocks originally built in the 1950s in Ludwigshafen in south-west Germany (Figure 3.28) was given a thorough thermal modernization including:

■ At least 200 mm thickness of foam insulation on the roof and in the walls (see Figure 3.29)

■ Triple-glazed windows with argon filling and low-emissivity coatings on the glass to cut the heat loss

■ Mechanical ventilation with heat recovery (MVHR); this uses heat recovered from outgoing air to preheat incoming fresh air

■ A fuel cell based combined heat and power (CHP) unit (see Chapter 11 for details of fuel cells).

Figure 3.28 The Brunck Estate in Ludwigshafen, Germany

Figure 3.29 Laying blocks of foam insulation on the roof

Monitoring showed that the net space heating energy use (i.e. that to be supplied by the heating system) fell by a factor of *seven* from 210 kWh per square metre of floor area per year to only 30 kWh m^2 y^{-1}. 30 kWh is equivalent to the energy content of 3 litres of heating oil – hence the project name, the 3 Litre House (Luwoge, 2008).

General passive solar heating techniques

There are some basic general guidelines for optimizing the use of passive solar heating in buildings.

(1) They should be well-insulated to keep down the overall heat losses.

(2) They should have a responsive, efficient heating system.

(3) They should face south (anywhere from south-east to south-west is fine). The glazing should be concentrated on the south side, as should the main living rooms, with little-used rooms, such as bathrooms, on the north.

(4) They should avoid overshading by other buildings in order to benefit from the essential mid-winter sun.

(5) They should be 'thermally massive' to avoid overheating in summer.

These guidelines were used, broadly in the order above, to design some low-energy, passive solar-heated houses on the Pennyland estate in Milton Keynes in central England in the late 1970s. The design steps (see Figure 3.30) were carefully costed and the energy effects evaluated by computer model.

The resulting houses had a form that was somewhere between the Wallasey School building and the Wates house. The houses faced south, there was not too much glazing, but not too little, and the main living rooms were concentrated on the south side (see Figures 3.31–3.33).

An entire estate of these houses was built and the final product carefully monitored. At the end of the exercise it was found that the steps 1–5 listed

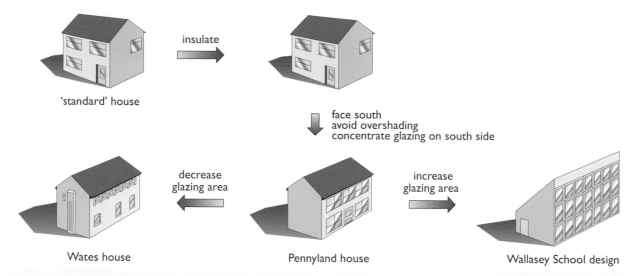

Figure 3.30 Design steps in low-energy housing

Figure 3.31 Passive solar housing at Pennyland – south elevation – the main living rooms have large windows and face south

Figure 3.32 Passive solar housing at Pennyland – the north side has smaller windows

Figure 3.33 Pennyland floor plans

on page 87 produced houses that used only half as much gas for low-temperature heating as 'normal' houses built in the preceding year. The extra cost was 2.5% of the total construction cost and the payback time was four years.

Here we come back to the difference between the 'broad' and 'narrow' definitions of passive solar heating. In its broad sense, it encompasses all the energy-saving ideas (1–5 above) put into these houses. In its narrow sense, it covers only the points that are rigidly solar based (3–5).

In this project, insulation and efficient heating saved the vast bulk of the energy, but approximately 500 kWh per year of useful space heating energy came from applying points 3–5 (Chapman et al., 1985).

Put another way, this 500 kWh is the difference in energy consumption between a solar and a non-solar house of the same insulation standard. We can call this figure the **marginal passive solar gain**. As we saw in Figures 3.26 and 3.27, even non-solar houses have some solar gains. What we are doing is trying to maximize them.

It is rather difficult to calculate the extra cost involved in producing marginal passive solar gains. After all, the passive solar 'heater' is an integral part of the building, not a bolt-on extra. Careful costing studies of different building designs and layouts have shown that modest marginal solar gains can be had at minimal extra cost.

Essentially, in its narrow sense, passive solar heating is largely free, being simply the result of good practice. In the 'wider' sense of integrated low-energy house design, the energy savings have to be balanced against the cost of a whole host of energy conservation measures, some of which involve glazing and perhaps have a solar element, and others which do not.

The balance between insulation savings and passive solar gains is also highly dependent on the local climate. In practice the most ambitious passive solar buildings are built in climates that have high levels of sunshine during cold winters.

Conservatories, greenhouses and atria

Direct gain design is really for new buildings: it cannot do much for existing ones. However, for many old buildings, conservatories or greenhouses could be added on to the south sides, just as they can be incorporated into new buildings (Figure 3.34).

Add-on conservatories and greenhouses are expensive and cannot normally be justified on energy savings alone. Rather, they are built as extra areas of unheated habitable space. A strong word of caution is necessary here. A conservatory only saves energy if it is not heated like other areas of the house. There is a danger that it will be looked on as just another room and equipped with radiators connected to the central heating system. One house built like this can easily negate the energy savings of ten others with unheated conservatories.

Figure 3.34 Conservatory on a Victorian terraced house

The costs can be reduced for new buildings if they are integrated into the design. The five houses in the Hockerton Housing Project in Nottinghamshire in the UK, completed in 1998 (see Figure 3.35), combine a 'Wallasey school' type design with a full-width conservatory (IEA. n.d). Not only do the walls and roof have 300 mm of insulation, but the rear of the houses is also built up with earth, giving extra thermal mass and protection from the worst winter weather. This is known as **earth sheltering**.

Figure 3.35 These low energy houses at Hockerton in Northamptonshire, completed in 1998, feature a full width conservatory, thick insulation and earth sheltering

Glazed atria are also becoming increasingly common. At their simplest, they are just glazed-over light wells in the centre of office buildings. At the other extreme, entire shopping streets can be given a glazed roof, creating an unheated but well-lit circulation space. Again a strong word of caution is necessary. 'Heated' shops may have wide-open doors, or even entire frontages, onto the 'unheated' circulation space, giving rise to unnecessary heat losses.

Avoiding overshading

One important aspect of design for passive solar heating is to make sure that the mid-winter sun can penetrate to the main living spaces without being obstructed by other buildings. This will require careful spacing of the buildings.

There are many design aids to doing this, but a useful tool is the **sunpath diagram** (see Figure 3.36). For a given latitude, this shows the apparent path of the Sun through the sky as seen from the ground.

In practice, the contours of surrounding trees and buildings can be plotted on it to see at what times of day during which months the Sun will be obscured. The Pennyland houses were laid out so that the midday sun in mid-December just appeared over the roof-tops of the houses immediately to the south.

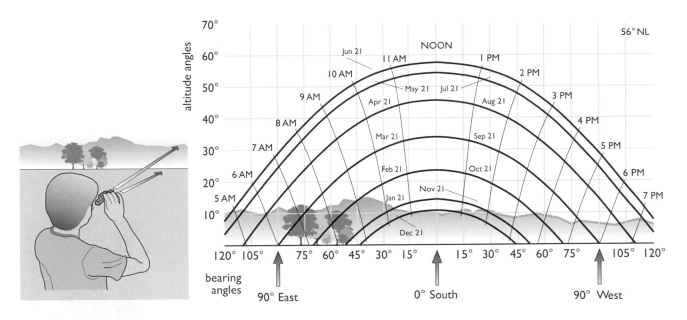

Figure 3.36 Plotting the skyline on a sunpath diagram can give important information about overshading. The sunpath diagram shown is for 56° N, which is approximately the latitude of Glasgow or Edinburgh

However, we need to ask whether it is advisable to cut down all offending overshading trees in the area to let the Sun through. To obtain maximum benefit from passive solar heating in its broad sense, it is necessary to follow another guideline:

Houses should be sheltered from strong winter winds.

Computer modelling suggests that, in houses such as those built at Pennyland, sheltering can produce energy savings of the same order of magnitude as marginal passive solar gains, approximately 500 kWh per year per house.

Where do the winter winds come from? This is immensely dependent on the local micro-climate of the site. In large parts of the UK, the prevailing wind is from the south-west. It would thus be ideal if every house could have a big row of trees on its south-west side. But is it possible to provide shelter from the wind without blocking out the winter sun? This is where housing layout becomes an art. Every site is different and needs solutions appropriate to it.

3.8 Daylighting

As well as providing heat, the Sun provides daylight. This is a commodity that we all take for granted. Replacing it with artificial light was, before the middle of the 20th century, very expensive (a topic discussed in Everett et al., 2012). Large mirrors were used in narrow London city streets to gather valuable daylight (Figure 3.37). With the coming of cheap electricity and efficient fluorescent lighting, daylight has been neglected and most modern office buildings are designed to rely heavily on electric light.

Figure 3.37 Mirrors used to catch valuable daylight in narrow London streets before the Second World War

Figure 3.38 Modern deep-plan office buildings, such as those at Canary Wharf in London, require continuous artificial lighting in the centre, which may create overheating in summer

Houses are traditionally well-designed to make use of natural daylight. Indeed, most of those that were not have long ago been designated slums and duly demolished. In the UK in 2013 domestic lighting accounted for under 3% of the sector's delivered energy use (BEIS, 2016a).

In some commercial buildings, however, lighting can account for up to 30% of the delivered energy use. Modern factory units and hypermarket buildings are built with barely any windows. Modern 'deep-plan' office buildings, such as those at Canary Wharf in London (Figure 2.41), have plenty of windows on the outside but there are many offices, central corridors and stairwells on the inside that require continuous lighting, even when the Sun is shining brightly outside.

Although in winter the heat from lights can usefully contribute to space heating energy, in summer (when there is most light available) it can cause overheating, especially in well-insulated buildings. Making the best use of natural light saves both on energy and on the need for air conditioning.

Daylighting is a combination of energy conservation and passive solar design. It aims to make the most of the natural daylight that is available. Many of the design details will be found in the better quality 19th century buildings. Traditional techniques include:

- shallow-plan design, allowing daylight to penetrate all rooms and corridors
- light wells in the centre of buildings
- roof lights
- tall windows, which allow light to penetrate deep inside rooms
- the use of task lighting directly over the workplace, rather than lighting the whole building interior.

Other experimental techniques include the use of steerable mirrors to direct light into light wells, and the use of optical fibres and light ducts.

An extreme example is in the small Norwegian town of Rjukan, set deep in a mountain valley, which has installed large steerable mirrors on an adjacent mountain-top to reflect a small, but very welcome, patch of sunlight into the town square.

When artificial light has to be used, it is important to make sure that it is used efficiently and is turned off as soon as natural lighting is available. Control systems can be installed that reduce artificial lighting levels when photoelectric cells detect sufficient natural light. Payback times on these energy conservation techniques can be very short and savings of 50% or more are feasible.

In designing new buildings, there is a conflict between lighting design and thermal design. Deep-plan office buildings have a smaller surface area per unit volume than shallow-plan ones. They will need less heating in winter. As with all architecture, there are seldom any simple answers and compromises usually have to be made.

3.9 **Solar thermal engines and electricity generation**

So far, we have considered only low-temperature applications for solar energy. If the Sun's rays are concentrated using mirrors, high enough temperatures can be generated to boil water to drive steam engines. These can produce mechanical work for water pumping or, more commonly nowadays, for driving an electric generator. This solar thermal-electric generation is known as **concentrating solar power (CSP)**.

The systems used have a long history and many modern plants differ little from the prototypes built 100 years ago. Indeed, if cheap oil and gas had not appeared in the 1920s, solar engines might have developed to be commonplace in sunny countries.

Concentrating solar collectors

Legend has it that in 212 BC Archimedes used the reflective power of the polished bronze shields of Greek warriors to set fire to Roman ships besieging the fortress of Syracuse. Although long derided as myth, Greek navy experiments in 1973 showed that 60 men each armed with a mirror 1 m by 1.5 m could indeed ignite a wooden boat at 50 m.

If each mirror perfectly reflected all its incident direct beam radiation squarely on to the same target location as the other 59 mirrors, the system could be said to have a **concentration ratio** of 60. Given an incident direct beam intensity of, say, 800 W m^{-2}, the target would receive 48 kW m^{-2}, roughly equivalent to the power density of a boiling ring on an electric cooker.

This use of steerable mirrors forms the basis of the modern 'power tower' systems described later.

The most common method of concentrating solar energy is to use a parabolic mirror. All rays of light that enter parallel to the axis of a mirror formed in this particular shape will be reflected to one point, the focus (Figure 3.39(a)). However, if the rays enter slightly off-axis, they will not pass through this point. It is therefore essential that the mirror tracks the Sun.

In the *line focus or trough collector* the Sun's rays are focused onto a pipe running down the centre of a trough (Figure 3.39(b)). The pipe is likely to carry a high temperature heat transfer fluid such as a mineral oil. Such systems are mainly used for generating steam for electricity generation. The trough can be pivoted to track the Sun up and down (i.e. in **elevation**) or east to west. A line focus collector can be oriented with its axis in either a horizontal or a vertical plane.

In the *point focus or dish collector Figure* 3.39(c), the Sun's image is concentrated on a steam boiler or a Stirling engine in the centre of the mirror. For optimum performance, the axis must be pointed directly at the Sun at all times, so it needs to track the Sun both in elevation and in **azimuth** (that is, side to side).

Most mirrors are assembled from sheets of curved or flat glass fixed to a framework.

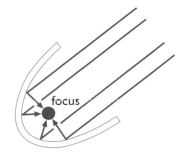

(a) Parabolic mirror brings light to a precise focus in centre

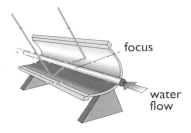

(b) **Line focus, 200 – 400 °C**

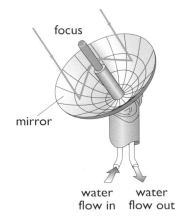

Figure 3.39 Parabolic mirrors for high-temperature applications (a) principles of focusing (b) a line focus or trough collector (c) a point focus or dish collector

The *fresnel* collector is a variant of the line focus collector in which a series of long strips of flat mirror carry out the same function.

There are trade-offs between the complexity of design of a concentrating system and its concentration ratio. A well-built and well-aimed parabolic dish collector can achieve a concentration ratio of over 1000. A line focus parabolic trough collector may achieve a concentration ratio of 50, but this is adequate for most power plant systems. The ratio required depends on the desired target temperature.

Unless the incident solar energy is carried away by some means, the target, be it a boat or a boiler, will settle at an equilibrium temperature where the incoming radiation balances heat losses to the surrounding air. The latter will be mainly by convection and re-radiation of infrared energy and will be dependent on the surface area of the target and its exposure to wind. A line focus parabolic trough collector can produce a temperature of 200–400 °C. A dish point focus system can produce a temperature of over 1500 °C.

What is important to appreciate is that no concentrating collector can deliver in total any more energy than falls on it, but what it does receive is all concentrated into one small area.

The first solar engine age

The process of converting the concentrated power of the Sun into useful mechanical work started in the 19th century. When, in the 1860s, France lacked a supply of cheap coal, Augustin Mouchot, a mathematics professor from Tours, had the answer: solar-powered steam engines. In the 1870s and 1880s, Mouchot and his assistant, Abel Pifre, produced a series of machines ranging from the solar printing press shown in Figure 3.40 to solar wine stills, solar cookers and even solar engines driving refrigerators.

Figure 3.40 Abel Pifre's solar-powered printing press

Their basic collector design was a parabolic concentrating collector with a steam boiler mounted at the focus. Steam pipes ran down to a reciprocating engine (like a steam railway engine) on the ground.

Although these systems were widely acclaimed, they suffered from the fundamental low-power density of solar radiation and a low overall efficiency.

In order to understand some of the problems of engines powered by solar energy, it is necessary to consider the Second Law of Thermodynamics (see Box 3.3).

BOX 3.3 Carnot efficiency of solar heat engines

Most concentrating solar systems use the same basic Rankine cycle as fossil fuelled power stations as described in Chapter 2. Most use water as the working fluid boiling it to produce high-pressure steam. This then goes to an 'expander', which was a reciprocating piston engine in Mouchot's engines, but today is likely to be a steam turbine. This extracts work which today is usually used to drive an electric generator.

As explained in Chapter 2, the maximum thermal efficiency of heat engines is limited by the **Second Law of Thermodynamics**. They all produce work by taking in heat at a high temperature, T_{in}, and rejecting it at a lower one, T_{out}. In the ideal case, the maximum efficiency they could be expected to achieve is given by:

$$\text{maximum efficiency} = 1 - \frac{T_{out}}{T_{in}}$$

where T_{in} and T_{out} are expressed in the Kelvin temperature scale (or degrees Celsius plus 273). This ideal efficiency is known as the **Carnot efficiency**.

For example, in a modern CSP plant well-designed parabolic trough collectors might produce steam at 350 °C. This would be fed to a steam turbine and low-temperature heat would be rejected in cooling towers at 30 °C. The theoretical efficiency of the system would therefore be:

   $1 - (30 + 273)/(350 + 273) = 0.51$ or 51%.

Its practical efficiency is more likely to be about 25%, due to various losses.

Normally, to boil water, its temperature must be raised to at least 100 °C. but as explained in Chapter 2, Organic Rankine Cycle (ORC) systems have been developed that use refrigerants with lower boiling points. These can be used with low-temperature solar engines, for example in Ocean Thermal Energy Conversion (OTEC) systems, described later.

Such low-temperature systems are likely to have poor efficiencies. For example, the theoretical Carnot efficiency of an OTEC engine heated with warm seawater at 25 °C and cooled by water at 5 °C from the bottom of the sea would be under 7%.

The early French solar steam engines were not capable of producing steam at really high temperatures and as a result their thermal efficiencies were poor. It required a machine that occupied 40 m² of land just to drive a one-half horsepower engine (less than the power of a modern domestic vacuum cleaner!)

By the 1890s, it was clear that this was not going to compete with the new supplies of coal in France, which were appearing as a result of increased investment in mines and railways.

At the beginning of the 20th century, in the USA, an entrepreneur named Frank Shuman applied the principle again, this time with large parabolic trough collectors. He realized that the best potential would be in really sunny climates. After building a number of prototypes, he raised enough financial backing for a large project at Meadi in Egypt. This used five parabolic trough collectors, each 80 m long and 4 m wide. At the focus, a finned cast iron pipe carried away steam to an engine.

In 1913, his system, producing 55 horsepower, was demonstrated to a number of VIPs, including the British government's Lord Kitchener. The payback time would have been only four years since the alternative fuel in Egypt at the time was coal, which had to be imported from the UK.

By 1914, Shuman was talking of building 20 000 square miles of collector in the Sahara, which would 'in perpetuity produce the 270 million horsepower required to equal all the fuel mined in 1909' (see Butti and Perlin, 1980, for the full story). Then came the First World War and immediately afterwards the era of cheap oil. Interest in solar steam engines collapsed and lay dormant for virtually half a century.

The new solar age

Solar engines revived with the coming of the space age. When, in 1945, a UK scientist and writer, Arthur C. Clarke, described a possible future 'geostationary satellite', which would broadcast television to the world, it was to be powered by a solar steam engine. In fact, by the time such satellites materialized, 25 years later, photovoltaics (see Chapter 4) had been developed as a reliable source of electricity.

Elsewhere, space rockets, guided missiles and nuclear reactors needed facilities where components could be tested at high temperatures without contamination from the burning of fuel needed to achieve them. The French solved this problem in 1969 by building an eight-storey-high parabolic mirror at Odeillo in the Pyrenees. This faced north towards a large field of **heliostats:** steerable flat mirrors, which, like those held by Archimedes' warriors, track the Sun. This huge mirror could produce temperatures of 3800 °C at its focus, but only in an area of 50 cm^2.

Power towers

In the early 1980s, the first serious, large, experimental solar thermal electricity generation schemes were built to make use of high temperatures. These are now known as concentrating solar power (CSP) plants. Several were of the 'power tower' type, using a large array of heliostats on the ground which focus the Sun's rays onto a central receiver at the top of a tower (Figure 3.41). This is a chamber where either steam can be produced directly, or a heat transfer fluid such as mineral oil or molten salt can be raised to a high temperature to be pumped away to generate steam at ground level. The steam is then used to drive a turbine to generate electricity.

A 10 megawatt (MW) plant, *Solar One*, was built at Barstow in California in 1981. Initially the Barstow plant used high-temperature synthetic oils to carry away the heat to a steam boiler. In 1995, it was rebuilt as Solar Two, and included the important feature of thermal energy storage (or TES) allowing it to potentially produce electricity on a 24-hour basis. This involves storing a molten salt mixture of sodium nitrate and potassium nitrate with a melting point of over 200 °C and heating it to 500 °C. This is used to produce steam later in the day.

More recently power tower technology has been taken up near Seville in southern Spain, where two plants were commissioned, the 11 MW *Planta Solar* 10 (PS10) completed in 2007 and the adjacent 20 MW *Planta Solar* 20 (PS20) in 2009 (Figure 3.42). These have limited steam heat storage and can use natural gas as back-up. The newer *Gemasolar* 20 MW plant also near Seville and completed in 2011 has increased thermal storage using molten salt.

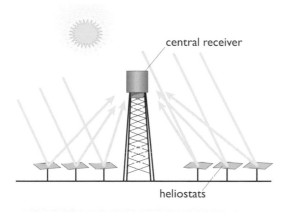

Figure 3.41 The central receiver on a power tower is heated by a large array of steerable heliostat mirrors on the ground

Figure 3.42 The PS10 (completed 2007) and PS20 (completed 2009) power tower plants near Seville in Spain have been followed by many similar projects.

This technology has since been taken up at a larger scale in the USA. The 377 MW Ivanpah plant in California, opened in 2014, has three power towers and covers an area of 14 km^2. It is designed to use natural gas as a back-up, but has no thermal storage (Brightsource, n.d.). In 2015, it produced 650 000 MWh of solar generated electricity.

Another 100 MW plant with a single power tower has been constructed at Crescent Dunes, near Las Vegas in Nevada, entering operation in 2015. This uses molten salt heat storage which potentially could supply full load output for 10 hours (Solar Reserve, 2016).

Parabolic trough concentrating collector systems

Until 2007 most of the world's solar thermal generation came from several large solar power stations developed by Luz International in the Mojave desert in California. Between 1984 and 1990, Luz constructed nine Solar Electricity Generating Systems (SEGS) of between 13 and 80 MW rating and totalling 354 MW. These are essentially massively uprated versions of Shuman's 1913 design, using large fields of parabolic trough collectors (see Figures 3.43 and 3.44). Each successive project has concentrated on increasing economies of scale in purchasing mirror glass and the use of commercially available steam turbines. The last 80 MW SEGS to be built (SEGS IX) has 484 000 m² of collector area.

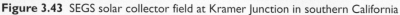

Figure 3.43 SEGS solar collector field at Kramer Junction in southern California

Figure 3.44 SEGS solar collector field – aerial view

The collectors heat synthetic oil to 390 °C, which can then produce high-temperature steam via a heat exchanger. The plants can also burn natural gas to top up their output and run into the evenings. The five plants at Kramer Junction (SEGS III to VII) have recorded annual plant efficiencies of 14% and peak efficiencies of up to 21.5% (Solarpaces, 2010). This is competitive with commercially available PV systems.

The SEGS plants were intended to compete with fossil fuel generated electricity to feed the peak afternoon air-conditioning demands in California, and for several years this objective was successfully achieved. In 1992, reductions in the price of gas, to which the price paid for electricity from the plant was tied, brought financial difficulties for the Luz company and the construction of new plants ceased. However those already built have continued to operate very reliably and cheaply for over 25 years and this is now regarded as a 'mature technology'. In 2015 these SEGS plants produced over 600 000 MWh of solar electricity (California Energy Commission, 2016).

Although almost no major concentrating solar power projects were built anywhere in the world between 1992 and 2007, the years since then have shown a surge in construction. In Spain, in addition to the power tower projects shown earlier, over 40 parabolic trough schemes have been constructed between 2008 and 2013, nearly all of a standard 50 MW size (NREL, 2016).

The construction of large solar parabolic trough power plants also resumed in the USA in 2007 with the commissioning of *Nevada Solar One*, a 64 MW parabolic trough plant outside Boulder City, Nevada. This has been followed by two 250 MW projects in California, another 250 MW project in Arizona and many smaller ones.

Most of the currently proposed plants are solar-fossil fuel hybrids where the steam turbine is powered by the Sun during the day, by stored heat in the evening and by natural gas at night. There are good thermodynamic reasons for combined gas and solar operation, since it allows the steam turbine to be run at its maximum operating temperature and thermal efficiency.

Parabolic dish concentrator systems

Instead of conveying the solar heat from the collector down to a separate engine, an alternative approach is to put the engine itself at the focus of a mirror. This has been tried both with small steam engines and with Stirling engines.

Stirling engines are described in Everett et al., 2012, and have a long history (they were invented in 1816). Although steam engines have fundamental difficulties when operating with input temperatures above 700 °C, Stirling engines, given the right materials, can be made to operate at temperatures of up to 1000 °C, with consequent higher efficiencies. Current experimental solar systems using these have managed very high overall conversion efficiencies, approaching 30% on average over the day.

A pilot scheme of 60 dishes each driving a 25 kW Stirling engine was constructed in Imperial Valley in Arizona in 2010 (Figure 3.45) but closed down and sold off a year later. The reasons for this are not clear, but may be a combination of fears of competition from photovoltaics and a legal dispute about land use (CSP Today, 2011). Another project of 430 dishes each with a 1.5 kW Stirling engine was installed in 2013 at the Toole US Army Base in Utah (Williams, n.d.). This, too, seems to have run into financial difficulties most likely due to competition from photovoltaics.

Figure 3.45 An array of 60 parabolic dishes each with a 25 kW Stirling engine constructed at Maricopa in Arizona in 2010

World concentrating solar power growth

Since 2007 the world has seen a surge of interest in concentrating solar power. Initially this was in Spain, which now has over 2 GW of installed CSP generating capacity. However, the 2008 global financial crisis meant that funding for new plants in Spain ceased. Much of the expertise moved to the USA where the installed capacity increased by about 1.3 GW between 2006 and 2015 (see Figure 3.46).

More recently, projects have been announced in a whole range of sunny countries including Australia, Chile, China, India, Israel, Mexico, Saudi Arabia and South Africa. An estimated 1 GW of plants were under construction in 2016. These include both power tower and parabolic trough schemes and almost all incorporate thermal energy storage (REN21, 2016).

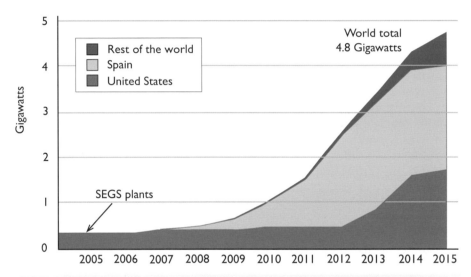

Figure 3.46 Growth in concentrating solar power installed capacity 2005–2015 (Source REN21, 2016)

However, the rate of CSP construction in the USA has slowed and it is obvious that there is serious competition from solar electricity generated by photovoltaics (PV). In 2015 PV supplied six times as much solar electricity to the State of California as did CSP plants. This raises the question of whether the continuing development of PV (to be described in the next chapter) will mean that large scale concentrating solar power will fail to compete. Box 3.4 lists some advantages and disadvantages of each technology. These will be examined a little further in Chapter 11.

BOX 3.4 Which is the best, CSP or Photovoltaics?

As pointed out in Chapter 1, between 2010 and 2015, the installed capacity of concentrating solar power increased at a rate of nearly 10% per annum, while the installed capacity of photovoltaics has been surging ahead, roughly doubling every two years. Table 3.5 sets out some advantages and disadvantages of each technology.

Table 3.5 Comparison of Concentrating Solar Power and Photovoltaics

Utility scale concentrating solar power	Photovoltaics
Advantages	**Advantages**
■ 'low tech' use of conventional steam turbine technology and conventional glass production for mirrors	■ basic simple systems have no moving parts and very low maintenance costs
■ can use natural gas as a 'top-up' and to generate in the evening	■ can operate under clear and cloudy skies
■ proven performance for over 25 years for parabolic trough systems	■ rapidly falling prices have encouraged investors
■ compatible with molten salt energy storage with potential to produce electricity day or night.	■ can be deployed in small project sizes as finance allows
	■ proven performance at a small scale.
Disadvantages	**Disadvantages**
■ requires clear skies	■ 'high-tech' PV panels have to be made in specialist factories
■ utility scale projects require large up-front capital investments.	■ any energy storage has to be electrical.

Low-temperature systems

The systems above all rely on producing high temperatures. As described in Box 3.3 this is essential to maximize generation efficiency. It also minimizes the land area required for a given power output. However, other low-temperature systems have been tried and remain of interest.

Solar ponds

Solar ponds use a large, salty lake as a kind of flat plate collector. If the lake has the right gradient of salt concentration (salty water at the bottom and fresh water at the top) and the water is clear enough, solar energy is absorbed at the bottom of the pond.

The hot, salty water cannot rise, because it is heavier than the fresh water on top. The upper layers of water effectively act as an insulating blanket and the temperature at the bottom of the pond can reach 90 °C. This is a high enough temperature to run an Organic Rankine cycle (ORC) engine. However, the thermodynamic limitations of the relatively low temperatures mean low solar-to-electricity conversion efficiencies, typically less than 2%. Nevertheless, a system of 5 MW peak electrical output, fed from a lake of over 20 hectares area, was demonstrated in Israel in the 1980s. The large thermal mass of the pond acts as a heat store, and electricity generation can go on day or night, as required. Their best location is in the large areas of the world where natural flat salt deserts occur.

In practice, the system has disadvantages. Large amounts of fresh water are required to maintain the salt gradient. These can be hard to find in the solar pond's natural location, the desert. Indeed, the best use for solar ponds may be to generate heat for water desalination plants, creating enough fresh water to maintain themselves and also supply drinking water.

Ocean thermal energy conversion (OTEC)

Ocean thermal energy conversion essentially uses the sea as a solar collector. It exploits the small temperature difference between the warm surface of the sea and the cold water at the bottom (Figure 3.47). In deep tropical waters, 1000 m deep or more, this can amount to 20 °C. This is sufficient to drive an ORC engine, using ammonia or an ammonia/water mixture as the working fluid.

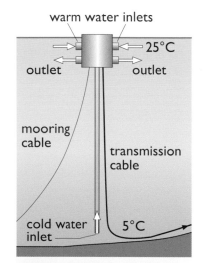

Figure 3.47 Ocean Thermal Energy Converter (OTEC) floating platform

Although the efficiency is likely to be low and the ORC system used needs to be finely adjusted to boil at just the right temperature, there is an extremely large amount of water available. Also it can operate for 24 hours a day. The technology is of considerable interest on tropical islands that have to import diesel fuel for electricity generation.

Initial experiments made on a ship in the Caribbean in the 1930s were only marginally successful. Water had to be pumped from a great depth to obtain a significant temperature difference, and the whole system barely produced more energy than it used in pumping.

The engineering difficulties are enormous. An OTEC plant producing 10 MW of electricity would need to pump nearly 500 cubic metres per second of both warm and cold water through its heat exchangers, whilst remaining moored in sea 1000 metres deep. Typically the pumps consume 20–30% of the electricity generated. Work on experimental systems has proceeded since the 1970s in Hawaii with encouragement from the US Navy. Although plans by the US Lockheed Martin Company for 10 MW plants in Hawaii and China suggested in 2013 have been shelved, a 100 kW prototype plant was connected to the Hawaii electricity grid in August 2015. (IRENA, 2014; Makai, 2015).

Solar updraft tower devices

Figure 3.48 This prototype 50 kW solar updraft tower at Manzanares in Spain operated between 1981 and 1989

These exploit warm air produced in a very large greenhouse. The air is allowed to rise through a tall chimney. Solar updraft towers are sometimes referred to as 'solar chimneys', but this term may also be used to refer to devices using rising warm air to ventilate buildings.

The updraft is used to turn an air turbine at the base of the chimney, driving a generator to produce electricity. This sounds simple enough. What is not so simple is the scale of construction. A 50 kW prototype built in Manzanares in Spain in 1981 (Figure 3.48) and operational until 1989 used a greenhouse collector 240 m in diameter feeding warm air to a chimney 195 m high.

In a very sunny region of the world, a cost-optimum plant might have an output of 100 MW using a collector 3.6 km in diameter and feeding a chimney 950 m tall (Schlaich, 1995). However, because such a design only produces warm air (a 35 °C temperature rise has been assumed in the calculations) the overall generation efficiency would be low, around 1.3%. Such a system would thus require considerably more land area than one of equivalent output using high-temperature concentrating collectors. However the system does not need clear skies and the ground beneath the collector provides a limited amount of energy storage, extending the output by a couple of hours into the evening.

A 200 kW prototype was constructed in 2010 at Jinshawan in Inner Mongolia in China and there are proposals for multi-megawatt plants in a number of countries.

3.10 Solar thermal process steam

Many industrial processes require high temperature steam and, where they take place in sunny countries, concentrating solar power is an ideal source. One such application is **solar thermal enhanced oil recovery**. High pressure steam is injected deep underground into oil fields to melt and flush out high-viscosity oil. Such systems have traditionally used natural gas, which may be a by-product of the oil field, as a fuel.

Using renewable energy to extract more fossil fuel from the ground is not perhaps totally in the interests of curbing global warming, but it is giving rise to the construction of one of the largest arrays of concentrating mirrors in the world.

Following the successful operation of a pilot project constructed in 2012 (see Figure 3.49), the Middle Eastern state of Oman announced in 2015 the construction of Miraah ('mirror' in Arabic), a US$600 million, 1 gigawatt solar thermal facility. This is adjacent to their Amal West oil field. The plant, when complete, will cover 3 square kilometres (Glasspoint, 2015).

The parabolic trough collectors are similar to those shown earlier in Figures 3.43 and 3.44 except that they are housed inside large greenhouses to protect them from dust storms. The greenhouses have automatic machines to clean their outside surfaces at night.

Figure 3.49 The pilot solar thermal enhanced oil recovery plant at Amal West in Oman. The steam-generating parabolic concentrating solar collectors are housed inside a protective greenhouse. The new project is estimated to save nearly 6 PJ of natural gas, enough to generate electricity for over 200 000 Oman residents.

3.11 Economics, potential and environmental impact

Although the motto of Bailey's 'Day and Night' Solar Water Heater company was 'Solar energy, like salvation, is free' the company still had to charge its customers money for the collection equipment. Fortunately, the assessment of the economics of solar thermal systems doesn't pose any great problems. Most solar thermal heating systems can be regarded in the same manner as conventional heating plant or building components and simple notions of 'payback times' often give an adequate assessment (see Appendix B). Concentrating solar thermal electricity plants have similar ratings and life expectancies to gas or diesel power stations. The difference is that they have no fuel costs and reasonably low operating and maintenance costs.

Domestic active solar water heating

In terms of collector area, domestic water heating for residential buildings makes up the bulk of the world solar thermal market. In 2014 the UK had an estimated '**solar park**' (i.e. total installed area) of approximately 578 000 m^2 of glazed and evacuated tube collectors (ESTIF, 2015). This represents a fivefold increase since 2001. However the figure contrasts with the 18 million m^2 installed in Germany.

At the time of writing (2016), sales of solar water heaters in the UK are low and prices are high. According to a 2011 report by the Energy Saving Trust (EST), the typical cost of an installed solar water heater in the UK was £4800. They estimated that, based on the field trial results (EST, 2011), typical savings from a well-installed and properly used system would be £55 per year when replacing gas heating and £80 per year when replacing electric immersion heating. They also estimate the typical carbon savings at around 230 kg CO_2 y^{-1} when replacing gas, higher when replacing electric immersion heating.

As pointed out earlier, solar collectors are in competition with other energy saving devices such as heat pumps (although these do require considerable amounts of electricity for their operation). The UK Committee on Climate Change (CCC) estimates that the levelized cost of heat from solar thermal panels for domestic use is nearly 27 p kWh^{-1} (see Appendix B for an explanation of 'levelized cost'). This compares with values of 6.8–10.5 p kWh^{-1} for heat from a conventional gas boiler (including the costs of the boiler itself) and 10.9–18.6 p kWh^{-1} for heat from heat pumps (CCC, 2011).

Both solar thermal panels and heat pumps in England, Wales and Scotland are currently eligible for subsidy payments under the Renewable Heat Incentive (RHI) scheme. Payments for solar thermal systems are currently (2016) nearly 20 p kWh^{-1} based on estimated system performance.

Looking further afield, the economics are more promising. In central and southern Europe solar thermal heat is considered cost competitive with both gas and electricity (IRENA, 2015). In southern Europe there is more sunshine, so a system may produce twice as much energy per square metre of collector as in the UK. Austria, France, Greece, Italy and Spain all have large areas of installed solar panels. Even in Denmark, economies of scale mean that large solar thermal arrays are competitive with gas-supported district heating systems.

Looking further afield, in 2014 China had about 440 million m² of solar collectors, over 70% of the world's installed capacity. Most of these are thermosyphon systems and use evacuated tubes (see Figure 3.50) (REN21, 2016).

Figure 3.50 A typical Chinese solar water heater using evacuated tube collectors

As for environmental impact, that of solar water heating schemes in the UK is likely to be small. The materials used are those of everyday building and plumbing. Pumped solar collectors can be installed to be visually almost indistinguishable from normal roof lights, with storage tanks hidden inside the roof space. Elsewhere, the use of free-standing thermosyphon systems on flat roofs can be highly visually intrusive. It is not so much the collector that is the problem but the storage tank above it. (Bailey's 'Day and Night' Solar Water Heater Company also had to face these problems. It offered to disguise the storage tank as a chimney.)

The situation is perhaps a little different for the kind of large district heating array shown in Figure 3.24 which obviously takes up a significant amount of urban or suburban ground area that could be used for other purposes.

Passive solar heating and daylighting

In its narrow sense of producing an increase in the amount of solar energy directly used in providing useful space heating, passive solar heating is highly economic, indeed possibly free. It should be borne in mind that to some extent passive solar heating and the use of daylight are already features of normal buildings, reducing the UK's energy demand by an estimated 0.5 EJ (BEIS, 2016b). This is a figure that does not appear in national renewable energy statistics.

Buildings specifically designed to use passive solar heating have been generally well-received by their occupants and are of interest to architects. However, the potential is limited by the low rate of replacement of the building stock.

In its broader, *Passivhaus,* sense of reducing overall heating energy demand, including retrofit projects, the potential is enormous. Yet, perversely, the superinsulation of buildings that contain large numbers of heat-producing appliances is likely to reduce the heating season to the dull mid-winter months, making the use of solar energy less important.

Designing buildings to take advantage of daylighting involves both energy conservation and passive solar heating considerations. In warmer countries, daylighting may be far more important, since cutting down on summer electricity use for lighting can also save on air conditioning costs.

Designing and laying out buildings to make the best use of sunlight has been part of the architectural tradition for centuries. It is generally seen as environmentally beneficial and has already shaped many towns and cities. For example, when in 1904 the city council of Boston, Massachusetts, USA, was faced with proposals for a 100 m high skyscraper, it commissioned an analysis of the shading of other buildings that this would cause. It was not pleased with the results and imposed strict limits on building heights.

However, a word of caution is necessary. In the UK, the tradition of new town development has been based partly on Victorian notions of the health aspects of 'light and air' in contrast to the overcrowded squalor of existing cities. This has been beneficial in terms of better penetration of solar energy into buildings. On the other hand, the encouragement of low building densities has led to vast tracts of sprawling suburbs and the consumption of enormous quantities of energy in transportation.

Solar thermal engines and electricity generation

As the original pioneers realized, it pays to build concentrating solar power systems in really sunny places. In order to generate the high temperatures necessary for thermodynamically efficient operation, the local climate has to have plenty of direct solar radiation – diffuse radiation will not do. Although there was a surge of interest in solar thermal electricity in the 1980s, low fossil fuel prices around the world dampened interest during the 1990s – in contrast to continued enthusiasm for photovoltaics. It is only since about 2005 that construction of new solar thermal power plants has revived.

There is currently fierce competition between CSP and photovoltaics. The US Department of Energy estimate that the cost of CSP electricity in the USA has fallen from US\$0.21 kWh^{-1} in 2010 to US\$0.15 kWh^{-1} in 2015 (Note: In late 2016 US\$1.00 = £0.78). Under their 'Sunshot' initiative they aim to reduce this to \$0.06 kWh^{-1} (including thermal energy storage) by 2020, making it competitive with conventional gas-fuelled electricity generation (NREL, 2016b). Each kilowatt-hour of solar electricity is likely to save on the emission of 0.4 kilograms of CO_2 from a gas-fuelled plant.

The Sunshot target price for photovoltaic electricity in 2020 is only US\$0.03 kWh^{-1}, the higher price for CSP electricity reflecting the ability to potentially operate 24 hours a day.

With modern concentrating solar power plant and a suitably sunny climate, 1 km^2 of land is enough to generate as much as 100–130 GWh of electricity per year (Greenpeace/ESTELA/Solarpaces, 2016). Ignoring the relative availability over the year, 3000 km^2 would thus be adequate to supply all of the UK's electricity and 35 000 km^2 (about 12% of the area of the state of Nevada) sufficient to supply all of the USA's electricity.

Sunny deserts, within striking distance of large urban electricity demands, are needed. In California, the Mojave desert is ideal. In Europe, many plants have now been built in central and southern Spain. Sites in other southern Mediterranean countries are possibilities, with proposals for grid links to future plants in North Africa.

A major problem is conveying the output to the loads. Although long-distance electricity transmission can be used, the manufacture and distribution of hydrogen might be another possibility. The possibilities of a future 'hydrogen economy' are described in Chapter 11.

The environmental consequences of solar thermal power stations are somewhat mixed. A major problem is the sheer quantity of land required.

Although this may be 'desert' there may still be problems about its use. In California, for example, solar projects have run into a range of problems including:

■ sites being part of the traditional lands of a native American tribe

■ sites being a habitat of rare desert tortoises

■ solar glare for passing aircraft

■ the deaths of birds killed by flying too close to the power towers.

CSP plants also require a certain amount of water for washing dust off the mirrors. This may be hard to come by in desert locations.

The costs of OTEC systems are not particularly clear, given the current lack of development. Figures of between US$0.07 and US$0.22 kWh^{-1} have been quoted (IPCC, 2011). Their environmental consequences may be mixed. On the one hand, it is claimed that the vast amounts of water being pumped circulate nutrients and can increase the amount of fish life. On the other, dissolved carbon dioxide can be released from the deep sea water, thereby negating some of the benefits of renewable energy generation. Only further experiments will resolve these issues.

3.12 **Summary**

There are four basic types of system for using solar thermal energy:

■ **Active solar heating** - using **solar collectors**

■ **Passive solar heating** – in its 'narrow sense' this means using **solar energy directly into buildings** for **space heating** (or to heat the interior spaces). In its broader sense it means the whole process of reducing a building's heat demand to the point where solar gains and other 'free heat' gains make a significant contribution in winter. Extremely low energy building design may be called **Passivhaus** technology.

■ **Daylighting** involves making the best use of natural daylight in buildings to avoid the use of artificial lighting.

■ **Solar thermal engines** use solar thermal heat to generate electricity, mostly by the use of concentrating solar collectors. Note that this is not the same as **photovoltaic electricity** (see the next chapter) where electricity is generated directly from sunlight.

Rooftop solar water heaters are used to supply domestic hot water. There are two basic types:

■ The **pumped solar water heater** uses a roof-mounted solar collector above a hot water storage tank, usually located inside the building. Water is circulated between them using an electric pump. This is the most common type in the UK and northern Europe.

■ The **thermosyphon solar water heater** has the storage tank mounted above the solar collector and uses natural convection to circulate water between the two. It is commonly used in Mediterranean countries and in China.

This chapter has also discussed the nature and availability of solar radiation.

The Sun radiates solar energy to the Earth at a range of frequencies.

- **'Light'** includes most of the **short-wave radiation**.
- **Short-wave infrared radiation** is also useful for solar thermal collectors.
- **'Heat'**, or **long-wave infrared radiation** may be emitted by warm surfaces.

Direct radiation, the unobstructed rays of the Sun, is important for concentrating collectors.

Diffuse radiation, light from the Sun reflected from the clouds or the sky is important for natural lighting and can be collected by both active and passive solar thermal systems. The total amount of solar radiation available over a year varies from country to country and is higher in those at a lower latitude (i.e. closer to the equator). Optimizing the amount of radiation falling on a flat plate collector requires an understanding of the appropriate tilt, which will depend on the latitude of the site.

Glass and other plastic glazing materials, have the ability to transmit light but block the re-radiation of long-wave infrared radiation (heat). There are three basic mechanisms of heat loss through a double glazed window: *conduction*, *convection* and *radiation*.

- **Conduction** can be reduced by the use of insulation.
- **Convection** in the gap between the panes of double glazing can be reduced by filling it with a heavy gas such as argon.
- **Radiation** of long wave infra-red radiation can be reduced by the use of **low-emissivity (low-e) coatings**.

The **U-value** of a piece of building fabric (such as a window) is a measure of its heat loss, the lower the *U*-value, the better.

A third of UK end-use fuel use is for low temperature applications. Solar energy is in principle ideal for supplying this heat.

There is competition from:

- **district heating fed by waste heat** from power stations or industrial processes
- **small-scale combined heat and power generation (CHP) plant**
- **heat pumps**

Domestic water heating uses about 4% of UK delivered energy use. A solar water heater will save more on emissions of CO_2 where it is used to substitute for oil or electric water heating than for gas water heating.

Although active solar heating can be used to contribute to space heating, the 'heating season' in the UK is in midwinter when there is little solar energy to collect. Solar space heating is best done in parts of Europe which are sunny and cold.

There are many types of solar heating system:

- **the rooftop active solar water heater** – usually with a glazed flat plate collector
- **swimming pool solar water heater** – usually large and unglazed
- **the conservatory or sunspace**

- **the Trombe wall** – using a narrow glazed space in front of a wall
- **direct gain** – the most common type of passive solar heating where solar energy penetrates windows into buildings.

There are four basic types of **active solar collector**:

- **Unglazed panels** – mostly used for swimming pool heating
- **Flat plate air collectors** – relatively uncommon and may be used for crop drying
- **Glazed flat plate water collectors** – commonly used for domestic solar water heating
- **Evacuated tube collectors** – used for domestic solar water heating and the main type in use in China. These may include **heat pipes** to carry heat from the collector to the water circuit.

A **selective surface** may be used to improve the solar collection of the black surface of solar panels. This has a high absorptivity in the visible light region but a low emissivity in the long wave infra red to cut heat losses.

Interseasonal storage of heat from summer through to winter is difficult. It requires large volumes of water and very good insulation. It pays to make stores very large to reduce the ratio of surface area to volume.

Solar district heating uses large arrays of solar collectors for whole towns. These can include very large heat stores and are becoming increasingly common in Denmark.

For **passive solar heating**, buildings are heated by a mixture of:

- **free heat gains** – heat given off by the occupants, lights and appliances
- **passive solar gains** – usually through windows
- **heat from a heating system**, usually fossil fuelled.

An optimum low energy house design may require a mixture of 'solar design' and 'superinsulation'.

Passivhaus design concentrates on reducing the heat demand of a building to the point where it may not need a conventional heating system.

Five basic rules for optimizing the use of passive solar heating in buildings are:

1 **They should be well-insulated** to keep down the overall heat losses.
2 **They should have a responsive, efficient heating system**.
3 In the northen hemisphere **they should face south** (anywhere from south-east to south-west is fine). **The glazing should be concentrated on the south side**, as should the main living rooms, with little-used rooms, such as bathrooms, on the north.
4 **They should avoid overshading by other buildings** in order to benefit from the essential mid-winter sun.
5 **They should be 'thermally massive'** to avoid overheating in summer.

A **sunpath diagram** shows the path of the sun through the sky at different times of year and can be used to assess solar overshading.

Daylighting, the avoidance of excessive use of artificial light, includes:

- shallow-plan design, allowing daylight to penetrate all rooms and corridors
- light wells in the centre of buildings
- roof lights and tall windows, allowing light to penetrate deep inside rooms
- the use of task lighting directly over the workplace
- steerable mirrors to direct light to where it is needed.

This chapter discussed solar thermal engines and electricity generation. There are two basic forms of these, *high temperature* and *low temperature*.

High temperature systems use solar thermal energy to raise steam driving engines and particularly to generate electricity, in what is now known as **concentrating solar power (CSP)**.

There are two basic forms of concentrating solar collector:

- the **line focus** or **parabolic trough**, which focuses the Sun's rays onto a horizontal pipe and the **point focus** or **dish collector** which focuses the Sun onto a central steam boiler or Stirling engine. Both types need to be moved to follow the Sun throughout the day.
- '**Power tower systems**' use an array of steerable flat mirrors or **heliostats** to concentrate the Sun's rays on a central receiver at the top of the tower.

High temperatures are needed to get a high thermal efficiency of electricity generation. **Clear skies** and plenty of **direct solar radiation** are required for CSP, diffuse radiation is not adequate. **Natural gas** is often used in large CSP systems as a top-up fuel and to run into the evening.

High temperature **thermal energy storage (TES)** using a molten salt mixture of sodium nitrate and potassium nitrate is often incorporated into new systems.

The large **California SEGS plants** have been operational for over 25 years.

There has been a twelve-fold increase in global CSP capacity between 2006 and 2015, mainly in Spain and the USA. These systems face stiff competition from photovoltaic electricity, but with thermal energy storage CSP has the advantage of possible 24 hour operation

Low temperature systems use Organic Rankine Cycle (ORC) engines and suffer from low thermal efficiencies. The most promising application is Ocean Thermal Energy Conversion (OTEC) using the small temperature difference between the top and bottom of the oceans.

Solar thermal process steam employs concentrating trough collectors to produce steam for industrial purposes.

Domestic solar water heaters are only marginally cost effective against gas water heating in the UK, but are supported by the UK government's Renewable Heat Incentive scheme. They have been heavily promoted in Germany. They are considered cost effective against gas and electric water heating in central and southern Europe.

Over 70% of the world's solar water heaters are in China.

Solar district heating is considered cost-effective in Denmark.

Passive solar design and **daylighting** are part of normal good practice in building design and it is difficult to assign 'costs' to them.

Passivhaus design has been shown to dramatically reduce the space heating demand of buildings.

The cost of **concentrating solar power** electricity generation is estimated to have fallen by nearly 30% between 2010 and 2015. It could be cost-effective against conventional gas-fuelled electricity generation in the USA by 2020.

CSP on an area equivalent to 12% of the State of Nevada could produce all of the USA's electricity.

References

Achard, P. and Gicquel, R. (eds) (1986) *European Passive Solar Handbook* (preliminary edition), Commission of the European Communities, DG XII.

Arcon-Sunmark (2016) *Solar Heating World Champ* [Online]. Available at http://arcon-sunmark.com/cases/vojens-district-heating (Accessed 29 November 2016).

BEIS (2016a) *Energy Consumption in the United Kingdom: data tables*, Department for Business, Energy and Industrial Strategy [Online]. Available at http://www.gov.uk (Accessed 9 October 2016).

BEIS (2016b) *Digest of UK Energy Statistics*, Department for Business, Energy and Industrial Strategy [Online]. Available at http://www.gov.uk (Accessed 9 October 2016).

BP (2016), *BP Statistical Review of World Energy*, British Petroleum [Online]. Available at http://www.bp.com (Accessed 9 October 2016).

BRE (2013) *SAP 2012 – the Government's Standard Assessment Procedure for Energy Rating of Dwellings*, Watford: Building Research Establishment [Online]. Available at http://www.bre.co.uk/filelibrary/SAP/2012/SAP-2012_9-92.pdf (Accessed 9 October 2016).

BRE (2008) *Passivhaus Primer: Introduction, An aid to understanding the key principles of the Passivhaus Standard*, Watford: Building Research Establishment [Online]. Available at http://www.passivhaus.org.uk (Accessed 1 November 2016).

Butti, K. and Perlin, J. (1980) *A Golden thread: 2500 Years of Solar Architecture and technology*, London: Marion Boyars.

California Energy Commission (2016) *California Solar Energy Statistics & Data* [Online]. Available at http://www.energy.ca.gov/almanac/renewables_data/solar/ (Accessed 1 December 2016).

CCC (2011) *the Renewable Energy Review*, London: Committee on Climate Change [Online]. Available at http://theccc.org.uk (Accessed 5 December 2016).

CEC (1994) 'Solar Radiation', *European Solar Radiation Atlas Volume 1*: Report EUR 9344, Directorate for General Science.

Chapman, J., Lowe, R. and Everett, R. (1985). *The Pennyland Project*, Energy Research Group, Milton Keynes, UK: Open University [Online]. Available at http://oro.open.ac.uk/19860/ (Accessed 29 November 2016).

CSP Today (2011) *What happened to Tessera Solar's projects?* [Online], Concentrating Solar Power Today. Available at http://social.csptoday.com/markets/what-happened-tessera-solars-projects (Accessed 30 November 2016).

EST (2011) *Here Comes the Sun: a Field trial of Solar Water Heating Systems*, Energy Saving Trust [Online]. Available at http://www.bre.co.uk/filelibrary/nsc/Documents%20Library/Not%20for%20Profits/NFP_Report_Energy-saving-trust-solar-thermal-test-report_0911.pdf (Accessed 28 November 2016).

ESTIF (2015) *Solar thermal Markets in Europe*, European Solar Thermal Industry Federation [Online]. Available at http://www.estif.org (Accessed 9 October 2016).

Everett, B., Boyle, G. A., Peake, S. and Ramage, J. (eds) (2012) *Energy Systems and Sustainability: Power for a Sustainable Future* (2nd edn), Oxford, Oxford University Press/Milton Keynes, The Open University.

Greenpeace/ESTELA/Solarpaces (2016) *Concentrating Solar Power: Global Outlook 2016*, Greenpeace/European Solar Thermal Electricity Association/ IEA Solarpaces [Online]. Available at http://www.greenpeace.org (Accessed 5 December 2016).

Glasspoint (2015) *Miraah project factsheet* [Online]. Available at http://www.glasspoint.com (Accessed 2 November 2016).

IEA (n.d.) Hockerton Housing Project, UK, International Energy Agency [Online]. Available at http://www.ecbcs.org/docs/Annex_38_UK_Hockerton.pdf (Accessed 29 November 2016).

IEA (2016) *CO_2 emissions from fuel combustion: highlights*, International Energy Agency, Paris [Online]. Available at http://www. iea.org (Accessed 9 October 2016).

IPCC (2011) *Special Report on Renewable Energy*, Intergovernmental Panel on Climate Change [Online]. Available at http://www.ipcc.ch (Accessed 5 December 2016).

IRENA (2014) *Ocean Thermal Energy Conversion: Technology Brief*, International Renewable Energy Agency [Online]. Available at http://www.irena.org/DocumentDownloads/Publications/Ocean_Thermal_Energy_V4_web.pdf (Accessed 30 November 2016).

IRENA (2015) *Solar Heating and Cooling for Residential Applications: Technology Brief*, International Renewable Energy Agency [Online]. Available at http://www.irena.org/DocumentDownloads/Publications/IRENA_ETSAP_Tech_Brief_R12_Solar_Thermal_Residential_2015.pdf (Accessed 6 December 2016).

Luwoge (2008) *Das 3-liter-Haus*, (in German) [Online]. Available at http://www.luwoge.de/fileadmin/user_upload/innovationen/media/3LH_Broschuere_d_EV_07_08_08.pdf (Accessed 30 November 2016).

Makai (2015) *Makai connects world's largest ocean thermal plant to US grid*, Makai Ocean Engineering, Oahu, Hawaii [Online]. Available at http://www.makai.com/makai-news/2015_08_29_makai_connects_otec/ (Accessed 30 November 2016).

NREL (2016a) Concentrating Solar Power Projects: Spain, National Renewable Energy Laboratory [Online]. Available at http://www.nrel.gov/csp/solarpaces/by_country_detail.cfm/country=ES (Accessed 2 December 2016).

NREL (2016b) *On the path to Sunshot Advancing Concentrating Solar Power Technology, Performance, and Dispatchability*, National Renewable Energy Laboratory [Online]. Available at http://www.nrel.gov/docs/fy16osti/65688.pdf (Accessed 5 December 2016).

REN21 (2016) *Renewables 2016 Global Status Report*, Renewable Energy Policy Network for the 21st Century, Paris [Online]. Available at http://www.ren21.net (Accessed 28 November 2016).

Schlaich, J. (1995) *the Solar Chimney – Electricity from the Sun*, Edition Axel Menges, Stuttgart.

Solarpaces (2010) *Annual Report, IEA Solar Power and Chemical Energy Systems* [Online]. Available at http://www.solarpaces.org (Accessed 1 December 2016).

SolarReserve (2016) *Crescent Dunes,* Santa Monica, California, [online] available at http://www.solarreserve.com/en/about (accessed November 29 2016)

State of Green (n.d.) *Marstal District Heating's Solar Thermal System* [Online]. Available at https://stateofgreen.com/en/profiles/aeroe/solutions/marstal-district-heating-s-solar-thermal-system (Accessed 29 November 2016).

Williams, J. (n.d.) *Infinia: Distributed Solar Power*, Infinia Corporation, Ogden, Utah [Online]. Available at https://www.esmap.org/sites/esmap.org/files/ESMAP_IFC_RE_Training_INFINIA_Letendre.pdf (Accessed 4 December 2016).

Chapter 4

Solar photovoltaics

by Godfrey Boyle and Bob Everett

4.1 Introduction

In Chapter 3 we saw how Concentrating Solar Power (CSP) can be used to generate electricity by producing high-temperature heat to power an engine, which then produces mechanical work to drive an electrical generator. This chapter is concerned with a more direct method of generating electricity from solar radiation, namely **photovoltaics**: the conversion of solar energy *directly* into electricity in a solid-state device.

If one were asked to design the ideal energy conversion system, it would be difficult to devise something better than the solar photovoltaic (PV) cell. In it we have a device which harnesses an energy source that is by far the most abundant of those available on the planet: as pointed out in Chapter 1 the net solar power input to the Earth is more than 6000 times humanity's current rate of use of fossil and nuclear fuels.

The PV cell itself is, in its most common form, made almost entirely from silicon, the second most abundant element in the Earth's crust. It has no moving parts and can therefore in principle, if not yet in practice, operate for an indefinite period without wearing out. Furthermore its output is electricity, probably the most useful of all energy forms.

This chapter starts with a brief history of photovoltaics, introduces the basic principles of the PV effect in silicon and describes various ways of reducing the cost and increasing the efficiency of crystalline silicon PV cells. It then examines non-crystalline PV cell designs using thin films, and discusses various innovative and emerging PV technologies. This is followed by a brief description of the electrical characteristics of PV cells and modules. The chapter then discusses PV systems for remote power supply, grid-connected PV systems for buildings, and large-scale PV power plants. The costs and environmental impact of energy from PV are then examined. The chapter concludes with a look at how electricity from PV can be integrated into electrical power systems, and some thoughts on how the enormous future growth potential for PV might be realized.

4.2 A brief history of PV

The term 'photovoltaic' is derived by combining the Greek word for light, *photos*, with *volt*, the name of the unit of potential difference (i.e. voltage) in an electrical circuit (see Chapter 1). The volt was named after the Italian physicist Count Alessandro Volta, the inventor of the battery. Photovoltaics thus describes the generation of electricity from light.

The discovery of the **photovoltaic effect** is generally credited to the French physicist Edmond Becquerel (Figure 4.1) who in 1839 published a paper

Figure 4.1 Edmond Becquerel, who discovered the photovoltaic effect

describing his experiments with a 'wet cell' battery, in the course of which he found that the battery voltage increased when its silver plates were exposed to sunlight (Becquerel, 1839).

The first report of the PV effect in a solid substance was made in 1877 when two Cambridge scientists, Adams and Day, described in a paper to the Royal Society the variations observed in the electrical properties of selenium when exposed to light (Adams and Day, 1877). Selenium is a non-metallic element similar to sulfur.

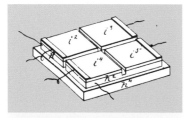

Figure 4.2 Diagram from Charles Edgar Fritts' 1884 US patent application for a solar cell

In 1883 Charles Edgar Fritts, a New York electrician, constructed a selenium solar cell that was, superficially, similar to the silicon solar cells of today (Figure 4.2). It consisted of a thin wafer of selenium covered with a grid of very thin gold wires and a protective sheet of glass. The efficiency of this cell, i.e. the proportion of the solar energy falling on its surface that is converted into electrical energy, was poor, less than 1%. Nevertheless, the response of selenium cells is well matched to the spectrum of visible light, and they came into widespread use in photographic exposure meters.

The underlying reasons for the inefficiency of these early devices were only to become apparent many years later, in the early decades of the 20th century, when physicists like Max Planck provided new insights into the fundamental properties of materials.

It was not until the 1950s that the breakthrough occurred that set in motion the development of modern, high-efficiency solar cells. It took place at the Bell Telephone Laboratories (Bell Labs) in New Jersey, USA, where a number of scientists, including Darryl Chapin, Calvin Fuller and Gerald Pearson (Figure 4.3), were researching the effects of light on **semiconductors**. These are non-metallic materials, such as germanium and silicon, whose electrical characteristics lie between those of conductors, which offer little resistance to the flow of electric current, and insulators, which block the flow of current almost completely. Hence the term *semi*conductor.

A few years before, in 1948, three other Bell Labs researchers, Bardeen, Brattain and Shockley, had produced another revolutionary device using semiconductors – the transistor. Transistors are made from semiconductors (usually silicon) in an extremely pure crystalline form, into which tiny quantities of carefully selected impurities, such as boron or phosphorus, have been deliberately diffused. This process, known as **doping**, dramatically alters the electrical behaviour of the semiconductor in a very useful manner which will be described in more detail later.

In 1953 the Chapin–Fuller–Pearson team, building on earlier Bell Labs research on the PV effect in silicon (Ohl, 1941), produced 'doped' silicon slices that were much more efficient than earlier devices in producing electricity from light.

By the following year they had produced a paper on their work (Chapin et al., 1954) and had succeeded in increasing the conversion efficiency of their silicon solar cells to 6%. Bell Labs went on to demonstrate the practical use of solar cells, for example in powering rural telephone amplifiers, but at that time they were too expensive to be an economic source of power in most applications.

In 1958, however, solar cells were used to power a small radio transmitter in the second US space satellite, Vanguard I. Following this first successful demonstration, the use of PV as a power source for spacecraft has become almost universal (Figures 4.4 and 4.5).

Figure 4.3 Bell Laboratories' pioneering PV researchers Pearson (left), Chapin (centre) and Fuller (right) measure the response of an early solar cell to light

The electrical power output of PV cells is obviously dependent on the intensity of light falling on them. The **power rating** of a particular cell or module is normally specified in peak watts (Wp), the output when illuminated by bright sunlight with an intensity of 1000 watts per square metre.

Rapid progress in increasing the efficiency and reducing the cost of PV cells has been made over the past few decades. Their terrestrial uses are now widespread and fall into three main categories:

- Off-grid applications where they provide power for telecommunications, lighting and other electrical uses.
- Small grid-connected power supplies for domestic, commercial and industrial buildings which may supply a substantial proportion of their electricity needs.
- Utility scale grid-connected electricity generation: multi-megawatt sized PV power plants.

The efficiency of the best crystalline silicon solar *cells* has now reached 26% in standard test conditions (see Box 4.1). Solar cells are actually sold packaged into solar panels or *modules*. At present a kilowatt of peak power

Figure 4.4 A model of the second US space satellite, Vanguard I, launched in 1958, whose radio transmitter powered by six very small solar PV cells operated for six years

Figure 4.5 The International Space Station is powered by large arrays of PV panels with a combined output of around 130 kW

typically requires about seven square metres of panels. The best silicon PV modules now available commercially have an efficiency of around 24% (Green et al., 2016) which means that a peak kilowatt could potentially be produced from just over four square metres of panels.

Between 2005 and 2015, the total global installed power capacity of PV systems increased at an extraordinary average rate of 28% per annum to about 230 GW_p (see Figure 4.6).

The selling prices of PV modules have decreased dramatically, particularly with the start of volume manufacture in China, falling to under US$1.00 per peak watt (Figure 4.7).

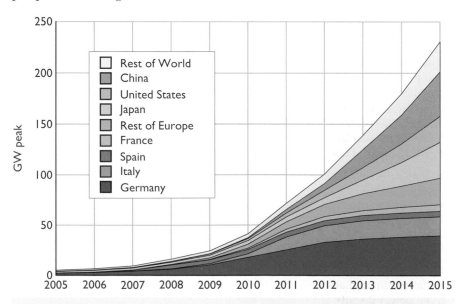

Figure 4.6 Growth of total PV installed capacity in various countries and regions of the world 2005–2015 (source BP, 2016)

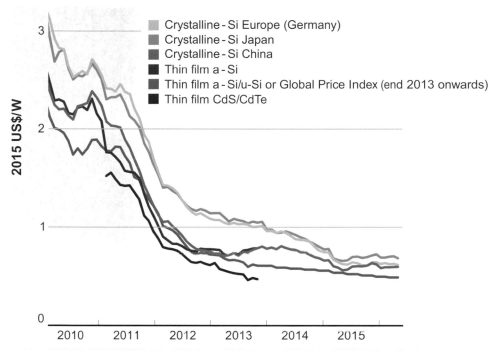

Figure 4.7 Average quarterly solar PV module prices by technology and manufacturing country sold in Europe, 2010-2016 (Source: IRENA, 2016). Note US$1.00 = £0.77 (September 2017)

BOX 4.1 Standard test conditions for PV cells and modules

PV cells and modules are traded globally, so it is important that their specified performance is measured under standard test conditions.

Chapter 3 has described how solar *thermal* collectors are tested under a light intensity of '1 sun', 1000 watts per square metre (W m^{-2}). These simply absorb light at different wavelengths as heat. However, as will be described later, the performance of a PV cell is critically dependent on the way in which the power contained in the solar radiation varies over the spectrum of wavelengths. Moreover, this precise spectrum will vary according to the height of the sun in the sky, becoming more 'red' at dawn and sunset and more 'blue' towards midday.

The concept of '**Air Mass**' (sometimes called 'Air Mass Coefficient') relates to the way in which the spectral power distribution of radiation from the Sun is affected by the distance the Sun's rays have to travel though the atmosphere before reaching a PV module.

In space, there is no atmosphere to affect the Sun's radiation. Immediately above the atmosphere, it has a power density of 1365 watts per square metre. The characteristic spectral power distribution of solar radiation as measured in space is described as the **Air Mass 0 (AM0)** distribution. This is used for testing PV cells for use in space.

At the Earth's surface, the various gases of which the atmosphere is composed (particularly ozone, water vapour and carbon dioxide) attenuate the solar radiation selectively at different wavelengths. This attenuation increases as the distance that the Sun's rays have to travel through the atmosphere increases.

When the Sun is at its zenith (i.e. directly overhead), the distance which the Sun's rays have to travel through the atmosphere to a PV module is at a minimum. The characteristic spectral power distribution of solar radiation observed under these conditions is known as the **Air Mass 1 (AM1)** distribution.

When the Sun is at a given angle θ to the zenith (as perceived by an observer at sea level), the Air Mass is defined as the ratio of the path length of the Sun's rays under these conditions to the path length when the Sun is at its zenith. By simple trigonometry (Figure 4.8), this leads to the definition:

$$\text{Air Mass} \approx \frac{1}{\cos \theta}$$

When the angle $\theta = 48°$, the Air Mass = 1.5. The spectral distribution corresponding to this path length through the atmosphere is used for testing

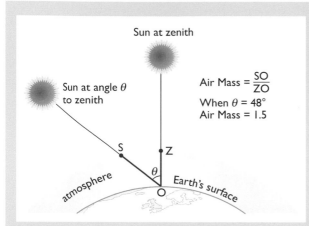

Figure 4.8 Air Mass is the ratio of the path length of the Sun's rays through the atmosphere when the Sun is at a given angle (θ) to the zenith (SO), to the path length when the Sun is at its zenith (ZO)

PV cells for terrestrial use. The overall intensity is taken to be 1000 W m^{-2}. The standard distribution for flat plate panels is called AM 1.5 Global, the term 'Global', denoting the inclusion of a small amount of diffuse radiation reflected from the sky (ASTM, 2012).

The spectral power distributions for Air Masses 0 and 1.5G are shown in Figure 4.9, together with that which would be received if the Sun was an ideal 'black body' radiator with a surface temperature of 6000°C. Visible light only makes up a small part of the spectrum. Much of the spectrum lies in the infra-red (heat) region where there are deep absorption bands produced by various atmospheric gases.

Also shown are the photon energies of different wavelengths of light, as described later in Box 4.2.

In practice the power rating of a PV cell or module, expressed in peak watts (Wp) of power output, is determined by measuring the maximum power it will supply at a temperature of 25 °C when exposed to radiation from lamps designed to reproduce the AM1.5 Global (AM 1.5G) spectral distribution at a total power density of 1000 W m^{-2} (1 'sun'). Cells for concentrating PV applications are tested at higher intensities, up to 500 suns.

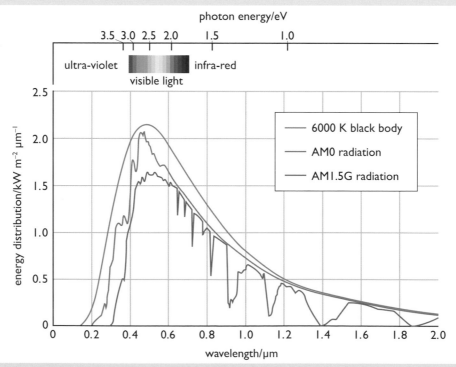

Figure 4.9 The spectral power distributions of solar radiation corresponding to Air Mass 0, Air Mass 1.5G and that of a perfect 'black body' radiator at 6000 °C. Also shown are the colours of the visible light range and the photon energies of different light wavelengths: Note: 1 μm = 1000 nm. (source: Green, 1982, ASTM, 2012)

4.3 The PV effect in crystalline silicon: basic principles

Semiconductors and 'doping'

Most metals are good conductors of electricity. Other materials may be good electrical insulators. Yet, at the sub-atomic level, both are made up of protons, neutrons and electrons. One key difference between them lies in the energy levels of the electrons. Those electrons that normally hold the atoms of a material together are described by physicists as occupying a lower-energy state known as the **valence band**. In certain circumstances, some electrons may acquire enough energy to move into a higher energy state, known as the **conduction band**, in which they can move around within the material and thus conduct electricity.

In metals, most electrons naturally lie in a conduction band. In good insulators the electrons are permanently bound up in the valence bands. Semiconductors, such as silicon, have most of their electrons in a valence band, but it only requires a small amount of energy to 'promote' an electron to a higher conduction band. This energy is known as the **band gap**.

Although PV cells can be made from one of a range of semiconductors, we will initially just consider silicon.

A crystal of pure (or **intrinsic**) silicon has a cubic structure, shown in Figure 4.10(a) in two dimensions for simplicity. The silicon atom has four 'valence' electrons. Each atom is firmly held in the crystal lattice by sharing two electrons (red dots) with each of four neighbours at equal distances from it. Occasionally thermal or light energy will provide enough energy to promote one of the electrons from the valence band into the conduction band, where the electron (red dot) is free to travel through the crystal and conduct electricity. When the electron moves from its bonding site, it leaves a '**hole**' (white dot). This 'absence of an electron' can be considered as a positively charged particle, which is also free to move through the crystal.

The 'car parking' analogy shown in Figure 4.11 may be helpful in visualizing the processes involved.

Conventional PV cells consist, in essence, of a junction between two thin layers of dissimilar semiconducting materials, known respectively as 'p' (positive)-type semiconductor, and 'n' (negative)-type semiconductor.

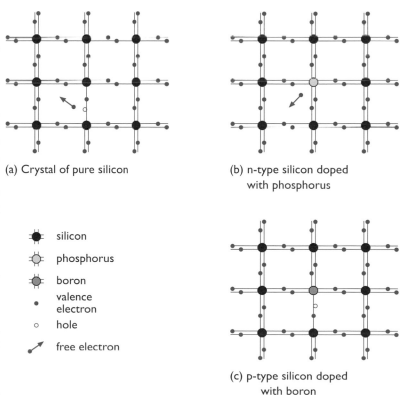

(a) Crystal of pure silicon

(b) n-type silicon doped with phosphorus

● silicon
◐ phosphorus
◉ boron
• valence electron
○ hole
➹ free electron

(c) p-type silicon doped with boron

Figure 4.10 Atomic structure of (a) a crystal of pure silicon; (b) n-type silicon doped with phosphorus; and (c) p-type silicon doped with boron (Source: adapted from Chalmers, 1976)

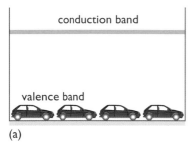

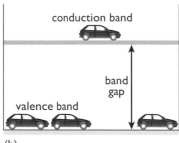

Figure 4.11 'Car parking' analogy of conduction processes in a semiconductor: A car park has two floors. The ground floor represents the valence band and the first floor the conduction band. (a) The ground floor of a car park is full and the cars there (representing electrons) cannot move around, but the first floor is empty. (b) A car (electron) is 'promoted' to the first floor (the 'conduction band'), where it can move around freely. This leaves behind a 'hole' that also allows cars on the ground floor (valence band) to move around (source: Green, 1982)

Note: λ = greek letter lambda

n-type semiconductors are made from crystalline silicon that has been 'doped' with tiny quantities of an impurity such as phosphorus (Figure 4.10(b)). Each phosphorus atom (shown in green) has five valence electrons, so that not all of them are taken up in the crystal lattice. **Electrons** are sub-atomic particles with a negative electrical charge, so silicon doped in this way has a *surplus of free electrons*.

p-type semiconductors are also made from crystalline silicon, but are doped with very small amounts of a different impurity such as boron. Each boron atom (shown in blue in Figure 4.10(c)) has only three valence electrons, so that it shares two electrons with three of its silicon neighbours and only one electron with the fourth. This causes the material to have a *deficit of free electrons*. Since the absence of a negatively charged electron can be considered equivalent to a positively charged particle, silicon doped in this way is known as a **p (positive)-type** semiconductor.

We can create what is known as a **p–n junction** by joining these dissimilar semiconductors. This is not a simple mechanical junction: the silicon crystal structure continues across it but the doping level changes from 'p' to 'n' gradually, not abruptly, across the junction. This sets up an **electric field** in the region of the junction, which promotes the flow of electrons and holes, i.e. an electric current.

The silicon solar cell

A silicon solar cell is a wafer of p-type silicon with a thin layer of n-type silicon on one side. What happens when light falls on it? An important factor is the wavelength of the light and the energy contained in the individual **photons** that make it up (see Box 4.2).

BOX 4.2 Light, photon energy and band gap

Light can be thought of as an electromagnetic wave travelling through space. In 1900 the physicist Max Planck suggested that at the atomic level, its energy could only be released in small discrete packets. It is thus also useful to describe light as consisting of a stream of tiny particles of energy, called photons.

The energy of a photon of light (E) is dependent on its wavelength λ:

$E = hc / \lambda$

where h = Planck's constant and c = velocity of light

At this atomic level, energy is normally described in terms of electron-volts (eV), the energy required to raise the charge on a single electron through one volt. This is 1.6×10^{-19} joules.

$E \text{ (eV)} = 1240 / \lambda \text{ (nm)}$

The photon energies of different wavelengths of light have been are shown earlier in Figure 4.9. They range from about 1 eV in the far infrared to 3.5 eV in the ultraviolet.

An important consequence of this theory is that a photon of blue light with a wavelength of 410 nm will have twice as much energy, 3 eV, as one of infrared radiation with a wavelength of 820 nm, which only has an energy of 1.5 eV.

These energy levels are important for understanding two mechanisms that limit the basic efficiency of PV cells:

(1) In order to promote an electron from the valence band to the conduction band an incoming photon *must have an energy greater than the band gap of the particular semiconductor*. For pure crystalline silicon this band gap is 1.1 eV. This corresponds to the energy of a photon of infra-red radiation with a wavelength of about 1130 nm. *Only light with a shorter wavelength than this will have any effect on a silicon PV cell.* The energy of the remaining low frequency infra-red radiation is effectively wasted.

(2) Even though a photon promoting an electron to the conduction band may have an energy considerably *greater* than the band gap, the useful electrical energy that it contributes to the cell is only *equal* to the band gap. Any surplus energy will be dissipated as heat and again wasted.

These two mechanisms of energy loss have led to the development of cells with different band gaps and multi-junction cells capable of absorbing light over a range of light frequencies.

Figure 4.12 shows a small section of a basic crystalline silicon PV cell. The p-n junction is at the top. Although the cell may be 10 cm or more square, the actual light collection junction is likely to be less than 200 μm (microns or millionths of a metre) thick.

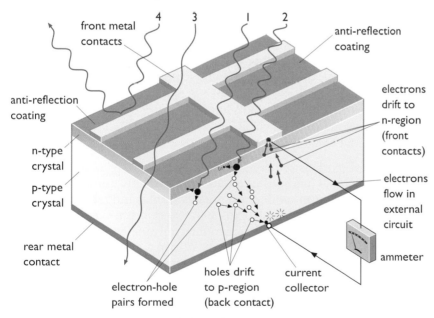

Figure 4.12 A basic silicon PV cell and four possible interactions with photons of light

There are four possible interactions with incident photons of light:

(1) If a photon of light with energy *equal* to the band gap of the semiconductor penetrates the cell near the p-n junction and encounters a silicon atom, it dislodges one of the valence electrons (red dot), leaving behind a hole (white dot). The electron thus promoted into the conduction band migrates into the layer of n-type silicon, and the hole migrates into the layer of p-type silicon. The electron then travels to a current collector on the front surface of the cell, generates an electric current in the external circuit and then re-emerges in the layer of p-type silicon, where it can recombine with waiting holes.

(2) If a photon with an energy *greater* than the band gap strikes a silicon atom, it again gives rise to an electron–hole pair and a current flow, but the difference in energy between the photon energy and the band gap will be converted into heat.

(3) A photon with an energy *smaller* than the band gap will not interact electrically with the silicon atoms and is likely simply to be absorbed as heat.

(4) Some photons will simply be reflected from the front surface of the cell or blocked from reaching the crystal by the current collectors that cover part of the front surface.

4.4 Crystalline PV cells

Monocrystalline silicon cells

The first commercial silicon solar cells were made from extremely pure **monocrystalline** silicon – that is, silicon with a single, continuous crystal lattice structure having virtually no defects or impurities. In 2015, this type still made up nearly a quarter of global PV cell production (Fraunhofer ISE, 2016). Silicon is widely found in nature as silicon dioxide in the form of quartz. The various steps of turning this into saleable PV modules are shown in Figure 4.13.

First, the silicon dioxide is reduced to 'metalurgical grade' silicon (MG-Si) by heating it with coke (carbon) in an arc furnace. This grade is typically only 98–99% pure. For the semiconductor electronics industry, the silicon must be *extremely* pure, at least 99.9999999% (i.e. 9 nines or 9N). The requirements for solar cells are less demanding, only 99.9999% or 6N purity. Large polycrystalline blocks of high purity silicon can be produced by the Siemens process, developed in the 1960s. This involves dissolving the impure MG-Si in hydrochloric acid to produce the chemical trichlorosilane and then reducing this with hydrogen back to very pure silicon.

Monocrystalline silicon is produced by the **Czochralski process**, initially developed for manufacturing 'silicon chips' for the electronics industry. In this, a small seed crystal is slowly pulled out of a molten mass, or 'melt', of high purity polycrystalline silicon to produce a large single crystal of silicon. The single crystals are then mechanically sawn into thin wafers, polished and doped to produce the necessary p-n junction. Pure silicon is reflective, so a rough anti-reflection coating is added. Then the front and rear contacts are added and the cell is packaged into a saleable module.

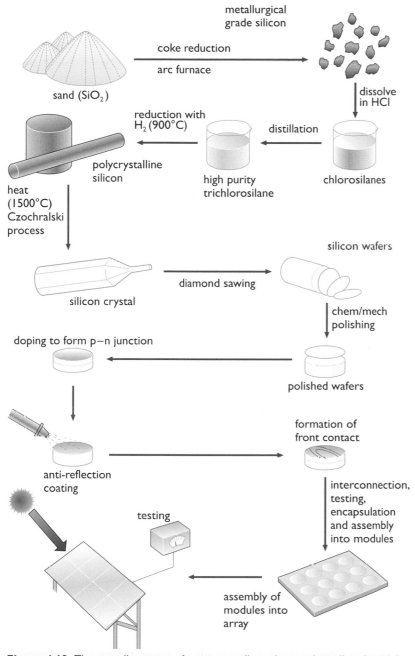

Figure 4.13 The overall process of monocrystalline silicon solar cell and module production. Note that most cells are finally trimmed to be square, or semi-square with rounded corners

Although monocrystalline silicon PV modules are highly efficient, they are also expensive because the Czochralski process is slow and is both labour- and energy-intensive.

A number of approaches to reducing the cost of crystalline PV cells and modules have been under development during the past 30 years.

The first approach involves improving the energy efficiency of pure silicon production. For example, the Fluidized Bed Reactor, developed by REC Silicon in the USA, is a modified form of the Siemens process, producing silicon granules at a 'solar grade' purity but using less energy and as a continuous flow process (REC Silicon, 2016).

The second approach is to use polycrystalline rather than single-crystal material, thus eliminating the slow Czochralski process.

The third is to minimize the amount of silicon needed by growing silicon wafers in ribbon form, or using it in extremely thin films.

Polycrystalline (or multicrystalline) silicon cells

Polycrystalline (or multicrystalline) silicon essentially consists of small grains of monocrystalline silicon (Figure 4.14). The term *multicrystalline* is likely to refer to cells with crystals larger than 1 mm square. Solar cell wafers can be made directly from polycrystalline silicon in various ways. These include the controlled casting of molten polycrystalline silicon into cube-shaped ingots that are then cut, using fine wire saws, into thin square wafers and fabricated into complete cells in the same way as monocrystalline cells (Figure 4.13).

Figure 4.14 Multicrystalline silicon consists of 'grains' of monocrystalline silicon

Another approach to polycrystalline silicon PV manufacture involves drawing a thin 'ribbon' of polycrystalline silicon from a silicon 'melt'. The main process used is known as 'edge-defined, film-fed growth' (EFG), and was originally developed by the US firm Mobil Solar.

Polycrystalline PV cells are easier and cheaper to manufacture than their monocrystalline counterparts, but they tend to be less efficient (21% compared to 26% (Green et al., 2016)). This is because light-generated charge carriers (i.e. electrons and holes) can recombine at the boundaries between the grains within polycrystalline silicon. However, their price advantage has meant that their share of the market has been steadily increasing, to about 70% of global PV production in 2015.

Polycrystalline silicon film

Conventional silicon solar cells need to be around 150–200 μm (millionths of a metre) thick in order to ensure that most of the photons incident upon them can be absorbed. However, 'light trapping' techniques involving giving the front surface of the cell a fine rough texture can maximize the interaction of photons with the material allowing much thinner 'films' of silicon to be used. A cell using a polycrystalline film of silicon only 35 μm thick has demonstrated an efficiency of 21% (Green et al., 2016). Other types of 'thin film' PV cells, using films 2 μm or less thick, are discussed later.

Gallium arsenide

Silicon is not the only crystalline material suitable for PV applications. Another is gallium arsenide (GaAs), a so-called **compound semiconductor**. Gallium is a far less common element than silicon, but is widely used in the electronics industry for high speed integrated circuits and light emitting diodes. GaAs has a crystal structure similar to that of silicon, but consisting of alternating gallium and arsenic atoms.

It is highly suitable for use in PV applications because it has a higher light absorption coefficient than silicon, so only a thin layer of material is required (less than 10 μm). GaAs cells also have a band gap wider than that of silicon, one close to the theoretical optimum for absorbing the energy in the terrestrial solar spectrum (see Box 4.3). Cells made from GaAs are therefore more efficient than those made from monocrystalline silicon. They can also operate at higher temperatures than silicon cells without appreciable performance degradation. This makes them well suited to use in *concentrating* PV systems (see Section 4.6).

On the other hand, cells made from GaAs are substantially more expensive than silicon cells. This is because they need to be produced in a monocrystalline form using **epitaxial** crystal growth techniques. These essentially involve growing monocrystalline GaAs by deposition of gallium and arsenic onto a supporting base material, in this case a single GaAs crystal. This support defines the orientation of the new crystal growth while the doping of the GaAs layer being deposited can also be controlled. GaAs cells have often been used when high efficiency is required – as in many space applications.

BOX 4.3 Semiconductors band gaps and PV cell efficiency

Different elements may have similar chemical properties. Often they lie in the same group of the Periodic Table of Elements. Table 4.1 shows the positions of some elements commonly used in PV semiconductors.

Table 4.1 Elements in the Periodic Table used in PV semiconductors

Group I	Group II	Group III	Group IV	Group V	Group VI
		Boron (B)	Carbon (C)		
		Aluminium (Al)	Silicon (Si)	Phosphorus (P)	Sulfur (S)
Copper (Cu)	Zinc (Zn)	Gallium (Ga)	Germanium (Ge)	Arsenic (As)	Selenium (Se)
	Cadmium (Cd)	Indium (In)	Tin (Sn)	Antimony (Sb)	Tellurium (Te)

Silicon lies in Group IV. It can be doped with boron from Group III to produce a p-type semiconductor or with phosphorus from Group V to produce an n-type semiconductor. Gallium arsenide (GaAs) uses one element from Group III and one from Group V and is termed a III-V semiconductor. Gallium indium phosphide is another III-V example.

For maximum efficiency of conversion of light into electric power, it is important that the band gap energy of the material used for a PV cell is well matched to the solar spectrum. Silicon has a band gap of 1.1 eV. As explained in Box 4.1, photons with energies *less* than the band gap (i.e. in the infra-red region of the spectrum) will not be absorbed. Photons with energies *greater* than the band gap will be absorbed but will only contribute an amount of useful energy *equal* to the band gap.

The optimum band gap for maximum efficiency for a single PV cell is about 1.4 eV. This can be achieved with several options shown in Table 4.2.

Table 4.2 Semiconductor materials and their band gaps

Semiconductor Material	Band Gap/eV
Germanium (Ge)	0.7
Crystalline silicon (Si)	1.1
Copper indium gallium diselenide (CIGS)	1.0 – 1.7 (depending on relative proportions of In and Ga)
Cadmium Telluride (CdTe)	1.4
Gallium Arsenide (GaAs)	1.4
Amorphous silicon (a-Si)	1.7
Gallium Indium Phosphide (GaInP$_2$)	1.9

The maximum theoretical conversion efficiency attainable in a *single-junction* silicon PV cell has been calculated to be about 30% (Wenham et al., 2007). However, *multi junction* cells (described later) have also been designed in which each junction is tailored to absorb a particular portion of the incident spectrum. A semiconductor with a wide band gap is used to absorb the blue part of the spectrum, another with an intermediate band gap is used to absorb the yellow part of the spectrum and yet another is used to absorb the red and infra radiation. These have achieved efficiencies of 45% or more (Green et al. 2016).

In practice, the highest efficiency achieved in commercially available single-junction monocrystalline silicon PV *modules* (as distinct from individual PV *cells*) is currently around 24% (Green et al., 2016). The efficiency of PV modules is usually lower than that achieved by cells in the laboratory because:

- it is difficult to achieve as high an efficiency consistently in mass-produced devices as in one-off laboratory cells under optimum conditions

- laboratory cells are not usually glazed or encapsulated

- in a PV module there are inactive areas, both between cells and due to the surrounding module frame, which are not available to produce power

- there are small resistive losses in the wiring between cells and in the diodes used to protect cells from short circuiting

- there are losses due to mismatching between cells of slightly differing electrical characteristics connected in series.

4.5 Thin film PV

Crystalline silicon wafers currently make up over 90% of the world market (see Figure 4.15). However, they are not the only materials suitable for photovoltaics. PV cells can also be made from 'thin films' of various kinds, usually only a few μm thick. The most common of these are amorphous silicon (a-Si), copper indium (gallium) diselenide (CI(G)S) and cadmium telluride (CdTe). These are often called 'second generation' solar cells.

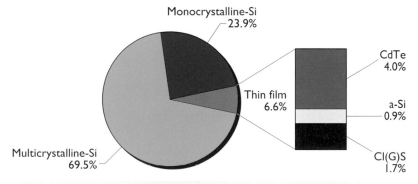

Figure 4.15 Percentage of global production for different PV technologies, 2015 (Source: Fraunhofer ISE, 2016)

Amorphous silicon

Solar cells can be made from very thin films of silicon (typically only 1 μm thick – i.e. only two wavelengths of green light) in a form known as **amorphous silicon (a-Si)**. The structure of a typical cell is shown in Figure 4.16(a). It is built up by depositing a thin layer of silicon dioxide onto a glass substrate, followed by the front electrical contact, a thin transparent layer of tin oxide (SnO_2). Then follows a layer of amorphous silicon, produced by electrically decomposing silane gas (SiH_4). In this layer the silicon atoms are much less ordered than in the crystalline forms described above. Not every silicon atom is fully bonded to its neighbours, which leaves so-called 'dangling bonds' that could potentially damage the overall efficiency of the cell. The hydrogen in the silane combines with some of these bonds, passivating them and making them electrically inactive.

The first part of the a-Si layer is made p-doped by including boron with the silane. The main part of the a-Si layer is pure intrinsic (or undoped) silicon and the final part is n-doped, made by including phosphorus with the silane. This creates not an abrupt 'p-n' junction, but a so-called 'p-i-n' junction. However, the overall electrical effect is similar. Finally a rear electrical contact layer of aluminium is added.

The operation of an a-Si cell is similar to that of a crystalline silicon one, except that the band gap is larger (1.7 eV rather than 1.1 eV).

The amount of silicon required for an a-Si cell is less than 1% of that for a crystalline cell. Also, the manufacturing process operates at a lower temperature, requiring less energy. They are thus much cheaper to produce.

However, a-Si cells are less efficient than crystalline silicon: maximum module efficiencies are currently around 10% (Green et al, 2016). Moreover, their efficiency can degrade by about 20% over the first few months, although this then stabilizes. Manufacturers sell modules with power ratings that correspond to their estimate of the degraded, stabilized performance.

Various approaches have been made to improve the performance. One of the most promising approaches involves combining an a-Si layer with a layer incorporating very small crystals of silicon (microcrystalline silicon or μ–Si) devices, which has resulted in reduced degradation and an improved module efficiency of about 12%.

a-Si cells are used in applications where the requirement is not so much for high efficiency as for low cost.

Copper indium (gallium) diselenide

Other thin film technologies include those based on compound semiconductors, in particular copper indium diselenide ($CuInSe_2$, usually abbreviated to CIS), and copper indium gallium diselenide (CIGS).

The addition of gallium in a CIGS cell means that the precise band gap can be chosen according to the proportions of indium and gallium used.

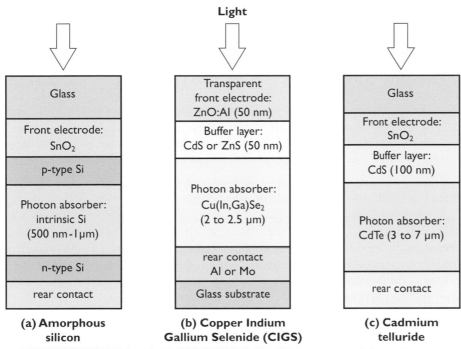

Figure 4.16 Structures of three different types of thin film PV cells (Note: 1 mm = 1000 µm = 1 000 000 nm)

The structure of a CIGS cell is shown in Figure 14.6(b). The cell is usually built up on a soda-lime glass substrate. A layer of molybdenum metal forms the rear contact. A p-doped CIGS absorber layer, 2–2.5 µm thick is then formed on top of this. A very thin buffer layer of n-doped cadmium sulphide (CdS) is then added. Then follows the front electrical contact – a thin layer of conductive aluminium doped zinc oxide.

Thin film CIGS cells have attained laboratory efficiencies of around 21%, and CIGS modules with stable efficiencies of 17% are available. The modules do contain small amounts of cadmium, a toxic heavy metal, and production can also involve the use of toxic hydrogen selenide gas, so care in production and eventual recyling is essential.

Cadmium telluride

Thin film PV modules can also be made using cadmium telluride (CdTe) using a relatively simple and inexpensive electroplating-type process. The cell structure is shown in Figure 4.16(c). A glass surface is covered with a thin tin oxide film which acts as an electrode. This is followed by a buffer layer of n-doped cadmium sulfide (CdS) and then the main active layer of p-doped CdTe, 3 µm to 7 µm thick. Finally, another electrode is added at the back.

The band gap of CdTe, 1.4 eV, is close to the optimum, and module efficiencies of over 18% have been achieved (Green et al., 2016). However, the modules contain cadmium, so precautions need to be taken during their manufacture, use and eventual disposal or recycling (see Section 4.11).

CdTe modules have now been deployed in some of the world's largest PV schemes.

4.6 Other PV technologies

Multi-junction PV cells and modules

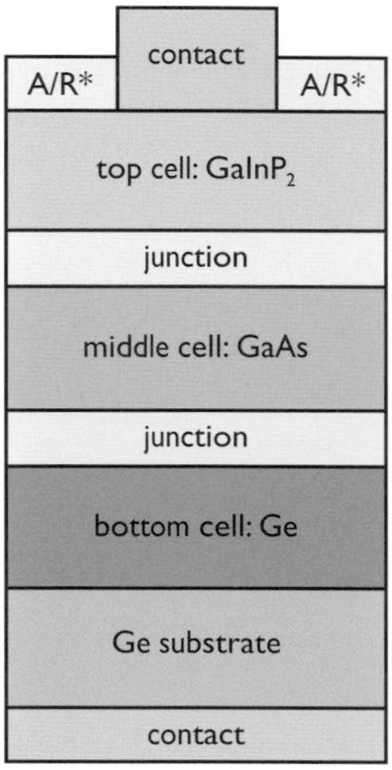

One way of improving the overall conversion efficiency of PV cells and modules is the 'stacked' or **multi-junction** approach, in which two (or more) thin PV junctions are layered one on top of the other, each layer extracting energy from a particular portion of the spectrum of the incoming light.

Such a cell, consisting of three layers, is shown in Figure 4.17. The top layer is made of gallium indium phosphide ($GaInP_2$), which has a band gap of nearly 1.9 eV. This absorbs light in the blue part of the spectrum. The remaining light passes through this cell to a gallium arsenide (GaAs) layer in the middle with a band gap of 1.4 eV that absorbs the yellow part of the spectrum. Finally the unabsorbed light passes to a germanium (Ge) layer at the bottom. This has a band gap of 0.7 eV, and absorbs light in the red portion and much of the infra-red spectrum.

Such cells are very expensive as their manufacture requires a slow epitaxial crystal growth process, but they are very efficient. Modules have achieved efficiencies of over 30% under terrestrial lighting conditions.

They have been demonstrated in applications such as long-distance solar car races (Figure 4.18), and used in commercial concentrating solar power applications (see below).

Figure 4.17 Structure of Spectrolab's 'ultra triple junction' III–V PV cell designed for space and concentrator applications

A/R*: anti-reflective coating

Figure 4.17 diagram labels (top to bottom): contact · A/R* · A/R* · top cell: $GaInP_2$ · junction · middle cell: GaAs · junction · bottom cell: Ge · Ge substrate · contact

Figure 4.18 The winner of the 2009 World Solar Challenge race across Australia was the Tokai Challenger, a solar car designed and tested by students from Tokai University. It covered almost 3000 km at an average speed (during daylight) of over 100 km per hour. The car was driven by electric motors powered by an 1.8 kW$_p$ array of Sharp triple-junction III–V PV cells of a type used in space applications, with an efficiency of 30% (source: Sharp Solar, 2009)

Concentrating PV (CPV) systems

The conversion efficiency of PV cells is not necessarily at its maximum at an illumination of '1 sun'. For example, the efficiency of a monocrystalline silicon cell may rise from 20.7% at '1 sun' to 26.5% at '500 suns' intensity, as long as the cell can be kept cool (McConnell and Fthenakis, 2012). Module efficiencies of almost 40% have been measured with multilayer gallium arsenide cells (Fraunhofer ISE and NREL, 2015).

Concentrating the Sun's rays onto cells using mirrors or Fresnel lenses also has the added advantage that only a small area of cells is required. (The approach is obviously similar to that described in Chapter 3, Section 3.9 on solar thermal engines.) A Fresnel lens is a thin sheet of glass containing stepped sections that carries out the same function as a conventional lens made of thicker solid glass.

The concentrating system must have an aperture equal to that of an equivalent flat plate array to collect the same amount of incoming energy, but the key factor is the concentration ratio, the ratio of the aperture to the actual cell area.

Low concentration systems with ratios up to ten are likely to use parabolic mirrors (as shown in Figure 4.19) and crystalline silicon cells. High concentration ratio systems (typically 400 times) use gallium arsenide cells with Fresnel lenses (Fraunhofer ISE, 2016). In both cases the cells need to be specially designed for adequate cooling.

One disadvantage is that such systems require clear skies and plenty of direct solar radiation; they cannot concentrate diffuse radiation. These systems need to track the Sun. For low concentration systems tracking in one axis is adequate, while high concentration systems need to track it in two axes (azimuth and elevation), ensuring that the cells always receive the maximum amount of direct solar radiation.

In 2014 there were 340 MWp of grid-connected concentrating PV projects in the world, including multi-megawatt schemes in China, South Africa and the USA. Although there has been much development work in the USA, competition from simpler flat plate PV systems has created severe financial problems for manufacturers. At the time of writing (early 2017) it is not clear whether or not this sector of photovoltaics will share in the large growth of other parts of the industry.

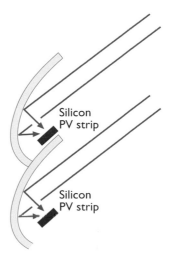

Figure 4.19 Schematic of part of a low concentrating ratio PV system using mirrors (source: adapted from SunPower Corporation, 2012)

Future and emerging PV technologies

The photovoltaic cells described so far have used p-n junctions in a rather restricted range of chemical elements, which are usually required to be produced in exceptionally high (and expensive) purity. Current commercial crystalline silicon cells have set expectations of performance: efficiencies of more than 20%, a guaranteed life expectancy of 25 years or more, and prices under US$1 per peak watt.

However electricity can also be produced by the action of light on a range of other materials, which may have future potential.

Perovskite solar cells

The term 'perovskite' refers to a particular crystal structure. Perovskite solar cells are primarily based on organic compounds of lead and halogen elements such as chlorine, bromine or iodine. Methylammonium lead halides are cheap to produce and simple to manufacture. The measured efficiency has increased from under 4% efficiency in 2010 to 20% in 2016 (Green et al., 2016). The material has a high light absorption coefficient so a film thickness of only about 0.5 µm is needed. The main research challenge is to achieve an adequate cell lifetime (Arthur D. Little, 2015).

Organic solar cells

These consist of layers of polymers or, more simply, plastics. These can potentially be produced at low production costs and in high volumes. Despite having been actively researched since 2001 the best research cell efficiency is only 13%. Commercial cells are being produced as a semi-transparent film with an efficiency of 6%.

Dye-sensitized solar cells (DSSC)

These can be based on a semiconductor such as titanium dioxide (commonly used as a white pigment in paint) together with a layer of a 'sensitizer' dye only one molecule thick, made of a proprietary 'transition metal complex' based on ruthenium or osmium (Grätzel, 2004). Such cells have demonstrated an efficiency of nearly 12%, but there are still ongoing research challenges about the ability to withstand bright sunlight without degrading.

Quantum dots (QDs)

This is PV technology based on 'nanotechnology' that aims to manipulate semiconducting materials at extremely small scales, measured in billionths of a metre, or nanometres (nm). Nanoparticles consisting of extremely small collections of atoms (up to around 10 000) are called 'quantum dots' (QDs). The band gaps of QDs can be 'tuned' to different parts of the solar spectrum according to their size. By incorporating QDs with a variety of sizes, more of the energy in the spectrum can be absorbed.

Although this technology is theoretically capable of a very high efficiency, of up to 66% (NREL, 2013), the efficiency of cells measured in laboratory tests is currently low – around 7%.

4.7 Electrical characteristics of silicon PV cells and modules

One very simple way of envisaging a typical silicon PV cell is as a solar powered battery that produces a voltage of around 0.5 V and delivers a current, proportional to cell size and sunlight intensity, of several amps.

In order to use PV cells efficiently we need to know a little more about how they behave when connected to electrical loads. Figure 4.20 shows a single silicon PV cell connected to a variable electrical resistance R, together with an ammeter to measure the current (I) flowing in the circuit and a voltmeter to measure the voltage (V) developed across the cell terminals. Let us assume the cell is being tested under standard conditions with an illumination of 1000 W m^{-2}.

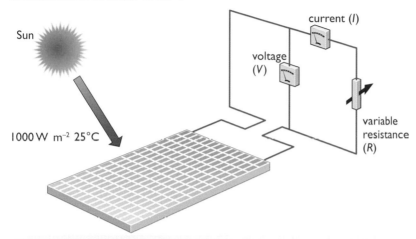

Figure 4.20 PV cell connected to variable resistance, with an ammeter and voltmeter to measure variations in current and voltage as the resistance varies

When the resistance is infinite (i.e. when the cell is 'open circuited') the current in the circuit is at its minimum (zero) and the voltage across the cell is at its maximum, known as the '**open circuit voltage**' (VOC). At the other extreme, when the resistance is zero, the cell is in effect 'short circuited' and the current in the circuit then reaches its maximum, known as the '**short circuit current**' (I_{SC}) (see Figure 4.21).

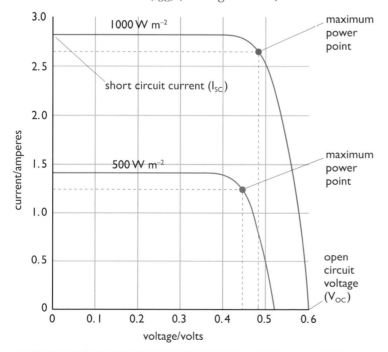

Figure 4.21 Current–voltage (I–V) characteristics of a small silicon PV cell under standard test conditions under an illumination of 1000 W m^{-2} and 500 W m^{-2}

If we vary the resistance between zero and open circuit, the current (I) and voltage (V) will be found to vary as shown in Figure 4.21, which shows the 'I–V characteristic' or 'I–V curve' of the cell. The cell will deliver maximum power (i.e. the maximum product of voltage and current) when the external resistance is adjusted so that the cell is working at the **maximum power point** (MPP) on the I–V curve.

The short circuit current is directly proportional to the intensity of solar radiation, whilst the open circuit voltage will only change slightly. If the level of solar radiation is reduced to 500 W m^{-2} then the general shape of the I–V characteristic stays the same, but the short circuit current is likely to be halved and the open circuit voltage will decrease slightly. The maximum power point will have changed.

The open circuit voltage also decreases linearly as the cell temperature increases.

When PV modules are delivering power to electrical loads in real world conditions, the intensity of solar radiation usually varies substantially over time. Many PV systems therefore incorporate a 'maximum power point tracking' circuit that automatically matches the output of the PV module to around the maximum power point.

As shown in Figure 4.21, a typical silicon PV cell produces a voltage of around 0.5 volts and, in the example shown, a current of 2.6 amps at its maximum power point.

PV modules are usually assembled as a string of cells in series. The number of cells in a module will depend on the peak operating voltage required. The total voltage will be then be equal to the number of cells × 0.5 V, but the current produced will only be that of an individual cell.

Multiple modules are normally connected in parallel to form an array and the total current will be the sum of that produced by each module (Figure 4.22).

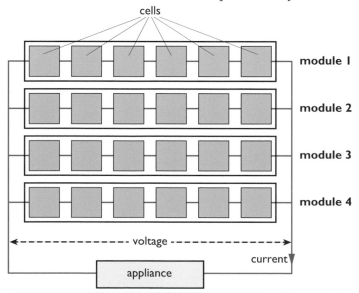

Figure 4.22 Cells are usually connected in series within modules and multiple modules can be connected in parallel

Current crystalline PV modules are around 1.4 to 1.7 square metres in area (though some are larger), usually incorporate 60–72 PV cells connected in a series – parallel combination, and have a peak power output of some 120–300 watts, depending on the design and technology.

4.8 **PV systems for remote power**

PV modules are now widely used in developed countries to provide electrical power in locations where it would be inconvenient or expensive to connect to conventional grid supplies. They usually charge batteries to ensure continuity of power. Some examples are shown in Figure 4.23.

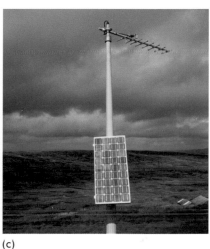

(a)   (b)   (c)

Figure 4.23 (a) PV parking meter (b) PV navigation buoy (c) PV telemetry

Off-grid PV systems may also be used to provide a basic electricity supply for rural houses. These are likely to require substantial batteries which may be the most expensive part of the system. Designing such systems for minimum overall cost requires that the system designer be able to answer the following questions:

- What are the daily, weekly and seasonal variations in the electrical demand of the house?
- What are the daily, weekly and seasonal variations in the amount of solar radiation in the area where the house is situated?
- What are the options for the orientation and tilt angle of the PV array?
- For how many sunless days will the battery need to be able to provide back-up electricity?

Computer programs have been developed to help engineers calculate the size and cost of PV systems to meet clearly specified energy requirements in given locations and climatic conditions.

In many 'developing' countries electricity grids are often non-existent or rudimentary, particularly in rural areas. In such countries, especially those with high solar radiation levels, PV can be highly competitive with other forms of electricity supply, such as petrol or diesel generators, and its use is growing very rapidly. Applications include: water pumping; refrigerators to help keep vaccines stored safely in health centres; providing energy for lights, radios, audio and video systems for homes and community centres; telecommunications systems; and street lighting.

In India and large parts of Africa, rural households use kerosene lamps and candles at night. These are very inefficient light sources, a source of smoke inside the home and can cause accidental fires. To address this problem, in 2008 the Energy and Resources Institute in Delhi, India set up the 'Lighting a Billion Lives' project (TERI, 2016). This aims to replace the traditional light sources with battery-powered solar lanterns using efficient compact fluorescent lamps (CFL) or light-emitting diode (LED) lamps. The solar lantern batteries can be recharged at solar PV charging stations in villages, run by selected local entrepreneurs (Figure 4.24).

In some villages they have also set up PV-powered micro-grids feeding multiple houses. By 2016, the scheme had distributed nearly 170 000 solar lanterns and provided illumination to over 870 000 households. Another project, SolarAid, sells small solar lights in a number of African countries, and by 2015 had distributed over 1.6 million of them.

Figure 4.24 PV-powered lanterns being charged in India, as part of the 'Lighting a Billion Lives' project

4.9 Grid-connected PV systems

PV systems for homes

In most parts of the developed world, grid electricity is easily accessible as a convenient backup to PV or other variable renewable energy supplies. Here it makes sense for PV energy systems to use the grid as a giant 'battery'. The grid can absorb PV power which is surplus to current needs (say, on sunny summer afternoons), making it available for use by other customers and reducing the amount that has to be generated by conventional means; and at night or on cloudy days, when the output of the PV system is insufficient, it can provide backup energy from conventional sources.

In these grid-connected PV systems, a so-called 'grid commutated inverter' (or 'synchronous inverter') transforms the direct current (DC) power from the PV arrays into alternating current (AC) power at a voltage and frequency that can be accepted by the grid. In many countries, 'Feed-in Tariffs' (FiTs) have been introduced. These provide for premium payments to be made for power produced by grid-connected PV arrays.

In the UK, surveys have suggested that about half of all domestic roofs are oriented in a direction sufficiently close to south to enable them to be used for solar collection purposes. PV arrays can be added to the roofs of existing dwellings, normally by mounting them on rails on top of the roof. In the roofs of new dwellings they can in some circumstances replace all or part of the conventional roof (Figure 4.25).

Figure 4.25 Solar PV roofs at the Solar Settlement, Freiburg, Germany (source: Rolf Disch Solar Architektur)

Electricity yield from PV systems

The annual amount of electricity that will be produced by a PV system depends on various factors, including:

- The annual total quantity of solar radiation available at the site. This is perhaps the most important factor and as described earlier, in Chapter 3 section 3.3, it is strongly influenced by the site latitude. In typical UK conditions, it is around 1000 kWh m^{-2} y^{-1} but in very sunny countries the figure can rise to well over 2000 kWh m^{-2} y^{-1}.

- The orientation (azimuth) and tilt (elevation) of the PV array. This influences both the month-by-month solar availability and the overall annual total. For maximum annual energy output, arrays in the northern hemisphere should be oriented close to south and with a tilt roughly equal to the latitude of the site (see Chapter 3, Figure 3.12), although departures from these optima do not have a serious effect. A higher tilt angle may be desirable for off-grid applications to maximize collection in winter. Arrays in the southern hemisphere should, of course, face north.

- The peak power rating of the arrays. The actual physical area in square metres will depend on other factors, particularly the efficiency of the PV modules. Typically a 1000 peak watt array of polycrystalline cells will occupy 5–7 m^2.

- The power-reducing effects of array shading by trees, nearby buildings, etc. Shading may be assessed using a sunpath diagram such as that shown in Chapter 3, Figure 3.36. PV modules are highly sensitive to shading and it is very important to make sure that all of the cells of individual modules are evenly illuminated by the Sun.

- Various energy losses. These include variations of panel efficiency with temperature and different lighting conditions, the effects of dirt and ageing of the panels, and the limited efficiency of any electrical inverter used. Overall these losses may be taken into account by including a 'performance ratio' into any energy calculation, assuming that the actual performance is only 75–80% of the peak theoretical performance.

The European Union's Joint Research Centre at Ispra, Italy, has produced PVGIS (PV Geographical Information System), a very useful online tool giving solar radiation data and estimated PV outputs. One version is available for Europe and surrounding countries. Another covers all of Europe, Africa and much of Asia. Users can input the geographic location, type of module, its power rating, array orientation, etc., and the software will calculate the expected monthly and annual energy yield (see EC JRC, 2012). Key aspects of this information have been summarized in a Solar Radiation Map of Europe (Figure 4.26).

Photovoltaic Solar Electricity Potential in European Countries

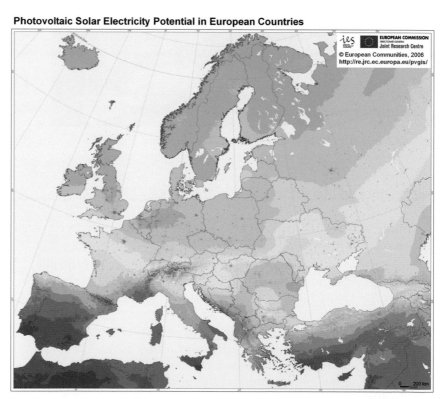

Yearly sum of global irradiation incident on optimally-inclined south-oriented photovoltaic modules

Global irradiation [kWh/m²]

| <600 | 800 | 1000 | 1200 | 1400 | 1600 | 1800 | 2000 | 2200> |

| <450 | 600 | 750 | 900 | 1050 | 1200 | 1350 | 1500 | 1650> |

Solar electricity [kWh/kWp]

Yearly sum of solar electricty generated by 1 kWp system with optimally-inclined modules and performance ratio 0.75

Figure 4.26 Solar radiation map of Europe, showing annual energy yields of optimally oriented PV arrays in various European locations. 'Performance Ratio' is the ratio of actual to the theoretical maximum PV array output

PV systems for non-domestic buildings

PV arrays can also be integrated into the roofs and walls of commercial, institutional and industrial buildings, replacing some of the conventional wall cladding or roofing materials that would otherwise have been needed and reducing the net costs of the PV system.

Commercial and industrial buildings are normally occupied mainly during daylight hours, which correlates well with the availability of solar radiation.

There are now many examples of non-domestic buildings incorporating grid connected PV systems, in countries like Germany, Japan, the Netherlands, Italy, the UK and the USA (see Figures 4.27 and 4.28).

Figure 4.27 This solar office building at Doxford, near Sunderland in the UK, has a 73 kW PV array integrated into its south-facing façade

Figure 4.28 A grain store near Warminster, Wiltshire, UK – fitted in 2012 with a 225kW rooftop PV array

Large, grid-connected PV power plants

Large, centralized PV power systems, at the multi-megawatt scale, have also been built to supply power for local or regional electricity grids in many countries, including the UK, Germany, Switzerland, Italy, China, India and the USA.

Compared with building-integrated PV systems, large stand-alone PV plants can take advantage of economies of scale in purchasing and installing large numbers of PV modules and associated equipment. On the other hand, the electricity they produce is not used onsite and has to be distributed by the grid, thus involving transmission losses.

Such systems usually use conventional fixed panels. Some may incorporate single axis tracking of the Sun, increasing the output and extending the performance from the morning right through to the evening.

Large plants also require substantial areas of land, which has to be purchased or leased, adding to costs. However, in many sunny countries there are suitable areas of desert, semi-desert or degraded land not used for agriculture or wildlife conservation. The 48 MW$_p$ plant outside Boulder City, Nevada, USA, shown in Figure 4.29, is such an example.

Figure 4.29 A 48 MW$_p$ solar PV facility in Boulder City, Nevada, USA commissioned in 2011. It uses cadmium telluride cells (Source: Sempra generation).

Elsewhere considerable ingenuity has gone into finding sites. Examples are:

- Low-grade agricultural land, including that prone to shallow flooding.
- Floating PV arrays on drinking water reservoirs.
- Commercial land such as car parks
- Noise barriers alongside motorways.

There has also been a proposal from a Chinese company to build a 1 GW$_P$ power plant on the radioactively contaminated land surrounding the Chernobyl nuclear power station in the Ukraine.

Large PV arrays do not necessarily cover all of the land area. This is particularly true at higher latitudes where an appreciable gap may be left between successive rows of panels to prevent mid-winter overshading of panels. The ground beneath can thus be sown with grass or 'wild-life' friendly plants and used for sheep grazing (Figure 4.30).

Figure 4.30 Sheep may graze immediately around and even under PV panels. (Source: Dr. Jonathan Scurlock, National Farmers' Union)

Large PV power plants are most economical in areas of the world with high levels of solar radiation, such as north Africa or southern California. These areas have clearer skies meaning that the majority of the radiation is direct, also allowing the use of concentrating PV systems.

The maximum size of schemes has increased considerably over the last ten years. Table 4.3 shows the details of the five largest in 2016.

Table 4.3 World's largest PV power plants, 2016

Location	Power / MW$_p$	Area /km^2
Longyangxia Dam Solar Park, Qinghai province, China	850	14
Kamuthi Solar Power Project, Tamil Nadu, India	648	10
Solar Star Projects, California, USA	579	13
Desert Sunlight Solar Farm, California, USA	550	9.5
Topaz Solar Farm, California, USA	550	16

(source: Wikipedia)

Satellite Solar Power System

Probably the most ambitious – and some would say the most fanciful – proposal for a 'grid-connected' PV plant is the Satellite Solar Power System (SSPS) concept, first suggested over 40 years ago (see Glaser, 1972 and 1992). The basic idea was to construct huge PV arrays, each over 50 km^2 in area and producing several GW of electrical power, in geostationary orbit around the Earth (Figure 4.31). The power would be converted to microwave radiation and beamed from 1 km diameter transmitting antennas in space to 10 km^2 receiving antennas on Earth. The PV panels could theoretically be in almost permanent bright sunshine.

Alas, there are many objections to such a scheme, two main ones being:

- The economics, given that it costs roughly $20 000 to put 1 kg of payload into space by rocket.
- The CO_2 emissions of all the rocket fuel necessary for such a project.

Even constructing a PV array with an area of 50 km^2 on Earth may have sounded like science fiction in 1972, but the combined areas of the five schemes listed in Table 4.3 exceed this.

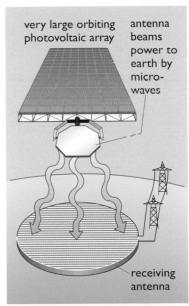

Figure 4.31 The Satellite Solar Power System (SSPS) concept

very large orbiting photovoltaic array

antenna beams power to earth by micro-waves

receiving antenna

4.10 **Costs of electricity from PV**

As with conventional electricity generation, the cost per kilowatt-hour of power from PV cells consists essentially of a combination of the costs of repaying the initial *capital cost* and the *running cost*.

The capital cost of a grid connected PV system is usually considered to be proportional to its rated power output. It is thus quoted in £ per peak kW (£ kWp $^{-1}$). It includes not only the cost of the PV modules themselves, but also the **'balance of system' (BOS)** costs. These include the costs of the interconnection of modules to form arrays, the array support structure, the costs of cabling, switching, metering and inverters, and, if the array is not building-mounted, the land and foundations.

As shown earlier in Figure 4.7, the costs of PV modules have been falling. Figure 4.32 shows the estimated cost breakdown for a 10 MW$_p$ ground mounted system in 2011 and with projections for the future. The falling module prices mean that BOS costs are likely to make up 50% or more of future system prices.

Small off-grid systems would also have to include extra battery capital costs which can add appreciably to the overall price.

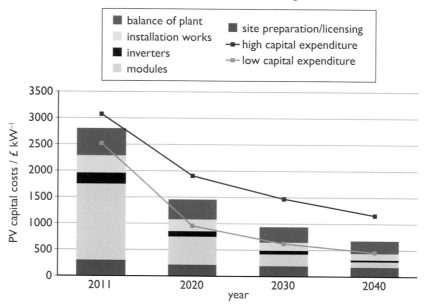

Figure 4.32 Capital cost breakdown for a 10 MW$_p$ ground-mounted PV system in 2011 and projected costs for 2020, 2030 and 2040 (source: CCC, 2011)

Figure 4.32 shows a total system price of nearly £3000 per kW$_p$ in 2011. Yet global average costs for utility-scale PV projects completed in 2015 had fallen to under £1400 per kW$_p$ (IEA, 2016a). Those for solar PV in buildings were around £1900 per kW$_p$. The prices for some large agricultural projects completed in the UK in 2016/7 had fallen to £800 per kW$_p$ (Scurlock, 2017).

The running costs of grid-connected PV are very low. There are no fuel costs and maintenance and insurance costs are typically £10–£15 per kW$_p$ of capacity per year.

Estimates of the annual electricity production from a kilowatt peak system have been shown earlier in Figure 4.26. This figure varies from about 1000 kWh per year in the southern UK up to 1500 kWh yr^{-1} in central Spain.

PV schemes are likely to have a long lifetime. Modules are normally sold with guarantees that they will meet over 95% of their specified power rating on delivery, but only 85% after 25 years. These figures are included in calculations as part of the 'performance ratio' described earlier.

In 2016 the domestic price of electricity of a kilowatt-hour electricity in the UK was about 17p. This contains a large price component for electrical distribution. The price of electricity as sold to medium sized industrial consumers was much lower, around 9p.

How does the cost of electricity from PV compare with these figures? In order to answer this we need to calculate the Levelized Cost of Electricity (LCOE) (see Box 4.4).

BOX 4.4 Levelized cost of electricity (LCOE) from a grid connected PV system

A simple, general approach to costing electricity from power plants is described in Appendix B. As described there the key cost elements of any electrical power plant are:

- Fuel costs (zero in the case of PV)

- Operation and maintenance (O&M) costs (low for PV)

- Initial capital costs (high in the case of PV)

- Final decommissioning costs (likely to be very low)

Let us consider a 10 MW_p (10 000 kW_p) scheme situated in the south-west of the UK with:

- A capital cost of £800 per kW_p

- An operating and maintenance (O&M) cost of £10 per kW_p per year

- A project lifetime of 25 years financed with a loan over that amount of time at a real interest rate of 8%.

Total capital cost = 10 000 × £800 = £8 000 000

The cost of borrowing £1000 at 8% per annum over 25 years can be found in Table 1 in Appendix B and is £94 per year.

Annual cost of repaying £8 000 000 over 25 years = 8 000 × £94 = £752 000

Annual operating and maintenance cost = 10 000 × £10 = £100 000

Total annual cost = £752 000 + £100 000 = £852 000

Figure 4.26 suggests that for a PV system in the south-west UK we can assume that each kW_p of capacity produces 1000 kWh of electricity each year:

Total annual electricity production = 10 000 × 1000 kWh = 10 000 000 kWh

Levelized Electricity Cost = £852 000 × 100 / 10 000 000 = 8.52 p kWh^{-1}

This is well below the domestic consumer price and slightly lower than the UK industrial electricity price. If this plant (with the same costs) was situated in central Spain with 50% more solar radiation, then the average electricity price would only be only 5.7p kWh^{-1}, highly competitive with fossil-fuelled electricity generation.

In many countries the PV market has been stimulated by Feed-in Tariffs (FITs) giving PV system owners and operators high prices for their output electricity. In the case of large schemes these payments are based on metered output. For small domestic users they are based on estimates from the system size and orientation.

The price of electricity from PV projects around the world has been falling and appears to be reaching 'grid parity' in many sunny countries, i.e. a price competitive with conventional electricity generation (see Figure 4.33). This has led to calls to reduce and even withdraw FITs for new schemes.

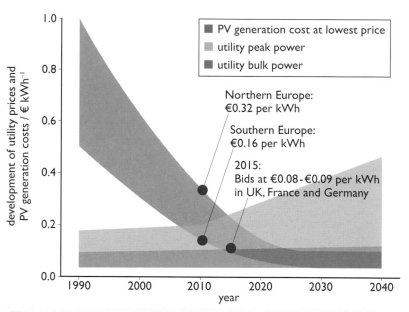

Figure 4.33 Progress towards 'grid parity': convergence of utility electricity prices and PV costs, 1990 to 2010 and projection to 2040 (source: EPIA/Greenpeace, 2011; Bid data from Irena, 2016b).

In many countries contracts for very large PV systems >500 MWp are currently being sold in auctions for the lowest 'bid price'. This has resulted in very low prices of under 4p kWh⁻¹ in Mexico, Brazil, the USA, Saudi Arabia and the United Arab Emirates (IRENA, 2017). These prices are truly competitive with conventional fossil-fuelled generation from oil and gas.

In order for full grid parity to be reached in less sunny countries there still need to be capital cost reductions, particularly the BOS and installation costs.

4.11 Environmental impact and safety

PV systems

The environmental impact of PV is probably among the lowest of all renewable or non-renewable electricity generating systems.

In normal operation, PV electricity systems emit no gaseous or liquid pollutants and no radioactive substances. Crystalline silicon PV modules contain relatively safe materials and pose little hazard either during their working life or in their eventual disposal. CdTe modules (and some CIS modules) contain cadmium, a toxic heavy metal. Although this poses little risk during operation, it is important that these cells are eventually recycled properly.

Even though PV arrays are potentially very long-lived devices, eventually they will come to the end of their useful life and will have to be disposed of – or, preferably, recycled (IRENA, 2016c). A kilowatt peak PV array is likely to contain 6 kg or more of very highly purified silicon.

Many European manufacturers are already beginning to recycle PV modules at the end of their working lives. EU recycling regulations under the Waste Electrical and Electronic Equipment (WEEE) directive now make it compulsory for manufacturers to take back and recycle at least 85% of their PV modules free of charge (Solar Waste, 2013).

Most PV modules have no moving parts, so they are also safe in the mechanical sense, and they emit no noise. However, as with other electrical equipment, there are some risks of electric shock – especially in larger systems operating at voltages substantially higher than the 12–48 volts employed in most small PV installations. But the electrical hazards of a well-engineered PV system are, at worst, no greater than those of other comparable electrical installations. The installation of roof mounted systems also involves some risks, but again no more than other maintenance work carried out at heights. Fires in PV systems are a possible problem and guidance for fire-fighters has been compiled in a US report (Grant, 2013).

PV arrays do, of course, have some visual impact. Rooftop arrays will normally be visible to neighbours, and may or may not be regarded as attractive, according to aesthetic tastes. Several companies have produced special PV modules in the form of roof tiles which blend into roof structures more unobtrusively than conventional module designs.

Large, multi-megawatt PV arrays require large amounts of land. The largest of such projects can now cover several square kilometres. In the UK there have been objections to their visual intrusiveness and schemes may be hidden by careful site selection and (non-shading) rows of trees. There have also been objections to their construction on agricultural land that could be used for food production.

PV module production

The environmental impact of manufacturing silicon PV cells is unlikely to be significant. Some of the issues are the same as for plants producing semiconductor integrated circuits. However, it is important that these are appreciated so that PV manufacture can be seen as truly creating 'green jobs' (SVTC, 2009).

The basic material from which the vast majority of PV cells are made, silicon, is not intrinsically harmful. However, the process of sawing crystals into wafers produces large amounts of dust. It is important that workers are not exposed to this and that this high purity material is recycled back into the production process.

The chemicals used in silicon production must also be treated with care. Silane gas, SiH_4, from which pure silicon is produced is inflammable and waste silicon tetrachloride ($SiCl_4$), which is highly toxic, can also be produced. Sulfur hexafluoride and nitrogen trifluoride, which are both potent greenhouse gases, are used for cleaning purposes and must not be released into the atmosphere.

Cadmium is obviously used in the manufacture of CdTe modules and very small amounts of cadmium may also be used in manufacturing CIS and CIGS modules – although zinc may potentially be substituted.

As in any chemical process, careful attention must be paid to plant design and operation, to ensure the containment of any harmful chemicals in the event of an accident or plant malfunction.

The energy balance of PV systems and potential materials constraints

A common misconception about PV cells is that as much energy is used in their manufacture as they generate during their lifetime. This might have been true in the early days of PV, when the refining of monocrystalline silicon and the Czochralski process were very energy-intensive, and the efficiency of the cells produced was relatively low.

However, modern cells are more electrically efficient and the use of modern PV production processes and thin film cells has made the energy balance of PV much more favourable. The energy payback time for new multicrystalline silicon PV rooftop systems in UK conditions is about 2.1 years. This falls to only 1.2 years for systems in southern Europe (Fraunhofer ISE, 2016).

Concerns have also been expressed over the availability of 'rare earth' elements such as indium and tellurium, used in thin film PV. PV cell manufacture was, in 2015, using about 5% of the world's silver production for cell contacts (Fraunhofer ISE, 2017). Given that this is expensive, there is commercial pressure to reduce the amounts used.

4.12 PV integration and future prospects

Integration

PV cells obviously only generate electricity during the day. Depending on their latitude, they are likely to produce more electricity in summer and less in winter. Locally their output may vary rapidly with passing clouds. Their average output may also vary from day to day as weather conditions change. All of these considerations raise issues for the integration of PV electricity into the present-day electricity grid, which already has to deal with large variations in electrical *demand*. Extra variations in electrical *supply* are unlikely to raise new problems as long as the proportion of PV in the generation mix is relatively small.

In 2016 PV provided about 3% of the UK's electricity supply. Elsewhere the proportions have been higher: 6% in California and over 7% in Germany (both estimates for 2016). Both California and Germany have set ambitious renewable energy targets involving continued expansion and investment in PV. Researchers have been investigating the integration problems of having more than 10% of PV in the electricity generation mix.

There are three main issues:

- Ramp rate – on a sunny day PV may supply a large proportion of the total electrical load, but this will disappear at sunset requiring that an equivalent amount of electrical power from other sources be 'ramped up' quickly. This has been recognized as a particular problem in California and may eventually tip the balance away from future investment in PV solar generation towards thermal Concentrating Solar Power (CSP). This, as described in the previous chapter, has the potential for some energy storage, allowing generation to continue into the evening.

■ Maintaining the voltage and frequency stability of the grid under high levels of input from PV.

■ Dealing with surplus PV generation on sunny days.

All of these issues are general problems shared with other technologies, such as wind, and are discussed later in Chapter 11 on integration.

The growing world photovoltaics market

Between 2005 and 2015, as shown in Figure 4.6, world installed PV capacity grew extremely rapidly, from some 5 GW$_p$ in 2005 to over 230 GW$_p$ in 2015, a rate of increase of over 40% per annum (BP, 2016).

Much of this growth has resulted from the rapid expansion of PV manufacturing capacity in China and Taiwan, which together in 2015 produced two thirds of the worlds cells (Figure 4.34). Nearly all of this production capacity has been put in place since 2010.

As shown earlier in Figure 4.15, multicrystalline silicon dominates the world market. Sales of both mono and multicrystalline silicon, as well as CdTe cells, have been growing.

In 2015 PV supplied only 1% of the world's electricity. Can the industry keep up its high growth rate into the future?

There are many reasons to say that it can:

■ PV electricity costs have now (early 2017) reached grid parity in many sunny countries.

■ Many countries have included PV targets as part of their climate change CO_2 reduction commitments. China could have well over 100 GW$_p$ of PV installed by 2020.

■ Both India and China have severe local pollution problems of smog from coal-fired power stations and are looking to PV for a (rapid) solution.

■ Countries such as India have committed to improving their standards of rural access to electricity, and PV may well be the cheapest solution.

■ Some Middle-Eastern countries have realized that oil and gas will not supply revenue for ever and that diversification into solar is necessary.

In view of the past high growth rate, future projections for world PV capacity (shown in Figure 4.35) are perhaps rather modest:

■ The International Energy Agency (IEA), in its *Renewable Energy: Medium Term Market Report,* has projected world PV capacity to grow to over 500 GW$_p$ by 2021 (IEA, 2016b). In its *World Energy Outlook, 2016*, its most ambitious 450 Scenario (i.e. roughly what is needed to keep global warming to 2°C) sees capacity rising to nearly 1300 GW$_p$ by 2030 (IEA, 2016a).

■ The International Renewable Energy Agency (IRENA), in their *REthinking Energy*, 2017 is more optimistic and suggests a PV capacity of 1760 GW$_p$ by 2030, generating 7% of the world's electricity (IRENA, 2017).

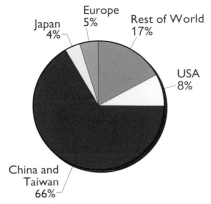

Figure 4.34 PV module production by region, 2015 (source:, Fraunhofer ISE, 2016)

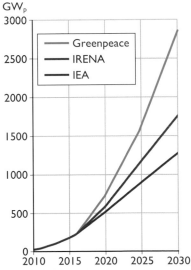

Figure 4.35 Projections of future world PV capacity (sources: IEA Renewable Energy: Medium term market report (IEA, 2016b), IEA World Energy Outlook, 2016 (IEA, 2016a) and IRENA, REthinking energy, 2017 (IRENA, 2017) Greenpeace Energy Revolution 2015 scenario (Greenpeace, 2015))

- Greenpeace, in their *Energy Revolution* 2015 scenario, do suggest a continuing high growth rate, reaching a PV capacity of over 2800 GW_p by 2030 providing over 11% of the world's electricity (Greenpeace, 2015).

It is worth pointing out that most past projections of global PV capacity (including those in previous editions of this book) have been *underestimates*.

Finally, an inspiring – and literally uplifting – symbol of the rising hopes of the global PV industry is the 'Solar Impulse' project. This has involved the design and construction of a solar PV-powered single-seater aircraft which successfully completed the world's first PV-powered round-the-world flight (Figure 4.36).

The leaders of the project, Bertrand Piccard and André Borschberg, do not see PV-powered aircraft replacing conventional aircraft in the foreseeable future. Piccard indicated that the project's goal was not to cause a revolution in aviation, rather, 'Our ambition for Solar Impulse is for the worlds of exploration and innovation to make a contribution to the cause of renewable energies. We want to demonstrate the importance of clean technologies for sustainable development; and to place dreams and emotions back at the heart of scientific adventure' (Solar Impulse, 2016).

Figure 4.36 The Solar Impulse 2 PV-powered aircraft flying over San Francisco. It completed a 40 000 km round-the-world trip in 17 stages starting in Abu Dhabi in March 2015 and eventually returning in July 2016. The plane was powered by over 17 000 monocrystalline silicon PV cells, mounted on the wings and tail. Four 7.5 kW electric motors drove the propellers and electrical storage was provided by lithium polymer batteries (source: Solar Impulse, 2016)

4.13 Summary

The photovoltaic effect – the direct conversion of light into electricity – has been known about since the 19th century. Silicon PV cells were developed in the 1950s for use in space satellites and have been used for Earth-based electricity generation since the 1980s.

Large scale PV module production since the 1990s has driven down prices and world installed capacity increased by over 40% per annum between 2005 and 2015.

Silicon dominates PV production – in 2015 it made up over 90% of the world's PV cell production.

PV cells only need to be thin – a crystalline silicon cell only needs to be 200 μm (one fifth of a millimetre) thick.

Doping produces an electric field in a silicon crystal – the addition of phosphorus to silicon creates *n-type* silicon which has a surplus of electrons. The addition of boron creates *p-type* silicon which has a deficit of electrons (or surplus of 'holes' – a lack of an electron). The active part of a PV cell is the junction between p-type and n-type silicon, a *p-n junction* which has an electric field across it.

Light as photons – at the atomic level light can be considered as consisting of individual particles called photons. The energy of a photon depends on its wavelength: blue light photons have more energy than photons of red light.

Silicon is a semiconductor – the electrons in a crystal of silicon are mainly in the *valence band* and help bond the silicon atoms together. However, if they are given enough energy from an incoming photon of light they can move into the *conduction band* and help create an electric current in an external circuit.

Band gap – this is the energy necessary to move an electron from the valence band to the conduction band. It is specified in *electron-volts (eV).* Silicon has a band-gap of 1.1 eV.

The efficiency of a PV cell depends on the band gap – light photons with less energy than the band gap have no effect on the cell. Photons with energies greater than the band gap only supply a quantity of energy equal to the band gap.

Test standards – the performance of cells and modules is specified in *peak watts (W_p)* under lamps which simulate sunlight. For Earth-based applications this has an intensity of '1 Sun' or 1000 W m^{-2}.

PV cells require very high purity silicon – but less pure than that required by the semiconductor electronics industry.

Monocrystalline and polycrystalline silicon cells made up nearly 70% of world PV cell production in 2015 – monocrystalline cells are slightly more efficient than polycrystalline ones. Polycrystalline cells are produced from blocks of refined silicon – monocrystalline cells require a further process to grow individual large crystals. Large blocks or crystals of silicon are sawn up to produce thin silicon wafers for cells.

Gallium arsenide (GaAs) cells – this is an alternative semiconductor consisting of alternate atoms of gallium and arsenic.

Band gap and cell efficiency – the optimum semiconductor band gap for a PV cell is about 1.4 eV. GaAs cells have this band gap and higher efficiencies than silicon.

PV cells can be produced using semiconductor films of less than 10 μm thickness.

Amorphous silicon cells use a film of silicon only 1 μm thick. Although cheap to produce they have relatively low efficiency.

Copper Indium Gallium Selenide (CIGS) cells use a film of a compound semiconductor less than 3 μm thick. The cells have good efficiency.

Cadmium Telluride cells also use a thin film of a compound semiconductor. They have good efficiencies and have been used in large power plant projects.

Other **PV technologies** include:

- **Multi-junction PV cells** can be constructed using stacked layers of different semiconductors with different band gaps, each one absorbing a different part of the light spectrum. These are expensive to produce and are used for space and concentrating PV applications.

- **Concentrating PV (CPV) systems** – higher cell efficiencies can be produced by concentrating the Sun's rays onto cells using mirrors or lenses. Although this requires less area of PV cell for a given power output it involves extra module complexity and they also need to be steered to track the Sun.

- **Emerging PV technologies** – there are many other possible approaches to PV cell design, though they will need to compete with the now well-developed technology based on silicon. *Perovskite* cells are based on lead, *dye-sensitized cells* are based on titanium dioxide and *organic photocells* are based on plastics. *Quantum dots* involve the 'fine-tuning' of the microscopic texture of more conventional PV cells to absorb a full spectrum of light.

An individual silicon PV cell can be thought of as a solar-powered battery with a voltage of about 0.5 volts and whose current is proportional to the light intensity on it.

PV modules usually consist of many individual cells connected in series to give a higher voltage. A PV array is likely to consist of multiple modules connected in parallel each one supplying more current.

A maximum power point tracking circuit may be included in a system to match the electrical load to the array's output so that the cells are always working to produce maximum power.

PV cells can also be used for remote power. For example, for **applications where it is difficult or expensive to supply mains electricity** – such as parking meter ticket machines, navigation buoys and remote telemetry stations.

PV charged batteries may also be the cheapest solution to **supplying lighting and basic electricity in rural areas in developing countries.** The technology is being widely deployed in India and Africa.

Grid electricity can be used as backup to PV or other variable renewable energy supplies.

Domestic PV systems typically use about 5 m² of PV panels and are connected to the grid via a DC to AC inverter. The electricity generated can be used by the home owner and any surplus exported to the grid.

The electricity output of a PV array is strongly dependent on the available solar radiation, which is highly dependent on the site latitude (the closer to the equator the better). Detailed estimates of the output can be made using online software from the EU's Joint Research Centre.

Large grid-connected PV power plants can use economies of scale in purchasing and installing arrays. They require large areas of land. The largest in the world each cover several square kilometres and have power ratings of hundreds of megawatts.

Factors affecting the costs of **PV electricity** include:

Levelized cost of PV – this is mainly made up of paying back the large initial capital cost over the project lifetime, typically 25 years. Annual operation and maintenance costs are small.

Capital costs – in the past the PV module costs have dominated the total, but falling modules prices now mean that 'balance of system' (BOS) costs may make up more than 50%.

Feed-in-Tariffs – many countries offer guaranteed high prices for PV generated electricity to encourage installations.

Grid parity – recent low bid prices for large utility-scale PV projects in many sunny countries are competitive with electricity from fossil-fuelled electricity generation.

Of all renewable and non-renewable electricity generating systems, PV probably has one of the **smallest impacts on the environment**. The main environmental concerns include:

- **Visual intrusion** – the visual impact of large areas of panels on the landscape is probably the main environmental impact.
- **Potentially hazardous chemicals used in manufacture** – PV manufacture uses inflammable gases, strong acids and some chemicals which are potentially strong greenhouse gases. These must be treated with proper care.
- **Toxic heavy metals** – cadmium telluride and most CIGS cells contain cadmium, so appropriate care must be taken in panel manufacture and in their eventual recycling.
- **Energy payback time** – overall the energy required to manufacture a new PV panel will be recovered in one to two years of operation.
- **Need for recycling as electronic waste** – large scale PV deployment creates a need for similar large scale decommissioning and recycling at the end of the project life. This should enable recovery of the materials used (high purity silicon and/or toxic cadmium).

Finally, this chapter looked at **PV integration** and future prospects.

Problems of PV integration – PV only produces electricity during the day. Having more than 10% of electricity from PV on the grid may create problems of:

- Ramp rate – having to bring on other forms of electricity generation rapidly in the evening when the Sun sets.
- Maintaining the stability of the grid under high levels of PV input.
- Dealing with surplus PV generation on sunny days.

Future prospects – in 2015 two thirds of the world's PV production was in China and Taiwan and most of that production capacity had been put in place since 2010.

Forecasts suggest that global PV capacity will continue to rise and it could be producing 7% of the world's electricity by 2030.

References

Adams, W. G. and Day, R. E. (1877) 'The action of light on selenium', *Proceedings of the Royal Society of London*, series A, vol. 25, p. 113.

Arthur D.Little (2015) *Emerging technologies in Solar PV: identifying and cultivating potential winners* [Online], Brussels, Arthur D. Little. Available at http://www.adlittle.com/downloads/tx_adlreports/ADL-Renewable-Energy-Emerging-PV-Technology.pdf (Accessed 18 February 2017).

ASTM (2012) *ASTM G173 - 03(2012) Standard Tables for Reference Solar Spectral Irradiances: Direct Normal and Hemispherical on 37° Tilted Surface* [Online], West Conshohocken, Pennsylvania, American Society for Testing and Materials. Available at https://www.astm.org/Standards/G173.htm, (Accessed 13 February 2017).

Becquerel, A. E. (1839) 'Recherches sur les effets de la radiation chimique de la lumière solaire au moyen des courants électriques' and 'Mémoire sur les effets électriques produits sous l'influence des rayons solaires', *Comptes Rendus de l'Académie des Sciences*, vol. 9, pp. 145–9 and pp. 561–7.

BP (2016) BP Statistical Review of World Energy [Online], London, The British Petroleum Company. Available at https://www.bp.com/content/dam/bp/pdf/energy-economics/statistical-review-2016/bp-statistical-review-of-world-energy-2016-full-report.pdf (Accessed 28 March 2017).

Chapin, D. M., Fuller, C. S. and Pearson, G. L. (1954) 'A new silicon p–n junction photocell for converting solar radiation into electrical power', *Journal of Applied Physics*, vol. 25, pp. 676–7.

Chalmers, R. (1976) 'The photovoltaic generation of electricity', Scientific American, October, pp. 34–43.

CCC (2011) *The Renewable Energy Review* [Online], London, The Committee on Climate Change. Available at https://www.theccc.org.uk/publication/the-renewable-energy-review (Accessed 22 February 2017).

EC JRC (2012) *Photovoltaic Geographical Information System* (PVGIS) [Online], Available at http://re.jrc.ec.europa.eu/pvgis/index.htm (Accessed 18 February 2017).

EPIA/Greenpeace (2011) *Solar Generation 6: Solar Photovoltaic Electricity Empowering the World* [Online], Brussels, European Photovoltaic Industry Association / Amsterdam, Greenpeace International. Available athttp://www.greenpeace.org/international/Global/international/publications/climate/2011/Final%20SolarGeneration%20VI%20full%20report%20lr.pdf (Accessed 22 February 2017).

Fraunhofer ISE/NREL (2015) *Current status of Concentrator Photovoltaic (CPV) Technology* [Online], Germany, Fraunhofer-Institut für Solare Energiesysteme ISE / USA, National Renewable Energy Laboratory. Available at http://www.nrel.gov/docs/fy16osti/65130.pdf (Accessed 6 March 2017).

Fraunhofer ISE (2016) *Photovoltaics Report* [Online]. Available at https://www.ise.fraunhofer.de/content/dam/ise/de/documents/publications/studies/Photovoltaics-Report.pdf (Accessed 17 February 2017).

Fraunhofer ISE (2017) *Recent Facts about Photovoltaics in Germany* [Online], Germany, Fraunhofer-Institut für Solare Energiesysteme ISE. Available at https://www.ise.fraunhofer.de/content/dam/ise/en/documents/publications/studies/recent-facts-about-photovoltaics-in-germany.pdf (Accessed 28 March 2017).

Glaser, P. (1972) 'The case for solar energy', paper presented at the *Annual Meeting of the Society for Social Responsibility in Science, Conference on Energy and Humanity*. Queen Mary College, London, 3 September.

Grant, C.C. (2013) Fire Fighter Safety and Emergency Response for Solar Power Systems [Online], Massachusetts, Fire Protection Research Foundation. Available at https://www1.maine.gov/dps/fmo/documents/firefighter_tactics_solar_power.pdf (Accessed 9 September 2017).

Grätzel, M. (2004) 'Conversion of sunlight to electric power by nanocrystalline dye-sensitized solar cells', *Journal of Photochemistry and Photobiology A: Chemistry*, vol. 164, pp. 3–14.

Green, M. (1982) *Solar Cells*, New York, Prentice-Hall.

Green, M. A., Emery, K., Hishikawa, Y., Warta, W., Dunlop, E., Levi, D. and Ho-Baillie, A. (2016) 'Solar Cell Efficiency Tables (Version 49)', *Progress in Photovoltaics: Research and Applications*, vol. 25, no. 1, pp. 3–13 [Online] Available at http://onlinelibrary.wiley.com/doi/10.1002/pip.2855/full (Accessed 12 February 2017).

Greenpeace, (2015) *Energy Revolution 2015 scenario*, Greenpeace International [online]. Available at http://www.greenpeace.org/international/en/publications/Campaign-reports/Climate-Reports/Energy-Revolution-2015/ (accessed 11 June 2017)

IEA (2016a) *World Energy Outlook 2016*, Paris, International Energy Agency.

IEA (2016b) *Renewable Energy: Medium-Term Market Report 2016*, Paris, International Energy Agency.

IRENA (2016a) *Solar PV in Africa: Costs and Markets* [Online], Abu Dhabi, International Renewable Energy Agency. Available at https://www.irena.org/DocumentDownloads/Publications/IRENA_Solar_PV_Costs_Africa_2016.pdf (Accessed 28 March 2017).

IRENA (2016b) *Letting in the Light* [Online], Abu Dhabi, International Renewable Energy Agency. Available at https://www.irena.org/DocumentDownloads/Publications/IRENA_Letting_in_the_Light_2016.pdf (Accessed 28 March 2017).

IRENA (2016c) *End-of-life Management: Solar Voltaic Panels* [Online], Abu Dhabi, International Renewable Energy Agency. Available at http://www.irena.org/DocumentDownloads/Publications/IRENA_IEAPVPS_End-of-Life_Solar_PV_Panels_2016.pdf (Accessed 26 March 2017).

IRENA (2017) *REthinking Energy 2017* [Online], Abu Dhabi, International Renewable Energy Agency. Available at http://www.irena.org/DocumentDownloads/Publications/IRENA_REthinking_Energy_2017.pdf (Accessed 28 March 2017).

McConnell, R. and Fthenakis, V. (2012) 'Concentrated Photovoltaics', in Fthenakis, V. (ed.) *Third Generation Photovoltaics*, Croatia, InTechOpen. DOI: 10.5772/39245 (Accessed 6 March 2017).

NREL (2013) *Quantum Dots Promise to Significantly Boost Solar Cell Efficiencies* [Online], Golden, Colorado, National Renewable Energy Laboratory. Available at http://www.nrel.gov/docs/fy13osti/59015.pdf (Accessed 18 February 2017).

Ohl, R. S. (1941) *Light Sensitive Device*, US Patent No. 2402622; and *Light Sensitive Device Including Silicon*, US Patent No. 2443542: both filed 27 May.

REC Silicon (2016) *REC Silicon's Fluidized Bed Reactor (FBR) process* [Online], Fornebu, Norway, REC Silicon. Available at http://www.recsilicon.com/technology/rec-silicons-fluidized-bed-reactor-process (Accessed 14 February 2017).

Scurlock, J. (2017) Private communication, National Farmers' Union, Scurlock. J.

Sharp Solar (2009) *Sharp* [Online]. Available at http://sharp-world.com/corporate/news/091029.html (Accessed 15 February 2017).

Solar Impulse (2016) *Solar Impulse: Exploration to Change the World* [Online]. Available at http://www.solarimpulse.com (Accessed 28 March 2017).

Solar Waste (2013) *Solar Waste and WEEE directive* [Online]. Available at http://www.solarwaste.eu/wp-content/uploads/2013/06/Solar-Waste-WEEE-Directive-Information-sheet.pdf (Accessed 21 February 2017).

SVTC (2009) *Towards a Just and Sustainable Solar Industry* [Online]. Available at http://svtc.org/our-work/solar/ (Accessed 21 February 2017).

TERI (2016) *Lighting a Billion Lives* [Online], New Delhi, The Energy and Resources Institute.). Available at http://labl.teriin.org/files/LaBL_publication/files/downloads/LaBL_Publication.pdf (Accessed 18 February 2017).

Wenham, S., Green, M., Watt, M. and Corkish, R. (2007) *Applied Photovoltaics* (2nd edn), London, Earthscan.

Wikipedia (2017) *List of photovoltaic power stations* [Online]. Available at https://en.wikipedia.org/wiki/List_of_photovoltaic_power_stations (Accessed 18 February 2017).

Chapter 5

Bioenergy

By Jonathan Scurlock, Caspar Donnison, Astley Hastings, Kevin Lindegaard and Hazel Smith

The authors are most grateful also to Prof. John Clifton-Brown (Aberystwyth University), Prof. Angela Karp (Rothamsted Research) and Dr Jeremy Woods (Imperial College London) for their comments and suggestions.

5.1 Introduction

Bioenergy is the general term for energy derived from organic materials such as wood, straw, oilseeds or animal wastes which are, or were recently, living matter, referred to collectively as **biomass**. Wood pellets, charcoal, bioethanol and biodiesel are all examples of energy-rich materials derived from biomass.

All the Earth's living matter, its total biomass, exists in the thin surface layer called the **biosphere**. It forms only a tiny fraction of the total mass of the Earth, but represents an enormous store of chemical energy (see Chapter 1). Although the majority of this is unavailable for human use, it is a store which is continually replenished by the flow of energy from the Sun, through the process of **photosynthesis** (from *photo*, to do with light, and *synthesis*, putting together). This in effect takes in carbon dioxide from the air and uses it to make living material, releasing oxygen in the process (see Figure 1.9 in Chapter 1). Although only a small fraction of the solar energy reaching the Earth each year is 'fixed' into living organic matter this way, the amount fixed annually as chemical energy in biomass is nevertheless equivalent to between two and five times the world's total primary energy consumption, as defined in Chapter 1.

It is important to appreciate that biomass is also vital in maintaining the Earth's atmosphere as we know it. If something were to sweep away all the plant life on Earth, the resulting loss of mass would be no more than one part in a billion – like blowing the dust off a school globe. Yet the physical consequences of this infinitesimal change would be enormous. There would no longer be a supply of oxygen to the atmosphere, and it is the particular mixture of nitrogen, oxygen and trace gases such as CO_2 in the atmosphere which maintains the surface conditions, including temperature, for life on Earth.

In nature, the energy that has been stored in the biomass of plants is dissipated through a series of conversions. These involve **metabolic** processes such as **respiration** (effectively the reverse of photosynthesis) in living matter, and physical processes such as re-radiation of heat energy and the **evaporation** of water (Figure 5.1). These processes transfer energy to the surrounding atmosphere and eventually the energy is radiated away from the Earth as low-temperature heat. Some newly-formed biomass will be metabolized within a year, but it can also accumulate over decades in the wood of trees and as the **humus** component of the soil organic matter. A small fraction may accumulate over centuries as **peat**, traditionally burned for heating, and over millions of years a tiny proportion has become the major fossil fuels: coal, oil and gas.

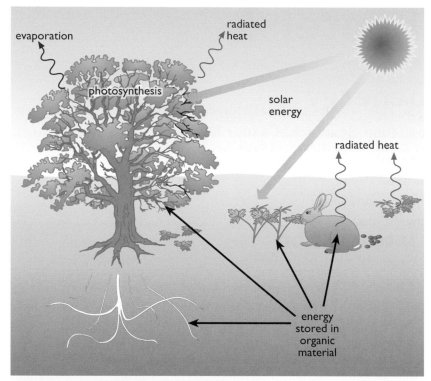

Figure 5.1 The bioenergy cycle on a local scale.

The significance of these processes is that if we can intervene and 'capture' some of the biomass at the stage where it is acting as a store of chemical energy, we can use it as a **fuel**, that is, a material that can release useful energy through a change in its chemical composition, usually through combustion. Material such as firewood, rice husks and other plant or animal residues can simply be burned in air to produce heat, and in many developing countries this **traditional biomass** use continues to account for a large part of energy consumption for cooking, heating and light. Biomass can also be burned to raise steam for electricity generation, or it can be converted into intermediate **bio-based fuels** such as charcoal, biogas, bioethanol and biodiesel. These many forms of bioenergy can be used to replace conventional fossil heating fuels or to power some form of engine for motive power or electricity generation.

Provided our anthropogenic consumption does not exceed the natural level of production (estimated over appropriate temporal and spatial scales), the burning of biomass *should generate no more heat and create no more carbon dioxide* than would have been formed in any case by natural processes. So it seems that here we have a truly sustainable energy source, with no deleterious global environmental effects. However, this is not always the case in practice, particularly where fossil fuels are used in the production process, and there may be other potential impacts to consider.

Box 5.1 summarizes some very approximate basic data about the quantities of biomass worldwide and related flows of energy. Freshly harvested biomass has a moisture content (m.c.) of around 50–80% of the total mass. If left to air-dry (as when turning fresh grass into hay, or seasoning logs) the moisture level drops naturally to between about 12% and 25%. To avoid the confusion caused by materials having varying moisture contents, biomass yields are often quoted as 'dry tonnes' or 'oven-dry tonnes' (odt), which assumes 0% m.c. as if the biomass had been dried in an oven.

BOX 5.1 Biomass – basic data

Note that almost all the data here is subject to considerable uncertainty.

World totals

Total mass of living matter[1]	Approx. 1800 billion dry tonnes
Total mass in land plants[1]	1800 billion dry tonnes
Total mass in oceans[1]	4 billion dry tonnes
World population (2016)	7.4 billion
Per capita terrestrial plant biomass	243 dry tonnes
Energy stored in terrestrial biomass[2]	25 000 EJ
Net annual production of terrestrial biomass[2]	130 billion dry tonnes y^{-1}

World energy comparisons

Rate of energy storage by land biomass[2]	2400 EJ y^{-1} (76 TW)
Rate of global primary energy consumption (2015)[3]	567 EJ y^{-1} (15.9 TW)
Biomass energy consumption[4]	57.7 EJ y^{-1} (1.6 TW)
Energy consumed as food[5]	29 EJ y^{-1} (0.9 TW)

Sources: 1 Slesser and Lewis, 1979; 2 Haberl et al., 2007; 3 IEA, 2015a; 4 IEA Bioenergy, 2009; 5 derived from WHO, 2003.

5.2 **Bioenergy past and present**

From wood to coal

Until recent times, the history of human energy use was essentially the history of bioenergy. Although there is evidence of coal-burning as early as 3000 years ago, its contribution remained relatively small until about 1800. Indeed, bioenergy was still dominant in many areas of life well into the Industrial Revolution, with wood used for heat, whale oil or tallow candles from animal fats for light, and grass and other agricultural crops as 'fuel' for the main means of land transport and haulage, horses and oxen.

The move from bioenergy to fossil fuel was a key feature of the Industrial Revolution. For many centuries, the high temperatures needed for iron smelting could be achieved only in furnaces using charcoal made from wood. The impurities and variable nature of coal made it unsuitable for smelting and attempts to convert it to a type of charcoal had little initial success. But in the early 1700s an effective 'coal charcoal' was produced and within a few decades this new material, now called coke, was replacing charcoal throughout the growing industrial sector.

The increased demand for coal led to deeper mines, and the need to pump flood water from great depths led to the first steam engines – powered of course by coal. By the end of the nineteenth century, coal was dominant in the world's **industrialized countries**. The twentieth century saw the rise of oil and natural gas, but it is worth noting that coal consumption also increased seven-fold between 1900 and 2008 (EIA, 2010).

Will the twenty-first century see the reversal of this process with a transition from coal to biomass? The data in Box 5.1 show that this is theoretically possible since the annual energy storage in biomass is approximately 4–5 times annual world total primary energy consumption. However, there are many practical, economic and environmental factors that may limit the uptake of modern bioenergy processes.

Present biomass contributions

Obtaining a reliable estimate of the total worldwide energy contribution from the many sources of bioenergy is a task fraught with difficulties. Unlike the fossil fuels, bioenergy has few global companies producing detailed reports on production and consumption. Indeed, the use of traditional biomass often involves no financial transaction at all, or the trading is local, informal and largely unrecorded. In recent years, the International Energy Agency has endeavoured to collect national data based on agreed categories of renewable energy, but they warn that their figures are still subject to a great deal of uncertainty. Recent estimates of the annual contribution to world primary energy from traditional biomass fall in the range 40–60 EJ, dominated by Asian and African countries (IEA Bioenergy, 2009; WBA, 2016).

Although the details may be uncertain, biomass is the fourth largest source of energy over much of the world, accounting for at least 10% of total human energy use (El Bassam, 2010; WBA, 2016). It accounts for about a third of total primary energy consumption in the developing countries (up to 90% in some of the poorest countries such as Uganda), and its total annual contribution continues to rise. Its percentage contribution to world primary energy is currently steady or falling slightly as developing countries industrialize and move towards fossil fuel use, but it remains important even in the more advanced of these, accounting for up to 10% of primary energy in China and 30% in India (El Bassam, 2010; WBA, 2016).

In the industrialized world, the energy contribution from biomass is significant and rising, particularly in countries with large forestry industries or well-developed technologies for processing residues and wastes – such as Sweden, Finland and Latvia, where biomass contributes 22%, 26% and 32% of energy consumption respectively (Eurostat, 2014). Advanced systems for domestic heating, district heating and CHP (combined heat and power) have helped Sweden towards a fivefold increase in the use of bioenergy since the 1980s, reaching nearly 50 GJ per head of population in 2006 (Björheden, 2006). In the UK, the bioenergy contribution rose at about 7% per year during the 1990s and had reached a total of 451 PJ, or 7.0 GJ per capita in 2015 (BEIS, 2016a, see also Figure 1.6 in Chapter 1).

5.3 **Biomass as a solar energy store**

The key mechanism in the use of bioenergy is photosynthesis, in which plants take in carbon dioxide and water from their surroundings and use energy from sunlight to convert these substances into plant biomass.

The essential features of the process can be represented by the chemical equation:

$$6CO_2 + 6H_2O + \text{light energy} \rightarrow C_6H_{12}O_6 + 6O_2$$

The first product on the right of the equation is glucose ($C_6H_{12}O_6$), a **carbohydrate** (see Box 5.2). The second product in the equation is oxygen which is released to the atmosphere. Although glucose is not necessarily the final 'vegetable matter' it is the crucial building block in subsequent biomass producing reactions. Thus a plant *grows* by using solar energy to convert carbon dioxide and water into carbohydrate or similar material, with a release of oxygen into the atmosphere.

BOX 5.2 **Carbon-containing compounds in biomass**

Carbon is a key element in the biosphere and is found combined with a range of other elements to form a vast array of simple and complex molecules that make up the bulk of all living tissues.

The term carbohydrate refers to a compound consisting only of carbon, hydrogen and oxygen having a general form $C_x(H_2O)_y$ (x and y can be different but crucially there is a hydrogen : oxygen ratio of 2:1). Of importance to the discussions in this chapter are carbohydrates such as the following.

Glucose – a simple sugar, which has the chemical formula $C_6H_{12}O_6$; these small carbohydrate molecules can be linked together into polymeric chains (polysaccharides).

Starch – a polymer consisting of many glucose units, it can be represented by the formula $(C_6H_{10}O_5)_n$, where the subscript n indicates that there are many identical $C_6H_{10}O_5$ units joined together.

Cellulose – another polymer with the formula $(C6H10O5)n$ but having a different overall structure that is an important component of plant cell walls, allowing the formation of fibrous structures.

Hemicellulose – a complex polymer containing a variety of different sugar units (i.e. a polymer consisting of more than just glucose base units), found with cellulose in plant cell walls.

Other organic (carbon containing) compounds discussed include the following:

Lipids – a class of carbon, hydrogen and oxygen containing compound: the generic term for the oils and fats contained in living tissues.

Proteins – large, highly complex organic molecules which also contain nitrogen and may consist of a number of polymeric chains. Proteins have complex folded structures.

In living organisms, these different types of compound are synthesized and broken down by the operation of specific **enzymes** (specialist proteins that act as catalysts) that enable the reactions to occur at ambient temperatures and pressures.

In an energy context, **hydrocarbons** which comprise only carbon and hydrogen are also important. They are mostly formed from previously living material by reactions occurring at high temperatures and pressures below the Earth's surface.

The glucose molecule contains more chemical energy than the sum total in the molecules of carbon dioxide and water from which it was formed. Within the plant, glucose can be converted into more complex carbohydrates including starches and cellulose, or it can be combined with nitrogen and other elements to form proteins and other components..

Some of the energy-rich carbohydrate formed is broken down within the plant through the process of respiration, releasing energy to power the synthesis of proteins and other components. Respiration is in effect the reverse of photosynthesis, with oxygen taken in, carbon dioxide given off and energy released. Plant material that has been consumed by an animal or allowed to decay also undergoes respiration with the energy released being used by the different organisms involved. All the energy involved in processes within living organisms ultimately appears as heat. So we have a continuous process, with energy from the Sun being stored in the form of carbon-based biomass which may persist for a time, or be consumed by another organism. Over time, nearly all this biomass is respired, releasing energy which is ultimately re-radiated back to outer space, and carbon dioxide which returns to the atmosphere (Figure 5.2). So carbon cycles around the biosphere while energy flows through it.

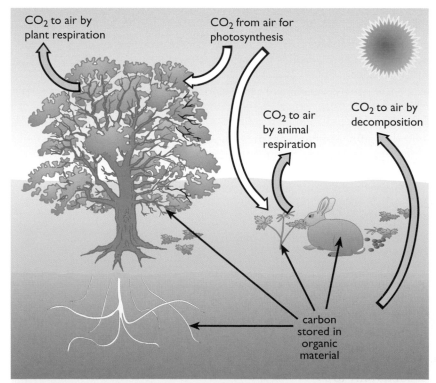

Figure 5.2 The carbon cycle on a local scale. In addition, a small fraction of the carbon from decomposing animal and plant residues (including roots) is transformed into soil organic matter, a significant carbon store.

Conversion efficiencies

The **yield** of a crop is the mass of (usually above-ground) biomass produced per hectare per year. For an **energy crop**, that is, one grown specifically for biomass energy, yield is evidently the primary determinant of the energy we can obtain from that crop. Yields depend on many factors: the location, climate and weather, the nature of the soil, supplies of water, nutrients, etc., and the choice of plant. For energy crops, the air-dry mass of plant matter produced annually on an area of one hectare can be as little as one dry tonne or, in favourable circumstances, as much as thirty dry tonnes. In energy terms, this represents a range from perhaps 15 GJ to 540 GJ per hectare per year.

However, these yields represent an extremely low conversion efficiency from solar energy to energy in biomass. Consider, for instance, northern Europe, where the average solar energy delivered in a year is about 1000 kWh per square metre (see Chapter 2). A well-managed energy crop in this region might yield perhaps 200 GJ per hectare per year: a solar-to-bioenergy conversion efficiency of about two-thirds of 1% (see Box 5.3; from Hall et al., 1993).

BOX 5.3 Conversion of solar energy

Consider one hectare (ha) of land, in an area such as southern England where the annual energy delivered by solar radiation is 1000 kWh m^{-2} y^{-1}.

1000 kWh is 3.6 GJ and 1 ha is 10 000 m^2,
so the total annual energy is   36 000 GJ

After losses about an eighth of this reaches the crop at the right time.
Say

12% of the annual energy reaches growing leaves	4320 GJ
50% of this is photosynthetically active radiation	2160 GJ
85% of which is captured by the growing leaves	1836 GJ
21% of which is converted into stored chemical energy	386 GJ
40% of which is consumed in respiration to sustain the plant	
or lost in photorespiration leaving	231 GJ

This is about 5.3% of the solar radiation reaching the growing plant, and only 0.64% of the original total annual energy.

The first significant point is that much of the annual solar radiation may be ineffective for photosynthesis. Some misses the plant altogether or arrives during the wrong season. There may not be enough water or nutrients for the plant to use the solar energy that does reach it, then there are the effects of diseases, pests (reducing the leaf area of the crop able to intercept light) and weeds (competing for light or nutrients). The remaining losses shown in Box 5.3 are specific to the interaction between the plant and sunlight. A leaf responds only to a part of the solar spectrum – the pattern of absorption is shown in Figure 5.3. Note the dip in the green region: a leaf looks green because it absorbs less green light than red or blue. Only about 85% of the 'useful' radiation that falls on a leaf surface is captured in the leaf for use in photosynthesis. The dominant plant types of the temperate regions

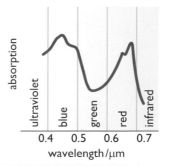

Figure 5.3 Relative absorption of different parts of the solar spectrum by a leaf

are referred to as **C3 plants**, because the first products of photosynthesis actually contain three carbon atoms rather than the six shown for glucose. in the general reaction described above. The fundamental biochemistry of C3 photosynthesis means that 21% of the energy of absorbed light is converted into the energy content of fixed carbohydrate. Finally, 40% of this fixed energy is used in respiration within the plant to drive the synthesis of proteins and other components of its biomass.

In the C3 plants, too much light relative to the CO_2 available in the leaf can lead to damage caused by the main photosynthetic enzyme producing a toxic molecule called **phosphoglycolate** rather than useful three-carbon compounds. The process of photorespiration protects the plant from damage by using previously stored energy to break down the phosphoglycolate, releasing carbon dioxide but effectively wasting as much as 30% of the stored energy.

Plants from tropical areas are at greater risk from such light-induced damage because of the higher light levels. Some tropical plants, especially those in the grass family (e.g. maize, sugar cane and miscanthus) effectively concentrate CO_2 in the cells where photosynthesis is carried out, so there is never a shortage and no risk of damage. They produce a four-carbon molecule as the first product of photosynthesis. While the fundamental efficiency of photosynthesis in these **C4 plants** is lower than the C3 process, such plants can potentially produce higher yields of biomass than C3 plants by avoiding the need for wasteful photorespiration. (For a clear explanation in more detail, see the classic textbook by Hall and Rao (1999)).

For C4 plants grown in tropical regions, the 5.3% of Box 5.3 might increase to perhaps 6.7%, giving an overall efficiency closer to 1%. Furthermore, good crop management can increase the fraction of solar radiation that is captured and used. Nevertheless, plants remain much less efficient solar energy converters than, for example, photovoltaic (PV) cells which can achieve solar to electrical energy conversions of from 10 to over 25%. However, plants are currently much cheaper per square metre of light intercepting surface!

5.4 Biomass as a fuel

What are fuels?

Fuels are those materials from which useful energy can be extracted. In using biomass as fuel, this release of energy usually involves burning (combustion). Some essential features of combustion are:

- it needs air – or to be more precise, oxygen
- the fuel undergoes a major change of chemical composition
- heat is produced, i.e. energy is released.

Consider, as an example, methane – the principal component of the fossil fuel natural gas and also one form of gaseous biofuel. Each methane molecule consists of one carbon and four hydrogen atoms: CH_4. Atmospheric oxygen has molecules consisting of two atoms (O_2), so in full combustion each methane molecule reacts with two oxygen molecules:

$$CH_4 + 2O_2 \rightarrow CO_2 + 2H_2O + \text{energy}.$$

The heat energy released in this process is the *difference* between the chemical energy of the original fuel plus oxygen, and the chemical energy of the resulting carbon dioxide plus water. In practice, it is usual to refer to it as the **energy content** (or *heat content*, or *heat value*) of the fuel – the methane in this case.

The reaction shown above shows the essential features of the combustion of many common fuels which contain carbon and hydrogen. The fuel interacts with oxygen from the air to produce carbon dioxide and water, the latter usually as water vapour in the form of steam. The process of respiration in living organisms is similar, but takes place at near-ambient temperature. If we know the composition of the fuel and the relative masses of the chemical elements making up its molecules, we can predict how much carbon dioxide will be produced in burning a given amount of fuel (see Box 5.4 and Table 5.1).

BOX 5.4 CO_2 from fuel combustion

As an example, consider the combustion of methane (CH_4). The masses of the atoms of carbon and oxygen are respectively 12 times and 16 times the mass of a hydrogen atom, so we can associate masses with the items in the combustion equation:

$$CH_4 \; + \qquad 2O_2 \qquad \rightarrow \qquad CO_2 \; + \qquad 2H_2O$$
$$12 + (4 \times 1) \; + \; 2 \times (2 \times 16) \quad \rightarrow \quad 12 + (2 \times 16) \; + \; 2 \times (2 \times 1 + 16)$$

We can see, therefore, that burning 16 tonnes of CH_4 releases 44 tonnes of CO_2.

The energy content of methane is 55 GJ t^{-1} if the water vapour produced is condensed but only 50 GJ t^{-1} if it isn't (see Box 5.5). This is the amount of heat produced by burning one tonne of methane and thus releasing 2.75 tonnes (2750 kg) of CO_2.

Other fossil fuels, apart from coal, are chemically more complex than methane (although many are hydrocarbons like methane in that they consist almost entirely of hydrogen and carbon), but their combustion follows a similar process. The heat produced per tonne is rather less, however, and they also produce more CO_2 per tonne because they have a higher ratio of carbon to hydrogen atoms, so the CO_2 per unit of heat output is greater (see Table 5.1). Coal, which is largely carbon, produces one of the highest outputs of CO_2 per unit of heat output. Coal and oil may also contain sulfur which burns to sulfur dioxide. This sulfur can be refined out or collected from flue gases and used in fertilizers.

Table 5.1 Heat content (net calorific value, see Box 5.5) and CO_2 emissions

Fuel	Heat content/GJ t^{-1}	CO_2 released/kg GJ^{-1}
Coal	24	94
Fuel oil	41	79
Natural gas	48	57
Air-dried wood	~15	~80*

Note that the composition of coal, oil and wood and the moisture content can vary significantly, so the figures are typical values.

* If the wood is grown sustainably, so that trees are planted to replace those harvested and combustion is complete, its life cycle CO_2 emission should be close to zero.

The energy released by burning one tonne or one cubic metre of various biological materials is shown in Table 5.2, with the main fossil fuels for comparison. Note again that the composition of coal and most biomass fuels is variable (Box 5.5) and the energy content per tonne can differ significantly from the averages shown here. The energy *per cubic metre* depends on the density of the material and can vary even more widely.

Table 5.2 Typical heat energy content (net calorific value, see Box 5.5) per unit of dry mass and volume of various forms of biomass and fossil fuel

Fuel	Energy content	
	GJ t^{-1}	GJ m^{-3}
Solid wood (green, 60% moisture)	6	7
Solid wood (air-dried, 20% moisture)	15	9
Solid wood (oven-dried, 0% moisture)	18	9
Charcoal	30	*
Miscanthus bales (0% moisture)	18	2.4
Dung (dried)	16	4
SRC chip (30% moisture)	13	2.9
Maize grain (air-dried)	19	14
Straw (as harvested, baled)	13	1.3
Sugar cane residues (bagasse)	17	10
Domestic refuse (as collected)	7	1.5
Commercial wastes (UK average)	15	*
Domestic heating oil	43	36
Coal (domestic heating, average)	28	50
Natural gas (at supply pressure)	48	0.04

* Indicates dependence on specific types of material.
Data from various sources including former UK Biomass Energy Centre (Forest Research, 2017) and BEIS (2016b)

BOX 5.5 Energy, moisture content and calorific values

We have seen that water is an inevitable combustion product of most common fuels. It results from the combustion of hydrogen in a fuel and also from the vaporization of moisture contained in the fuel. It appears as water vapour at a high temperature (steam) and some extra useful heat can be obtained by cooling and condensing it.

Fuels are often quoted with two different calorific values giving their potential heat output when burned. Values for biomass, industrial wastes, coal and oil are normally quoted as the lower heat value (LHV) or net calorific value (NCV) of the fuel. This assumes that any water vapour produced is not condensed. However, if it is condensed, then a higher heat value (HHV), also called the gross calorific value (GCV) is appropriate. For fuels such as coal and oil the difference between NCV and GCV is small enough to be ignored in rough calculations, but this is not the case for natural gas. This is largely methane whose combustion in described in Box 5.4. In this case the GCV is about 10% higher than the LCV. For hydrogen (to be discussed in Chapter 11) the difference is 17%.

Unlike fossil fuels, biomass often has a high water content prior to combustion, which adds to the mass but contributes no energy. This water also absorbs energy for heating and evaporation. In general, each 10% increase in moisture content reduces the NCV by roughly 11%. Condensing all the water can increase the heat output by 50% or more when burning fresh green plants, but is difficult in practice.

Moisture content also affects the rate at which biomass materials decay naturally. Drier materials such as straw or air-dried wood can be stored for relatively long periods, whereas wet materials will decay rapidly unless treated.

Making use of biomass

An important issue in the use of biomass as an energy source is the need to process it into a suitable form. The input to these processes may be purpose-grown plants in the form of **energy crops**, but can also be organic **wastes**, **residues** or **byproducts/coproducts** including straw, animal manures and human sewage. The output may be useful heat, or one of a range of solid, liquid or gaseous biofuels. Figure 5.4 shows a very general outline of the primary sources of biomass, the conversion processes and products involved.

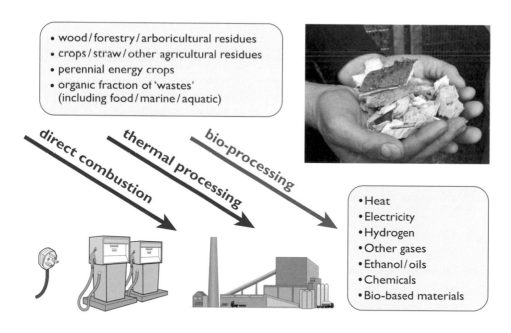

Figure 5.4 Biomass offers a high degree of versatility with respect to input resources (feedstocks), conversion processes and product outputs

The next section will describe the types of materials considered primary or virgin biomass energy sources, and those which are secondary sources or co-products. Then Section 5.6 will discuss the physical, thermochemical and biochemical processing of biomass, which may be required to enable the release of energy.

5.5 Biomass resources

Biomass resources can be categorized as either primary materials, derived directly from plants, or secondary 'wastes', the otherwise low-value, unwanted or unutilized co-products and residues from human activities. These secondary materials can only be classified as biomass if they themselves have been produced from renewable plant-based materials.

Primary biomass energy sources: plant materials

Biomass resources have attracted increased attention over the past few decades for a number of reasons: as an alternative to fossil fuels to reduce net greenhouse gas emissions; as a way of enhancing national energy security by domestic production or having a more diverse range of suppliers; and as a way of adding value to indigenous natural resources, subject to the constraints of local climate, soils, competing land use, etc.

In addition to microorganisms, primary biomass sources can be sub-divided into four general categories according to the type of plant-derived material used:

- Starch/sugary biomass – Starch is a relatively simple carbohydrate which is abundant in cereal grains and in the tubers of crops such as potatoes. Sugar beet roots and sugar cane stems have high levels of simple sugars, which can be converted to bioenergy.
- Cellulosic biomass – Terrestrial plant and seaweed cell walls are primarily made up of cellulose and related hemicelluloses. Bioenergy grasses and cereals also contain lignin and lignocellulose, although this is at a much lower level than in woody plants.
- Woody biomass – Trees are the main source of woody biomass, which is a complex mixture of carbohydrates, lignin, cellulose and hemicellulose.
- Oily biomass – Oils form the basis of the storage material in some seeds such as oilseed rape, soya and oil palms.
- Microorganisms – A novel 'energy crop' is represented by microalgae, which may have high concentrations of oils in their cells.

This general categorization is expanded in Table 5.3, which also gives typical yields and structural information (important for processing and storage). Although primary bioenergy sources can be described this way, the categories may be overlapping. For example, a crop may be grown primarily for its sugar or starch content, but its supporting lignocellulosic structural materials may also be utilized.

Table 5.3 A broad generalized classification of primary bioenergy sources

Category	Major energy-rich components	Structural strength /resistance to natural decay	Examples	Typical yields of dry matter /t ha^{-1} y^{-1}
Woody	Lignin/lignocellulose (complex carbohydrates)	High	Trees (deciduous or hardwoods)	10 (temperate) to 20 (tropics)
Cellulosic	Cellulose/lignocellulose (complex carbohydrates)	Medium	Grasses, water hyacinth, seaweeds	10 (temperate) to 60 (tropical aquatics)
Starchy/sugary	Relatively simple carbohydrates	Low	Cereals, sugar cane, tubers/roots	10 (temperate cereals) to 60 (sugar-cane)
Oily	Lipids (i.e. oils/fats)	Low	Oilseeds (rape, sunflower, oil palm, jatropha)	8 to 15
Microorganisms	Oils	Low	Microalgae	Not yet known at scale

The gross biomass yield per hectare is given as tonnes of oven-dry mass of all above-ground matter in Table 5.3, although the energy harvest index of the different energy crop types is diverse. For example, the yield of particular components, such as starch or oil, will be less than the figures presented and will depend on the processing route from crop to product. Bioenergy crop yields can vary depending not only on crop type but also on the growing system (Table 5.4).

Many non-woody plants are currently grown for food, fodder or fibre, yet could also be used for bioenergy, by providing starch/sugar, oils or cellulosic biomass. As an indication of the spectrum of possible bioenergy crop types, El Bassam (2010) lists over 100 species that could be used for **ethanol** production. Reflecting the desire to identify crops which can be grown on marginal land, minimizing competition between food and fuel production, he also lists about 70 species that are particularly suited to arid and semi-arid areas. The various species have inherently diverse characteristics (water and nutrient requirements, nitrogen-fixing ability, etc.) which could make them useful candidates for specific situations.

Table 5.4 Yield components and annual energy production of commonly grown bioenergy crops in selected research trials, grown in regions suitable to their plant type. Adapted from Long et al., 2015

Category	Feedstock	Region of measured growth	Total dry biomass yield /t ha^{-1}	Grain/ seed/ sugar yield /t ha^{-1}	Easily accessible biofuel /GJ ha^{-1}	Cellulosic biomass fuel /GJ ha^{-1}	Combustion of residue for bioenergy /GJ ha^{-1}	Easily accessible + cellulosic + residue bioenergy /GJ ha^{-1}
Oil	Rapeseed (*Brassica napus*)	Europe	5.6	2.8	33.2	12.3	8.4	53.9
	Oil palm (*Elaeis guineensis*)	Indonesia	34.0	17.0	128.8	149.4	50.9	329.2
Starchy/ Sugaryl	Wheat (*Triticum aestivum*)	Europe	8.8	5.3	34.9	19.4	13.2	67.6
	Maize (*Zea mays*)	USA	18.4	9.2	72.8	40.4	27..6	140.8
	Sugarcane (*Saccharum officinarum*)	Brazil	38.0	12.0	156.8	167.0	113.9	437.7
Cellulosic	Reed Canary Grass (*Phalaris arundinaceae*)	Denmark	12.0	0.0	0.0	105.4	71.9	177.3
	Napier Grass (*Pennisetum purpureum*)	El Salvador	84.0	0.0	0.0	738.2	503.5	1241.7
Woody	SRC Willow *Salix* hybrids	Sweden	10.0	0.0	0.0	43.9	30.0	73.9
	SRC Poplar *Populus* hybrids	Italy	14.0	0.0	0.0	61.5	42.0	103.5
	SRF Eucalyptus *Eucalyptus* hybrids	Brazil	18.2	0.0	0.0	80.0	54.5	134.5

Woody biomass

Well-managed forests can provide a sustainable fuel source, reducing atmospheric CO_2 as the trees grow, storing it for up to a few decades to centuries, providing a substitute for fossil fuel when logged and burned. Historically, energy production from conventional forestry has generally been incidental, although some countries use the forest and process residues (branches, bark, sawdust, etc.) to provide heat and power (see 'Secondary sources' below). More recently, new technologies have raised the prospect of producing liquid biofuels from woody crops and conventional forestry production techniques have been intentionally altered to allow the simultaneous production of energy alongside traditional timber.

Domestic use of wood fuel, logs and pellets is well developed in central Europe, Scandinavia and other regions and wood is still a vital fuel for cooking and heating in many developing countries. For example, the production process can include planting at a higher than normal density so that periodic thinning of the trees to produce wood chips during the growing cycle is possible. An advantage of woody biomass is that it is possible, through cultivation, to take multiple harvests without the need to replant the site and this can enable the production of a sustainable woody fuel source. In recent years there has been increased interest in woody crops planted and harvested entirely for energy production and this is generally achieved through **short rotation coppicing or forestry (SRC/SRF)**.

It has been estimated that using short-rotation woody crops for energy will reduce greenhouse gas emissions by 80–90% compared to the fossil fuel baseline (Djomo et al., 2011). SRC involves planting cuttings at 10–15 000 per hectare, which are cut back to a stool at regular intervals of approximately 2–4 years, allowing regrowth of multiple stems. SRF, on the other hand, follows a longer cycle with trees being felled at 8–20 year intervals. Both SRC and SRF species are generally able to be grown on poorer-quality agricultural land, thus limiting competition with food production. Modern short-rotation cycles can be repeated for up to 30 years, although traditional UK coppice woodlands of ash, hazel or willow were often harvested over much longer periods. Annual yields of 10 dry tonnes per hectare should be achievable in northern Europe and up to 30 dry tonnes in more favourable regions – in all instances subject to sufficient water (typically >600mm/year), nutrients (e.g. as waste-water treatment sludge or digestate from anaerobic digestion), good land preparation and weed control (Table 5.5). Modern harvesting machines can produce chipped wood or longer billets, whole rods for chipping when dry, or bales for transport, storage and future use.

Table 5.5 Woody biomass yields of poplar, willow and Eucalyptus when grown in the European Union

Species	Yield (dry matter) /t ha^{-1} y^{-1}	Rotation period* /years	References
Poplar	10–33.7	8–21	Manzone et al., 2009; Havlicková and Weger, 2009; Fiala and Bacenetti, 2012
Willow	4.15–12	10–22	Mola, 2011; Rosenqvist and Dawson, 2005
Eucalyptus	10–20.5	8 – 15	Iriarte et al., 2010; Jiménez et al., 2013

*Shorter for SRC/SRF as described above.

Short-rotation woody crops for renewable energy production have been successfully applied in Sweden, where around 11 000 ha of willow coppice supplies energy for district heating, although the total area of SRC across the EU remains modest – only about 50 000 ha.

Figure 5.5 Short rotation coppice willow

Despite around 30 years of supportive policy measures in many European countries, it has proven difficult to persuade farmers to plant these novel crops (Lindegaard et al., 2016). Nevertheless, modelling by the UK Energy Technologies Institute has suggested that planting 1–1.8 Mha of energy crops could reduce the cost of meeting Britain's 2050 carbon target by more than 1% of GDP (Adams and Lindegaard, 2016). Beyond their role in energy security, SRC crops have a very favourable energy balance compared to conventional agricultural land use because there is no annual cultivation cycle. The multiple environmental benefits of converting agricultural land to short-rotation woody cropping systems include soil carbon storage, enhanced biodiversity and pollination services, increased water and nutrient retention and decreased run-off (King et al., 2004; Rowe et al., 2009).

Outside Europe, there has been some development of SRC in the Americas, India, Australia and New Zealand, where *Eucalyptus* species are typically grown. This is partly due to its potential for improving degraded land as well as its value as an energy source. In Australia, millions of hectares of cereal lands have been going out of production due to rising salt levels in soils as a result of poorly-managed irrigation during arable cultivation. Growing *Eucalyptus* in strips around 100 m apart for SRC can help lower the water table, since the trees can tolerate the salt and have a high water demand. Lowering the water table carries the salts back down to lower levels. However, in other countries such as Portugal, there are concerns about the crops' high water usage in areas where water supplies are limited.

Cellulosic materials

The structural integrity of non-woody (i.e. herbaceous) plants is provided mainly by cellulose. Humans are unable to digest cellulose, but it forms a major food source for sheep, cattle and other ruminant livestock. It is a major component of the straw and other supporting tissues of all the major food crops, and can be used directly as a fuel or potentially converted to bioethanol. Herbaceous (cellulosic) bioenergy crops offer advantages over woody energy crops in that they can be managed with relatively conventional agricultural equipment and they may also offer greater flexibility in land use.

The most commonly considered cellulosic bioenergy crop suitable for temperate climates is **miscanthus** (Figure 5.6), a tall perennial grass from eastern Asia. This is found naturally in a wide geographical area from the equator to eastern Russia, with many genotypes growing well in temperate and continental northern Europe. Ongoing research and breeding programmes have produced and identified varieties with yields up to 20 dry tonnes per hectare per year under UK conditions. *Miscanthus x giganteus*, a highly productive sterile interspecies hybrid, is propagated vegetatively by means of underground stems called **rhizomes**. These can be split mechanically when the plant is dormant to create propagules used for planting new crops. The stems of miscanthus, ranging from 8 to 12 mm diameter, are suitable for direct combustion since they have a low water content (12–30%) when harvested in late winter or early spring. As a grass, miscanthus naturally propagates by airborne seeds, but research has developed varieties that can be mechanically seed-propagated under European conditions but are either sterile or do not set seed, thereby avoiding the likelihood of becoming an alien invasive weed.

(a)   (b)

(c)   (d)

Figure 5.6 Cellulosic bioenergy crops which are currently being deployed or developed: (a) miscanthus, (b) giant reed, (c) switchgrass/reed canary grass and (d) water hyacinth.

Miscanthus and other energy grass crops, such as reed canary grass (*Phalaris arundinacea*), Napier grass (*Pennisetum purpureum*) and switchgrass (*Panicum virgatum*), have so far been grown for bioenergy using mostly agricultural land, equally suitable for food crops, although the trend now is to develop genotypes and management techniques for production on more marginal land (Zegada-Lizarazu et al., 2010; Clifton-Brown et al., 2010; Clifton-Brown et al., 2016).

For example, giant reed (*Arundo donax*) is a weedy C3 grass which has been proposed as a valuable bioenergy source as it is high yielding and requires low inputs (Angelini et al., 2005; Nassi o Di Nasso et al., 2013). Although *A. donax* is known to use large quantities of water, it has also been suggested to be tolerant of drought. Additionally, it is possible to irrigate the crop using wastewater unsuitable for food production. Moreover, despite *A. donax* being a C3 crop, it has been reported to have a higher water use efficiency when compared to the C4 miscanthus (Mantineo et al., 2009).

Table 5.6 A comparison of common herbaceous bioenergy crops. All are harvested annually.

Crop	Yield (dry matter) /t ha^{-1} y^{-1}	References
Miscanthus	10–40	Heaton et al., 2004
Switchgrass	10–25	Venturi and Venturi, 2003
Giant reed (A. donax)	7–61	Mantineo et al., 2009

Other possible non-agricultural resources in this general category include bracken (*Pteridium aquilinum*), regarded in the UK as a weed, although the dry fronds can be used as a horticultural mulch, livestock bedding, or processed into briquettes or pellets. Such weedy species are inherently hardy and may be able to tolerate the stresses presented by being cultivated on marginal land, alongside the challenges posed by a changing climate. However, this pervasiveness has presented its own problems, with species such as *A. donax* becoming invasive in some parts of Europe. Another example of this is water hyacinth (*Eichornia crassipes*), a very fast growing aquatic weed that causes problems in tropical reservoirs and irrigation channels. Control of water hyacinth by harvesting it for bioenergy use has been proposed and attempted in a variety of locations worldwide, providing a dual benefit. The spread of weedy species needs to be controlled but, if managed successfully, they may become productive bioenergy crops that are able to be cultivated without affecting food production through diversion of land and resources from conventional agricultural crops.

Lastly, some agricultural crops that were primarily grown for their sugar or starch content, such as maize and new varieties of sugar beet (see below) are increasingly used whole for bioenergy, along with fodder beet and grasses. When used as animal feed these crops are often conserved as **silage** by storing the fresh material in near-airtight conditions, promoting the production of organic acids. Such silages are also used as high-energy feedstocks for anaerobic digestion (Section 5.6) together with animal manures.

Figure 5.7 Harvesting miscanthus using conventional agricultural machinery

Starchy/sugar crops

The major biomass energy component of these crops is their sugar or starch, which, as shown in Figure 5.4, can be fermented to produce ethanol as a liquid fuel. Their supporting cellulosic tissues may also be used in the same way as the mainly cellulose crops. Globally, at the time of writing, the most important crops for bioenergy purposes are sugar cane and maize. Both are high yield C4 crops, with the former having a particularly favourable energy balance (that is, a good ratio of energy input to final energy output – see Section 5.7). Sugar cane, as its name implies, is widely grown for the sugary sap (12–16% sugar content) which is stored in its 5 cm diameter stems.

Sugar cane is restricted to areas of the world between 37° north and 31° south, with Brazil producing 768 million tonnes (fresh weight of cane stems) out of a world total of around 1 905 million tonnes in 2013 (FAO, 2013). India also produces a substantial quantity of sugar cane (about 340 million tonnes per year). This is primarily used for edible sugar with only the molasses (the liquid remaining after crystallization of the sugar) used for ethanol production. The cane is a perennial grass, propagated vegetatively using above ground stem sections called nodes. Several successive harvests of stems can be taken before yield declines and the crop has to be replanted (due to the stems being harvested before senescence). Replanting may take place every 5–10 years, so sugar cane is sometimes called a 'semi-perennial' crop.

In contrast, almost all the other crops in this category are annuals that provide a single harvest from each sowing. Maize can be grown over a wide range of latitudes, with the USA growing over 350 million tonnes of grain (air-dry; 8–15% m.c.) out of a world production of 1 017 million tonnes in 2013 (FAO, 2013). Other cereals such as wheat (total world production 729 million tonnes in 2014), rice, barley and sorghum could potentially be used to produce starch, as could the starchy tubers of potatoes. World production of potatoes in 2013 was 376 million tonnes (fresh weight), with China the largest producer. Sweet sorghum, which is also a sugar-rich crop, requires less water and nitrogen than sugar cane and can grow in semi-arid regions (Woods, 2001). Two crops per year are possible and there is genetic potential for improving yields or developing hybrids suitable for the European climate. Another sugar-rich crop of interest in Europe is sugar beet, where the swollen, fleshy roots contain 15–20% sugar.

Sugar-cane bagasse (the sugar cane's fibrous residue) and the straw from maize (stover) and other cereals can also be used for energy in the same way as cellulosic crops. Straw and bagasse are considered in the next section as co-products.

(a)   (b)   (c)

Figure 5.8 Starch/sugar crops: (a) sugar cane, (b) maize harvesting, (c) sugar beet

Oilseed crops

Sunflowers, oilseed rape and soya beans are grown widely for the oil in their seeds, while in tropical areas, oil palm (*Elaeis guineensis*) is a major crop. The worldwide total production of vegetable oils in 2013 was around

160 million tonnes, of which 54 million tonnes came from oil palm and 42 million tonnes from soya (FAO, 2013).

The oils from these various crops are extracted by pressing, with or without additional chemical treatments and can be converted to a diesel substitute known as biodiesel (Section 5.6). As with the starchy/sugar crops considered above, there is a protein-rich residue left from these processes that is currently often used as animal feed, but which could otherwise be a bioenergy feedstock. The oil from all these crops has a vast number of uses, in foods, cosmetics, etc. However, major environmental concerns have been raised about the conversion of mature forests and tropical peatlands to oil palm plantations in South East Asia and West Africa, and increased use of palm oil for bioenergy in the absence of rigorous sustainability standards has attracted much criticism.

The vivid yellow of oilseed rape (*Brassica napus*, also known as canola in Canada) has become familiar in many countries over recent decades. There are two general types – winter varieties that are sown in the autumn to overwinter and spring varieties, sown and harvested in the same year to avoid frost damage but with a shorter total growing period and slightly lower yield. World production of rapeseed oil was about 25 million tonnes in 2013 (FAO, 2013) with a well-developed biofuel market in Europe. Sunflower (*Helianthus annuus*) is another crop which originated in the Americas, but which is widely grown in southern Europe for its oil and seed. Some non-edible oil yielding plants such as *Jatropha curcus* and *Pongamia pinnata* have been recently used in government programmes in India. These can be grown with relatively fewer inputs, are also used as fencing material, and were responsible for 3.6 million tonnes of biodiesel production in 2013.

Soya beans are extensively grown in the middle latitudes, with the USA, Brazil and Argentina together producing some 153 million tonnes out of a global total of 278 million tonnes in 2013 (FAO, 2013). Soya oil, like palm oil, is in widespread use in a whole range of products and the crop is also a major source of protein for human and livestock feed. It has an advantage over other herbaceous crops in that it is a *leguminous crop*. It is therefore less dependent on external supplies of nitrogen fertilizer since it can fix nitrogen from the air through the agency of *Rhizobium* bacteria present in its roots. The extent of soya cultivation worldwide is limited by the requirement for a particular day length for the crop to flower at the correct time (although this could potentially be overcome in a breeding programme). Soya was one of the first crops to undergo direct genetic modification (GM), incorporating a bacterial gene that gives the crop resistance to the widely used herbicide glyphosate (commercial name: Roundup). This simplifies cultivation, but has been a continuing source of controversy mainly in the UK and Europe with regard to use of GM crops in foodstuffs.

Most of the cellulosic, starchy and oilseed crops considered here are familiar in current agriculture, requiring no new agricultural technology for their exploitation. They have in the past been bred primarily for food production and to respond to high levels of manufactured fertilizer. However, the production of fertilizers is a relatively energy-intensive process, requiring 30–40 MJ per kilogram of the major plant nutrients applied. This, together with the recurrent energy costs for soil preparation, sowing and harvesting, affect the *net* energy yield of these crops (Section 5.7). Their existing widespread cultivation also means that pests and diseases have had time to co-evolve with them and could spread relatively easily between food and biomass crops. Pesticide production and use will therefore also have to be taken into account. There may be advantages in considering more exotic plants that are not currently in widespread use, such as algae.

Microalgae and other microorganisms

Seaweeds are one form of large algae or macro-algae, but there are also single-celled aquatic microalgae and cyanobacteria that photosynthesize. These could potentially make a very attractive bioenergy source as:

- they grow in water, and are tolerant of wide ranges of salinity and temperature
- they do not occupy land that could be used for other products
- under appropriate conditions, the cells of the algae can contain high percentages of oils, and the cyanobacteria can actually excrete oil into their surroundings: the microbial residues after oil extraction can also be used as an energy source
- there are suggestions they can be used simultaneously to clean up waters polluted with plant nutrients and the resulting biomass, including the oil, can be used for energy (Clarens et al., 2010)
- some forms are also seen as candidate material for the capture of carbon dioxide from power plants.

Research in this area has taken place over several decades especially in the USA, with major oil companies investing many millions of dollars, and in China, with strong government support. In one prominent example, so-called 'green crude oil' is being produced on a continuous basis equivalent to 100 barrels of oil per day from algae grown in 40 hectare open ponds, using seawater, CO_2 from the atmosphere and sewage as a nutrient source (Sapphire Energy, 2016). This 'green crude' has been used as a feedstock to demonstrate that it can be processed into gasoline to power cars without modification and into jet engine fuel for aviation, with testing in flights by a Boeing 737-800 of Continental airlines and a 737-300 of Japan Airlines. There are many designs of smaller-scale bio-reactors which are the subject of ongoing research.

Figure 5.9 A possible bioreactor design for use of microalgae as an energy source

Secondary biomass sources: wastes, residues and co-products

Materials such as wheat straw, rice husks or sawdust, resulting from 'non-energy' uses of biomass are sometimes discarded as 'wastes' (at which point they may become subject to environmental regulations) but they are also potential sources of biomass for energy. Since they are co-products and often collected and transported as part of the primary product supply chain, they tend to be much cheaper than purpose-grown wood or energy crops.

Another possible biomass source includes the many kinds of mixed urban and industrial waste, much of which is carbon-containing material and will release energy if burned. However, it is arguable whether this industrial waste material should be regarded as a truly renewable resource, especially if it contains more than a minimal proportion of matter of non-biological origin, for instance, polymeric materials ('plastics') that have been synthesized from fossil fuels.

This section looks first at the co-products and residues arising from existing uses of biomass in forestry, arable agriculture and animal husbandry. It then moves on to household, or more generally, municipal wastes, and finally to the specific wastes associated with industrial processes.

Wood residues

Around 15% of the standing tree crop is left behind as **forestry residues** during operations such as thinning plantations and de-limbing felled trees. These residues are often left to rot or burned on site, with some being used as a form of matting to prevent soil damage by the harvesting machinery. This has the environmental merit of retaining nutrients and a relatively slow release of carbon dioxide as the residues decay, given that their bulk and inconvenient form makes transporting the residues for wider use economically unattractive. However, with the development of integrated harvesting techniques, some fraction of the residues is increasingly being used for heat and/or power generation in many countries. Sweden has been using fuel chips from forest residues for some 30 years and the rate of utilization has been claimed to be increasing at 10% per annum, towards an estimated potential of 20 million m^3 per year. Most of the residues (more than 71%) are derived during the final felling of the trees (Kuiper and Oldenburger, 2006).

The quantity of residues used as firewood is uncertain, but was thought to be about a million dry tonnes per year in the UK in 1999 (ETSU, 1999). For comparison, total UK sales of **roundwood** (tree trunks and limbs) for fuel increased from around 0.4 million tonnes fresh weight in 2005 to over 2 million tonnes in 2015 (Forestry Commission, 2016). About half a million tonnes of forestry residues are used by the wood-burning power stations at Wilton on Teesside and Steven's Croft near Lockerbie (Ofgem, 2016).

Possibly the largest use of wood residues as fuel is in the pulp and paper industries, where the production plant has ready access to such wastes. Some 3% of the USA's total energy demand is supplied in this way (Haq, 2002). In Britain, the Iggesund paperboard mill in Cumbria installed a 50 MW combined heat and power plant in 2013, which annually uses over 0.5 Mt of wood chip, including 0.2 Mt derived from sawmill waste. It also consumes a small quantity of SRC willow, but in 2014/15 this amounted to only 1.4% of the feedstock (Ofgem, 2016).

Temperate crop co-products

Worldwide, residues from wheat and maize (corn), the two main temperate cereal crops, amount to more than a billion tonnes per year, with an estimated energy content of 15–20 EJ. The residues have many uses including as bedding and feed and hence may be better described as co-products rather than residues, but in major cereal-growing regions substantial quantities are ploughed back into the soil. In the 1970s much of the straw in Britain was burned in the field, but air pollution concerns led to a ban on field burning from the end of 1992. There has been similar legislation across Europe and elsewhere. China experienced a similar pollution problem in the 1990s when residues that had been used for heating and cooking were replaced by 'modern' fuels that left a surplus of residues that was burned in the fields. In this case, the solution was the introduction of biomass-fuelled village-scale gasifiers (see Section 5.6) distributing gas to households. Emissions were reduced and the conversion efficiency was better than direct combustion in individual straw-fuelled stoves.

For use as a bioenergy source, straw must be baled, removed from the fields, stored in a dry atmosphere and transported to its point of use. Although straw has a reasonable energy density of up to 15 GJ per tonne, it has a relatively low mass density. One tonne of bales can occupy a volume of up to 6 cubic metres, which makes transport and storage relatively expensive.

Several European countries already have wide experience of straw burning, and Denmark (Box 5.6) has a programme to use 1.2 Mt per year in combined heat and power (CHP) plants with district heating. It was previously estimated that straw could provide up to 3.2% of UK energy demand by 2020 (DTI, 2004) and Britain's first straw-fired power station, using 0.2 Mt per year, was the world's largest when it was commissioned in 2000. As of 2016, three UK straw plants are operational, consuming about 0.75 Mt per year, with at least one more in development.

Co-firing of straw with other solid fuels occurs in the USA with, for example, a 35 MW plant in Tacoma that derives 40% of its energy from biomass (Haq, 2002).

BOX 5.6 The Avedøre-2 straw-fired combined heat and power station

In 1996 the Danish authorities banned the burning of coal on its own in power stations. The 570 MW Avedøre-2 power plant in Denmark (Figure 5.10) is designed to burn biomass and is located adjacent to a former coal-fired 250 MW CHP plant (itself converted to 100% wood pellet firing in 2016). As a condition of project approval, SK Power had to decommission three older coal-fired power plants to reduce net emissions. The Avedøre-2 plant uses a combination of gas turbines, a fossil fuel boiler and a biomass boiler. The unit can supply district heat to about 180 000 homes and provide electricity consumption for 800 000 households.

When the plant was originally built, natural gas was expected to contribute 85% of total fuel consumption. However, the price of natural gas then rocketed, so in early 2001 it was decided to switch to biomass as the main fuel source. The system requires 300 000 tonnes of straw pellets, at least half of which come from local sources. After transport to the plant, the pellets are re-ground to a small particle size (the improvement in logistics and transport costs for pelleted feedstocks more than compensates for the energy and economic cost of this process).

The bioenergy plant comprizes the straw storage facility, a boiler, an ash separator and ash and fly-ash handling equipment. A flue-gas filter restricts particle discharge into the atmosphere. Bottom ash is recycled as a fertilizer.

Figure 5.10 The Avedøre-2 straw-fired power plant

Tropical crop residues

The total energy content of the annual residues of two major tropical food crops, **sugar cane** and **rice**, is estimated to be about 18 EJ (similar to the total for temperate crops). Significant quantities of tropical crop residues are already being used as fuels.

Bagasse, the fibrous residue of sugar cane after juice extraction, is used in sugar factories as a fuel for raising steam and to produce electricity for use in the plant, in much the same way as wood residues are used in pulp and paper mills. The leaves and stem tops, known as trash, are either burned or left in the field as mulch (the available energy in the biomass is divided roughly equally between the sugar juice, bagasse and trash).

In Brazil, during the 6 to 7 month sugar cane processing period, around 600 MW of electricity is generated to run the crushing mills, and sugar extraction and ethanol production processes, with a surplus of 100 MW being exported to the grid. There is currently a surplus of bagasse, some of which is utilized in other industries although much is used as mulch in the fields. Electricity generation in 2015 was 20 TWh and by 2023 with more efficient boilers or biomass **gasification**, continued liberalization of the electricity market and use of all bagasse, this is expected to increase to 75 TWh. If all cane trash was used as well, total production could be as much as 170 TWh (UNICA 2016; EPE 2016). Such an increase in the renewable contribution to Brazilian power production should be feasible, since the sugar harvesting season corresponds to periods of reduced rainfall and lower hydropower output.

Likewise, in India, a bagasse co-generation programme for producing surplus power from sugar mills by upgrading the boilers has operated successfully. The estimated potential for bagasse co-generation in India is about 5000 MW, out of which about 1500 MW had been installed by 2007 (Purohit and Michaelowa, 2007).

Rice husks are among the most common agricultural residues in the world, making up about one-fifth of the dry weight of unmilled rice. Around 40 million tonnes of rice husks were produced in China in 1999 (Jin et al., 2002), out of a world total of some 140 million tonnes. Although they have a high silica (ash) content that can lead to combustion problems compared with other biomass fuels, their uniform texture makes them suitable for technologies such as gasification. Rice-husk power plants have been deployed in China, India, Thailand and other rice-growing countries.

Other important agricultural residues include ground-nut shells, palm kernel expeller, olive pits, cotton stalks, mustard stalks, jute sticks, etc., some of which are currently used inefficiently in developing countries like India for cooking, and could be utilized better by deploying technologies such as gasification (Kishore et al., 2004).

Animal wastes

The management of **animal manures** and slurries can be a major source of greenhouse gases. Manure from grazing livestock that is deposited in the field decomposes **aerobically** through respiration by a whole range of organisms that feed on it, releasing mainly carbon dioxide. However, manure and slurry from housed livestock that is stored in bulk decomposes **anaerobically** (in the absence of air) releasing methane rather than carbon dioxide. Manure management has been estimated to account for 8% of methane emissions in the USA (EPA, 2010). It can also result in emissions of nitrous oxide, another powerful greenhouse gas. However the increased housing of livestock, together with stricter environmental controls on odour and water pollution, and incentives for renewable energy production, are encouraging farmers to invest in controlled **anaerobic digestion** (Section 5.6) where the biogas generated can be collected and used on-farm or for export of energy services.

Poultry litter, a mixture of chicken droppings and material such as straw, wood shavings etc., has a relatively low moisture content compared to other manures, and an energy content in the range 9–15 GJ t^{-1}, enabling its direct combustion for electricity generation (Box 5.7).

BOX 5.7 **Power from poultry litter**

The world's largest biomass power plant to run exclusively on poultry manure was opened in 2008 in Moerdijk, in the Netherlands. The €150 million plant converts roughly 440 000 tonnes of chicken manure into energy annually. Running at a capacity of 36.5 MW it can generate more than 270 GWh of electricity per year. The plant is intended to convert one third of the Netherlands' chicken waste into energy while also reducing the environmental problems associated with chicken manure in the country.

The first UK poultry manure power plant, with a 12.7 MW output capacity, started operating in 1992 at Eye in Suffolk, using some 140 000 tonnes of litter per year from surrounding poultry farms. A second 13.5 MW plant was commissioned a year later, followed by a 38.5 MW plant, the largest in the UK, in 1998, which consumes 420 000 tonnes of litter each year. Residues from combustion are sold as fertilizer, and one plant has since been modified to burn material from animal carcass disposal. All three projects were originally supported by the NFFO (the Non-Fossil Fuel Obligation, a mechanism of Government incentives for producing low-carbon electricity which ran from 1990 to 1998).

At time of writing (Jan 2017) £1 = €1.15

Sewage sludge, the semi-solid residue from the initial settlement of raw sewage, can be treated anaerobically, as has been done in the UK since the first large 'sewage farms' were built in the last century. Originally, much of the resulting methane-containing biogas was simply flared (burnt off), but an increasing proportion is now used for heat and electricity production on-site. The sludge residue may then be landspread, or alternatively dewatered with mechanical presses and incinerated, producing further heat and electricity. In 2005–6, the water industry in the UK generated approximately 500 GWh of electricity (POST, 2007), although this only represents about 6% of the total energy use in that industry.

Municipal solid waste

The average household in an industrialized country generates rather more than a tonne of solid waste per year. Up to 20 million tonnes of food waste is generated in the UK each year, with each household throwing away 25% of all food purchased (WRAP, 2009). Households in the USA throw away some 40 million tonnes of food annually (Kantor et al., 1997). Most household waste is collected as Municipal Solid Waste (MSW) with an energy content of about 9 GJ per tonne (Figure 5.11). In principle, therefore, the average UK household throwing away a tonne of waste could supply one tenth of its total annual energy consumption of about 90 GJ from its own wastes.

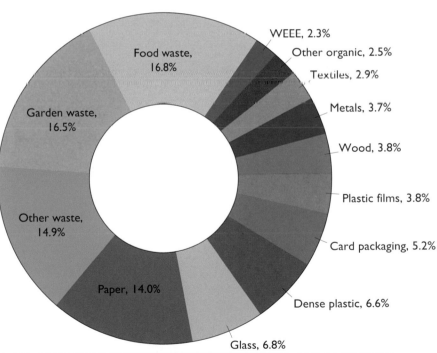

Figure 5.11 Composition of local authority collected waste, England 2010/11 (Defra, 2015). Roughly one-third is food and garden wastes, and one-fifth is paper and card, out of a total of 62% organic material. WEEE = Waste Electrical and Electronic Equipment.

Global waste generation is expected to increase from around 1.4 billion tonnes per year currently to over 3 billion tonnes per year by 2050. Whilst waste represents a major and increasing feedstock for energy provision, its physical nature and energy content is also changing worldwide as recycling increases and dietary preferences change. In continental Europe, where there are now nearly 400 MSW incinerators in operation, and elsewhere, refuse incineration with heat recovery, or energy-from-waste (EfW), is an important part of waste management. The heat may be used directly for district heating or for power production (often in CHP plants). Countries with successful recycling and composting programmes have often seen parallel growth in EfW, which accounts for 30–60% of MSW disposal in most western European countries. In 2015 around 2.79 TWh of electricity was generated in the UK by incinerating 4.5 million tonnes of municipal waste (BEIS, 2016a). Some MSW combustion plants in the UK now also make use of their heat output (see, for instance, Figure 5.12). Worldwide installed capacity is over 10 GW, mostly located in China, the USA and Europe.

Landfilling of waste, involving compaction and burial, remains a common disposal method for MSW in a number of countries, including the USA and the UK. Other countries landfill a much smaller fraction, or even none at all, and in continental Europe up to 60% or so of MSW is incinerated. Legislation and taxation in the UK are gradually decreasing the fraction of MSW going to landfill (47% in 2014), as local authorities diversify into other methods of waste treatment including materials recovery and energy recovery. However, the extent to which MSW can be regarded as a renewable energy source due to its variable content of non-organic and fossil-fuel derived components is debatable.

The composition and characteristics of MSW may be quite different in many of the emerging economies of South Asia, South East Asia and Africa. Here, the majority of wastes like paper, plastic and metals are removed by informal rubbish-pickers and recycled, leaving mainly wet organic wastes and construction debris, for which incineration may not be a suitable option.

Figure 5.12 The SELCHP (South-East London Combined Heat and Power) plant, commissioned in 1994, was designed to incinerate 420 000 tonnes of MSW per year and produce steam for a 31 MW turbo-generator and heat for a local district heating scheme (however the latter was not implemented until 20 years later in 2014)

Commercial and industrial wastes

Commercial and industrial wastes of organic origin can be used as fuel. The UK generated about 28 Mt of specialized wastes in 2014, a decrease from 33 Mt in 2012, but a significant proportion of this is combustible, such as waste from the furniture and construction industries (Defra, 2016).

The wastes from commercial food processing facilities must be treated before discharge to prevent water pollution. Much food waste was traditionally used as animal feed, particularly for pigs, but problems of disease transmission have led to restrictions on this use in Europe and elsewhere. However, the nature of general food waste makes it a useful substrate for anaerobic digestion.

Fats, either in the form of used cooking oils or unwanted fatty tissue removed during meat processing, have a high energy density and are potentially suitable either for direct combustion or conversion to biodiesel.

Hospital wastes, all of which must be incinerated, are increasingly subject to energy recovery as health authorities upgrade their waste-handling equipment. Approximately 200 000 tonnes of healthcare waste is produced annually in the UK, of which 26 450 tonnes, with an energy content of some 15 GJ tonne^{-1}, requires high-temperature treatment (Defra, 2007).

The majority of the 55 million tyres discarded in Britain every year are unsuitable for reuse. With an energy content of 32 GJ per tonne, they constitute a significant fuel resource – although they contain only a modest proportion of renewable biomass (natural rubber). Under EU legislation, whole tyres have been banned from landfill sites since 2003 and chipped tyres from 2006. Lafarge Cement / Sapphire Energy Recovery consumed over 75 000 tonnes of waste tyres in 2008, using them as a partial substitute for coal or coke.

5.6 **Bioenergy processing and conversion technologies**

All the raw bioenergy feedstocks discussed in the previous section require some kind of processing and/or conversion technology to release their energy in a useful form. The nature of this processing depends on the type of biomass and the energy product required. Upstream pre-processing may also be required before subsequent conversion technologies are employed. These processes of densification, concentration and conversion can be categorized broadly into physical, thermochemical and biochemical, which may be used individually or in combination (Figure 5.13).

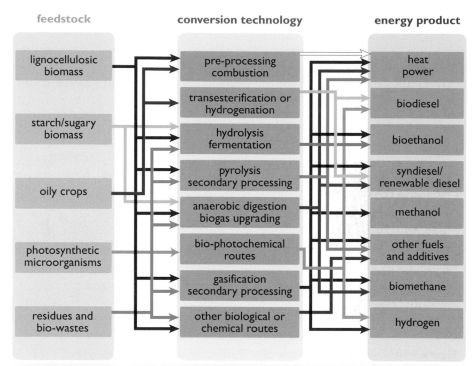

Figure 5.13 Biomass conversion routes, where pre-processing may involve any densification process

Solid biomass fuels

Firewood and chipped wood

Despite the availability of more sophisticated bioenergy conversion routes, almost 40% of the world's population still relies on firewood for energy provision (IEA, 2013). Although direct combustion of timber avoids the losses involved in making more refined bioenergy products, the former is not always convenient to use, transport or store in its raw form. Chipping of firewood may be undertaken to homogenize the fuel, making it easier to handle in large quantities without incurring significant energy or financial costs. Wood chips can be used alone as a power station fuel, or in a 15–30% co-firing mix with pulverized coal to yield an energy conversion of 33–37% (DoE, 2000). Although fuel delivery systems need to be adapted or retro fitted, only modest changes to existing coal boilers are required for co-firing, so the practice is considered cost-effective. In 2012, the USA generated 1.4% of its total electricity using wood chips (IEA, 2013). However, when wood chips are used in co-firing at fractions above 30%, boilers can experience problems due to the different properties of wood ash compared with coal ash (Guo et al., 2015).

Pelleting

Wood chips can be further refined to produce pellets, a process which requires further mechanical grinding before the resultant sawdust is compressed and extruded through 6–10 mm dies (Figure 5.14). This compression causes a temperature spike which plasticizes the lignin content so that the pellet bonds upon cooling. Pellet energy density ranges from approximately 9 MJ m^{-3} for unrefined 'white' wood pellets to 18 MJ m^{-3} for fully torrefied 'black' pellets (IEA, 2011) – see below for an explanation of **torrefaction**. In 2015, 28 million tonnes of wood pellets for heating and industrial purposes were produced worldwide (FAO, 2017), a figure which various authorities expect to increase to between 45 and 60 million tonnes by 2020 (World Wide Recycling Group, n.d.; Wood Pellet Association of Canada, n.d.).

(a)

(b)

Figure 5.14 A wood pelleting plant (a) and a sample of the resultant pellets (b)

Conventional 'white' wood pellets have a low moisture content of approximately 5–10 %, with a high physical durability and packing density of approximately 650 kg m^{-3}. This stability and uniformity makes wood pellets suitable for automatic feeding and they are more easily packed, stored and transported than wood chips or logs. Pelleting can also be applied to biomass feedstocks such as miscanthus, sugarcane bagasse and even coffee grounds (Box 5.8), and there is some evidence that agricultural biomass and energy crops need less energy for processing than wood. For example, size reduction and pelleting requires energy inputs of 0.1–0.4 GJ per tonne for miscanthus (Jannasch et al., 2002) and 0.7 GJ per tonne for wood (Risović et al., 2008). However, these energy inputs are variable and depend on factors such as drying requirements (Mani et al., 2006).

BOX 5.8 Bio-bean

There is considerable potential for turning a wide variety of agricultural residues and wastes into densified biomass fuels. Among the most innovative of these are the biomass pellets and briquettes produced from waste coffee grounds in the UK by bio-bean, a clean technology company established in 2014 (Figure 5.15).

Britain produces 500 000 tonnes of waste coffee grounds each year, costing the coffee service industry over £70 million in disposal costs. bio-bean is the first company in the world to industrialize the processing of coffee grounds into a range of biomass fuels. Working within the existing waste disposal infrastructure, bio-bean collects waste coffee grounds from coffee shops, offices, restaurants and factories and transports them to their production plant in Cambridgeshire – which can process up to 50 000 tonnes of grounds a year, or 10% of UK coffee waste production.

The gross calorific value of bio-bean pellets and briquettes is 22.6 GJ/tonne, higher than typical wood pellets and briquettes.

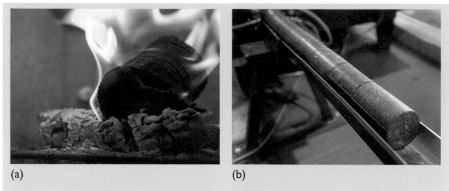

(a)   (b)

Figure 5.15 Combustion (a) and manufacture (b) of briquettes produced from waste coffee grounds (Courtesy of bio-bean Ltd.)

The UK is currently the world's largest importer of industrial wood pellets, mainly due to demand from Drax power station in Yorkshire, which in 2012 embarked on a programme of converting three of its six previously coal-fired generating units to biomass firing. Altogether, the 2000 MW of biomass capacity at Drax can consume annually 7–8 million tonnes of pellets, most of which are imported from North America. The conversion of Lynemouth power station in Northumberland to biomass power in 2017 will account for an additional 1.4 million tonnes/year, and a further one million tonnes or more are burned in the thousands of small and medium-sized domestic and industrial heating boilers installed since 2011 under the UK government's Renewable Heat Incentive scheme (Box 5.9).

Further examples of large recently-converted pellet-fired power stations are the 250 MW Avedøre-1 plant in Denmark (see Box 5.6), and the 205 MW Atikokan Generating Station in Ontario, Canada, currently the largest of its kind in North America. Dedicated biomass power plants using wood chips and agricultural residues as fuel include Engie's Polaniec station in south-east Poland, which consumes 1.1 million tonnes/year of biomass to replace 205 MW of capacity within an existing coal power plant complex, and the 49 MW Mikawa biomass plant in Omuta, Japan – likely to be fitted with carbon capture technology (see Section 5.9) – while the older 265 MW Alholmens Kraft power station in Finland is mostly fired with forest residues. See also Box 5.10.

BOX 5.9 The Renewable Heat Incentive scheme in Britain

The Renewable Heat Incentive (RHI) was introduced by the UK Government, among other renewable energy policy measures, to encourage users to install renewable energy technologies for space and water heating in buildings or for industrial process heat. It was the world's first financial incentive of its kind, and includes solid biomass used in biomass boilers and combined heat and power plants, as well as biogas. The aim of the RHI was to help attain the UK's National Renewable Energy Action Plan goal of 12% of heat generated from renewable sources by 2020, under the EU Renewable Energy Directive. The scheme was anticipated to create 13 000 installations in industry and 110 000 installations in the commercial and public sector.

The RHI scheme was launched for non-domestic users e.g. offices, hotels and schools in November 2011 and extended to domestic buildings in April 2014. The 'non-domestic' RHI pays rebates for 20 years to the project developer based on the capacity of the biomass system and its quarterly meter readings.

In order to be eligible for the scheme the project developer has to go through an accreditation process with Ofgem, which is completed after the boiler is installed.

By November 2016 there were over 16 000 non-domestic RHI installations, with solid biomass making up nearly 94% of the total. The vast majority of installations fell into the 'small biomass' category with capacities of under 200 kW. Here, the rebate levels were initially quite high (more than the price of heating oil) in order to encourage uptake. However, the popularity of the RHI scheme led to the introduction of progressive quarterly reductions in the tariffs from July 2014. Installations hit their peak in March 2015 at around 1800 permonth, but by January 2017 the 'small biomass' main tariff had dropped to 2.95p/kWh compared with 7.9p/kWh initially, with a consequent major impact on the number of installations. At the time of writing (November 2016) total installations across all sizes of biomass boilers had fallen to just 100 per month.

The domestic RHI was designed as a 'boiler replacement scheme' for households (open only to existing properties and not newly built homes), with payments made over seven years. About half of all installations have been air source heat pumps but biomass boilers account for around a further 30% of the total. Again, the initial popularity of the scheme brought about frequent tariff reductions and a decrease in biomass installations to about 60 per month at the time of writing. 73% of domestic RHI-funded renewable heat systems have been for households off the gas network.

Since September 2013 all newly accredited biomass installations have been required to have emissions certificates or environmental permits, requiring the recipients of RHI rebates to demonstrate that combustion of the specific biomass fuel in their boiler did not exceed a maximum level of 30 grams per gigajoule (g/GJ) of particulate matter (PM) and 150 g/GJ for oxides of nitrogen (NOx).

From October 2015 all biomass projects receiving RHI rebates have had to demonstrate that the fuels used were sustainably sourced. Biomass fuel has to meet a lifecycle greenhouse gas (GHG) emissions target of 34.8g CO_2 equivalent per MJ of heat, or 60% GHG savings against the EU fossil fuel average. In addition, the biomass needs to satisfy land criteria, complying with rules on the type of land on which the biomass was produced. Compliance schemes such as the Biomass Supplier List (BSL) and the Sustainable Fuel Register (SFR) have been set up to help RHI applicants to meet these obligations.

BOX 5.10 Iggesund and Brigg power stations

At the time of writing (November 2016) there were only two major outlets for UK produced perennial energy crops. The Iggesund paperboard mill operates a 50 MW combined heat and power (CHP) plant in Workington, Cumbria. Completed in 2013 at a cost of £108 million, the plant requires 500 000 green tonnes of woodfuel per year and aims to source 10% of its fuel needs from short rotation coppice (SRC) grown in northwest England, southern Scotland and across the north of England.

The plant used 8 400 tonnes of willow in 2014/15 which equates to 1.4% of the total feedstock consumed. Local planting of SRC is proceeding at around 100 hectares per year. In addition, some larger growers in Yorkshire and the East Midlands who previously sold their crop to Drax Power Station for co-firing have agreed to supply the Iggesund facility following the end of their Drax contract.

The main buyers for miscanthus are Terravesta, who previously processed the crop into pellets for supply to Drax. The latter used over 31 000 tonnes of miscanthus in 2014/15. With the cessation of this contract, Terravesta will instead be supplying 25 000 tonnes of whole miscanthus bales to the 40 MW Brigg Renewable Energy Power Plant in Lincolnshire that was completed in January 2016. Agreements are index-linked and vary from five to ten years. The 2016 price for a long-term miscanthus supply contract was £74.57/tonne, delivered to Brigg, equivalent to around £63–£66/tonne before haulage costs.

Charcoal and torrefaction

The need to increase the energy density and combustion performance of traditional biomass fuels has historically driven the development of charcoal through **pyrolysis**. To achieve this process, wood materials are heated to 400 °C in the absence of air to make a product which is almost pure carbon. The consequent charcoal has an energy content of approximately 30 MJ kg_{-1}, can be burnt without flame or smoke and can reach temperatures of 2700 °C (Antal and Gronli, 2003). However, charcoal-making (traditionally carried out in the forest where the wood is cut) releases volatile matter, volatile tars and oils and the products of incomplete combustion into the atmosphere, and this can make charcoal production seem an environmentally damaging process. The efficiency of conversion varies widely with practice, from about 6 to 12 tonnes of air-dry wood per tonne of charcoal product, and the level of government regulation and industry good practice differs markedly between countries. Regrettably, in many instances, harvesting of wood as a 'free' resource has undermined the economic case for improving kiln efficiency. Yet charcoal is relatively clean burning and this is particularly important in the developing world where otherwise food may be cooked over an open fire or in an inefficient stove inside a home with poor ventilation. Worldwide, various projects to promote improved charcoal kilns and to introduce locally-manufactured improved charcoal stoves, such as the Kenyan 'jiko', are considered to have been successful (UNDP, 2000 and Anderson et al., 1999). Globally, about 52 million tonnes of wood charcoal was produced in 2015, led by Brazil and several African countries (Nigeria, Ethiopia, D.R. Congo, Mozambique) (FAO, 2016).

Solid biomass fuel for industry can be similarly enhanced and stabilized through the thermochemical process of torrefaction, where partial pyrolysis is achieved through heating to 240–300 °C in the absence of air. In this mild pyrolysis process, the biomass (particularly the hemicellulose) only partly decomposes, releasing some of the component combustible volatiles. The resulting solid torrefied biomass (sometimes referred to as biocoal) has approximately a 30% higher heat energy content, increased density and reduced content of undesirable elements such as chlorine in comparison to the original biomass. Torrefied biomass is also **hydrophobic**, that is, it does not absorb moisture from the air. It is particularly suitable for co-firing with coal because it can be stored in the open for long periods in the same way as coal. It can also be transported more easily and cheaply than unrefined wood chips or pellets. Although the economics of the process have yet to be fully demonstrated, torrefaction pilot plants, demonstration projects and small commercial units are being commissioned in many parts of the world (e.g. the 200 000 tonnes/year proposed Scandinavian Biopower Oy biocoal plant in Mikkeli, Finland, announced in 2016). There are significant prospects for large-scale production and trade in torrefied biomass for co-firing in the world's existing fleet of coal power stations.

Combustion of solid biomass

Most biomass is initially solid, and can be burned in this form to produce heat for use *in situ* or in close proximity, although it may first require relatively simple physical processing, sorting, chipping, compressing and/or air-drying, as discussed above.

Direct combustion is a simple process to release useful energy. Unfortunately, it can also be very inefficient (as described in Box 5.11) and can potentially produce a wide range of pollutants.

BOX 5.11   Boiling a litre of water over a wood fire

How much wood is needed to bring one litre of water to the boil?

The starting point for answering this question is to apply the following equation:

Heat energy required = mass × temperature rise × specific heat capacity of water

Note: the **specific heat capacity** (sometimes called the specific heat) of a substance is the amount of energy in joules that has to be transferred to 1 kg of material to raise its temperature by 1 kelvin.

Data

Specific heat of water	$= 4200 \text{ J kg}^{-1} \text{ K}^{-1}$
Mass of 1 litre of water	$= 1 \text{ kg}$
Heat value of air-dry wood (Table 5.1)	$= 15 \text{ MJ kg}^{-1}$
Density of air-dry wood	$= 600 \text{ kg m}^{-3}$
1 cubic centimetre (1 cm^3)	$= 10^{-6} \text{ m}^3$

Calculation

Heat energy needed to heat 1 litre of water from 20 °C to 100 °C	$= 80 \times 4200 \text{ J}$
	$= 336 \text{ kJ}$
Heat energy released in burning 1 cm^3 of wood	$= 15 \times 600 \times 10^{-6} \text{ MJ}$
	$= 9.0 \text{ kJ}$
Volume of wood required	$= 336 \div 9.0 \text{ cm}^3$
	$= 37 \text{ cm}^3$. (say, two 200 mm sticks)

Experience suggests that on an open fire many more than two thin 200 mm sticks would be needed. However, a well-designed enclosed stove using small pieces of wood could boil the water with as little as four times this 'input' – implying an efficiency of 25%.

Designing a stove or boiler that will make good use of valuable fuel requires an understanding of the series of processes involved in combustion. The first process, which consumes rather than produces energy, is the evaporation of any water in the fuel. More energy must then be supplied to raise the temperature of the material to its ignition point, when it starts to burn.

Once it starts burning, there are two stages because any solid fuel contains two combustible constituents. The **volatile matter** is released as a mixture of gases and vapours as the temperature of the fuel rises. The combustion of these produces the little spurts of flame seen around burning wood or coal. The solid matter consists of the **char** together with any inert matter. The char, mainly carbon, burns to produce CO_2, whilst the inert matter becomes ashes, slag or clinker.

It is essential that the design of any stove, furnace or boiler ensures that the vapours which are released burn and don't just disappear up the chimney. Air must also reach all the solid char, which is best achieved by burning the fuel in small particles. This can raise a different problem because finely divided fuel may produce finely divided fly ash or **particulates** that must be removed from the flue gases. The air flow should also be controlled: too little oxygen means incomplete combustion and leads to the production of poisonous carbon monoxide, while too much air is wasteful because it carries away heat in the flue gases. This is the reason that burning wood in an open fire is only about 10% efficient.

Modern systems for burning solid biomass fuels are as varied as the fuels themselves, ranging in size from small stoves in domestic space and water heating systems to large boilers producing megawatts of heat. Combustion efficiencies (heat output/heat content of fuel) of over 85% can be achieved in well-designed modern systems.

Liquid biofuels

In order to at least supplement the world's dwindling supply of cheap crude oil, the principal resource for the production of gasoline, diesel and jet fuel for passenger transportation, there is a need to develop renewable sources of substitute liquid fuels which fit the existing supply systems. At present, biomass is the only renewable feedstock which can be used to yield liquid biofuel, so this is a primary focus. The market share of biofuels in road transport remains modest, at around 3.5–4.0% worldwide, but it is notably higher in some European countries, in the United States, and especially in Brazil—where it exceeds 20% (REN21, 2015).

At the time of writing, the EU Renewable Energy Directive obliges all member states to attain a 10% contribution of renewable energy in transport by 2020, but progress in most countries has been slow, reaching an average of 5.7% by 2014, mostly from biofuels. Sweden was the first EU member state to exceed the 10% goal, with a 2013 share reaching 16.7%. Overall, half of member states had achieved at least 5% or higher share of renewable energy in transport by 2013 (European Commission, 2017).

Bioethanol

A widely-used commercial practice is to utilize the sugars which can be extracted from biomass feedstocks and convert them to ethanol through fermentation. Fermentation is an anaerobic biological process where the simple sugars from the biomass feedstock are converted to alcohol and carbon dioxide via the action of microorganisms, such as yeasts. The ethanol product (C_2H_5OH) is then separated from other components using heat to distil the mixture; the ethanol boils off and can then be cooled and condensed back to liquid. Although it is possible to generate bioethanol from all plant materials, which are universally made up of cellulose, hemicellulose and lignin, there are technical difficulties in separating these compounds and depolymerising cellulose and hemicellulose into simple sugars. Research and development are on-going in order to devise effective pre-treatment methods to improve the efficiency of bioethanol production from lignocellulosic feedstocks and other complex plant materials, but starch/sugar-based crops are the most commonly used (Table 5.7). Sugar cane, sugar beet, sweet sorghum, potato, cassava, corn, wheat and barley are all target crops for bioethanol production given their high sugar/starch contents. The choice of crop type is generally dependent on growing regions; the USA dedicated 42% of its grain crops to bioethanol production in 2012, while Brazil favours sugar cane (Figure 5.16), Europe uses mainly wheat and sugar beet and China depends upon corn, wheat and cassava (Guo et al., 2015).

Anhydrous bioethanol is most commonly used as an extender in gasohol, which is petrol (gasoline) containing a percentage of ethanol. Higher blends (typically 85% ethanol, 15% gasoline – designated E85) are used in flexible-fuel vehicles (FFVs) in countries such as Sweden, the USA and Brazil, the latter of which has many cars which also run on 100% hydrous bioethanol. Such high ethanol blends can be burned directly in suitably modified petrol engines. Most FFVs involve minor changes to the fuel-injection system, generally using a fuel mixture sensor communicating with the engine control unit.

(a)   (b)

Figure 5.16 Sugar cane stems (a) and a Brazilian bioethanol plant (b)

In order to produce bioethanol, a number of steps are necessary, which typically include: milling, liquefication, saccharification, fermentation, distillation, drying and denaturing. For example, wheat is milled to a fine powder before the addition of water produces a slurry, which is converted to a mash after it is heated to just over 100 °C for approximately two hours. Subsequently, α-amylase and glucoamylase are used in the liquefication and saccharification processes, before yeast is added to ferment the sugars to alcohol. The liquid resulting from fermentation contains typically 8–12%

ethanol, after which distillation is required to concentrate it: this requires a considerable heat input, which may be supplied by crop residues. The energy content of ethanol is about 30 GJ per tonne or 0.024 GJ per litre (Guo et al., 2015). Improving the efficiency of yeast strains which can be implemented commercially for sugar fermentation is a key focus in improving this process, and progress has been made at the scale of pilot plants (Foust and Bratis, 2013).

Table 5.7 Ethanol yield varies depending on feedstock type

Raw material	Litres per tonne*	Litres per hectare per year#
Sugar cane (stem)	70	400–12 000
Maize (grain)	360	250–2000
Cassava (roots)	180	500–4000
Sweet potatoes (roots)	120	1000–4500
Wood (stem)	160	160–4000

* This depends mainly on the proportion of the raw material that can be fermented
# The ranges reflect worldwide differences in yield. The upper figure is a theoretical maximum.

These sugar and starch-based biomass feedstocks generally also function as food crops and the ethanol product is termed a 'first generation' biofuel: it competes for resources with both human food and animal feed production, but also offers increased diversity and a degree of flexibility in the market for producers, arguably stimulating agricultural investment. For example, sugar beet was the feedstock for the first bioethanol plant in the UK, with an annual capacity of 70 million litres, while two subsequent British wheat-to-ethanol plants each have a capacity of around 400 million litres. Brazil's PRO-ALCOOL programme, established in 1975, produces ethanol from sugar residues and was the world's largest commercial biomass system until surpassed in 2006 by USA bioethanol production, mostly from maize (27 billion and 56 billion litres, respectively, in 2015). World production has grown to over 100 billion litres per year, equivalent to more than one million barrels of oil per day.

To reduce competition with food and fibre production, much research effort is being invested in developing 'second generation' biofuels, which rely on non-food crops and agricultural residues as feedstock, but worldwide production capacity for cellulosic ethanol has stalled at around 200 million litres per year (less than 0.2% of the industry total), and actual output in 2015 was barely 10 million litres. Extracting sugars from lignocellulosic biomass has proven technically and economically challenging, and overcoming this bottleneck is the subject of intense research effort (Bomgardner, 2013). Higher process costs and operational problems with feedstock quality have also held back commercialization.

Pre-treatment of lignocellulosic feedstocks to improve the efficiency of sugar release and subsequent bioethanol yield may involve acid or enzymatic hydrolysis (Badger, 2002). When **acid hydrolysis** is employed to treat crushed biomass, energy yields are variable with a balance needing to be found between speed and efficient sugar recovery. For example, treating biomass rapidly with dilute H_2SO_4 under high pressure and temperatures tends to yield only about 50% of the cellulose as sugars because the released sugars are rapidly degraded. To improve the efficiency of sugar recovery (up to 80%), milder conditions may be employed to release pentoses before conditions are intensified to release the remaining hexoses (Lenihan et al., 2010). However this is time-consuming and it can be difficult to separate the sugar products from the acids.

Alternatively, enzymes can be employed to convert lignocellulosics into simple sugars, although this is not straightforward as cellulose in biomass is protected by hemicellulose and lignin. To expose cellulose for enzymatic hydrolysis, the feedstock is first treated through methods such as freezing, radiation or autohydrolytic hydrothermal deconstruction (Kumar et al., 2009). **Steam explosion** may also be used to remove lignin from the biomass in order to expose the cellulose and other polymeric carbohydrates (such as hemicellulose) for subsequent breakdown. Enzymes for treating steam-exploded biomass are usually derived from micro-organisms such as bacteria or fungi and break down both the cellulose and hemicellulose to simpler carbohydrates that can be used in traditional fermenters. More recently, thermostable cellulases, able to degrade cellulose at temperatures in excess of 70 °C, have been identified from mesophilic and thermophilic fungal strains, such as Thermoascus aurantiacus, Chaetomium thermophile and Myceliophthora thermophila. This is a major area of research effort which could greatly increase the efficiency with which sustainably-sourced feedstocks can be converted to biofuels (Viikari et al., 2012).

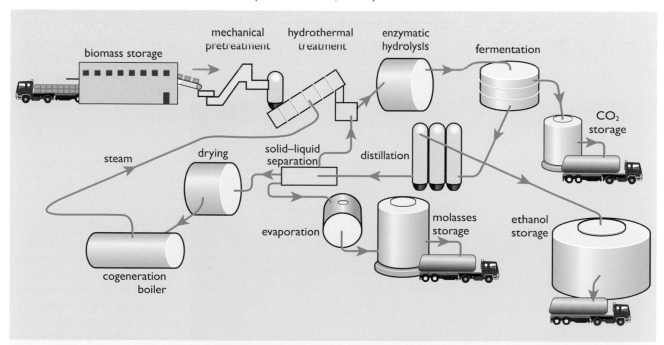

Figure 5.17 Schematic of the Inbicon integrated process for ethanol production from straw, where the co-generation boiler supplies steam for the hydrothermal treatment

A small demonstration second-generation plant was established by Inbicon in 2009 at Kalundborg, Denmark, and had completed 15 000 hours of operation by 2016 (Inbicon, 2017). This uses 4 tonnes of straw per hour as its feedstock, with a combination of mechanical and hydrothermal treatment to separate the cellulose and lignin. The cellulose is then subjected to enzymatic hydrolysis and fermentation to ethanol. After distillation to separate ethanol from the lignin and unfermented molasses, the lignin is dried for use as fuel in a co-generation boiler. Figure 5.17 is a schematic diagram of the process, which is designed to produce 4 300 tonnes of ethanol, 13 000 tonnes of lignin and 11 000 tonnes of molasses each year. A 50 million litres/year commercial plant, processing 195 000 tonnes of corn stover (stalks) and wheat straw annually using Inbicon technology, is, at the time of writing, under development in North Dakota, USA.

In 2013, the first of several commercial-scale cellulosic ethanol projects in the USA and Europe, the Crescentino Biorefinery, started production at Vercelli in north-western Italy. It can produce up to 75 million litres of ethanol annually using wheat and rice straw together with giant reed (Arundo donax). The conversion technology for these feedstocks consists of pre-treatment with steam and high pressure and subsequent enzymatic hydrolysis.

An alternative approach to ethanol production from lignocellulosic feedstocks involves the thermochemical gasification of biomass to **syngas** (a mixture of H_2 and CO) which is then either fermented to ethanol using micro-organisms, or thermochemically synthesized in a catalytic reactor. These methods can produce up to 50% ethanol yield from syngas input (Badger, 2002; Phillips et al., 2007; Ptasinski, 2016) but they are still relatively expensive and technically demanding, and therefore have not yet seen commercialization.

Vegetable oil and biodiesel

Following his invention of the compression ignition engine that bears his name, Dr Rudolf Diesel demonstrated in 1911 that it could run on a variety of vegetable oils, although petroleum-based diesel fuel subsequently dominated the commercial vehicle market in the late 20th and early 21st century.

Certain vegetable oils can indeed be burned directly in some modern diesel engines, either pure or blended with petroleum diesel, but most applications require minor modifications to the engine and fuel system (Figure 5.18). Alternatively, the upgrading of vegetable oils to biodiesel (fatty acid methyl ester - FAME) results in a fuel that can blend with, or replace, petroleum diesel in unmodified engines.

Figure 5.18 A car powered by vegetable oil against a backdrop of oilseed rape fields

The oil in crops such as rapeseed, soya and oil palm is contained within the seed, which also contains the plant embryo and may have a structural shell or coat for protection. The oil has to be separated from these surrounding tissues, either by pressing or by using a solvent to dissolve the oil, thus facilitating easier separation from the remaining material. Commercial oilseed extraction involves a number of steps including:

- seed cleaning – removal of foreign matter
- tempering – pre-heating of the seed to improve ease of oil extraction
- dehulling – removal of seed coat
- flaking – to increase surface area
- conditioning – re-heating the flaked seed
- mechanical extraction – by pressing and extrusion and/or expansion
- solvent extraction – for maximum extraction of oil.

A typical press for the mechanical extraction stage uses a metal screw to force the material over a sieve, which allows the oil to pass out but retains the fibrous material for separate ejection (Figure 5.19).

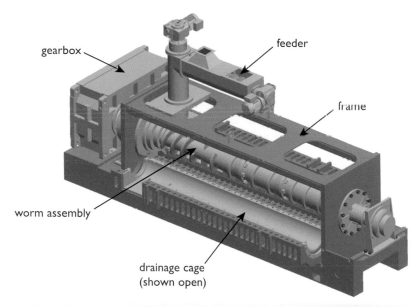

Figure 5.19 Schematic diagram of a screw press which can be used for oilseed extraction

Vegetable oils are **triglycerides**, a form of **ester** whose large molecules are effectively various organic acids combined with **glycerol** (an alcohol). In biodiesel the main components are esters where the organic acids are combined with other, lighter alcohols rather than glycerol. The conversion process from vegetable oil to biodiesel, called **transesterification**, involves adding the alcohols, usually methanol or ethanol (both potentially derived from biomass) to the vegetable oil in the presence of a catalyst. This converts the triglycerides into esters of methanol or ethanol, together with free glycerol. Global production of biodiesel, from a wide variety of plant and animal oils and fats, has increased steadily in recent years, with 23.6 billion litres produced in 2013 compared to only 0.8 billion litres in 2000 (EIA, 2014).

In countries with warmer climates, blends of up to 30% vegetable oil with diesel have been used without the need for transesterification. Coconut oil is used in tractors and lorries in the Philippines, palm and castor oil in Brazil and sunflower oil in South Africa. However, since the food and cosmetics industries can usually pay a higher price, these applications are usually limited to places where diesel fuel is expensive and in short supply, or where fuel standards are less stringent.

Bio-oil from pyrolysis

When plant biomass is subjected to complete pyrolysis (unlike the partial pyrolysis which produces charcoal), it yields three components: biochar, a black solid residue; **bio-oil**, the vapour condensate; and syngas, which is uncondensed vapour. The relative proportions of these three products can be manipulated according to the method of pyrolysis and the feedstock type, with **slow pyrolysis** typically producing 35%, 30%, and 35% of biochar, bio-oil and syngas respectively, while fast pyrolysis yields are in the region of 10–30%, 50–70% and 15–20% respectively (Laird et al., 2009). A characteristic of biomass is that the volatile matter carries more of the energy than the char, and so these volatile compounds have been the subject of much research on pyrolysis for bioenergy. Wood is the preferred feedstock for bio-oil production, with up to 75% conversion of the feedstock (Carpenter et al., 2014), yet the resultant bio-oil has a number of drawbacks: it is unstable, corrosive, immiscible with hydrocarbon fuels, viscous, difficult to ignite and has a low energy density (Czernik and Bridgwater, 2004; Xu et al., 2008). It is therefore preferable to further process bio-oil to remove moisture and reduce acidity, both of which improve its storage properties and energy value. Although many processing methods have been investigated in order to upgrade bio-oil efficiently, there is not yet a technology which is able to economically separate pure compounds from the mixture, making its use as a petroleum fuel alternative challenging (Vamvuka, 2011). Pilot plants have, however, been implemented in countries such as Canada, Italy, the USA, Brazil, Germany, the UK and Australia (Vamvuka, 2011), which convert woody feedstocks to bio-oil for use as a heating fuel and industrial feedstock.

Other liquid biofuels

Research and development have continued on potentially more convenient 'drop-in' biofuels, which, unlike bioethanol and biodiesel, would require no modifications at all to existing fuel supply systems. These fuels include hydro-treated vegetable oils, wood-based 'tall oil' and waste animal fats, as well as bio-butanol, sugar-based hydrocarbons and syngas complexes (AFDC, 2012). A number of conversion pathways have been developed, including fermentation or catalysis of lignocellulosic sugars, hydrothermal biomass liquefaction and transformation, or up-grading, of bioethanol and syngas. For example, a number of reactions may be carried out using catalysts to convert sugars to hydrocarbon fuels, including dehydrogenation and deoxygenation (Biddy and Jones, 2013). Commercial-scale production of bioethanol can be modified to produce **biobutanol** using the acetone-butanol-ethanol fermentation process, which utilizes solvenogenic clostridial bacterial strains (Schiel-Bengelsdorf et al., 2013). Alternatively, catalysts can be utilized to upgrade bioethanol or to convert syngas to synthetic gasoline, such as in Syngas to Gasoline Plus (STG+) (Primus, 2013). These emerging technologies offer potential alternatives to the more well-known liquid biofuel processes, yet most of them are still under development, requiring further research to bring them to market.

Gaseous biofuels

Anaerobic digestion for biogas

Biogas offers a renewable alternative to natural gas, and can be produced through anaerobic digestion (AD) of organic wastes; typically, it consists of 60–65 % methane and 30–35% of CO_2 plus smaller quantities of water vapour, H_2 and H_2S (Weiland, 2010). Although the majority of AD plants consume the biogas product on-site for heat and/or electricity production, an increasing number of larger plants upgrade most of their biogas to '**biomethane**' for direct injection into the local natural gas pipeline network and (less commonly) for motor vehicle use.

The AD of biomass comprizes four main steps, each of which is based on bacterial activity: (1) hydrolysis, (2) acidogenesis, (3) acetogenesis and (4) methanogenesis. Hydrolysis of large inorganic molecules such as carbohydrates, proteins and lipids yields relatively simple molecules of sugars, amino acids and fatty acids. These simple organic molecules are then subjected to acidogenic bacteria that break them down into their basic compounds, some of which are organic acids, which are in turn acted upon by acetogenic bacteria to form acetic acid, along with H_2, NH_4 and CO_2. Finally, methanogenic bacteria facilitate the decomposition of acetic acid to methane and CO_2.

The feedstock for AD may include dung or sewage, food processing wastes or discarded food, agricultural residues or specially grown silage crops. Wood-based biomass is generally excluded as most anaerobic bacteria are unable to degrade lignin (Weiland, 2010). Digestion can take place in either 'wet' or 'dry' (with much lower moisture content) systems. In wet systems, the raw feedstock is usually converted to a slurry with up to 80–95% water, and fed into a purpose-built **digester** with control of temperature, pH, etc. Less energy is required to mix the material in a wet system, but this also means large quantities of water are required, and large volumes of the resulting co-product, digestate, must then be handled. Digesters can range in size from perhaps one cubic metre (roughly 200 gallons) for a small 'household' unit in rural India or China, to around 100–1000 m^3 for a typical farm plant (Figure 5.20) and more than 10,000 m^3 for a large installation. The input may be continuous or in batches and digestion is allowed to continue for a period of between ten days and a few weeks. Approximately one tonne of dry biomass can produce 120 m^3 of biomethane or 200 kWh of electricity (European Bioplastics, 2011).

Figure 5.20 On-farm anaerobic digestion plant in Somerset, UK, in which manure and slurry from a large dairy farm is supplemented by small amounts of waste silage and animal feed

The bacterial action itself generates a small amount of heat, but in cooler climates additional heat is normally required to maintain the ideal process temperature for bacterial action, with temperatures of around 35–40 °C needed for **mesophilic digestion** and between 50–55 °C for more rapid **thermophilic** processing. If the ambient climate is extreme and temperatures drop below 10 °C or rise above 70 °C, AD can virtually cease (Kardos et al., 2011). In some small digesters, the majority of the biogas produced may be consumed 'parasitically' simply to maintain optimal plant operation, but, while the net energy output is then zero, the plant may still pay for itself in savings of fossil fuels and other costs associated with processing the wastes by other means. The residual co-product remaining after digestion (**digestate**) contains much of the nitrogen and other plant nutrients from the original feedstock, and can be used as a fertilizer and soil conditioner.

Figure 5.21 A small-scale biogas plant in India

Worldwide electricity production from biogas in 2012 was in the region of 47–95 TWh (Rajgor, 2012), with China and Germany dominating the sector. Germany is the largest European producer of biogas, with nearly 9000 AD plants in operation in 2016, generating around 30 TWh or 5% of German electricity consumption (IEA, 2015b). The majority of these AD plants have been developed in the agricultural sector using manures and silage crops such as maize. The UK has around 350 AD plants operating outside the water industry, of which 225 are fed predominantly with farm-based feedstocks – and has an aspiration of 1000 AD plants by the early 2020s (NFU, 2016).

China initiated large-scale programmes for small-scale domestic AD starting in the 1970s. A later drive involving more commercial enterprises resulted in some 5 million domestic plants operating successfully by the mid-1990s, and tens of millions by the 2010s. In India (Figure 5.21), the government actively encourages the growth of anaerobic digesters on all scales and supports each state with subsidies and incentives with a view to utilizing the majority of human and animal waste (Ministry of New and Renewable Energies, 2016). Biogas makes up a relatively small proportion of global bioenergy production, but if all bio-waste were processed by AD, a quarter of current natural gas consumption could be displaced, which equates to approximately 6% of primary energy demand (Guo et al., 2015).

Likely future developments in the AD industry include pre-processing of feedstocks to increase their biogas yield, and the post-processing of digestate to increase its value. Research interest includes the identification and development of microbes adapted to higher temperatures, such as extreme thermophiles, which offer potentially higher returns on capital investment. Other research involves the use of **psychrophiles**, which are organisms adapted to lower temperatures that would reduce the need for digester heating in cold climates.

Landfill gas

Over the period 2013–2015, Britain generated about 5 TWh/year of electricity from **landfill gas** (BEIS, 2016a), so the majority of the UK's energy from AD actually comes from the decomposition of organic waste in the landfill sector. A large proportion of municipal solid waste is material of biological origin (Figure 5.11) and its historic disposal in deep landfills has furnished suitable conditions for anaerobic digestion to occur naturally. It was known for decades that landfill sites produced methane and systems were fitted to burn it off safely, but the idea of collecting and using this landfill gas developed only in the 1970s. The natural digestion process in a landfill (Figure 5.22a) takes place over years, rather than the days or weeks of in-vessel systems. In developing a landfill gas site, each area is covered with a layer of impervious material after it is filled and the gas is collected by an array of interconnected perforated pipes placed at depths of up to 20 metres in the refuse (Figure 5.22b). In a large well-established landfill there can be several kilometres of pipes, with as much as 1000 m^3 per hour of gas being pumped out.

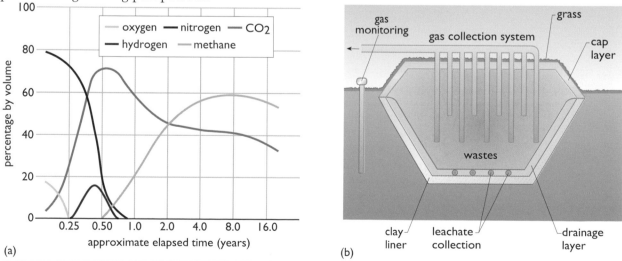

Figure 5.22 (a) The changing composition of evolved gas in a landfill site over time, (b) Extraction method for landfill gas

In theory, the lifetime yield per tonne of waste in a good site should lie in the range 150–300 m^3 of gas, with between 50% and 60% methane by volume. This suggests a total energy output of 5–6 GJ tonne-1 of refuse but, in practice, the heat energy output per tonne of wastes is less than 2 GJ, with extremely variable and often unpredictable yields dependent on waste composition and the site conditions. In the longer term, the prospects for energy from landfill gas appear less favourable, given the implementation across Europe of the 1999 EU Landfill Directive, which requires a reduction in the amount of waste going to landfill 'to prevent or reduce as far as possible negative effects on the environment'.

Gasification for syngas

As described above, gasification is a specialized form of thermochemical pyrolysis by which a gaseous fuel may be produced from a solid fuel. It is a commercialized process whereby carbon-rich materials are intensely heated to over 700 °C and partly combusted under a controlled flow of air, steam or oxygen. 'Coal gas' or 'town gas', the product of coal gasification, was widely used in the UK and elsewhere for many decades. 'Wood gas' was used for heating, lighting, and even as vehicle fuel during the coal shortages of World War II, but both were superseded in Europe by natural gas by the 1970s. Both syngas and lower-quality '**producer gas**' (containing a higher fraction of carbon dioxide and nitrogen) can be produced from biomass.

There are many different designs of modern gasifier but, during the process of gasification, woody biomass undergoes three distinct transitions. Firstly, wood is dehydrated rapidly before temperatures exceed 200 °C, at which point pyrolysis is initiated and char and vapours are produced as previously detailed. At this stage, the presence of O_2 partially oxidizes the char and produces CO and CO_2, while the vapours are combusted to form CO_2 and H_2O. After further chemical reactions, a gaseous mixture of CO, H_2 and CO_2 is recovered, this is then known as syngas. Although crude syngas contains impurities such as tars, ammonia and sulphur, these compounds can be condensed or chemically removed. The process of gasification, when applied to wood, gives an energy conversion rate of approximately 60%.

Syngas can be directly combusted for electricity generation but it is more common for it to be purified, as then almost any hydrocarbon compound may be synthesized from it, including premium fuels such as methanol, ethanol and methane. The first stage in this synthesis is to adjust the proportions of H_2 and CO to the ratio required in the desired product. For example, methanol is CH_3OH, and therefore needs two H_2 molecules for each CO converted. Using the **Fischer-Tropsch** process, named after the chemists who developed it in the 1920s, syngas from biomass has been successfully converted to synthetic diesel. Additionally, the hydrogen which can be isolated from syngas has been used to power hydrogen **fuel cells** for the generation of electricity and to power electric vehicles (E4tech, 2009). However, these purification steps are costly and energy-intensive, and efficiency needs to be improved if they are to become widely used. In addition to the challenges which face purification of syngas, the implementation of gasification is limited because of the initial feedstock requirements. Moisture and ash content, homogeneity, particle size, bulk density and energy content must all be tightly controlled if gasification is to be efficient and this has limited uptake of the technology. In 2010, only 0.5% of worldwide gasification relied on biomass, with the rest using coal, petroleum, natural gas and petcoke as feedstocks (NETL, 2013).

5.7 Environmental assessment of bioenergy

As we saw in the Introduction to this chapter, the world's biomass plays a very basic role in maintaining the environment and providing our food, animal feed and fibre. However, it is important to consider not only the *benefits* of using it for bioenergy but also the possible *deleterious* effects – global or local – of the harnessing of these natural processes. The following section concentrates on some of the more significant benefits and impacts of bioenergy, considering first atmospheric emissions and then land use and energy balance.

Atmospheric emissions

Emissions to the air, both locally and on a wider national or international scale, are a major concern for all energy systems, especially those involving combustion. Carbon dioxide, nitrous oxide and methane are greenhouse gases associated with climate change, while other oxides of nitrogen and sulfur are also significant air pollutants.

Carbon dioxide

The concept of 'fixing' atmospheric CO_2 by planting trees on a very large scale has attracted much attention. Halting deforestation would bring many environmental benefits, as would replanting large areas with trees. A new forest plantation absorbs atmospheric CO_2 while the trees and forested areas mature, taking between 30 and 300 years depending upon species. Following this CO_2 emissions by respiration may balance photosynthetic absorption, or the forest may decline, becoming a modest net source of CO_2 (see, for example, Tang et al. 2014 and references therein). A wider bioenergy strategy, concentrating on the substitution of biomass fuels for fossil fuels, is arguably a more effective lasting solution.

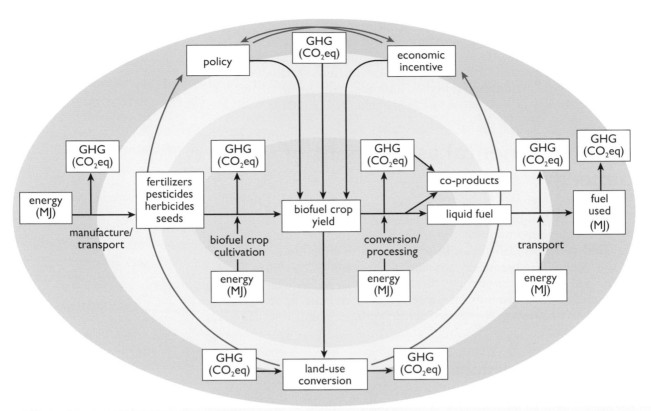

Figure 5.23 Different system boundaries for the life-cycle analysis (LCA) of biomass energy systems (Davies et al., 2009) (GHG = Greenhouse Gases, measured in terms of the amount of CO_2 that would have an equivalent greenhouse effect, CO_2eq)

To analyze the benefits of substitution, it is essential to assess all the effects in a **life-cycle analysis (LCA)**. This involves considering all the inputs and outputs of carbon dioxide from the whole bioenergy system, including growing the biomass, its conversion to useful product(s), transport (if necessary) to the site of end use, and effects associated with this end use. Figure 5.23 shows one possible interpretation of this, using various possible **system boundaries**. So, we could analyze the inputs and outputs of just the bioenergy crop system (the innermost coloured area in Figure 5.23), or we could widen the boundary to include the effects involved in producing the inputs for the crop, and in using the biofuel product. The widest boundary includes the changes in carbon dioxide emissions that may result from land use changes caused by the use of biomass for energy, but there is also 'credit' for the CO_2 *removed* from the atmosphere by the growing crop.

A life-cycle analysis tool of this kind, the Biomass Environmental Assessment Tool, was previously developed for use in bioenergy sustainability reporting in the UK (see, for example, Matthews et al., 2014), although this has now been superseded by the sustainability reporting criteria administered by the government regulator Ofgem.

In the introduction to this chapter, it was suggested that the utilization of plant biomass for energy should add no more carbon dioxide to the atmosphere than would have been formed in any case by natural processes as long as consumption does not exceed the natural level of production, estimated over appropriate temporal and spatial scales. Such assumptions about carbon budgets and accounting methods have been challenged by claims that, while ultimately renewable, the production of many biomass feedstocks (especially wood fuels from forest timber) would take decades to replace the carbon released in their combustion - and that such a 'carbon deficit' does not fit the urgent timescale for mitigating climate change (Searchinger et al., 2009; Haberl et al., 2012: RSPB et al., 2012).

However, in a well-managed system of forestry rotation, every tonne of biomass that is harvested and burnt will be balanced by a tonne of new biomass that is growing somewhere else nearby. If the rate of harvest is at least equalled by the rate of regrowth, when averaged over the whole supply region and measured on a timescale appropriate to forestry practice (e.g. 1–5 years), the carbon deficit may be non-existent or even reversed. Suitable data may be available from national forest inventories to assess whether overall standing biomass stocks are indeed static or increasing across a region. For annually harvested agricultural energy crops, the appropriate measurement period for regrowth would be just 12 months. The 'counter-factual' that forest regrowth alone would achieve a greater reduction in atmospheric CO_2 than the use of sustainably managed bioenergy feedstocks does not hold true, except under extreme and unlikely scenarios (Mackey et al., 2013). In reality, many biomass feedstocks (e.g. residues, thinnings, break crops) are co-produced alongside timber and food products of higher value, under the operation of diverse commercial markets.

Table 5.8 shows the emissions of carbon dioxide and also of the two main contributors to **acid rain**: sulfur dioxide and nitrogen oxides. The data is *life-cycle emissions per unit of electrical output*, taking into account all the processes involved. Even the best systems are not entirely carbon-neutral, but all the bioenergy systems, even MSW combustion, have lower CO_2 emissions than any of the fossil fuel pathways.

Table 5.8 Net life cycle gaseous emissions from electricity generation systems in the UK

	Emissions[1]/t GW h^{-1}		
	CO$_2$	**SO$_2$**	**NO$_x$**
Combustion, steam turbine			
Poultry litter	10	2.42	3.90
Straw	13	0.88	1.55
Imported wood pellets	50	0.10	1.58
Forestry residues	29	0.11	1.95
MSW (EfW)	364	2.54	3.30
Anaerobic digestion, gas engine			
Sewage gas	4	1.13	2.01
Animal slurry	31	1.12	2.38
Landfill gas	49	0.34	2.60
Gasification, BIGCC[2]			
Energy crops	14	0.06	0.43
Forestry residues	24	0.06	0.57
Fossil fuels			
Natural gas: CCGT	446	0.0	0.5
Coal: with minimal pollution abatement	955	11.8	4.3
Coal: Flue Gas Desulfurization and low NO$_x$[3] burner	987	1.5	2.9

[1] Note that 1 g kWh^{-1} is the same as 1 t GWh^{-1}

[2] Biomass Integrated Gasification Combined Cycle. Wood is gasified and the gas used to feed a gas turbine to generate electricity, with the hot exhaust gas used to raise steam to power a further, steam turbine.

[3] Flue gas desulfurization is a process for removal of sulfur compounds after combustion, and special forms of burners can be used that minimize emissions of oxides of nitrogen.

Methane

Methane is a significant greenhouse gas and is produced from the anaerobic breakdown of biomass (either naturally or by human intent). A molecule of CH_4 is about 28– 36 times as effective as a molecule of CO_2 in trapping the Earth's radiated heat, modelled over a 100-year period of progressive breakdown in the atmosphere. It is therefore said to have a Global Warming Potential (GWP) of 28–36. Collection of the methane released from anaerobic manure and slurry stores, and from landfills, and subsequent combustion effectively replaces each CH_4 molecule by a CO_2 molecule. The combustion of landfill gas was estimated to have reduced UK greenhouse gas emissions by the equivalent of some 70 Mt of carbon dioxide in 2002. Without this, total UK greenhouse gas emissions in that year would have been more than 10% higher. Likewise, the deployment of on-farm AD plants, processing about one-fifth of UK agricultural manures and slurries, could potentially reduce agricultural methane emissions by 0.6 Mt (NFU, 2017).

Table 5.8 shows that life-cycle CO_2 emissions are higher for Energy from Waste plants than for landfill gas combustion, although with careful storage and efficient furnaces the methane emissions from MSW combustion should be low. With landfill, it is rarely possible to collect all the gas and there are inevitably 'fugitive' methane emissions added to the atmosphere. Depending on the collection efficiency, these could add the equivalent of another 100–200 g kWh^{-1} to the actual CO_2 emissions shown in Table 5.8.

Nitrous oxide

Nitrous Oxide (N_2O) is a powerful greenhouse gas with a 100-year GWP of between 265 and 298. It is produced when nitrogen compounds in the soil, such as nitrates, urea and ammonia, are broken down by microbial action in an environment that is anoxic (lacking oxygen). It is also produced during the manufacture of some chemical fertilizers, and from the storage of manure. This greenhouse gas is an inevitable consequence of the soil nitrogen cycle and is stimulated whenever either organic or chemical fertilizers are used to grow biomass, for food or non-food purposes. It is generally estimated that 1% of the nitrogen applied is emitted as N_2O, with another 1– 2% being emitted during the production of the chemical fertilizer or the storage of manure.

Other emissions

Nitrogen oxides (NOx) are an inevitable product of the combustion, in air, of any fuel, because four-fifths of the air is nitrogen. High temperatures – in furnaces or internal combustion engines – increase NO_x production, and bioenergy systems need to meet the same 'clean-up' requirements as those using fossil fuels. This also applies to the removal of particulates. The emissions of sulfur oxides depend on the sulfur content of the biomass fuel, which will vary with the particular characteristics of the feedstock concerned. Many biomass feedstocks contain very little sulfur, so emissions of SO_2 may be greatly reduced compared with potentially high-sulfur fossil fuels such as coal and oil. There may also be liquid effluents, from flue gas cleaning, for instance, that require further treatment.

Land use

The production of most biomass feedstocks requires a certain area of land for plant growth, and it has been suggested that using this land for non-biological forms of renewable energy instead of biomass may do more to mitigate CO_2 emissions (Smith et al., 2000).

Consider, for instance, the electricity needs of 5000 households (such as a small European rural town), i.e. about 20 gigawatt-hours per year (GWh/year) of electricity, corresponding to the output of a small 3 MW biomass thermal power station operating with a 75% 'load factor', or its equivalent in other renewable energy technologies.

Assuming a modest energy conversion efficiency of about 20%, the fuel needs are 100 GWh per year. This is about 20 000 tonnes/year of oven-dry biomass or 34 000 tonnes 'as received'; i.e. the yield from 1340 hectares of miscanthus at 15 t/ha (dry matter basis) or 1800 ha of short-rotation coppice willow at 11 t/ha. A combined heat and power plant of this size should also be able to provide something like 50 GWh/year of useful heat to industry, or for domestic and commercial buildings.

This much electricity could also be generated by 3 MW of AD plants, for which 37% conversion efficiency is more typical, thereby reducing the fuel energy input required. Based on current yields and operating experience in the south of England, about 1200 hectares of land would be required to supply the total feedstock needs in the form of maize or rye, probably in rotation with other arable crops. The net land area required could be just 600 ha if 50% of the feedstock needs were met by manures, slurries, crop discards or food waste. Co-production of useful heat could be as much as 20 GWh/year (increased conversion to electricity leading to a lower heat figure than for the small thermal combustion plant).

The same electrical output would be achieved by a small 10 MW wind farm with a 23% load factor, which would have a 'footprint' of about 250 ha, based upon typical separation distances between turbines and their distance from nearby dwellings. The actual concrete footprint of the foundations and roadways would be a small fraction of this, with agriculture, wildlife or amenity use of the land in-between.

20 GWh/year is also the output of a 24 MW solar PV farm in Britain with a load factor of 9.5%, which would occupy about 60 ha based upon typical layout. Many solar farms in the UK also provide grazing land for sheep and/or agri-environmental features to create wildlife habitats.

In summary, the same annual amount of electricity could be generated using roughly 1500 ha for perennial energy crops, 600–1200 ha for AD feedstocks, a wind farm of 250 ha, or just 60 ha for a solar farm. The wind and solar options produce electricity intermittently (albeit in a reasonably predictable manner), while allowing multi-purpose land use; in contrast, the bioenergy plants generate dispatch-able power and heat but leave little space for other use of the land. In reality, these differing renewable energy options are more likely to be distributed throughout the landscape rather than competing for the same land, and they could even overlap (for example, miscanthus or maize could be grown within a wind farm).

In the case of 'first generation' transport biofuels from food crops such as maize and oilseeds, there have been widespread concerns about the diversion of crops for biofuel rather than for food. With the explosion of interest in the US use of maize for ethanol production in the first decade of this century, it was claimed that this had increased the price of maize flour as a staple food in Mexico, leading to food riots. Subsequently it turned out that the shortage of maize flour was worsened by hoarding and speculation by investors, but this case highlighted the tension over 'food versus fuel'. Since then, in a significant shift of opinion, the Director-General of the UN Food and Agriculture Organization, Jose Graziano da Silva, stated in 2015 that 'biofuels should be seen as a key part of the global agriculture complex...we need to move from the food versus fuel debate to a food and fuel debate.'

The concept of 'Indirect Land Use Change' (ILUC) is another possible impact. It has been suggested that ILUC could occur through a process whereby growing energy crops on agricultural land may displace existing food-crop production, causing land use change in another location (Searchinger et al., 2008). There has been much debate since about the methods and underlying assumptions used to estimate ILUC, recognizing that land use change is very complex and affected by a wide range of factors, including agricultural markets and farmer decision-making. Some authorities have cast doubts on the validity of ILUC models. A recent publication backed by the International Food Policy Research Institute (IFPRI) concluded that food and energy security are complementary goals and challenged the idea of 'food vs fuel' competition for resources (Kline et al., 2016). Likewise, in 2015 the European Commission concluded that the 2020 EU target for 10% renewable energy in transport would not significantly impact global food prices and food affordability in developing countries.

A further environmental concern is a reduction in biological diversity through conversion of existing vegetation to fuel crops. In theory, loss of biodiversity from land use change could be exacerbated by the use of pesticides in bioenergy crops, although such crops generally do not require such rigorous quality control as food crops. For example, Field et al. (2008) suggested that, to preserve biodiversity, only land that has been abandoned from agricultural use should be used for bioenergy production. However, some bioenergy systems such as short rotation forestry or coppice or energy grasses can actually increase biodiversity compared to conventional agriculture (Rowe et al., 2011; Milner et al., 2015; McCalmont et al., 2015). Furthermore, the impact on biodiversity of forest residue removal may be less than previously suspected (Fritts et al., 2016).

Effects on soil also need to be considered. Soils can contain large amounts of organic matter or humus, formed from the recalcitrant remains of vegetation that are hard to break down, including material such as lignin. The humus acts as a store of large amounts of carbon over long periods if the soil is undisturbed (tens to hundreds of years). Increasing the proportion of stable carbon-containing material in soils has been suggested as one means of reducing atmospheric CO_2 levels. Soil organic matter is also essential to the maintenance of **soil structure**. The humus fraction of soil includes gums and other compounds that bind together smaller soil particles and improve the drainage characteristics of the soil as well as acting as a store of plant nutrients. However, cultivating the soil exposes the organic material to the air, and it then begins to break down more rapidly, releasing carbon dioxide. This loss of carbon is more of a problem with annual crops than with perennials such as trees and grasses, because of the higher frequency of soil disturbance. This may be partly addressed by newer, reduced-tillage forms of cultivation. The emerging national and international sustainability standards for liquid transport fuels and solid bioenergy incorporate measures to exclude the use of land with previously high carbon stocks, whether in the form of soil organic matter (e.g. peat soils) or in vegetation (typically forest).

Another idea that has received considerable attention recently is the use of biochar, potentially produced as a co-product of pyrolysis or gasification of biomass. Rather than using this char as a fuel, it can be added directly to soils. There are claims that this both acts as a long-term store of carbon and improves soil fertility. The value and practicability of this approach are subject to continuing research (Shackley et al., 2011). However, at present there is little financial incentive to store carbon from biochar rather than to use it as an energy source.

Energy balance

The terms **energy balance**, **energy use efficiency**, **energy payback ratio**, or **fuel energy ratio** are used to describe the relationship between the energy output of a system and the energy inputs needed to operate it (usually from fossil fuels). The concept came to the fore when doubts arose concerning some of the early fuel-from-biomass projects introduced following the oil price increases of the 1970s. There were claims that, when a full life-cycle analysis was undertaken, the fossil-fuel energy input for some schemes was actually greater than their bioenergy output.

The ratio of output to input will of course depend on the type of system and the extent of the processing involved. In particular, ratios will normally be lower if the final 'output' is electricity because of the inherent limit on the Carnot efficiency (see Box 2.4) of heat engines used in generation.

Davies et al. (2009) and Hastings et al. (2012) examined the fuel energy ratio (FER) for a selection of bioenergy pathways reported in recent literature (Table 5.9). The FER is the ratio of the useful energy in the fuel to the total amount of fossil fuel used in producing that fuel, including that used to construct equipment. If this ratio is less than one, then more energy is used to make the bioenergy product than its fuel value. FER for transport biofuel production varied from over 5 to less than one, but Davies et al. (2009) also make the point that there is little consistency among LCA estimates of biofuel energy efficiency.

Over the full range of renewable sources, the energy ratio of output to input can vary from less than 1:1 to as much as 300:1 (this for some hydroelectric plants). Woody energy crops perform well, with ratios between 10:1 and 20:1 on a heat output basis, but biodiesel may achieve only 3:1. Ethanol from grain may barely break even, at just over 1:1 in the worst cases.

When wastes or residues are the input, the questions arise of how much of the energy input to the whole process should be attributed to the energy extraction system as opposed to alternative disposal pathways for wastes, and how energy should be allocated between primary products (such as food or timber) and crop or forest residues. This can result in high FER values, e.g. 30:1 typically for electricity from woody sawmill wastes.

Table 5.9 The range of fuel energy ratios for selected bioenergy systems reported in the literature to 2016

	Fuel energy ratio	
	Lowest	Highest
Lignocellulosic alcohol (generalized)	0.75	1.24
Switchgrass (combustion)	0.44	4.43
Corn (alcohol)	0.9	1.14
Miscanthus (rhizome to pellet combustion)	1.77	9.23

Only one set of data for miscanthus systems was available.
Source: derived from Davies et al., 2012 and Hastings et al., 2012

Energy ratios can be improved by good design of the production and processing systems to ensure that no energy is wasted. Co-products such as the straw or bagasse associated with biofuel feedstocks can be used to replace fossil fuels such as coal or natural gas in supplying process heat. The energy balance of a biomass energy system also affects its environmental impact. The greater the final energy outputs, the greater the quantity of fossil fuel displaced. The lower the fossil fuel inputs, the lower the extra demands put upon the environment by the biomass system.

5.8 Economics of bioenergy

Bioenergy is generated from a number of different feedstocks, processes and fuels with several established international bioenergy markets (Figure 5.24). The energy production costs of each of these options vary considerably and bioenergy economics is chiefly concerned with these costs and how they compare to fossil fuel energy costs. This is closely connected to the role of government policies which strongly influence the economic viability of bioenergy systems and are present in most countries generating bioenergy. The economics of bioenergy is also determined by barriers to market entry, with factors of feedstock availability, accessibility of conversion processes and market maturity varying across regions. Finally, energy and bioenergy economics concern externalities and whether and how these can be addressed. These four determinants of the economics of bioenergy will be discussed broadly first, before being applied to case study examples of existing bioenergy technologies in the UK and elsewhere in the world.

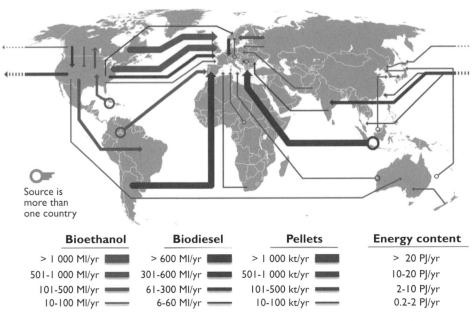

Bioethanol	Biodiesel	Pellets	Energy content
> 1 000 Ml/yr	> 600 Ml/yr	> 1 000 kt/yr	> 20 PJ/yr
501-1 000 Ml/yr	301-600 Ml/yr	501-1 000 kt/yr	10-20 PJ/yr
101-500 Ml/yr	61-300 Ml/yr	101-500 kt/yr	2-10 PJ/yr
10-100 Ml/yr	6-60 Ml/yr	10-100 kt/yr	0.2-2 PJ/yr

Ml = Million litres; kt = Thousand tonnes; PJ = Peta joules (1 Exa joule = 1 000 Peta joules)
Source: HLPE (2013)

Figure 5.24 Net trade flows of wood pellets, biodiesel and ethanol in 2011 source: HLPE (2013)

Bioenergy production costs

Energy is not one single commodity and is generated from a variety of sources, which are often good substitutes for each other, based upon price and availability. The ability to substitute energy sources is particularly evident in the electricity sector, where the end user's supply of power could have been generated from a wide variety of fossil fuel or renewable energy sources. For example, if a power station that co-fires biomass and coal experiences an increase in the price of coal, it may replace more coal in its energy mix with biomass, thus limiting the rise in input costs. In the USA, the blending of bioethanol from maize with gasoline from crude oil is another example where fuels can be substituted to different extents. When the oil price is high more ethanol can be blended into gasoline in order to minimize the cost to vehicle fuel users. The ethanol itself can be produced from several different feedstocks, each with differing costs; when using maize the minimum oil price that US ethanol producers require in order to achieve a reasonable margin is estimated at around US$50-$60 per barrel (Nersesian, 2016) whilst a price of over US$100 per barrel is required if using cellulosic (second generation) feedstocks (Tyner, 2008). As liquid biofuels are most often used to substitute for petroleum oil products, it may be argued that the greatest risk to the biofuel industry is low oil prices.

The four main factors that determine the cost of energy from any system are described in Appendix B. For most renewable energy systems, the initial capital cost (including the cost of borrowing the money) is a major component. Unlike many other renewable energy technologies, bioenergy systems can also have significant fuel costs.

The remaining two factors are common to most energy systems. Operation and Maintenance (O&M) costs are usually proportional to the output of the plant, and will depend on the type of fuel – in particular, the nature of its emissions and the residues it leaves. It is usually assumed that the decommissioning costs of bioenergy plant will be covered by the scrap value of equipment. It is also worth noting that for any electrical plant, the cost per kWh of output depends on the annual output, so it is important to maximize the 'load factor' or 'capacity factor' (Box 6.1 in Chapter 6).

The fuel costs for energy crops, for example, include planting them into prepared land, fertilizing, protecting against weeds and pests, harvesting and transporting. This cost may also be considered as the opportunity cost: the value that could have been achieved through using that land for some other agricultural purpose, such as food production. In the case of 'first generation' energy crops that opportunity cost may be quite high if the land is of the best and most versatile quality, but with 'second generation' energy crops, which may be produced on poorer, more marginal land, the opportunity cost is likely to be lower. Meanwhile, the feedstock costs associated with generating bioenergy from waste and by-products such as agricultural and forest residues may be very low and occasionally negative, depending upon alternative commercial applications and disposal costs.

Bioenergy policy support

Governments across the world operate policy mechanisms in order to support the substitution of fossil-fuel energy for renewable sources. Intervention in the market is justified for various reasons, chiefly because the fossil fuel energy systems produce external costs of pollution which are not borne by the polluter, whilst alternative renewable energy sources such as bioenergy can avoid these costs. However, markets for renewable energy sources are relatively immature and at present they require support in order to compete with fossil fuel energy. The IEA has stated that for as long as the external costs of fossil fuels are not accounted for, bioenergy systems will need some degree of financial assistance (IEA, 2012).

A number of different policy mechanisms are implemented in order to promote bioenergy. Some policies fund research and development of the biofuel or bioenergy technology (technology-push policies), whilst others directly intervene in the market to stimulate demand for the biofuel or bioenergy technology (market-pull policies). Market-pull policies may comprize feed-in tariffs and premiums; quota obligations (mandates) and tradable renewable certificates; incentives to invest in generating plants; and disincentives to non-renewable energy (Bahar et al. 2013).

Barriers to market entry

As well as production costs, and incentives and subsidies provided by government policy, the economics of bioenergy are also determined by factors external to costs and prices. Some fuels may be more suited to a particular end-use than others. For example, around 70% of sub-Saharan African households depend upon wood-based bioenergy for their cooking needs (Nersesian, 2016), particularly in rural areas where the cost of extending the electricity grid would be prohibitive. Meanwhile, in the UK, the production of second-generation perennial energy crops is thought to be held back not by the cost of production but by a lack of available market or local demand for the feedstock, which deters farmers from committing to investing in the crop (Sherrington and Moran, 2010). There is a perceived risk when investing in such a developing and immature sector, even if the potential economic returns appear competitive (Alexander & Moran, 2013). The risks may be a function of uncertainty over government policy, which this form of bioenergy relies upon, uncertainty and unfamiliarity about the technology, and also the reduced versatility of dedicated energy crops which cannot be diverted to food markets if bioenergy prices fall.

Externalities

An externality is a cost or benefit borne by a party which did not choose to receive that cost or benefit (Buchanan and Stubblebine, 1962). For the affected party, the externality can either have a positive or negative effect. Pollution from fossil fuel combustion is a negative externality, a social cost borne by the affected party which is not reflected in the prices charged for the provision of the fossil fuel. Because the costs of pollution are not reflected in market prices for fossil fuels the negative externality will be over-provided, i.e., provided above the socially efficient level. Negative externalities tend to be over-provided whilst positive externalities tend to be under-provided, unless they are 'internalized'. Environmental economists argue for the 'internalization' of externalities through mechanisms which put prices on the cost or benefit that is provided. Thus the externality caused by pollution from a fossil fuel power station can be internalized by imposing a tax on emissions. Policies which pay for ecosystem services are an example of internalizing positive externalities. Carbon taxes and emissions permits in a number of countries have begun internalizing some of the negative externalities associated with fossil fuels, but there is much progress to be made yet and the fossil fuel industry continues to receive large subsidies in many parts of the world (Bast et al, 2015).

Bioenergy systems and the policies which support them are associated with their own externalities. Researchers disagree over the exact impact of bioenergy on food prices (see Section 5.7 above), but the increased use of maize to produce ethanol in the US has been associated with price links between corn, ethanol and oil, possibly causing food price volatility (Wisner, 2009). However, no such price link was found between second-generation bioenergy feedstocks and oil (Alexander and Moran, 2013). It has also been argued that higher food prices could be a positive development by increasing the incomes of poor farmers worldwide and stimulating agricultural investment (Wisner, 2009).

Ecosystem services

Ecosystem services are the direct and indirect benefits that humans derive from ecosystem functions (Costanza et al., 1997). Many of these ecosystem services are positive externalities because they are not internalized into the market, they do not have a price, and they are typically under-provided by land managers. Second-generation energy crops such as Miscanthus and short-rotation coppice (SRC) provide a number of ecosystem services including carbon sequestration, higher levels of biodiversity, and improved soil quality (Holland et al, 2015; Milner et al, 2015). Such ecosystem service benefits are most positive when intensively managed arable land is converted to second generation bioenergy crops, which require less intensive management and reduced inputs such as fertilizer. Policies to support 'payment for ecosystem services' have been implemented by a number of governments around the world and could be applicable to bioenergy feedstock production.

Case studies

In order to demonstrate the economics of bioenergy in different scenarios, the above economic considerations of energy production costs are applied below to case studies in four bioenergy systems; 1) wood fuel; 2) waste and by-products 3) first-generation biofuels; and 4) second-generation biofuels.

Wood fuel in Africa

Some regions of the world are particularly reliant on bioenergy because no alternative energy sources are available. An estimated 1.5 billion people are not connected to an electricity grid and, as stated above, around 70% of households in sub-Saharan Africa reply upon wood fuel for cooking (Nersesian, 2016). Whilst the lack of alternative energy options has led to the widespread use of biomass in these regions, bioenergy from wood fuel and charcoal is also associated with relatively low costs. The fuel is either freely collected or inexpensive to purchase, and domestic cooking and heating systems require relatively basic infrastructure. Particularly an issue in many African countries, government renewable energy policies are either lacking or relatively inconsequential to the overall economics of wood fuel and charcoal, although aid funding and non-government organization initiatives are often targeted at supporting sustainable forestry initiatives and the efficiency of cooking equipment (see also Section 5.6 above).

Whilst the monetary costs associated with wood fuel burning in Africa are relatively low, the opportunity cost of collecting fuel does reduce time spent by people in other income-generating activities, or in education. In addition, the labour required to collect heavy loads of wood fuel can often be very strenuous, bringing health and wellbeing concerns (IEA, 2006). Concern has also been raised regarding the unsustainable use of wood fuel in parts of Africa (SEI, 2008), as well as the respiratory problems associated with burning wood-fuel without sufficient ventilation. Whilst these costs are not borne by the market (they are externalities), they are clearly important and need to be considered in government energy policy and international development assistance.

Wood fuel in the UK

Substantial quantities of wood fuel are also burned in large-scale electricity and combined heat and power (CHP) plants in North America and northern Europe. Biomass represented over 5% of UK electricity supply in 2015, the majority of which comprized wood pellets imported from North America. Despite the need for transportation across the Atlantic, they are a relatively inexpensive fuel, typically made from low-value thinnings, pulpwood and residues from the southeast US forestry industry. Wood pellets exported to Europe accounted for under 3% of total harvest removal volumes in the southeast US in 2014 (Dale et al., 2016). Set against these manageable fuel costs are the significant capital costs associated with building dedicated biomass power plants, or converting existing coal plants to biomass.

The combustion of wood pellets in UK power stations is supported by Renewable Obligation Certificates (ROCs), which are issued by the government per unit of renewable electricity generated, although the scheme was closed to new entrants in 2017. Introduced in 2002, the effect of this mandate mechanism has been to increase the value of energy produced from biomass and other renewables relative to fossil fuels. More recently, the UK government has committed to banning coal combustion in power stations by 2025. This policy may further stimulate demand for bioenergy, assuming that some coal plants choose to continue operating beyond 2025 by investing in conversion to biomass combustion.

Bioenergy from wastes in the UK

In Britain, some 'residual waste' from commercial and household sources is combusted to generate electricity or CHP. Agricultural and food wastes are more typically processed through AD to produce biogas. Waste incineration plants use a feedstock which has little or no alternative economic value; Energy from Waste (EfW) plants charged 'gate fees' of up to £131/tonne in 2015/16, marginally less than the fee for alternative disposal to landfill where up to £145/tonne was charged (WRAP, 2016). However, in order to meet strict environmental standards modern EfW plants are expensive, with capital cost estimates in the range of £145–£200m (Defra, 2013). In 2016, 36 such plants were in operation across the UK, and a further 7 approved for construction.

AD plants are much smaller in capacity than EfW installations, with capital costs typically in the range £0.5–10 m. The UK Government recently assumed a median cost of £4.4 m per MW of installed capacity (GIB, 2015). The feedstocks used in AD plants vary from agricultural and commercial organic waste to arable crops such as maize silage. The number of AD plants in the UK has increased dramatically in recent years, and the market for source-segregated food waste has become more competitive as capacity has increased. The result has been for gate fees to decline and an increased use of alternative feedstocks (GIB, 2015).

The British government's progressive introduction of landfill taxes as well as subsidies for renewable energy generation has incentivized the economics of generating energy from waste (EfW). UK and EU law requires that biodegradable municipal waste be diverted from landfill and, as such, significant landfill taxes apply. The Waste Infrastructure Grant programme provides some assistance towards capital expenditure for energy from waste infrastructure, and CHP incinerator plants are also eligible for Renewables Obligation Certificates (ROCs). AD plants in the UK are supported by renewable energy incentives in the form of ROCs or alternatively the Feed-in Tariff scheme for installations up to 5 MW.

Whilst environmental regulations are aimed at reducing landfill volumes, which carry externalities of their own including increased greenhouse gas emissions, there are also externalities (such as public health concerns) associated with EfW plants. However, Public Health England (PHE) has found the contribution of emissions from EfW plants towards health problems to be very small (Defra, 2014). Nevertheless, public acceptability can be an important determinant and whilst a smaller number of larger plants may be more cost-effective the public often values other factors more highly. For example, residents in Hampshire in southern England recently rejected the construction of one large incinerator in favour of three smaller ones (Defra, 2013).

Ethanol fuel from crops in the USA

In 2015, 14.8 billion US gallons of bioethanol were produced in the United States, up from just 1.6 billion gallons in 2000 (RFA, 2016). The US government has actively encouraged the bioethanol industry in order to promote energy security, protecting the economy from oil price shocks. Mostly produced from corn (maize), American bioethanol is blended with petroleum-derived gasoline and so the profitability of its production is closely connected to oil prices. Under a high oil price scenario it is estimated that bioethanol output would be sustained regardless of government incentives (Tyner, 2008). With reference to the US$50–60 per barrel break-even price described above, note that ethanol production increased by 43% year-on-year in 2008 when oil prices peaked at an average of over $90 per barrel (Figure 5.25). In contrast, when gasoline prices fell by around 50% year-on-year in 2015 to an average of US$42 per barrel, US bioethanol output edged up by just 3% (RFA, 2016).

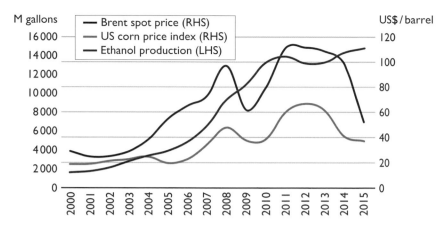

Figure 5.25 Oil price vs ethanol production in the US

The US government supports bioethanol production through a range of tax breaks and infrastructure measures, as well as the Renewable Fuel Standard (RFS) which stimulates demand through mandating the use of bioethanol in transport fuel. However, some argue that a developed industry such as US corn ethanol should no longer receive government support, and in 2015 the US government set lower than expected increases in the RFS mandate (EPA, 2015; Farm and Dairy, 2015).

'First generation' biofuels use crops for which there is already an established agricultural market as well as expertize and familiarity. Farmers can be seen as enjoying a 'win-win' situation because it makes little difference to their land management decisions whether they sell their crops for food or energy production. Such flexibility is foregone in the case of 'second generation' biofuel feedstocks, which typically have a more restricted range of end uses.

The most often discussed externality associated with first generation biofuels is their impact on food prices and food price volatility, but it remains inconclusive exactly what the influence of the corn ethanol industry has been on US domestic and international food prices. Some argue that other factors such as variation in crop yields and rising global demand for all agricultural products are more significant, pointing to recent divergent movements in food prices and ethanol production. Others argue that the bioethanol industry has established price links between corn, oil and ethanol which has increased the volatility of food markets (Wisner, 2009). Another important externality concerns the carbon balance of corn ethanol. The high energy inputs to feedstock production and processing have led some to question whether net greenhouse gas emissions are reduced at all relative to displaced fossil fuel (Lynd et al. 2006; Wilhelm et al. 2007).

Perennial energy crops in the UK

Much current research and development is focused on 'second generation' biofuels, which represent a small and immature bioenergy sector. Second generation bioenergy crops, or dedicated energy crops, such as Miscanthus, short-rotation coppice (SRC) and short-rotation forestry (SRF) can deliver high yields without intensive energy inputs. These crops may be grown on land where they are not directly competing with arable crops for food (Lemus and Lal 2005), whilst also delivering positive externalities in the way of ecosystem services.

Perennial energy crops of Miscanthus and SRC, which accounted for just 1.3% of UK arable land in the UK in 2014 (Defra, 2015a), are used as feedstock in electricity and CHP power stations or heating systems. Both of these crops are able to deliver relatively high yields on low grade and marginal agricultural land, resulting in a reduced opportunity cost compared to growing them on higher grade agricultural land. Under these circumstances, they can often offer improved returns for farmers compared to other arable crops. Miscanthus and SRC have been estimated to sell in the region of £60 odt^{-1} (NNFCC, 2010; Sherrington & Moran, 2010) and £40 odt^{-1} (Aylott et al., 2010) respectively, although these crops involve relatively high establishment costs and require specialist machinery for harvesting. After establishment, further input costs are relatively low with minimal fertilizer required (Aylott et al., 2008). See also Table 5.10 below.

Note: odt = oven-dry tonnes, i.e. fuel with minimal moisture content

To offset the relatively high establishment costs of perennial energy crops, the Defra-funded Energy Crop Scheme (ECS) subsidized up to 50% of these costs (initially up to £1,000/ha) from 2000 until the scheme was closed in 2013. The combustion of dedicated energy crops in either small-scale or large commercial heating or electricity systems is also supported through the Renewable Heat Incentive (RHI), Feed-in Tariff (FIT) or ROCs. Significant funding is currently being directed towards research into developing second generation biofuels, such as through the European Industrial Bioenergy Strategy Initiative , launched in 2009/10 with a budget of EUR €6–8 billion to fund demonstration plants with the goal of reducing production costs of biomass fuels to allow them to compete with fossil fuels.

Notwithstanding the reasonable long-term financial returns available, the modest level of uptake of perennial energy crops in the UK may reflect a numbers of barriers to their adoption, chiefly the lack of local offtakers and the resulting long distance for transport to available markets, the switch from familiar crop rotations to an unfamiliar one, and the overall perceived high risk (Alexander et al. 2014; Sherrington and Moran, 2010; Adams and Lindegaard, 2016; Lindegaard et al., 2016). Nevertheless they may be associated with long-term contracts which provide stable incomes, can be grown on land not otherwise suited to food production, and are independent of food crop prices (Alexander & Moran, 2013).

Well-managed second-generation energy crops can deliver significant ecosystem services in the form of improved soil quality, insect pollination and pest control, water quality, carbon sequestration, and biodiversity (Holland et al. 2015, Milner et al. 2015). Such positive externalities are not yet accounted for by the market or recognized by government policy. However, the UK government has agreed to protect and enhance the natural environment through a 25 year plan (Defra, 2015b). The concept of 'payment for ecosystem services' could be translated into domestic agricultural policy following the UK's departure from the EU and its Common Agricultural Policy, offering a potential opportunity to enhance the economics of perennial energy crops.

Table 5.10 Typical Miscanthus and SRC production costs

Item	Unit	*Miscanthus*	SRC willow
Establishment Cost	£ha⁻¹	1949	2183
Establishment Grant	£ha⁻¹	975	1092
Removal	£ha⁻¹	109	547
Fixed overhead	£ha⁻¹ yr⁻¹	95	95
Fertilizer	£ha⁻¹ per application	0	27
Harvesting Cost	£ha⁻¹ per harvest	219	141
Storage Cost	£ha⁻¹ per harvest	42	23

Source: Alexander et al. (2013) Note: energy crop establishment grants were discontinued in 2013

5.9 Future prospects for bioenergy

As described earlier (Section 5.2), bioenergy is the world's fourth largest source of energy and remains prominent among the renewables. For the European Union, as well as many other parts of the industrialized world, bioenergy is the principal source of renewable energy (as is the case within most EU member states), followed by hydropower, wind energy and solar power (WBA, 2016; Eurostat, 2014).

In the near future, much of the growth in renewable energy supply in many parts of the world may come from increased direct combustion of a wide variety of biomass feedstocks for heat and electricity generation, together with a substantial increase in the supply of 'first generation' liquid transport biofuels. In the longer term, newer biomass technologies, such as electricity from novel high-efficiency small power plants or fuel cells, and 'next generation' liquid biofuels, could play an increasing role – but by that time bioenergy might be overtaken in developed and emerging economies by faster-growing renewables such as wind and solar power. Some authorities predict that traditional use of biomass will be largely phased out by as early as 2030, to be replaced by modern renewables, although bioenergy will still account for half of all renewable energy use worldwide (IRENA, 2016). Others forecast that traditional biomass will remain the sole domestic fuel for 2.8 billion people in 2030 (IEA, 2010): this leads to the conclusion that the improvement of traditional wood and charcoal stoves should be one of the most important technical aims in the field of bioenergy today.

New sources of large-scale demand for biomass may also continue to be significant, in particular, expected demand for 'biocoal' formed by torrefaction or steam explosion, and pelleting of a wide range of biomass feedstocks (Thraen. et al., 2016). Europe and the USA are predicted to remain the dominant regions in the global biomass power generation market as it grows from 106 GW in 2016 to around 165 GW in 2025, with the UK's 2 GW of biomass capacity at Drax power station the largest plant worldwide for the time being (see also Kaltschmitt and Janczik, 2015). However, the Chinese 5-year plan for bioenergy development published in December 2016 included a goal for 2020 of 30 million tonnes per year of pellets, which could add significantly to the global pellet market over the next few years.

Biomass is also likely to remain an important source of process heat in certain energy-intensive industries around the world such as iron/steel and cement manufacture, as well as specific market segments for transport fuels (aviation, long-distance trucking, rural areas of emerging economies).

Table 5.11 summarizes the potential future global bioenergy contribution according to a wide range of authors. The most optimistic of these forecasts suggests over 1500 EJ per year, significantly above the total world energy use at the time of writing. A more widely accepted figure might be of the order of 100–200 EJ of bioenergy per year. As we have seen in Section 5.2 and Box 5.1, traditional biomass currently contributes at a minimum estimate 40 EJ (probably more) to world primary energy, with a further 9 EJ or so from modern bioenergy (mainly in the form of electricity). 200 EJ would therefore represent a 4-fold increase, and a new contribution equal to about a third of present world total primary energy consumption.

The future role of bioenergy in the UK energy system has been forecast by the Energy Technologies Institute (ETI, 2014), further to the current policy laid out in the Government's 2012 Bioenergy Strategy (DECC, 2012). Bioenergy could supply 10–12% of energy needs and is anticipated to be used for heat, power and transport as well as to create 'negative emissions' through carbon capture and storage (CCS). Both bioenergy and CCS are considered essential for the delivery of an affordable low carbon energy system in Britain.

Table 5.11 Potential future world roles for bioenergy

Source, date	Time frame (year)	Bioenergy contribution	
		EJ per year	Percentage of total global primary energy
UNDP[1], 2000	2050–2100	94–325	11–50%
Hoogwijk et al., 2003	2050	33–1130	5–190%
Moreira, 2006	2100	164	15%
Ladanai and Vinterbäck, 2009	2050	1135–1548	~200%
Beringer et al., 2011	2050	126–216	13 to 22%
IRENA 2014	2030	154–226	16 to 23%

[1] UNDP, 2000 contained estimates from a number of sources.

The future of bioenergy with carbon capture and storage (BECCS)

Coupling biomass or biocoal power plants to carbon capture and storage offers the prospect of carbon-negative electricity production, since the photosynthetic production of the biomass feedstock on a sustainable basis would remove carbon dioxide from the atmosphere. Following its first description in the scientific literature at the beginning of this century (Obersteiner et al., 2001), and after a decade or more of debate, BECCS was formally endorsed as a viable strategy by Working Group III of the UN Intergovernmental Panel on Climate Change in advance of the ground-breaking Paris Climate Agreement (IPCC, 2014). According to the IPCC, such 'additional strategies' may be required to actively remove CO_2 from the atmosphere and BECCS is a 'critical component' of most strong climate change mitigation scenarios.

The UK Energy Technologies Institute has recommended that the British government should support a demonstration of BECCS technology to allow 'negative emissions' to be delivered within the next decade (ETI, 2016). It claims that all the major components of a BECCS system have now been demonstrated or proven individually, making full deployment achievable by 2030. Its ESME energy systems modelling project concluded that BECCS could lead to the removal of up to 55 million tonnes per year of CO_2 emissions from the atmosphere, equivalent to half the UK emissions target in 2050, whilst meeting up to 10% of the UK's future energy demand. Similar conclusions about the potential of BECCS have been reached elsewhere, such as for electricity generation in the US western region (Sanchez et al., 2015).

Future perspectives on bioenergy feedstock production

In terms of the production of bioenergy crops, there are two main targets for improvement: (i) crop genetics and (ii) agronomic management practices. The first enables the development of better genotypes (varieties) of bioenergy crops, while the second optimizes the yields of these varieties. Improving bioenergy crops genetically relies on identifying key traits for improved quality and yield, which can then be selected for using either traditional or molecular breeding methods. Although biomass feedstock quality can have a significant impact on conversion efficiency and costs, yield is generally considered to be of greater importance when considering economic drivers for bioenergy production. If yield can be increased through sustainable intensification, then more biomass can be produced without expanding cropping area, transport costs can be minimized, and competition with food crops on better quality agricultural land will be reduced. Traditional breeding programmes have previously delivered substantial results in terms of bioenergy crop improvement with, for example, willow yields being doubled between 1975 and 2005 (Karp et al., 2011). Now, 'next-generation sequencing' technologies are rapidly advancing, which will enable molecular breeding approaches such as marker assisted breeding and genomic selection to be applied to traditional breeding programmes, in order to speed the pace of genetic improvements (Arnoult and Brancourt-Hulmel, 2015; Wang et al., 2016).

Yields of bioenergy crops are variable (Zegada-Lizarazu et al., 2010) and yield gaps exist (Allwright and Taylor, 2016), leaving room for substantial yield improvements. Given that agronomic management of energy crops is generally under-developed, this is a primary target for narrowing these yield gaps. In the first instance, crop suitability must be determined with agronomic management, climatic adaptability and biomass accumulation in mind. For example, in northern Europe, woody and grass perennials such as poplar, willow and miscanthus can be produced efficiently with high growth rates, whereas eucalyptus and giant reed (Arundo) are better suited to the Mediterranean climate of southern Europe. Furthermore, each bioenergy crop has its own set of requirements for growth including water and nutrition availability, weed and crop management and cropping patterns.

These two approaches to improving bioenergy yield are not stand-alone and show clear interactions. For example, varieties with high resource use efficiency may need to be developed in the future in order to optimize the carbon and energy balance of feedstock production.

5.10 Summary

Bioenergy is the general term for energy derived from organic materials such as wood, straw, oilseeds or animal wastes which are, or were recently, living matter, referred to collectively as **biomass**. Wood pellets, charcoal, bioethanol and biodiesel are all examples of energy-rich materials derived from biomass.

Material such as **firewood, rice husks and other plant or animal residues** can simply be burned in air to produce heat. In many developing countries this **traditional biomass** use continues to account for a large part of energy consumption for **cooking, heating and light**.

Biomass can also be burned to raise steam for **electricity generation**, or it can be converted into intermediate **bio-based fuels** such as **charcoal, biogas, bioethanol and biodiesel**. These many forms of **bioenergy can be used to replace conventional fossil heating fuels or to power some form of engine for motive power or electricity generation**.

Recent estimates of the annual contribution to world primary energy from traditional biomass fall in the range 40–60 EJ, dominated by Asian and African countries (IEA Bioenergy, 2009; WBA, 2016).

At a chemical level, **bioenergy refers to various types of carbohydrates**. The term carbohydrate refers to a compound consisting only of carbon, hydrogen and oxygen having a general form $C_x(H_2O)_y$ (x and y can be different but crucially there is a hydrogen:oxygen ratio of 2:1). The **yield** of a crop is the mass of (usually above-ground) biomass produced per hectare per year.

Biomass resources can be categorized as either **primary materials**, derived directly from plants, or **secondary 'wastes'**, the otherwise low-value, unwanted or unutilized co-products and residues from human activities.

These secondary materials can only be classified as biomass if they themselves have been produced from renewable plant-based materials.

All the raw bioenergy feedstocks discussed in the previous section require some kind of **processing and/or conversion technology to release their energy in a useful form**. The nature of this processing depends on the type of biomass and the energy product required. Upstream pre-processing may also be required before subsequent conversion technologies are employed. These processes of **densification, concentration and conversion** can be categorized broadly into **physical, thermochemical and biochemical**, which may be used individually or in combination.

To analyze the benefits of substitution of fossil fuels with bioenergy, it is essential to assess all the effects in a **life-cycle analysis (LCA)**. The utilization of plant biomass for energy should add no more carbon dioxide to the atmosphere than would have been formed in any case by natural processes as long as consumption does not exceed the natural level of production, estimated over appropriate temporal and spatial scales.

The production of most biomass feedstocks requires a certain area of land for plant growth and it has been suggested that using this land for non-biological forms of renewable energy instead of biomass may do more to mitigate CO_2 emissions. The same annual amount of electricity could be generated using roughly 1500 ha for perennial energy crops, 600–1200 ha for AD feedstocks, a wind farm of 250 ha, or just 60 ha for a solar farm. However, different renewable energy options are more likely to complement one another than to compete for the same land, and they may even overlap.

The term **fuel energy ratio** is used to describe the relationship between the **energy output of a system and the energy inputs needed to operate it** (usually from fossil fuels).

Over the full range of renewable sources, the energy ratio of output to input can vary from less than 1:1 to as much as 300:1 (this for some hydroelectric plants). **Woody energy crops** perform well, with ratios between 10:1 and 20:1 on a heat output basis, but biodiesel may achieve only 3:1. Ethanol from grain may barely break even, at just over 1:1 in the worst cases.

When wastes or residues are the input, the questions arise of how much of the energy input to the whole process should be attributed to the energy extraction system as opposed to alternative disposal pathways for wastes, and how energy should be allocated between primary products (such as food or timber) and crop or forest residues. This can result in high FER values, e.g. 30:1 typically for electricity from woody sawmill wastes.

Governments across the world operate policy mechanisms in order to support the substitution of fossil-fuel energy for renewable sources. The IEA has stated that for as long as the external costs of fossil fuels are not accounted for, **bioenergy systems will need some degree of financial assistance**.

A number of different policy mechanisms are implemented in order to promote bioenergy. Some policies fund research and development of the biofuel or bioenergy technology (technology-push policies), whilst others directly intervene in the market to stimulate demand for the biofuel or bioenergy technology (market-pull policies).

Market-pull policies may comprize feed-in tariffs and premiums; quota obligations (mandates) and tradable renewable certificates; incentives to invest in generating plants; and disincentives to non-renewable energy.

In the near future, much of the growth in renewable energy supply in many parts of the world may come from increased direct combustion of a wide variety of biomass feedstocks for heat and electricity generation, together with a substantial increase in the supply of 'first generation' liquid transport biofuels.

In the longer term, newer biomass technologies, such as electricity from novel high-efficiency small power plants or fuel cells and 'next generation' liquid biofuels, could play an increasing role – but by that time bioenergy might be overtaken in developed and emerging economies by faster-growing renewables such as wind and solar power.

Some authorities predict that traditional use of biomass will be largely phased out by as early as 2030, to be replaced by modern renewables, although bioenergy will still account for half of all renewable energy use worldwide. Others forecast that traditional biomass will remain the sole domestic fuel for 2.8 billion people in 2030 (IEA, 2010): this leads to the conclusion that the **improvement of traditional wood and charcoal stoves should be one of the most important technical aims in the field of bioenergy today**.

The most optimistic forecasts on the potential future global bioenergy contribution suggests over 1500 EJ per year, significantly above the total world energy use at the time of writing. A more widely accepted figure might be of the order of 100–200 EJ of bioenergy per year.

Coupling biomass or biocoal power plants to carbon capture and storage offers the prospect of carbon-negative electricity production, since the photosynthetic production of the biomass feedstock on a sustainable basis would remove carbon dioxide from the atmosphere.

In terms of the **production of bioenergy crops**, there are two main targets for improvement:

■ **crop genetics**

■ **agronomic management practices.**

The first enables the development of better genotypes (varieties) of bioenergy crops, while the second optimizes the yields of these varieties.

These two approaches to improving bioenergy yield are not stand-alone and show clear interactions. For example, varieties with high resource use efficiency may need to be developed in the future in order to optimize the carbon and energy balance of feedstock production.

References

Adams, P. W. R. and Lindegaard, K. (2016) 'A critical appraisal of the effectiveness of UK perennial energy crops policy since 1990', *Renewable and Sustainable Energy Reviews*, vol. 55, pp. 188–202.

AFDC (2012) *Drop-in biofuels*, Washington, DC, Alternative Fuel Data Center, Department of Energy.

Alexander, P., Moran, D., Smith, P., Hastings, A., Wang, S., Sunnenberg, G., Lovett, A., Tallis, M. J., Casella, E., Taylor, G., Finch, J. and Cisowska, I. (2014) 'Estimating UK Perennial Energy Crop Supply using farm-scale models with spatially disaggregated data', *Global Change Biology Bioenergy*, vol. 6, no. 2, pp. 132–155.

Alexander P. and Moran D (2013) 'Impact of perennial energy crops income variability on the crop selection of risk averse farmers'. *Energy Policy*, vol. 52, pp. 587–596.

Allwright, M. R. and Taylor, G. (2016) 'Molecular breeding for improved second generation bioenergy crops', *Trends in Plant Science*, vol. 21, pp. 43–54.

Anderson, T., Doig, A., Rees, D. and Khennas, S. (1999) *Rural Energy Services: A Handbook for Sustainable Energy Development*, London, Practical Action.

Angelini, L. G., Ceccarini, L. and Bonari, E. (2005) 'Biomass yield and energy balance of giant reed (Arundo donax L.) cropped in central Italy as related to different management practices', *European Journal Of Agronomy*, vol. 22, no. 4, pp. 375–389.

Antal, M. J. and M. Gronli (2003) 'The art, science, and technology of charcoal production', *Industrial & Engineering Chemistry Research*, vol. 42 , pp. 1619–1640.

Arnoult, S. and Brancourt-Hulmel, M. (2015) 'A review on Miscanthus biomass production and composition for bioenergy use: genotypic and environmental variability and implications for breeding', *BioEnergy Research*, vol. 8, pp. 502–526.

Aylott, M. J., Casella, E., Farrall, K. and Taylor G. (2010) 'Estimating the supply of biomass from short-rotation coppice in England, given social, economic and environmental constraints to land availability', *Biofuels*, vol. 1, no. 5, pp. 719–727.

Aylott M. J., Casella E., Tubby I., Street N. R., Smith P. and Taylor G. (2008) 'Yield and spatial supply of bioenergy poplar and willow short-rotation coppice in the UK', *New Phytol*, vol. 178, no. 2, pp. 358–370.

Badger P. C. (2002) Ethanol from cellulose: a general review, in Janick, J. and Whipkey, A., (eds) *Trends in new crops and new uses*, Alexandria, VA, ASHS Press, pp. 17–21.

Bahar, H., Egeland, J. and Steenblik, R. (2013) 'Domestic Incentive Measures for Renewable Energy With Possible Trade Implications', *OECD Trade and*

Environment Working Papers, no. 2013/01 [Online].. DOI: http://dx.doi. org/10.1787/5k44srlksr6f-en (Accessed 15th February 2017).

Bast E., Doukas A., Pickard S., Burg L. and Whitley, S. (2015) *Empty promises: G20 subsidies to oil, gas and coal production* [Online], London and Washington, Overseas Development Institute and Oil Change International. Available at https://www.odi.org/sites/odi.org.uk/files/odi-assets/publications-opinion-files/9957.pdf (Accessed 16th February 2017).

BEIS (2016a) *Digest of United Kingdom Energy Statistics (DUKES)* [Online], London, Department for Business, Energy & Industrial Strategy, [online]. Available at https://www.gov.uk/government/statistics/renewable-sources-of-energy-chapter-6-digest-of-united-kingdom-energy-statistics-dukes (Accessed 31 July 2016).

BEIS (2016b) *DUKES: calorific values* [Online], London, Department for Business, Energy & Industrial Strategy. Available at https://www.gov.uk/government/statistics/dukes-calorific-values (Accessed 1 Jan 2017).

Beringer, T. D., Lucht, W. and Schaphoff, S. (2011) 'Bioenergy production potential of global biomass plantations under environmental and agricultural constraints', *Global Change Biology Bioenergy*, vol. 3 [Online]. DOI: 10.1111/j.1757 1707.2010.01088.x (Accessed 4 April 2011).

Biddy, M. and S. Jones (2013) *Catalytic upgrading of sugars to hydrocarbons technology pathway*,. Golden, Colorado, National Renewable Energy Laboratory. NREL/TP-5100-58055).

Björheden, R. (2006) 'Drivers behind the development of forest energy in Sweden', *Biomass and Bioenergy*, vol. 30, no. 4, pp. 289–95.

Bomgardner, M. M. (2013) 'Seeking biomass feedstocks that can compete', Chemical and Engineering News, vol. 91, no. 32, pp. 11–15.

Buchanan, J. M. and Stubblebine, W.C. (1962) 'Externality', *Economica*, vol. 29, no. 116, pp. 371–384.

Carpenter, D., Westover, T. L., Czernik, S. and Jablonski, W. (2014) 'Biomass feedstocks for renewable fuel production: a review of the impacts of feedstock and pretreatment on the yield and product distribution of fast pyrolysis bio-oils and vapors', *Green Chemistry*, vol. 16, pp. 384–406.

Clarens, A. F., Resurreccion, E. P., White, M. A. and Colosi, L. M. (2010) 'Environmental life cycle comparison of algae to other bioenergy feedstocks', *Environmental Science & Technology*, vol. 44, no. 5, pp. 1813–19.

Clifton-Brown, J. C., Hastings, A., Mos, M., McCalmont, J. P., Ashman, C., Awty-Carroll, D., Cerazy, J., Chiang, Y-C., Consentino, S., Cracroft-Eley, W., Scurlock, J., Donnison, I. S., Glover, C., Golab, I., Greef, J. M., Gwyn, J., Harding, G., Hayes, C., Helios, W., Hsu, T-W., Huang, L. S., Jezowski, S., Kim, D-S., Kiesel, A., Kotecki, A., Krzyzak, J., Lewandowski, I., Lim, S. H., Liu, J., Loosely, M., Meyer, H., Murphy-Bokern, D., Nelson, W., Pogrzeba, M., Robinson, G., Robson, P., Rogers, C., Scalici, G., Schuele, H., Shafiei, R., Shevchuk, O., Schwarz, K-U., Squance, M., Swaller, T., Thornton, J., Truckses, T., Botnari, V., Vizir, I., Wagner, M., Warren, R., Webster, R., Yamada, T., Youell, S., Xi, Q., Zong, J. and Flavell, R. (2016) 'Progress in upscaling Miscanthus biomass production for the European bio-economy with seed-based hybrids', *Global Change Biology Bioenergy*, vol. 9, no. 1, pp. 6–17 [Online]. DOI: 10.1111/gcbb.12357 (Accessed 27April 2017).

Clifton-Brown, J. C., Renvoize, S., Chiang, Y-C., Ibaragi, Y., Flavell, R., Greef, J., Huang, L., Hsu, T. W., Kim, D-S., Hastings, A., Schwarz, K., Stampfl, P., Valentine, J., Yamada, T., Xi, Q. and Donnison, I. (2010) 'Developing Miscanthus for Bioenergy', in Halford, N. G. and Karp, A., (eds) *Energy Crops*, Cambridge, Royal Society of Chemistry, pp. 301–321.

Costanza, R., d'Arge, R., de Groot, R., Farber, S., Grasso, M., Hannon, B., Limburg, K., Naeem, S., O'Neill, R. V., Paruelo, J., Raskin, R. G., Sutton, P. and van den Belt, M. (1997) 'The value of the world's ecosystem services and natural capital', *Nature*, vol. 387, pp. 253–260.

Czernik, S. and Bridgwater, A. V. (2004) 'Overview of applications of biomass fast pyrolysis oil', *Energy Fuels*, vol. 18, pp. 590–598.

Dale V., Parish E. S. and Kline K. L. (2016) 'Lessons from the Forest', *World Biomass*, 2016/2017 [Online]. Available at http://www.dcm-productions. co.uk/flippages/flipbook/index.html?page=20 (Accessed 2 May 2017).

Davies, S. C., Anderson-Teixeira, K. J. and DeLucia, E. H. (2009) 'Life-cycle analysis and the ecology of biofuels', *Trends in Plant Science*, vol. 14, no. 3, pp. 140–146.

DECC (2012) *UK Bioenergy Strategy* [Online], London, Department for Energy and Climate Change. Available at https://www.gov.uk/government/ uploads/system/uploads/attachment_data/file/48337/5142-bioenergy-strategy-.pdf (Accessed 26 February 2017).

Defra (2007) *Waste strategy for England 2007 Annexes* [Online], London, Department for Environment Food and Rural Affairs. Available at http:// webarchive.nationalarchives.gov.uk/20130402151656/http://archive. defra.gov.uk/environment/waste/strategy/strategy07/documents/waste07-annexes-all.pdf (Accessed 26 February 2017).

Defra (2011) 2011 *Guidelines to Defra/DECC's Greenhouse Gas Conversion Factors for Company Reporting* [Online], London, Department for Environment Food and Rural Affairs. Available at https://www.gov. uk/government/uploads/system/uploads/attachment_data/file/69314/ pb13625-emission-factor-methodology-paper-110905.pdf (Accessed 26 February 2017).

Defra (2013) *Incineration of Municipal Solid Waste* [Online], London, Department for Environment Food and Rural Affairs. Available at https:// www.gov.uk/government/uploads/system/uploads/attachment_data/ file/221036/pb13889-incineration-municipal-waste.pdf (Accessed 15 February 2017).

Defra (2014) *Energy from waste: a guide to the debate* [Online], London, Department for Environment Food and Rural Affairs. Available at https:// www.gov.uk/government/publications/energy-from-waste-a-guide-to-the-debate (Accessed 15 February 2017).

Defra (2015) *Digest of Waste and Resource Statistics 2015* [Online], London, Department for Environment Food and Rural Affairs. Available at https://

www.gov.uk/government/collections/waste-and-recycling-statistics (Accessed 5 January 2017).

Defra (2015a) *Area of Crops Grown for Bioenergy in England and the UK: 2008-2014* [Online], London, Department for Environment Food and Rural Affairs. Available at https://www.gov.uk/government/uploads/system/uploads/attachment_data/file/483812/nonfood-statsnotice2014-10dec15.pdf (Accessed 15 February 2017).

Defra (2015b) *The government's response to the Natural Capital Committee's third State of Natural Capital report* [Online], London, Department for Environment Food and Rural Affairs. Available at https://www.gov.uk/government/uploads/system/uploads/attachment_data/file/462472/ncc-natural-capital-gov-response-2015.pdf (Accessed 16 February 2017).

Defra (2016) *UK Statistics on Waste* [Online], London, Department for Environment Food and Rural Affairs. Available at https://www.gov.uk/government/uploads/system/uploads/attachment_data/file/593040/UK_statsonwaste_statsnotice_Dec2016_FINALv2_2.pdf (Accessed 8 January 2017).

Djomo, S. N., Kasmioui, O. E. and Ceulemans, R. (2011) 'Energy and greenhouse gas balance of bioenergy production from poplar and willow: a review', *Global Change Biology Bioenergy*, vol. 3, no. 3, pp. 181–197 [Online]. DOI:10.1111/j.1757-1707.2010.01073.x (Accessed 28 April 2017).

DoE (2000) *Biomass co-firing: a renewable alternative for utilities* [Online], Washington, DC, U.S. Department of Energy. Available at http://www.nrel.gov/docs/fy00osti/28009.pdf (Accessed 28 April 2017)..

DTI (2004) *Renewables Innovation Review* [Online], London, Department of Trade and Industry. Available at http://webarchive.nationalarchives.gov.uk/20100809181248/http://www.decc.gov.uk/assets/decc/what%20we%20do/uk%20energy%20supply/energy%20mix/renewable%20energy/policy/file22017.pdf (Accessed 26 February 2017).

E4tech (2009) *Review of technologies for gasification of biomass and wastes*, London, E4tech Ltd, NNFCC project 09/008 final report..

EBTP (2017) *European Industrial Bioenergy Initiative* (EIBI) [Online]. Available at http://www.biofuelstp.eu/eibi.html (Accessed 3 May 2017).

EIA (2010) *World Coal Consumption 1980–2008* [Online], Washington D.C., US Energy Information Administration. Available at http://www.eia.doe.gov/aer/txt/ptb1115.html (Accessed 29 March 2011).

EIA (2014) *Monthly biodiesel production report*, Washington, DC, U.S. Energy Information Administration.

El Bassam, N. (2010) *Handbook of Bioenergy Crops: A Complete Reference to Species, Development and Applications*, London, Earthscan.

EPA (2010) *Methane: Sources and emissions* [Online]. Available at http://www.energybc.ca/cache/naturalgas/www.epa.gov/methane/sources.html (Accessed 5 May 2017).

EPA (2015) *Final Renewable Fuel Standards for 2014, 2015 and 2016, and the Biomass-Based Diesel Volume for 2017* [Online], Washington,

DC, United States Environmental Protection Agency. Available at https://www.epa.gov/renewable-fuel-standard-program/final-renewable-fuel-standards-2014-2015-and-2016-and-biomass-based#rule-history (Accessed 16 February 2017).

EPE (2016) *Cenários de Oferta de Etanol e Demanda do Ciclo Otto* [Online], Rio de Janeiro, Empresa de Pesquisa Energética,. Available at http://www.epe.gov.br/Petroleo/Documents/EPE-DPG-SGB-Bios-NT-02-2016_Cen%C3%A1rios%20de%20Oferta%20de%20Etanol%20e%20Demanda%20do%20Ciclo%20Otto.pdf (Accessed 25 February 2017).

ETI (2014) *The role of bioenergy in the future UK energy system* [Online], Loughborough, Energy Technologies Institute. Available at http://www.eti.co.uk/wp-content/uploads/2014/11/IMECHE_The-role-of-Bioenergy-in-the-future-UK_201114.pdf (Accessed 18 February 2017).

ETI (2016) *The Evidence for Deploying Bioenergy with CCS (BECCS) in the UK* [Online], Loughborough, Energy Technologies Institute. Available at http://www.eti.co.uk/news/uk-ccs-strategy-should-include-a-demonstration-of-bioenergy-with-ccs-technology-to-allow-negative-emissions-to-be-delivered-within-the-next-decade (Accessed 25 February 2017).

ETSU (1999) *New and Renewable Energy: Prospects in the UK for the 21st Century: Supporting Analysis*, Didcot, Energy Technology Support Unit, ETSU R-122.

European Bioplastics (2011) *Factsheet: Anaerobic digestion.* Berlin, European Bioplastics..

European Commission (2017), *Progress reports* [Online]. Available at https://ec.europa.eu/energy/en/topics/renewable-energy/progress-reports (Accessed 29 January -2017).

Eurostat (2014) *Renewable Energy Statistics for 2014* [Online], Luxembourg, European Union Statistical Office. Available at http://ec.europa.eu/eurostat/statistics-explained/index.php/Renewable_energy_statistics (Accessed 1 January 2017).

FAO (2013) FAO Statistical Yearbook 2013, World Food and Agriculture, [Online], Rome, Food and Agriculture Organization of the United Nations. Available at http://www.fao.org/docrep/018/i3107e/i3107e00.htm (Accessed 23 February 2017).

FAO (2016) 2015 Global Forest Products Facts and Figures, [Online], Geneva, Food and Agriculture Organization of the United Nations. Available at http://www.fao.org/3/a-i6669e.pdf (Accessed 16 January 2017).

FAO (2017) *FAOSTAT* [Online] Available at http://www.fao.org/faostat/en/#data/FO; http://faostat3.fao.org/download/F/FO/E (Accessed 16 January 2017).

Farm and Dairy (2015) 'US EPA scales back renewable fuel standard', Farm and Dairy, 30 May [Online]. Available at http://www.farmanddairy.com/news/u-s-epa-scales-back-renewable-fuel-standard/262784.html (Accessed 16 February 2017).

Fiala, M. and Bacenetti, J. (2012) 'Economic, energetic and environmental impact of short rotation coppice harvesting operations', *Biomass Bioenergy*, vol. 42, pp. 107–113.

Field, C. B., Campbell, J. E. and Lobell, D. B. (2008) 'Biomass energy: the scale of the potential resource', *Trends in Ecology and Evolution*, vol. 23, no. 2, pp. 65–72.

Forest Research (2017) Former UK Biomass Energy Centre Guidance and Information Sheets [Online], Bristol, Forest Research. Available at https://www.forestry.gov.uk/fr/beeh-9uxqlz / https://www.forestry.gov.uk/fr/beeh-ab6jy7 (Accessed 2 February 2017).

Forestry Commission (2016) *Forestry Statistics 2016 - UK-Grown Timber: Deliveries of UK-grown roundwood* [Online]. Available at http://www.forestry.gov.uk/website/forstats2016.nsf/0/8BE129C3F1FC3AEF80257FE0004B2CFF (Accessed 4 January 2017).

Foust, T. and Bratis, A., (2013) *Review of recent pilot scale cellulosic ethanol demonstration* [Online], Golden, Colorado, National Renewable Energy Laboratory, Department of Energy. Available at https://energy.gov/sites/prod/files/2014/05/f15/b13_foust_op-1.pdf (Accessed 27 April 2017).

Fritts, S., Moorman, C., Hazel, D., Homyack, J., Castleberry, S., Pollock, K., Farrell, C., and Grodsky, S. (2016) 'Do biomass harvesting guidelines influence herpetofauna following harvests of logging residues for renewable energy?', *Ecological Applications*, vol. 26, pp. 926–939.

GIB (2015) *The UK anaerobic digestion market* [Online], London, Green Investment Bank. Available at http://www.greeninvestmentbank.com/media/44758/gib-anaerobic-digestion-report-march-2015-final.pdf (Accessed 15 February 2017).

Guo, M., Song, W. and Buhain, J. (2015) 'Bioenergy and biofuels: history status and perspectives', *Renewable and Sustainable Energy Reviews*, vol. 42, pp. 715–725.

Haberl, H., Sprinz, D., Bonazountas, D., Cocco, P., Desaubies, Y., Henze, M., Hertel, O., Johnson, R. K., Kastrup, U., Laconte, P., Lange, E. and Novak, P. (2012) 'Correcting a fundamental error in greenhouse gas accounting related to bioenergy', *Energy Policy*, vol. 45, pp. 18–23.

Haberl, H., Erb, K. H., Krausmann, F., Gaube, V., Bondeau, A., Plutzar, C., Gingrich, S., Lucht, W. and Fischer-Kowalski, M. (2007) 'Quantifying and mapping the human appropriation of net primary production in Earth's terrestrial ecosystems', *Proceedings of the National Academy of Sciences of the USA*, vol. 104, no. 31, pp. 12 942–12 947 [Online]. Available at http://www.pnas.org/content/104/31/12942.full.pdf+html (Accessed 26 February 2017).

Hall, D. O. and Rao, K. K. (1999) *Photosynthesis (Studies in Biology)* (6th edn), Cambridge, Cambridge University Press.

Hall, D. O., Rosillo-Calle, F., Williams, R. H. and Woods, J. (1993) Biomass for energy: supply prospects in Johansson, T. B., Kelly, H., Reddy, A. K. N. and Williams, R. H. (eds) *Renewables for Fuels and Electricity*, Washington, D.C, Island Press..

Haq, Z. (2002) *Biomass for energy generation*, Washington, Energy Information Administration..

Hastings, A., Yeluripati, J., Hillier, J. and Smith, P. (2012) 'Biofuel crops and greenhouse gases', in Singh, B. P. (ed) *Biofuel Crop Production*, New Jersey, Wiley-Blackwell, pp. 383-406.

Havlicková, K.Weger, J. and Zanova, I. (2009) 'Short rotation coppice for energy purposes—economy conditions and landscape functions in the Czech Republic', in Goswami, D. Y. and Zhao, Y. (eds) *Proceedings of ISES World Congress 2007* (Vol. 1–5), Beijing and New York, Tsinghua University Press and Sringer Berlin Heidelberg, , pp. 2482–2487.

Heaton, E. A., Clifton-Brown, J., Voigt, T. B., Jones, M. B. and Long, S. P. (2004) 'Miscanthus for renewable energy generation: European union experience and projections for Illinois', *Mitigation Adaptation Strategies Global Change*, vol. 9, pp. 433–451.

HLPE (2013) *Biofuels and food security*, Rome, High Level Panel of Experts on Food Security and Nutrition

Holland, R. A., Eigenbrod, F., Muggeridge, A., Brown, G., Clarke, D. and Taylor, G. (2015) 'A synthesis of the ecosystem services impact of second generation bioenergy production', *Renewable & Sustainable Energy Reviews*, vol. 46, pp. 30–40.

Hoogwijk, M., Faaij, A., van den Broek, R., Berndes. G., Gielen, D. and Turkenburg, W. (2003) 'Exploration of the ranges of the global potential of biomass for energy', *Biomass and Bioenergy*, vol. 25, no. 2, pp. 119–33.

IEA (2006) *Energy for Cooking in Developing Countries* [Online], in IEA (ed) World Energy Outlook. Paris, IEA. Available at https://www.ica.org/publications/freepublications/publication/cooking.pdf (Accessed 15 February 2017).

IEA (2010) *Energy Poverty: How to make modern energy access universal?* [Online], Paris, International Energy Agency. Available at http://www.worldenergyoutlook.org/media/weowebsite/2010/weo2010_poverty.pdf (Accessed 18 February 2017).

IEA (2011) *Global wood pellet industry market and trade study, IEA bioenergy task 40*, Paris, International Energy Agency.

IEA (2012) *Technology Roadmap: Bioenergy for Heat and Power* [Online], Paris, International Energy Agency. Available at http://www.iea.org/publications/freepublications/publication/2012_Bioenergy_Roadmap_2nd_Edition_WEB.pdf (Accessed 15 February 2017).

IEA (2013) *World Energy Outlook 2013*, Paris, International Energy Agency..

IEA (2015a) *Key World Energy Statistics 2015* [Online], Paris International Energy Agency. Available at https://www.iea.org/publications/freepublications/publication/KeyWorld_Statistics_2015.pdf (Accessed 31 July 2016).

IEA (2015b) *Member Country Reports* [Online]. Available at http://www.iea-biogas.net/country-reports.html (Accessed 28 August 2016).

IEA Bioenergy (2009) *Bioenergy – a Sustainable and Reliable Energy Source (Executive Summary)* [Online], Paris, IEA Bioenergy. Available at http://www.ieabioenergy.com/publications/bioenergy-a-sustainable-and-reliable-energy-source-executive-summary-2/ (Accessed 25 February 2017).

Inbicon (2017) *Danish projects* [Online]. Available at http://www.inbicon.com/en/global-solutions/danish-projects (Accessed 25 January 2017).

IPCC (2014) *Climate Change 2014: Synthesis Report. Contribution of Working Groups I, II and III to the Fifth Assessment Report of the Intergovernmental Panel on Climate Change*, Geneva, IPCC..

IRENA (2014) *Global Bioenergy: Supply and Demand, A working paper for REmap 2030* [Online], Abu Dhabi, International Renewable Energy Agency. Available at http://irena.org/remap/IRENA_REmap_2030_Biomass_paper_2014.pdf (Accessed 30 August 2016).

IRENA (2016) *Remap: Roadmap for a Renewable Energy Future* [Online], Abu Dhabi, International Renewable Energy Agency. Available at http://www.irena.org/remap/ (Accessed 18 February 2017).

Iriarte, A., Rieradevall, J. and Gabarrell, X. (2010) 'Life cycle assessment of sunflower and rapeseed as energy crop under Chilean conditions', *Journal of Cleaner Production*, vol. 18, pp. 336–345 [Online].DOI:10.1016/j.jclepro.2009.11.004

Jannasch, R., Quan, Y. and Samson, R. (2002) *A process and energy analysis of pelletizing switchgrass* [Online], Québec, REAP-Canada. Available at http://www.reap-canada.com/online_library/feedstock_biomass/11%20A%20Process.pdf (Accessed 25 February 2017).

Jiménez, B. J. A., Perea, T. F., Lobo, G. J., Pavón, P.L. and Durán, Z. V. H. (2013) 'Evaluación del cultivo del eucalipto para la producción de biomasa en Andalucía', *Vida Rural*, vol. 366, pp. 62–66

Jin, J-Y., Wu, R. and Liu, R. (2002) 'Rice production and fertilization in China', *Better Crops International*, vol. 16, Special Supplement May, pp. 26–9.

Kaltschmitt, M. and Janczik, S. (2015) 'REMIPEG Report: Part 2 - Biomass to power is on the rise globally', Renewable Energy Focus, 24 November [Online]. Available at http://www.renewableenergyfocus.com/view/43303/remipeg-report-part-2-biomass-to-power-is-on-the-rise-globally/ (Accessed 25 February 2017).

Kantor, L. S., Lipton, K., Manchester, A. and Oliveira, V. (1997) 'Estimating and addressing America's food losses', *Food Review*, vol. 20, no. 1, pp. 1–12.

Kardos, L., Juhasz, A., Palko, G. Y., Olah, J., Barkacs, K. and Zaray, G. Y. (2011) 'Comparing of mesophilic and thermophilic anaerobic fermented sewage sludge based on chemical and biochemical tests', *Applied Ecology and Environmental Research*, vol. 9, pp. 293–302.

Karp, A., Hanley, S. J., Trybush, S. O., Macalpine, W., Pei, M., and Shield, I. (2011) 'Genetic improvement of willow for bioenergy and biofuels', *Journal of Integrative Plant Biology*, vol. 53, pp. 151–165.

King, J. A., Bradley, R. I., Harrison, R. and Carter, A. D. (2004) 'Carbon sequestration and¬ saving potential associate¬d with changes to the management of agricultural soils in England', *Soil Use Management*, vol. 20, pp. 394–402.

Kishore, V. V. N., Bhandari, P. M. and Gupta, P. (2004) 'Biomass energy technologies for rural infrastructure and village power – opportunities and challenges in the context of global climate change concerns', *Energy Policy*, vol. 32, no. 6, pp. 801–810.

Kline, K.L., Msangi, S., Dale, V. H., Woods, J., Souza, G. M., Osseweijer, P., Clancy, J. S., Hilbert, J. A., Johnson, F. X., McDonnell, P. C. and Mugera, H. K. (2016) 'Reconciling food security and bioenergy: priorities for action', *Global Change Biology – Bioenergy*, vol. 9, no. 3, pp. 485–661 [Online]. DOI: 10.1111/gcbb.12366 (Accessed 28 April 2017).

Kuiper, L. and Oldenburger, J. (2006) *The harvest of forest residues in Europe* [Online], Wageningen, Biomassa-upstream Stuurgroep. Available at http://www.probos.net/biomassa-upstream/pdf/reportBUSD15a.pdf (Accessed 25 February 2017).

Kumar, P., Barrett, D. M., Delwiche, M. J., and Stroeve, P. (2009) 'Methods for pretreatment of lignocellulosic biomass for efficient hydrolysis and biofuel production', *Industrial and Engineering Chemistry Research*, vol. 48, no. 8, pp. 3713–3729.

Ladanai, S. and Vinterbäck, J. (2009) *Global potential of sustainable biomass for energy* [Online]. Available at https://www.researchgate.net/publication/48343698_Global_potential_of_sustainable_biomass_for_energy (Accessed 28 April 2017).

Laird, D. A., Brown, R. C., Amonette, J. E. and Lehmann, J. (2009) 'Review of the pyrolysis platform for coproducing bio-oil and biochar', *Biofuels, Bioproducts and Biorefining*, vol. 3, pp. 547–562.

Lemus, R. and Lal, R. (2005) 'Bioenergy Crops and Carbon Sequestration', *Critical Reviews in Plant Sciences*, vol. 24, no. 1, pp. 1–21.

Lenihan, P., Orozco, A., O'Neill, E., Ahmad, M. N. M., Rooney, D. W. and Walke, G. M. (2010) 'Dilute acid hydrolysis of lignocellulosic biomass', *Chemical Engineering Journal*, vol. 156, pp. 395–403.

Lindegaard, K. N., Adams, P. W. R., Holley, M., Lamley, A., Henriksson, A., Larsson, S., von Engelbrechten, H-G., Lopez, G. E. and Pisarek, M. (2016) 'Short rotation plantations policy history in Europe: lessons from the past and recommendations for the future', *Food and Energy Security*, vol. 5, pp. 125–152.

Long, S. P., Karp, A., Buckeridge, M., Davis, S. C., Jaiswal, D., Moore, P. H., Moose, S. P., Murphy, D. J., Onwona-Agyeman, S. and Vonshak, A. (2015) 'Feedstocks for biofuels and bioenergy', in Souza, G. M., Victoria, R. L., Joly, C. A., and Verdade, L. M. (eds) *Bioenergy & Sustainability: bridging the gaps*, Paris, SCOPE, pp. 302–347.

Lynd, L., Greene, N., Dale, B., Laser, M., Lashof, D., Wang, M. and Wyman, C. (2006) 'Energy Returns on Ethanol Production', *Science*, vol.. 312, no. 5781, pp. 1746-1748.

Mackey, B., Prentice, I. C., Steffen, W., House, J. I., Lindenmayer, D., Keith, H. and Berry, S. (2013) 'Untangling the confusion around land carbon science and climate change mitigation policy', *Nature Climate Change*, vol. 3, pp. 552–557.

Mani, S., Sokhansanj, S., Bi, X. and Turhollow, A. (2006) 'Economics of producing fuel pellets from biomass', *Applied Engineering in Agriculture*, vol. 22, no. 3, pp. 1–6 [Online]. DOI: 10.13031/2013.20447 (Accessed 28 April 2017).

Mantineo, M., D'Agosta, G. M., Copani, V., Patane, C. and Cosentino, S. L. (2009) 'Biomass yield and energy balance of three perennial crops for energy use in the semi-arid Mediterranean environment', *Field Crops Research*, vol. 114, pp. 204–213.

Manzone, M., Airoldi, G. and Balsari, P. (2009) 'Energetic and economic evaluation of poplar cultivation for the biomass production in Italy', *Biomass Bioenergy*, vol. 33, pp. 1258–1264 [Online]. DOI:10.1016/j. biombioe.2009.05.024 (Accessed 28 April 2017).

Matthews, R., Mortimer, N., Mackie, E., Hatto, C., Evans, A., Mwabonje, O., Randle, T., Rolls, W., Sayce, M. and Tubby, I. (2014) *Carbon impacts of using biomass in bioenergy and other sectors: forests* [Online]. Available at https://www.gov.uk/government/uploads/system/uploads/attachment_ data/file/282812/DECC_carbon_impacts_final_report30th_January_2014. pdf (Accessed 28 April 2017). See also https://www.gov.uk/government/ uploads/system/uploads/attachment_data/file/65618/7014-bioenergy- strategy-supplementary-note-carbon-impac.pdf and http://www.forestry. gov.uk/fr/beeh-9uynmd.

McCalmont , J. P., Hastings, A., McNamara, N. P., Richter, G. M., Robson, P., Donnison, I. S. and Clifton-Brown, J. (2015) 'Environmental costs and benefits of growing *Miscanthus* for bioenergy in the UK', *Global Change Biology Bioenergy*, vol. 9, no. 3, pp. 489–507. DOI: 10.1111/gcbb.12294 (Accessed 28 April 2017).

Milner S., Holland, R. A., Lovett, A., Sunnenberg, G., Hastings, A., Smith, P., Wang, S. and Taylor, G. (2015) 'Potential impacts on ecosystem services of land use transitions to second generation bioenergy crops in GB', *Global Change Biology Bioenergy*, vol. 8, pp. 317–333.

Ministry of New and Renewable Energies (2016) *Ministry of New and Renewable Energy* [Online]. Available at http://mnre.gov.in/schemes/ decentralized-systems/schems-2/ (Accessed 28 August 2016).

Mola, Y. B (2011) 'Trends and productivity improvements from commercial willow plantations in Sweden during the period 1986–2000', *Biomass Bioenergy*, vol. 35, pp. 446–453 [Online]. DOI:10.1016/j. biombioe.2010.09.004 (Accessed 28 April 2017).

Moreira, J. R. (2006) 'Global biomass energy potential', *Mitigation and Adaptation Strategies for Global Change*, vol. 11, pp. 313–342.

Nassi o Di Nasso, N. N., Roncucci, N., Bonari, E. (2013) 'Seasonal dynamics of aboveground and belowground biomass and nutrient accumulation and remobilization in giant reed (Arundo donax L.): a three-year study on marginal land', *Bioenergy Research*, vol. 3 pp. 1–12.

Nersesian, R. L. (2016) *Energy Economics: Markets, History and Policy* London and New York, Routledge.

NETL (2013) *Syngas composition*, Albany, Oregon, National Energy Technology Laboratory.

NFU (2016) *Anaerobic Digestion: progress towards NFU's aspiration of 1000 on-farm plants* [Online], Stoneleigh, National Farmers' Union. Available at https://www.nfuonline.com/assets/80339 (Accessed 31 December 2016).

NFU (2017) *Greenhouse Gas Action Plan for the agricultural industry* [Online], Stoneleigh, National Farmers' Union of England and Wales. Available at https://www.cfeonline.org.uk/home/about-us/greenhouse-gas-action-plan/ (Accessed 2 February 2017).

NNFCC (2010) Crop Factsheet: *Miscanthus. National Non-Food Crops* Centre, York, UK.

Obersteiner, M., Azar, C., Kauppi, P., Möllersten, K., Moreira, J., Nilsson, S., Read, P., Riahi, K. and Schlamadinger, B. (2016) *Biomass Sustainability Dataset 2014-15* [Online], London, Ofgem. Available at https://www.ofgem. gov.uk/publications-and-updates/biomass-sustainability-dataset-2014-15 (Accessed 4 January 2017).

Phillips, S., Aden, A., Jechura, J., Dayton, D. and Eggeman, T. (2007) *Thermochemical Ethanol via Indirect Gasification and Mixed Alcohol Synthesis of Lignocellulosic Biomass*, Golden, Colorado, National Renewable Energy Laboratory, Technical Report NREL/TP-510-41168..

POST (2007) *Energy and Sewage* [Online], London, Parliamentary Office of Science and Technology. Available at http://www.parliament.uk/ documents/post/postpn282.pdf (Accessed 31 March 2011).

Primus (2013) *Introduction to Primus'* STG+ Technology, Hillsborough, New Jersey, Primus Green Energy.

Ptasinski, K. J. (2016) *Efficiency of Biomass Energy: An Exergy Approach to Biofuels, Power, and Biorefineries*, Chichester, John Wiley..

Purohit, P. and Michaelowa, A. (2007) 'CDM potential of bagasse cogeneration in India', *Energy Policy*, vol. 35, no. 10, pp. 4779–4798.

Rajgor, G. (2013) 'Renewable power generation — 2012 figures. Part six: Biomass', *Renewable Energy Focus*, 18 November [Online]. Available at http://www.renewableenergyfocus.com/view/35689/renewable-power-generation-2012-figures/ (Accessed 28 April 2017).

REN21 (2015) *Renewables 2015 Global Status Report*, Paris, REN21 Secretariat.

RFA (2016) *Fueling a high octane future: Ethanol Industry Outlook* [Online], Washington, DC, Renewable Fuels Association. Available at http://ethanolrfa.org/wp-content/uploads/2016/02/RFA_2016_full_final. pdf (Accessed 15th February 2017).

Risović, S., Djukić, I. and Vućković, K. (2008) 'Energy analysis of pellets made of wood residue', *Croatian Journal of Forest Engineering*, vol. 29, no. 1, pp. 95–108 [Online]. Available at http://www.crojfe.com/articles-635 (Accessed 25 February 2017).

Rosenqvist, H. and Dawson, M. (2005) 'Economics of willow growing in Northern Ireland', *Biomass Bioenergy*, vol. 28, pp. 7–14 [Online]. DOI:10.1016/j.biombioe.2004.06.001 (Accessed 28 April 2017).

Rowe, R. L., Street, N. and Taylor, G. (2009) 'Identifying potential environmental impacts of large-scale deployment of dedicated bioenergy crops in the UK', *Renewable and Sustainable Energy Reviews*, vol. 13, pp. 271–290.

Rowe, R. L., Hanley, M. E., Goulson, D., Clarke, D. J., Doncaster, C. P. and Taylor, G. (2011) 'Potential benefits of commercial willow Short Rotation Coppice (SRC) for farm-scale plant and invertebrate communities in the agri-environment', *Biomass Bioenergy*, vol. 35, pp. 325–336.

RSPB, Friends of the Earth and Greenpeace (2012) *Dirtier than coal? Why Government plans to subsidise burning trees are bad news for the planet* [Online] Sandy and Amsterdam, RSPB, Friends of the Earth and Greenpeace. Available at http://www.rspb.org.uk/Images/biomass_report_tcm9-326672.pdf (Accessed 28 April 2017).

Sanchez, D. L., Nelson, J. H., Johnston, J., Mileva, A. and Kammen, D. M. (2015) 'Biomass enables the transition to a carbon-negative power system across western North America', *Nature Climate Change*, vol. 5, pp. 230–234.

Sapphire Energy (2016) *The Sapphire Story* [Online]. Available at www.sapphireenergy.com (Accessed 11 May 2017).

Schiel-Bengelsdorf, B., Montoya, J., Linder, S., Dürre, P. (2013) 'Butanol fermentation', *Environmental Technology*, vol. 34, pp. 1691–1710.

Searchinger, T. D., Hamburg, S. P., Melillo, J., Chameides, W., Havlik, P., Kammen, D. M., Likens, G. E., Lubowski, R. N., Obersteiner, M., Oppenheimer, M., Robertson, G. P., Schlesinger, W. H. and Tilman, G. D. (2009) 'Fixing a critical climate accounting error', *Science*, vol. 326, pp. 527–528.

Searchinger, T. D., Heimlich, R., Houghton, R. A., Dong, F., Elobeid, A., Fabiosa, J., Tokgoz, S., Hayes, D. and Yu, T. (2008) 'Use of US croplands for biofuels increases greenhouse gases through emissions from land-use change', *Science*, vol. 319, pp. 1238–1240.

SEI (2008) *Annual Report 2008* [Online]. Available at https://www.sei-international.org/mediamanager/documents/Publications/SEI-AnnualReport-2008.pdf (Accessed 15 February 2017).

Shackley, S., Carter, S., Sims, K. and Sohi, S. (2011) 'Expert perceptions of the role of biochar as a carbon abatement option with ancillary agronomic and soil-related benefits', *Energy & Environment*, vol. 22, no. 3, pp. 167–88.

Sherrington, C. and Moran, D. (2010) 'Modelling farmer uptake of perennial energy crops in the UK', *Energy Policy*, vol. 38, no. 7, pp. 3567-3578.

Slesser, M. and Lewis, C. (1979) *Biological Energy Resources*, London, E. & F.N. Spon.

Smith, P., Powlson, D. S., Smith, J. U., Falloon, P. and Coleman, K. (2000) 'Meeting Europe's climate change commitments: quantitative estimates of the potential for carbon mitigation by agriculture', *Global Change Biology*, vol. 6, pp. 525–39.

Tang, J., Luyssaert, S., Richardson, A. D., Kutsch, W. and Janssens, I. A. (2014) 'Steeper declines in forest photosynthesis than respiration explain age-driven decreases in forest growth', *Proceedings of the National Academy of Sciences of the United States of America*, vol. 111, pp. 8856–8860.

Thraen, D., Witt, J., Schaubach, K., Kiel, J., Carbo, M., Maier, J., Ndibe, C., Koppejan, J., Alakangas, E., Majer, S. and Schipfer, F. (2016) 'Moving torrefaction towards market introduction - Technical improvements and economic-environmental assessment along the overall torrefaction supply chain through the SECTOR project', *Biomass and Bioenergy*, vol. 89, pp. 184–200 [Online]. DOI: http://dx.doi.org/10.1016/j.biombioe.2016.03.004 (Accessed 28 April 2017).

Tyner, W. E. (2008) 'The US Ethanol Biofuels Boom: Its Origins, Current Status, and Future Prospects', *Bioscience*, vol. 58, no. 7, pp. 646-653.

UNDP (2000) *World Energy Assessment*, New York, United Nations Development Programme. Available at http://www.undp.org/content/undp/en/home/librarypage/environment-energy/sustainable_energy/world_energy_assessmentenergyandthechallengeofsustainability.html (Accessed 25 February 2017).

UNICA (2016) *A Bioeletricidade da Cana em Números – Janeiro de 2016* [Online], Sao Paolo, Uniao da Industria de Cana-de-acucar. Available at http://www.unica.com.br/download.php?idSecao=17&id=28739779 (Accessed 25 February 2017).

Vamvuka, D. (2011) 'Bio-oil, solid and gaseous biofuels from biomass pyrolysis processes—an overview', *International Journal of Energy Research*, vol. 35, pp. 853–862.

Venturi, P. and Venturi, G. (2003) 'Analysis of energy comparison for crops in European agricultural systems', *Biomass Bioenergy*, vol. 25, pp. 235–255.

Viikari, L., Vehmaanpera, J. and Koivula, A. (2012) 'Lignocellulosic ethanol: from science to industry', *Biomass Bioenergy*, vol. 46, pp. 13–24.

Wang, Y., Fan, C., Hu, H., Li, Y., Sun, D., Wang, Y. and Peng, L. (2016) 'Genetic modification of plant cell walls to enhance biomass yield and biofuel production in bioenergy crops', *Biotechnology Advances*, vol. 34, pp. 997–1017.

WBA (2016) *WBA Global Bioenergy Statistics 2016* [Online], Stockholm, World Bioenergy Association. Available at http://www.worldbioenergy.org/content/wba-global-bioenergy-statistics-2016-0 (Accessed 5 May 2017)).

Weiland, P. (2010) 'Biogas production: current state and perspectives', *Applied Microbiology Biotechnology*, vol. 85, pp. 849–860.

WHO (2003) Diet, *Nutrition and the Prevention of Chronic Diseases: Report of a Joint WHO/FAO Expert Consultation*, Geneva, WHO, Technical Report Series 916.

Wilhelm, W., Johnson, J., Karlen, D., Lightle, D. (2007) 'Corn Stover to Sustain Soil Organic Carbon Further Constrains Biomass Supply', *Agronomy Journal*, vol. 99.

Wisner, R. (2009) *Corn, Ethanol and Crude Oil Prices Relationships – Implications for the Biofuels Industry* [Online], Ames, Iowa, Agricultural Marketing Resource Center,. Available at http://www.agmrc.org/renewable-energy/ethanol/corn-ethanol-and-crude-oil-prices-relationships-implications-for-the-biofuels-industry/ (Accessed 15 February 2017).

Wood Pellet Association of Canada (n.d.) *Global pellet market outlook in 2017* [Online]. Available at https://www.pellet.org/wpac-news/global-pellet-market-outlook-in-2017 (Accessed 2 May 2017).

Woods, J. (2001) 'The potential for energy production using sweet sorghum in Southern Africa', *Energy for Sustainable Development*, vol. 5, no. 1, pp. 31–38.

World Wide Recycling Group (n.d.) *The Wood Pellet Market* [Online]. Available at http://www.wwrgroup.com/en/biomass-market/the-wood-pellet-market (Accessed 2 May 2017).

WRAP (2009) *Household Food and Drink Waste in the UK* [Online]. Available at http://www.wrap.org.uk/sites/files/wrap/Household_food_and_drink_waste_in_the_UK_-_report.pdf (Accessed 25 February 2017).

WRAP (2016) *Gate Fees report 2016* [Online]. Available at http://www.wrap.org.uk/content/comparing-cost-alternative-waste-treatment-options-gate-fees-report-2016 (Accessed 18 February 2017).

Xu, J., Jiang, J., Sun, Y. and Lu, Y. (2008) 'Bio-oil upgrading by means of ethyl ester production in reactive distillation to remove water and to improve storage and fuel characteristics', *Biomass Bioenergy*, vol. 32, pp. 1056–1061.

Yamagata, Y., Yan, J., van Ypersele, J. P. (2001) 'Managing climate risk', Science, vol. 294, pp. 786-787.

Zegada-Lizarazu, W., Elbersen, H. W., Cosentino, S. L., Zatta, A., Alexopoulou, E. and Monti, A. (2010) 'Agronomic aspects of future energy crops in Europe', *Biofuels, Bioproducts and Biorefining*, vol. 4, pp. 674–691.

Chapter 6

Hydroelectricity

By Janet Ramage and Bob Everett

6.1 Introduction

> How to promote socio-economic development and eradicate poverty, whilst simultaneously halting environmental degradation, is one of the greatest challenges at the start of the 21st century. This challenge is most conspicuous in the policy for water and energy, as both are essential elements for human life.
>
> (Sustainable Hydropower, 2011a)

Water power, like most renewable energy sources, is indirect solar power, and like others such as the wind, it has been contributing to local energy supplies for many centuries. It is, however, unique in that it became a major 'modern' energy source over a hundred years ago, supplying the input for some of the earliest power stations. Hydroelectricity has become a well-established technology, delivering in 2016 about a sixth of the world's annual electricity supply (Figure 6.1).

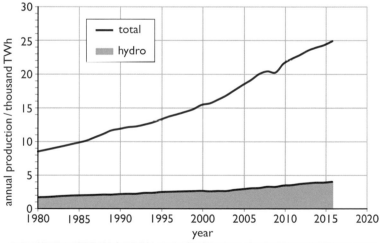

Figure 6.1 World total electricity production and hydro contribution, 1980–2016 (source: BP, 2017)

To introduce the terminology and main features of hydroelectric systems, this chapter starts with an account of one relatively modest hydro scheme. Commissioned over three-quarters of a century ago and still operating with much of its original plant, this group of power stations in Scotland exemplifies both the technical and economic aspects of hydroelectricity.

The chapter continues with a discussion of the nature of the hydro resource and its present contribution to world energy. This is followed by a summary of the basic science and a brief history of the development of water power, leading to the modern turbine systems that are the subject of Sections 6.7 and 6.8.

The remaining sections are concerned with the problems and the potential of hydroelectricity. We find the familiar issues of cost, firmness of supply and integration which arise for all renewable sources, but for large-scale hydroelectricity the questions are rather different: whether there are limits to growth, what determines these limits, and whether we are already reaching them.

6.2 The Galloway Hydros

The Galloway Hydroelectric Scheme on the River Dee in south-west Scotland makes an interesting study for several reasons. Initially commissioned in 1935, it was the first major UK scheme designed specifically to provide extra power at times of peak demand. Its six power stations are controlled as one integrated system. Several of its dams are on major salmon-fishing rivers, raising environmental issues common to many hydro schemes. It is also technically interesting because significant differences between the site conditions at the power station locations mean that across the system several types of turbo-generator are used.

Origins

The Galloway Hydros owe their origin to local pride and individual enthusiasm, and an Act of Parliament (ScottishPower, 2011). The first proposals to use the rivers and lochs of south-west Scotland for hydropower appeared in the 1890s, but the scheme became feasible only with the establishment of a National Grid in the 1920s. This meant that the great industrial conurbation of Glasgow became a potential customer, and also that the hydro system could meet the need for a plant that could quickly respond to daily and seasonal peaks in the otherwise coal-dominated grid system.

The scheme

The system (Figure 6.2) has three main elements. The first is Loch Doon, which provides the main long-term seasonal storage. Its natural outflow is to the north, but a dam now restricts this and the main flow is diverted eastwards through a 2 km tunnel into the upper valley of the Water of Ken. An interesting feature is the Drumjohn Valve: when demand for power is low, this directs the flow from two eastern tributaries of the

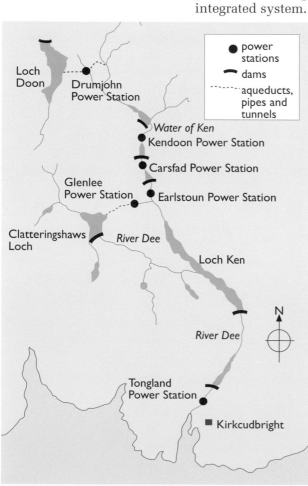

Figure 6.2 The Galloway Hydros (source: adapted from Hill, 1984)

Water of Ken through the tunnel in the 'reverse' direction *into* Loch Doon, adding to the stored volume. The level in the loch can vary by 12 metres, releasing 80 million cubic metres of water. Falling through the 200 metre height difference down to the final outflow at Tongland, this represents a gross release of some 150 million MJ of energy – over 40 million kWh.

The second element of the system, for fast response to short-term demand variations over the course of a day, is the series of dams and power stations along the course of the Water of Ken: Kendoon, Carsfad and Earlstoun, and the Tongland plant near the mouth of the River Dee.

Clatteringshaws Loch, the third element, is fed by the tributaries of the Dee, rising to the west of the main valley. This is the only completely artificial reservoir in the system, and its main outflow, through a tunnel nearly 6 km long and pipes with a fall of over 100 metres, supplies the 24 MW Glenlee power station before joining the Ken below Earlston. The remaining natural flow of the Dee and its tributaries is south-east into Loch Ken (into which the Ken also flows).

Table 6.1 gives key details for the different power plants in the scheme.

Figure 6.3 Carsfad power station and dam

Power

The essential characteristics of a hydro site are the **effective head** (H), the height through which the water falls, and the **flow rate** (Q), the number of cubic metres of water passing through the plant per second. As we shall see in Section 6.5, there is a simple approximate relationship between these two quantities and the power delivered by the water (P) at the *input to the turbine*, measured in kilowatts (kW):

$$P \text{ (kW)} = 10 \times Q \text{ (m}^3 \text{ s}^{-1}) \times H \text{ (metres)}$$

The conversion of energy carried by water into electrical energy is carried out by the **turbo-generator**: a rotating turbine driven by the water and connected by a common shaft to the rotor of a generator. (The electric power *output* will of course be rather less than this input, for reasons discussed in Section 6.5.)

Table 6.1 The Galloway power stations

Power station	Average head / m	Maximum flow / m³s⁻¹	Output capacity / MW	Number of turbines
Drumjohn	11	16	2	1
Kendoon	46	55	24	2
Carsfad	20	73	12	2
Earlstoun	20	71	14*	2
Glenlee	116	26	24	2
Tongland	32	127	33	3

*Although the *average* head and flow are similar to Carlsfad, a slightly higher dam at Earlstoun allows it to take advantage of seasonal variations in flow.
Source: ScottishPower, 2011

BOX 6.1 Capacity factor – a reminder

As we saw in Chapter 1, the annual capacity factor of any plant is equal to its actual annual output divided by its output if it were to generate continuously at its full rated power, expressed as a percentage.

So, if Carsfad, with its capacity of 12 000 kW, generates 30 million kWh of electricity in a year (remembering that there are 8760 hours in a year), its capacity factor will be:

$$30 \times 10^6 / (12\ 000 \times 8760) = 0.285, \text{ or } 28.5\%$$

Of course, in practice, the annual capacity factor of a plant depends on both the demand for its output and the power that it can produce at any given time.

The turbines

The head and the required power are critical in determining the most suitable type of turbine for a site. Glenlee's high head puts it at one extreme in the Galloway system, with Drumjohn's very low head and power rating at the other. Of the four river plants, Kendoon and Tongland have intermediate heads and fairly high power ratings whilst Carsfad and Earlstoun have almost identical low heads and powers.

All turbines consist of a set of curved blades designed to deflect the water in such a way that it gives up as much of its energy as possible. The blades and their support structure make up the turbine **runner**, and the water is directed on to this either by channels and guide vanes or through a jet, depending on the type of turbine. The Galloway plants include two types of runner – 'propellers' and Francis turbines (see Figure 6.17). As discussed in more detail later, 'propeller' types are most suitable for large flows at low heads and Francis turbines for medium to high heads. Comparison of Tables 6.1 and 6.2 shows that this is true for the Galloway scheme.

Table 6.2 The turbines

Power station	Turbine rating /MW	Turbine type	Rate of rotation* / rpm
Drumjohn	2	Propeller	300
Kendoon	12	Francis	250
Carsfad	6	Propeller	214.3
Earlstoun	7	Propeller	214.3
Glenlee	12	Francis	428.6
Tongland	11	Francis	214.3

* The significance of these rates is explained in Section 6.7.
Source: ScottishPower, 2011

The salmon

The principal environmental issue raised during the approval process was the possible effect on salmon fishing. Several dams blocked rivers below their salmon spawning pools, and concern was expressed about the fate of adult salmon making their way upstream and young smolt on the reverse journey. The response was the incorporation of fish ladders at four dams. These are a series of stepped pools with a constant downward flow of water to attract the fish, which leap up from pool to pool (Figure 6.4). The Doon dam had insufficient space for a long series of pools, so the fish ladder there is partly inside a round tower – it is claimed that the fish do not find this spiral staircase a problem. The issue does not arise at Glenlee since salmon do not use the man-made Clatteringshaws Loch. (It is worth noting that a much more detailed environmental impact statement would be required today.)

Economics

Figure 6.4 The fish ladder at Earlstoun

The Galloway scheme was built to supply extra power to the grid at times of peak demand. In other words, it was assumed that it would generate for only a few hours a day, resulting in an annual capacity factor of no more than perhaps 25%. This would suggest a poor return on the initial investment, which was in any case higher than the cost of a coal-fired plant of similar capacity. However, a number of circumstances made the scheme financially attractive:

- The company was able to assume a firm demand for the planned output.
- During its first three years the scheme received an annual treasury grant of £60 000 (about £4 million at present-day values), reducing the cost per kWh by about 20%.

From the start, demand, and the consequent economic performance, exceeded expectations, and only in a few years of serious drought has output fallen below the planned level. After more than 80 years, the original five plants are still generating power, joined in 1984 by the 2 MW plant at the Drumjohn Valve. The original construction costs were repaid many years ago.

In 2009, ScottishPower announced a £20 million investment programme to upgrade equipment at three hydroelectric sites, including the Galloways and Cruachan (Figure 6.8), which is intended to safeguard their operation until at least 2035 (ScottishPower, 2010).

6.3 The resource

For hydroelectricity, as for other renewables, the resource is basically an amount of *power* – a rate at which energy is delivered. However, for comparison with other major energy sources, where the resource is effectively a quantity of *stored energy* (tonnes of coal, barrels of oil, etc.), it is usual to express a hydro resource in terms of the total energy it delivers in the course of a year. The customary unit on the national or world scale is **terawatt hours per year, TWh y^{-1}** (1 TWh = 3.6 PJ, or 1 EJ = 278 TWh).

The world resource

We saw in Chapter 1, Figure 1.5 that about 20% of the 5.4 million EJ (1.1 billion TWh) of solar energy reaching the Earth's atmosphere each year is consumed in the evaporation of water. The water vapour in the atmosphere therefore represents an enormous, constantly replaced, store of energy. Unfortunately most of it is not available to us. When the water vapour condenses into water, most of its stored energy is released into the atmosphere as heat, and ultimately re-radiated into space. But a tiny fraction, about 200 000 TWh y^{-1}, reaches the Earth as rain or snow. Roughly one fifth of this precipitation falls on hills and mountains, descending ultimately to sea level as the world's streams and rivers. The 40 000 TWh y^{-1} of energy carried by this flowing water can be regarded as the world's **total hydro resource**.

It is obviously not possible – or desirable – to build hydro plants on every river or stream, so the usable fraction of this flow will be significantly lower, and, in 2017, the world's **technical hydro potential** is estimated to be about 16 000 TWh y^{-1}, or roughly two-fifths of the above total resource (WEC, 2010). This figure must be regarded as approximate, in part due to the different ways countries assess their resource. Some, for instance, include in their 'technical potential' only those sites that have been fully surveyed.

There remains one important question in the assessment of the realistic resource: 'How much hydroelectricity is available at a cost that is competitive with power from other sources?' In other words, what is the **economic potential** for hydroelectricity? The issues involved in financing hydro plants are discussed in Section 6.11, and Chapter 11 considers the relative costs of electricity from different sources; but the economic potential for any country or region must also take into account the social and environmental effects of dam construction and the likely flooding of large areas of land.

This obviously leaves many questions open, and we should not be surprised to find that estimates of the economic potential for hydropower in different countries and regions are generally regarded as much less reliable than the estimates of the total or technical potential. The following discussion of resources therefore considers only the *technical potential*, leaving the financial aspects to Section 6.11.

Regional resources

The first two columns of Table 6.3 show the estimated hydro technical potential for different regions of the world, together with their percentage share of the world's total.

Table 6.3 Regional and world hydro potential and generated output, 2016

Region	Technical potential		Output	
	Technical potential / TWh y⁻¹	% of world technical potential	Annual output / TWh y⁻¹	Percentage of technical potential used
North America	2420	15%	680	28%
South America	2840	18%	689	24%
Europe and Eurasia	2760	17%	892	32%
Middle East	280	1.7%	21	7%
Africa	1890	12%	114	6%
Asia Pacific	5820	36%	1627	28%
World	16 000		4023	25%

Sources: WEC, 2010; BP, 2017

The third and fourth columns show the extent to which the regions have actually developed their hydro potential.

Overall world hydro output in 2016 was a quarter of the technical potential, suggesting that a fourfold global increase in output might be possible in future. But there are marked regional differences. There would appear to be considerable scope for development in Africa. It has 12% of the world's total technical hydro potential, but only 6% of this has been developed. One major African scheme, Grand Inga, is described later in Section 6.12.

National resources

Table 6.4 shows data for the 13 countries whose hydro output in 2015 was greater than 50 TWh, together with three other European countries of interest. It is worth noting that several countries, for example Japan, Sweden, Italy and Switzerland, appear to have developed more than half their technical potential already; but, as pointed out above, some may limit their estimates of 'technical potential' to sites that have been studied in detail and possibly passed a full environmental assessment.

Table 6.4 National hydro potential and contributions, 2015

Country	Technical potential / TWh y⁻¹	Annual output / TWh y⁻¹	Installed capacity * / GW	Average capacity factor	Percentage of nation's electricity
China	2500	1115	296	43%	19%
Canada	830	378	79	55%	58%
Brazil	1250	360	92	45%	62%
USA	1340	247	79	36%	6%
Russia	1670	170	49	40%	16%
Norway	240	138	29	54%	95%
India	660	133	42	36%	10%
Japan	140	84	22	44%	8%
Venezuela	260	76	15	58%	59%
Sweden	130	75	16	54%	46%
Turkey	220	67	26	29%	26%
Vietnam	120	57	15	43%	36%
France	100	54	18	34%	9%
Italy	65	46	22	24%	16%
Switzerland	43	38	14	31%	54%
Austria	75	37	8	53%	57%

Note: * excluding pumped storage
Source: WEC, 2010; WEC, 2016; BP, 2017

The output from any hydroelectric plant will of course depend on the available flow of water, but the overall output for an individual country will also depend on the role played by hydroelectricity in the national power system. As Table 6.4 shows, the countries with the highest capacity factors tend to be those where hydropower makes a significant contribution, but where it is not the only major source of electricity – a situation that allows the relatively cheap hydropower to be used to its full potential, with an alternative source of electricity being used when hydropower cannot meet demand.

In general, if almost all of a nation's electricity comes from hydro plants, annual capacity factors are usually lower, because the installed capacity must be large enough to meet the maximum demand experienced during any day (or year). Conversely, it is interesting to note that countries with significant amounts of nuclear power, such as France and Switzerland, have low hydro capacity factors, a consequence of the use of hydroelectricity mainly for providing power at times of high demand.

Compared with the countries in Table 6.4, the hydro resource of the UK is small, only 14 TWh y⁻¹, and the installed capacity in 2015 was under 1.8 GW (BEIS, 2016a), mainly in Scotland. The annual output has varied in recent years between about 3.3 and 6.3 TWh – reflecting year-on-year weather variations throughout this relatively small area. In 2015, the average capacity factor was 40% and hydro supplied only 2% of the UK's electricity.

World output

In 1900, about two decades after the first commercial hydroelectric plants, world annual output had reached an estimated 3.7 TWh from an installed capacity of about 1.3 GW. Despite two world wars and the Great Depression of the 1930s, world hydro generation rose at a continuous annual rate of nearly 10% per year throughout the first half of the twentieth century. Figure 6.5 shows the trend since 1965, an average annual increase of about 60 TWh per year, which has been dominated since 2000 by hydro development in China. This has included the construction of two of the world's largest hydro schemes, the 22.5 GW Three Gorges Dam (described later) and the 13.9 GW Xiluodo Dam.

Total world hydro generation in 2016 was about 4000 TWh and the estimated total installed generation capacity at the end of that year was nearly 1100 GW (REN21, 2017).

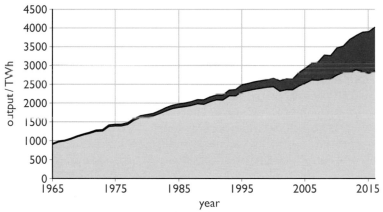

Figure 6.5 World annual hydroelectricity output, 1965–2016 (source: BP, 2017)

The noticeable drop in world output in 2001 was attributed mainly to exceptionally dry conditions in the Americas, the source of about a third of world output.

6.4 Small-scale hydro (SSH)

In the early days of electric power, generators with output ratings between a few kilowatts and a few megawatts were installed on streams or rivers, often using the dams and sluices of old watermills. As late as the mid-twentieth century, many towns in Europe and elsewhere were still served by hydro or other plants in the upper part of this range. However, with continually rising demand for electricity, and the growth of national transmission networks, capacities of several hundred megawatts have become the norm for modern power stations, and plant outputs below about 10 MW are now referred to as **small-scale**. However, this cut-off point is not universally accepted. The figure in the UK is only 5 MW. In 2015 small-scale systems supplied about 15% of the UK's hydro output (BEIS, 2016a). Other countries use higher capacity ratings, for example 25 MW in India, 30 MW in Brazil, 50 MW in Canada and China, and up to 100 MW in the USA (IPCC, 2012). A 'small' hydro scheme could thus include a plant capable of powering a small city.

The term **micro-hydro** is commonly used for plants with capacities below 100 kW and the term **pico** appears occasionally for very small plants; but again there is a lack of agreement between countries on these definitions.

The past few decades have seen growing interest in smaller power plants, for several reasons. In the industrialized countries, environmental issues have increasingly limited the potential for further major hydro development, whilst small-scale schemes (such as the Elan Valley scheme described in Box 6.2), are considered to produce fewer deleterious effects, and have received growing encouragement and financial incentives.

More recently, a market has begun to emerge for micro-hydro plants, generating a few tens of kilowatts for an isolated house or farm.

Small-scale hydro is currently more costly than electricity from large hydro or other conventional sources and in many European countries investment in electricity from renewables during the first decade of the twenty-first century has concentrated on wind and solar PV.

In an entirely different context, small-scale plants are a practicable independent option for electricity in developing countries without extensive grid systems. Large areas of Nepal, for instance, have no electricity supply and only mules or human porters for transport, but do have many mountain streams suitable for 'high-head' plants. This has led to the development of local industries producing extremely small-scale systems, transportable by a single person on foot. The Peltric (Pelton electric) turbo-generator set, for example, consists of a tiny Pelton wheel (see Section 6.8) driving a simple generator. Operating under heads of 50–70 metres, it produces a kilowatt or so of output. With cheap and relatively simple 'civil works', these tiny systems have proved sufficiently popular to be copied in other countries.

BOX 6.2 Elan Valley

The Elan Valley scheme in the Cambrian Mountains in mid-Wales (Figure 6.6) was financed in 1997 under the UK's Non-Fossil Fuel Obligation (NFFO) low carbon electricity incentive scheme. It supported the development of five power stations. Like the Galloway Hydros they are in sequence along two rivers; but these plants use the dams and reservoirs of an existing drinking water-supply system – and their total output capacity is only 4.2 MW.

Table 6.5 gives some details and the contribution from Foel Tower on the Garreg Ddu reservoir is worth noting: its Kaplan turbine (see Section 6.8) sits inside one of the 14 metre diameter pipes supplying water to the city of Birmingham.

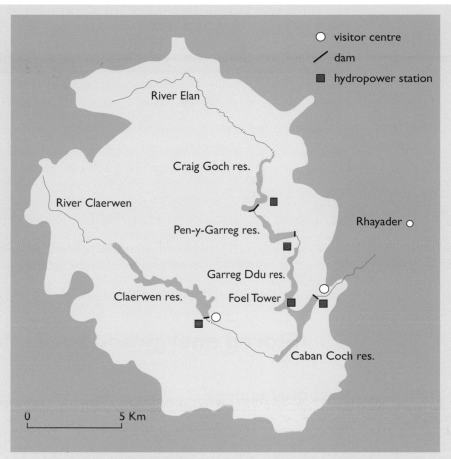

Figure 6.6 The Elan Valley hydro scheme

Table 6.5 The Elan Valley plants

Site	Craig Goch	Pen-y Garreg	Caban Coch	Foel Tower	Claerwen
Head / m	36.5	37.5	37	13.5	56
Capacity / kW	480	810	950	300	1680
Turbine	Francis	Francis	2 × Francis	Kaplan	Francis

All the reservoirs and dams date from the early twentieth century – the latter are listed historic monuments. Caban Coch was already a hydro plant. One of its old turbines was replaced some time ago and the other as part of the NFFO scheme. In a region of great natural beauty and with 90% of its area designated as Sites of Special Scientific Interest, the four new plants have required careful siting. To meet environmental constraints, their turbines make use of existing discharge pipes and their associated buildings are mainly below ground (Elan Valley Trust, n.d.).

It is generally agreed that world output from small-scale hydro plants is rising; but because many countries do not report their statistics separately it is impossible to estimate either the total output or the installed capacity with any reasonable precision. A World Energy Council survey in 2010 (WEC, 2010) suggests that the global total small-scale capacity (<10 MW) at the end of the 2009 was about 60 GW (including an estimated 33 MW from those plants in China that meet the <10 MW criterion). This is about

6% of the world's total hydro capacity, a proportion that seems to have remained much the same for several decades.

BOX 6.3 Direct uses of water power

Although the generation of electricity is by far the major use of water power today, other uses still remain. The kinetic energy of moving water (or the potential energy of water at a height) can be used directly to drive machines – indeed this was the sole use of water power until the mid-nineteenth century.

Old watermills do still operate in a few places in today's industrialized countries, grinding corn or sawing wood; and some mountain railways use a counter-balancing tank filled with water at the top and emptied at the foot of the hill. But these are now rarities.

In the developing countries, direct water power plays a slightly greater role. In Nepal, for instance, simple turbines which can be produced locally are used to drive machinery, and in the Middle East and Asia the use of a flowing stream to raise water for irrigation has by no means disappeared. Nevertheless, the worldwide energy contribution from direct use is negligibly small compared with the hydroelectric power output.

6.5 Stored energy and available power

Potential energy

Water held at a height represents stored energy – the gravitational potential energy discussed in Chapter 1. As we saw, if M kilograms are raised through H metres, the stored potential energy in joules is given by the following simple equation:

$$\text{potential energy (joules)} = M \,(\text{kg}) \times \text{g} \times H \,(\text{m})$$

where g here is the acceleration due to gravity (about 9.81 m s^{-2}, although for rough calculations, 10 m s^{-2} is often used). Thus about 10 joules of energy input are needed to lift one kilogram of anything vertically through one metre against the gravitational pull of the Earth.

However, here we are concerned with large reservoirs, whose capacities are almost always given in *cubic metres*, rather than kilograms. The necessary conversion is simple, because one cubic metre of fresh water has a mass of 998 kg, which is so close to 1000 kg that the tiny difference is not significant in most calculations. We can therefore say that, within this degree of precision, the energy stored by a volume of V cubic metres of water raised through a height H is:

$$\text{stored energy (joules)} = 1000 \times V \times \text{g} \times H \tag{1}$$

and this is therefore the energy that will be released when this volume of water *falls* through a vertical distance H.

The joule is a very small unit of energy. A more familiar unit of energy is the kilowatt-hour which is 3.6 MJ. Storing 1 kWh requires raising 1 cubic metre of water through a height of 360 metres. This is perhaps quite a sobering view of the amount of energy required to run a one bar electric fire for an hour.

Pumped storage

There has been growing interest in the *storage* of electrical energy in recent decades. One major reason for this that has been steadily emerging since the 1960s has been the growth in the size of power plants, both fossil and nuclear, which can now easily exceed 500 MW output. A failure of a turbine, or an electrical fault, could mean that this amount of generating capacity could suddenly disappear from the grid. It might take an hour or more for other conventional power plants to 'ramp up' their generation to provide extra power.

Similar problems can result from the synchronization of consumer electrical demand. The end of a popular TV show or football match can produce a rapid 'demand pickup' of hundreds of megawatts as thousands of electric kettles are turned on.

The need to provide rapid short-term back-up for a period of several hours is thus vital to prevent power cuts.

The second reason is the increasing use of renewables such as PV and wind for electricity generation. Here the problem is that the output may vary considerably from hour to hour. Back-up power is therefore needed which can be brought on stream quickly.

At present the most practicable and economically viable way to store electrical energy in large quantities is to use it to pump water up a mountain. Pumped storage has thus become increasingly important, with installed capacity worldwide having grown from 78 GW in 2005 to 150 GW in 2016 – equivalent to one seventh of the world's total hydro capacity (REN21, 2017).

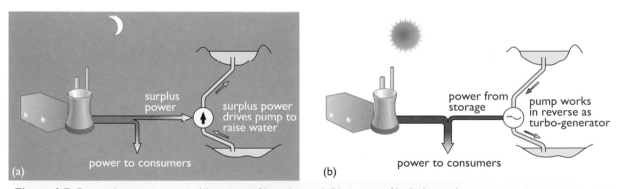

Figure 6.7 Pumped storage system: (a) at time of low demand, (b) at time of high demand

The principle is simple. Electrical energy is converted into gravitational potential energy when the water is pumped from a lower reservoir to an upper one, and the process is reversed when it is released to run back down, driving a turbo-generator on the way (Figure 6.7). The economic viability of the method depends on two nice technological facts.

- A suitably designed generator can be run 'backwards' as an electric motor: the machine which converts mechanical energy into electrical energy can perform the reverse process.
- A suitably designed turbine (Section 6.8) can also run in both directions, either extracting energy *from* the water as a turbine or delivering energy *to* the water as a pump.

The complete reversal is thus **turbo-generator** to **electric pump**. The machines must of course be designed for this dual role, but the cost saving is obviously significant.

There will, as always, be losses associated with the conversion processes, but turbines and generators are very efficient, and nearly 80% of the input electrical energy can be retrieved as electrical output when needed. The value of the system is enhanced by its speed of response: any of the six 300 MW Francis turbines of the Dinorwig storage plant in Wales (commissioned in 1984) can be brought to full power in just 12 seconds if initially spinning in air, and even from complete standstill the process takes only a few minutes (Dinorwig, n.d.). Pumped storage is thus particularly useful as back-up in case of sudden changes in generation or demand, or a failure elsewhere in a grid system.

The location must of course be suitable. A low-level reservoir of at least the capacity of the upper one must be available, or must be constructed. Sites such as Cruachan in Scotland (Figure 6.8), where the mountains rise from a large loch or lake, are obviously ideal. The high-level reservoir, behind a large dam, provides an operating head of 365 metres. Running the four 100 MW reversible machines for 20 hours at full capacity, as electric pumps or turbo-generators, raises or lowers the reservoir level by about 15 metres, storing or releasing about 8 million kilowatt-hours of energy (Cruachan, 2017).

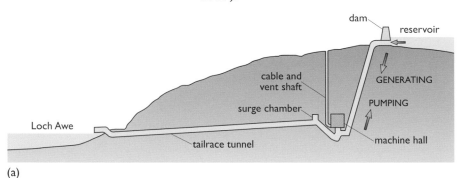

Figure 6.8 Cruachan pumped storage plant, commissioned between 1965 and 1967 (a) the installation, (b) the dam

Pumped storage can be combined with 'normal' hydroelectric generation in locations where the potential exists. The upper reservoir will in any case have a local catchment area, so there may be a positive net output from the plant.

There are however trade-offs. Very high heads have the advantage of needing smaller reservoirs for a given amount of stored energy, but the types of turbine most suited to high heads (see Section 6.8) cannot be run 'backwards' as pumps. At the other extreme, very low heads need much greater volumes of stored water, but the switch from pumping to generating may be achieved simply by reversing the pitch of the propeller-type Kaplan turbines used in such circumstances.

Power, head and flow rate

In estimating the value of any proposed hydroelectric plant, the *power* available at any time is probably the most important factor. The power supplied by a plant, the number of *watts*, is the rate at which it delivers energy: the *number of joules per second*. This will obviously depend on the **volume flow rate** of the moving water. Note that this is not just the speed of the water; it is the number of *cubic metres per second* passing through the plant, usually represented by the symbol Q (think quantity). It then follows from Equation (1) above that the power P (in joules per second or watts), will be

$$P \text{ (watts)} = 1000 \times Q \text{ (m}^3 \text{ s}^{-1}) \times g \times H \text{ (m)} \tag{2}$$

However, resource estimates must take into account energy losses. In any real system the water falling through a pipe will lose some energy due to frictional drag and turbulence, and the **effective head** will thus be less than the actual, or **gross head**. These flow losses vary greatly from system to system: in some cases the effective head is no more than 75% of the actual height difference, in others as much as 95%. Then there are energy losses in the plant itself. Under optimum conditions, a hydroelectric turbo-generator is extremely efficient, converting all but a few per cent of the input power into electrical output. Nevertheless, the **efficiency** – the ratio of the output power to the input power, usually expressed as a percentage – is always less than 100%. With these factors incorporated, the output power becomes:

$$P = 1000 \times \eta \times Q \times g \times H \tag{3}$$

where H is now the effective head and η (Greek letter eta) is the turbo-generator efficiency. (Note that although efficiency is often quoted as a percentage, in an equation like this it will be a number between 0 and 1. For example, 85% becomes 0.85.)

If we now express P in kilowatts, and use the approximation $g = 10 \text{ m s}^{-2}$, we obtain a very useful, simple expression:

$$P \text{ (kW)} = 10 \times \eta \times Q \text{ (m}^3 \text{ s}^{-1}) \times H \text{ (m)} \tag{4}$$

Box 6.4 shows how this can be used for rough calculations of power output.

BOX 6.4 Available power

As examples of power calculations, we can consider two systems, each with a plant efficiency of 83%, but of very different sizes.

The first site is a mountain stream with an effective head of 25 metres and a modest flow rate of 600 litres a minute, which is 0.010 cubic metres per second. Using Equation (4), we find that the power output will be

   $P = 10 \times 83\% \times 0.010 \times 25 = 2.075 \text{ kW}$

In contrast, suppose that the effective head is 100 m and the flow rate is 6000 cubic metres per second – roughly the total flow over Niagara Falls. The power is now

   $P = 10 \times 83\% \times 6000 \times 100 = 4.98 \text{ million kWh, or nearly 5 GW}$

6.6　A brief history of water power

The prime mover

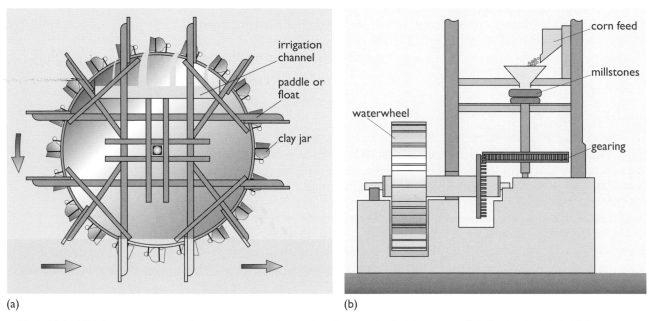

(a)　　　　　　　　　　　　　　　　　　　　　　(b)

Figure 6.9 (a) A noria – in this earliest waterwheel the paddles dip into the flowing stream and the rotating wheel lifts a series of jars, raising water for irrigation, (b) A Roman mill – this corn mill with its horizontal-axis wheel was described by Vitruvius in the first century BC (note the use of gears)

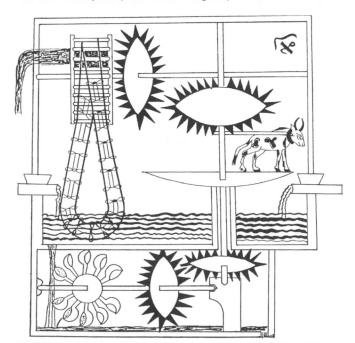

Figure 6.10 Medieval saqiya. The diagram comes from the Book of Knowledge of Ingenious Mechanical Devices of al-Jahazi, written in Mesopotamia 700 years ago. (The ox is a wooden cut-out designed to fool the public, who can't see the hidden waterwheel and gears below.)

Moving water was one of the earliest energy sources to be harnessed to reduce the workload of people and animals. No one knows exactly when the waterwheel was invented, but irrigation systems existed at least 5000 years ago and it seems probable that the earliest water power device was the *noria*, used to raise water for this purpose (Figure 6.9(a)). It appears to have evolved over a long period from about 600 BC, perhaps independently in different regions of the Middle and Far East (Strandh, 1989).

The earliest **watermills** were probably corn mills, which seemed to appear during the first or second century BC in the Middle East, and a few centuries later in Scandinavia. In the following centuries, increasingly sophisticated watermills were built throughout the Roman Empire and beyond its boundaries in the Middle East and Europe (Figures 6.9(b) and 6.10). In England, the Saxons are thought to have used both horizontal- and vertical-axis wheels. The first documented mill was in the eighth century, but three centuries later the Domesday Book of

1086 AD recorded about 5000, suggesting that every settlement of any size had a mill.

Raising water and grinding corn were by no means the only uses of the watermill, and during the following centuries the applications of this power source kept pace with the developing technologies of mining, iron working, paper-making, and the wool and cotton industries. Water was the main source of mechanical power, and by the end of the seventeenth century England alone is thought to have had some 20 000 working mills.

There was much debate on the relative efficiencies of different types of waterwheels (Figure 6.11), and the period from about 1650 until 1800 saw some excellent scientific and technical investigations of various designs. These revealed output powers ranging from about one horsepower to perhaps 60 for the largest wheels (in modern terms roughly 1–50 kW) with overshot wheels being in principle the most efficient. They also confirmed that for maximum efficiency the water should pass across the blades as smoothly as possible and fall away with minimal speed, having given up almost all its kinetic energy: two features that were to be important in modern turbines.

But then steam power entered the scene, putting the whole future of water power in doubt.

Nineteenth-century hydro technology

An energy analyst writing in the year 1800 would have painted a very pessimistic picture of the future for water power. The coal-fired steam engine was taking over and the waterwheel was fast becoming obsolete. However, like many later experts, this one would have been suffering from an inability to foresee the future. A century later the picture was completely different: the world now had an electrical industry and a quarter of its generating capacity was water powered.

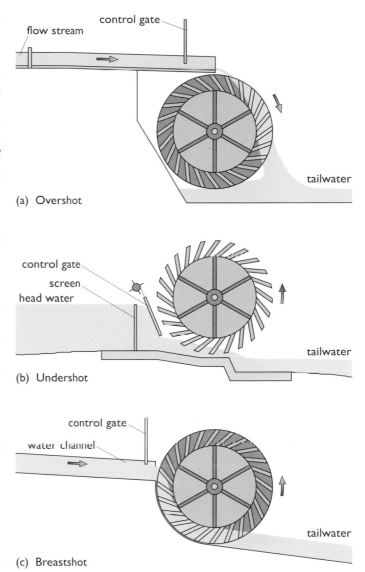

(a) Overshot

(b) Undershot

(c) Breastshot

Figure 6.11 Types of waterwheel (a) Overshot – water falls onto blades with closed sides, (b) Undershot – driven by water pressure against lower blades, (c) Breastshot – water strikes paddles at about the level of the wheel axle

The growth of the electric power industry was the result of a remarkable series of scientific discoveries and developments in electro-technology during the nineteenth century, but significant changes in what we might now call *hydro technology* also played their part. In 1832, the year of Faraday's discovery of the principle of the electric generator, a young French engineer, Benoît Fourneyron, patented a new and more efficient waterwheel: the first successful water **turbine**. (The name, from the Latin *turbo*: something that spins, was coined by Claude Burdin, one of Fourneyron's teachers.) The traditional waterwheel, essentially unaltered for nearly two thousand years, had finally been superseded.

Fourneyron's turbine (Figure 6.12) incorporated many new features. It was a vertical-axis machine, itself something of a novelty. But the important innovations were that the turbine ran *completely submerged* and that the water entered the turbine vertically, *along the axis*, and was directed outwards by **guide vanes** fixed on to the *blades* mounted on the *runner*, which was free to rotate. These are the features that ensured the smooth flow of water, which is essential for high efficiency. The water entering at the centre travelled horizontally outwards almost parallel to the faces of the runner blades as it reached them. Deflected as it crosses the faces, it exerted a sideways pressure that transmitted energy to the runner. Having given up its energy, it then falls away into the outflow.

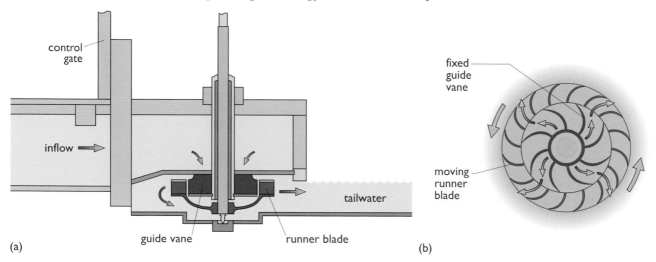

(a)

(b)

Figure 6.12 Fourneyron's turbine – the runner consists of a circular plate with curved blades around its rim and a central shaft (a) vertical section, (b) the flow across guide vanes and runner blades

Tests showed that Fourneyron's turbine converted as much as 80% of the energy of the water into useful mechanical output – an efficiency previously equalled only by the best overshot wheels. The rotor could also spin much faster, an advantage in driving 'modern' machines. The first pair of these turbines to come into use were installed in 1837 in the small town of St Blasien in the Grand Duchy of Baden (now part of southern Germany). Development did not stop there, and within a few years the American engineer James Francis started his experiments on *inward-flow radial* turbines which ultimately led to the modern machines known by his name (see Section 6.8).

These turbines were of course used to provide *mechanical* power. Half a century of development was needed before Faraday's discoveries were translated into a major electrical industry. In 1878 Lord Armstrong, a wealthy Victorian engineer and inventor, installed one of the first 'hydro-electric' plants to provide electric lighting in 'Cragside', his house in the north of England (National Trust, 2017). This house has now been re-equipped with a modern Archimedes Screw generator (see Section 6.8).

About three years later and 300 miles south, the small town of Godalming opened the world's first public electricity supply, powered by the

River Wey. Unfortunately, the water source proved unreliable and the waterwheel was soon replaced by a steam engine. Credit for the first *successful* public hydroelectricity scheme therefore usually goes to a little 12 kW plant on the Fox River in Wisconsin, USA, providing lighting for local paper mills from 1882.

From these primitive beginnings, the electrical industry grew during the final decades of the nineteenth century at a rate far exceeding that of any earlier technology. The capacities of individual power stations, including many hydro plants, were also rising, increasing tenfold, from about 100 kW to over 1 MW, during the 1890s. This growth was to continue. As described in Section 6.3 world total hydro output rose over a thousand-fold from 3.7 TWh in 1900 to over 4000 TWh in 2016. Over the same period, the outputs of the largest turbo-generators rose to over 780 MW. Of these, 18 have been deployed in the Xiluodu dam in western China, giving a total capacity of nearly 14 GW (Daily Fusion, 2013).

It is interesting to note, however, that not only the modes of operation but the main components of today's enormous turbo-generators (and today's small- or micro-hydro plants) are very similar to those of their much smaller precursors of over a hundred years ago. These components, mainly the turbines, are the subject of the next section.

6.7   Types of hydroelectric plant

Present-day hydroelectric installations range in capacity from a few hundred watts to more than 20 GW – a factor of some hundred million between the smallest and the largest output. We can classify installations in different ways:

- by the effective head of water
- by the capacity – the rated power output
- by the type of turbine used
- by the location and type of dam, reservoir, etc.

These categories are not of course independent of one another. The available head is an important determinant of the other factors, and the head and capacity together largely determine the type of plant and installation. We start therefore with the customary classification in terms of head, but shall soon see that it is really the fourth criterion that matters.

Low, medium and high heads

Two hydroelectric plants with the same power output could be very different: one using the huge volume flow of a slowly moving river and the other a relatively low volume of high-speed water from a mountain reservoir. Sites, and the corresponding hydroelectric installations, can be classified as *low*, *medium* or *high* head. The boundaries are fuzzy, and tend to depend on whether the subject of discussion is the civil engineering work or the choice of turbine; but **high head** usually implies an effective head of appreciably more than 100 metres and **low head** less than perhaps ten metres. Figure 6.13 shows the main features of the three types.

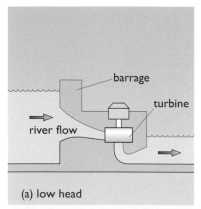
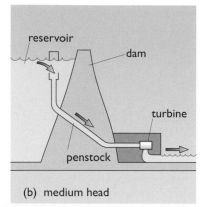
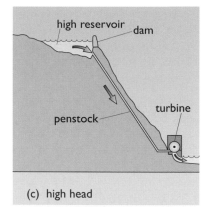

(a) low head   (b) medium head   (c) high head

Figure 6.13 Types of hydroelectric installation

The low dam or barrier of the installation in Figure 6.13(a) serves to maintain a head of water and also houses the plant. It may incorporate locks for ships (or as we have seen above, a fish-ladder for salmon). '*Run-of-river*' power stations of this type, having relatively little storage capacity, are dependent on the prevailing flow rate and can present problems of reliability if the flow varies greatly with the time of year or the weather. The large volume flow through a low-head plant means that the plant and the associated civil engineering works are likely to be massive, which means high capital cost, although this may be ameliorated where there is a second function such as flood control or irrigation.

The plant in Figures 6.13(b) and 6.14 is typical of the very large hydroelectric installations with a dam at a narrow point in a river valley. The large reservoir behind the dam provides sufficient storage to meet demand in all but exceptionally dry conditions. (It will also have flooded an extensive area and may not have been entirely welcomed by the population (see Section 6.10).) The USA has some of the world's largest dams of this type, including the Grand Coolee (170 m high), which when completed in 1942 had the distinction of being the first artificial bulk structure with a volume greater than the Great Pyramid! On this scale, the civil engineering costs are obviously considerable, but the large reservoir normally ensures a reliable supply. Systems of this type don't of course have to be on a gigantic scale, and quite small reservoirs can provide power for a hydroelectric plant located below their dams.

Figure 6.14 The Hoover Dam, 1936. This dam on the Colorado River (originally called the Boulder Dam) is 220 metres high and its reservoir, Lake Mead, holds 35 billion cubic metres of water. The 2.1 GW power plant is at the foot of the dam.

To call the 220 metre head of the Hoover Dam 'medium' may seem rather surprising, but it illustrates the fact that the distinction between this and high-head systems lies more in the type of installation. Figure 6.13(c) shows the difference. In the high-head plant the entire reservoir is well above the outflow, and the water flows through a long **penstock** – possibly passing through a mountain – to reach the turbine. (The penstock was originally the wooden gate or 'stock' which controlled the flow of 'penned-up' water. It later came to mean the channel, or the pipe, carrying the flow.)

With a high head, the flow needed for a given power is much smaller than for a low-head plant, so the turbines, generators and housing are more compact. But the long penstock adds to the cost, and the structure must be able to withstand the extremely high pressures below the great depth of water – as much as 100 atmospheres for a 1000 metre head (Box 6.5).

BOX 6.5 Height, depth and pressure

The pressure in a liquid (or gas) is the force with which it presses on each square metre of surface of anything submerged in it.

Atmospheric pressure, due to the weight of the air above us, is equivalent at sea level to the weight of a 10 tonne mass acting on each square metre of any surface. As you move up through the atmosphere, the pressure decreases, initially by about 1% per 100 metres vertically.

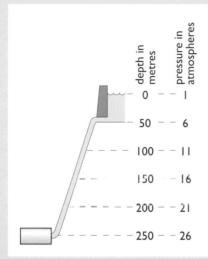

Figure 6.15 Depth and water pressure

As you move down through any body of water, the pressure increases due to the increasing weight of water above. Since water is several hundred times denser than air, the change is much more noticeable – at a depth of about 10 metres the pressure is twice that at the surface: a pressure of two atmospheres. This increase of about one atmosphere per 10 metres continues as the depth becomes greater (Figure 6.15).

Rates of rotation

The **rate of rotation** of a turbine is the number of completed revolutions per minute (rpm), and, as we saw for the Galloway plants in Table 6.2, this can vary appreciably depending on the site and the turbine type. However, if the turbine drives the generator directly, only certain rates of rotation are permitted, for the following reason.

The alternating voltages from all the power stations contributing to any grid system must have the same frequency. In European countries and many others, the agreed frequency is 50 Hz (hertz, or cycles per second).

So in a very simple generator, consisting of a magnet and a pair of coils, the magnet (or the coils) would need to spin at 50 Hz, that is 50 cycles per second, which is 50 × 60 = 3000 rpm. In the usual terminology, this simple system would be called a **two-pole generator.**

A large power-station generator will have more than one pair of coils on its spinning rotor (see Figures 6.16 and 6.22). A 20-pole machine, for instance, would need to rotate at only 300 rpm, a tenth of the above rate, to produce the required 50 Hz. A little arithmetic reveals that all the rates of rotation shown earlier in Table 6.2 are sub-multiples of 3000 rpm.

Note however that in the USA and other countries where the supply frequency is 60 Hz, the rates of rotation are sub-multiples of 3600 rpm. (Generators are discussed in more detail in Chapter 9, *Electricity*, of Everett et al., 2012.)

Estimating the power

Reliable data on flow rates and, equally important, their variations, is essential for the assessment of the potential capacity of a site. Stopping the flow and catching the water for a measured time is hardly practicable for large flows or as a routine method. The preferred techniques depend on establishing empirical relationships between flow rate and either water depth or water speed at chosen points. Simple depth or speed monitoring then provides a record of flow rates. For many major rivers, particularly in developed countries, such data has been accumulated for years.

Where such records are not available, an entirely different approach is to determine the annual precipitation over the catchment area. This gives the total flow into the system and is particularly suitable for large systems. However, allowance must be made for losses due to processes such as re-evaporation, take-up by vegetation or leakage into the ground, and as these could account for as much as three-quarters of the original total they are hardly negligible corrections.

Dealing with time variations adds further problems. In most areas there will be seasonal changes, but these at least come at known times. The more serious problems are with changes over very long or very short periods. Year-to-year variations can be large: the average annual precipitation on the catchment area of the River Severn in the UK, for instance, is 900 mm but it can range from as little as 600 mm to as much as 1200 mm. For countries which depend heavily on hydroelectric power a succession of dry years can mean a serious supply shortage.

At the other extreme, the installation must be designed to survive the '100-year flood', the sudden rush of water following unusually heavy rain. As in any power system, the need to guard against rare but potentially catastrophic events adds to the cost.

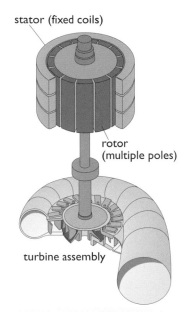

stator (fixed coils)

rotor (multiple poles)

turbine assembly

Figure 6.16 Multi-pole generator in a hydro power plant

6.8 Types of turbine

Present-day turbines come in a variety of shapes (Figure 6.17). They also vary considerably in size, with runner diameters ranging from as little as a third of a metre to some 20 times this. In the next four subsections we look at how they work, the factors that determine their efficiency, and the site parameters that determine the most suitable turbine.

Francis turbines

Francis turbines (Figures 6.17 to 6.20) are by far the most common type in present-day medium- or large-scale plants, being used in locations where the head may be as low as 2 m or as high as 200 m. They are radial-flow turbines, and although the water flow is inwards towards the centre instead of the outward flow of Fourneyron's turbine, the principle remains the same.

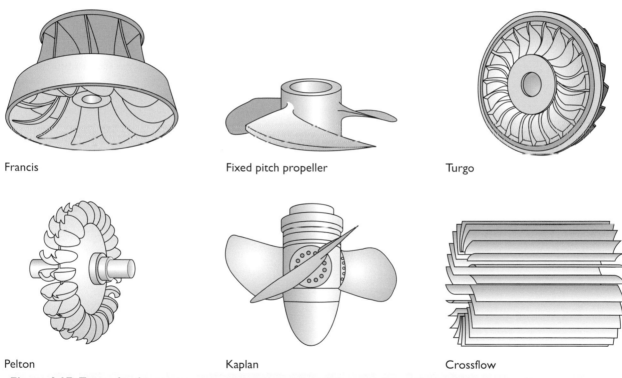

Francis

Fixed pitch propeller

Turgo

Pelton

Kaplan

Crossflow

Figure 6.17 Types of turbine runner

Figure 6.18 The photograph shows six Francis runners and behind them three Turgo runners. The different sizes and shapes reflect their outputs and also the 'head' under which each will operate. The largest of the Francis turbines, for instance, is designed for an output of 10 MW at a head of 280 m, whilst the smallest one generates only 600 kW at a head of 80 m. (The subsection Ranges of Application, below, discusses these factors.)

Action of the turbine

As the Francis turbine is completely submerged, it can run equally well with its axis horizontal (Figure 6.19) or vertical (Figure 6.20). In medium- or high-head turbines the flow is channelled in through a scroll case (also called the volute), a curved tube of diminishing size rather like a snail shell, with the guide vanes set in its inner surface. Directed by the guide vanes, the water flows in towards the runner. The shapes of the guide vanes and runner blades and the speed of the water are critical in producing the smooth flow that leads to high efficiency (see below). Francis turbines run most efficiently when the blade speed is only slightly less than the speed of the water meeting them.

Figure 6.19 The 450 kW horizontal-axis Francis turbine of a small-scale plant in Scotland, commissioned in 1993. The inflow (at lower right) is 2.1 m³ s⁻¹ at a head of 25 m. Part of the generator casing can be seen on the left and the adjustment mechanism for the guide vanes is also visible.

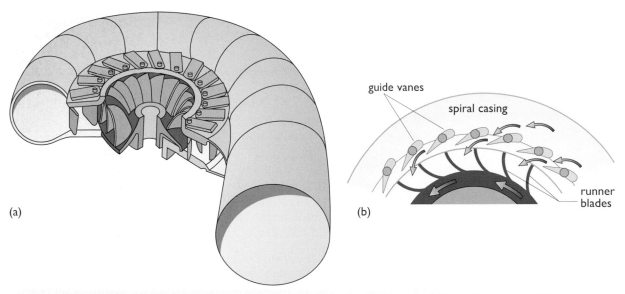

Figure 6.20 Structure of a Francis turbine, showing the central runner blades, the pivoting guide vanes and the surrounding volute

As it crosses the curved runner blades the water is deflected sideways, losing its whirl motion. It is also deflected into the axis direction, so that it finally flows out along the central draft tube to the tail race. That the water exerts a force on the blades is obvious because it has changed direction in passing through the turbine. In being deflected by the blades, it pushes on them in the opposite direction – the way they are travelling – and this reaction force transfers energy to the runner and maintains the rotation. For this reason, these are called **reaction turbines**. An important feature of this type is that the water arrives at the runner under pressure, and the pressure drop through the turbine accounts for a large part of the delivered energy.

Maximizing the efficiency

Although, as we saw above in Section 6.5, there are always energy losses, modern turbines can achieve efficiencies as high as 95% – but only under optimum conditions. Maintaining exactly the right speed and direction of the incoming water relative to the runner blades is important, and this leads to a problem. Suppose demand falls: the output power can be reduced by reducing the water flow, and in a Francis turbine this is done by turning the guide vanes; but this changes the angle at which the water hits the moving blades and the efficiency falls. This is a characteristic which must be accepted with this type of turbine. (As we'll see below, some 'propeller' types allow adjustment of the pitch, changing the angle of the blades to match the new conditions.)

A rather different cause of less than 100% efficiency is that the water flowing out carries away kinetic energy. A partial remedy is to flare the draft tube. If the tube becomes larger but the volume flow stays the same, the actual speed of the water must decrease, reducing the energy loss. It may seem strange that this change *after* the water has left the turbine makes any difference to its efficiency, but the effect of the deceleration is to reduce the pressure back at the exit from the turbine, increasing the pressure drop across it and therefore the energy it extracts.

Limits to the Francis turbine

The available head is an important factor in selecting the best turbine for a particular site. If the head is low a large volume flow is needed for a given power. But a low head also means a low water speed, and these two factors together mean that a much larger input area is required. Attempts to increase this area whilst adapting the blades to the reduced water speed and at the same time deflecting the large volume into the draft tube led to turbines with wide entry and blades which were increasingly twisted. Ultimately the whole thing began to look remarkably like a propeller in a tube, and this is indeed the type of turbine now commonly used in low-head situations (see below).

High heads bring problems too, because they mean high water speeds. As mentioned above, Francis turbines are most efficient when the blades are moving nearly as fast as the water, so high heads imply high speeds of rotation. A look at Tables 6.1 and 6.2 in Section 6.2 reveals that the turbine at Glenlee, with its much higher head, rotates at up to twice the speed of the other Francis turbines in the Galloway system. For sites with very high heads, the Francis turbine becomes unsuitable, and yet another type takes over, as we shall see.

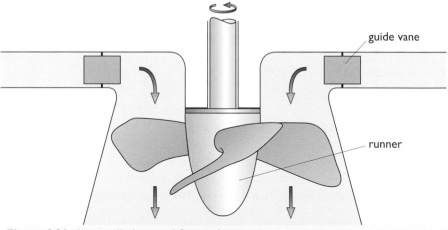

Figure 6.21 A 'propeller' or axial-flow turbine

'Propellers'

In the 'propeller' or **axial-flow** turbines shown in Figures 6.17 and 6.21, the area through which the water enters is as large as it can be: it is the entire area swept by the blades (these turbines are, again, reaction turbines). Axial-flow turbines are therefore suitable for very large volume flows and have become usual where the head is only a few metres. They have the advantage over radial-flow turbines in that it is technically simpler to improve the efficiency by varying the angle of the blades when the power demand changes. Axial-flow turbines with this feature are called **Kaplan turbines**. These are used in both hydro plants and in tidal barrage schemes, described in Chapter 7.

An important feature of 'propeller' turbines is that their optimum blade speed for maximum efficiency is appreciably greater than the water speed – as much as twice as fast. This allows the high rate of rotation needed by the generator even with relatively low water speeds. (Note that because the outer parts of the blade move faster than the more central parts, the blade angle needs to increase with distance from the axis. This is why a propeller has its familiar twisted shape.)

With axial flow there is no need to feed the water in from the side, and it is obviously simpler to let it flow in along the axis instead of being deflected through a right angle. However, this raises the problem of where to position the generator; if located directly along the axis of the turbine it will either get in the way and/or get wet! Several different solutions to this problem – rim generators and tubular turbines – are shown in Chapter 7.

Archimedes screw

The Archimedes screw can be thought of as an extended propeller in a trough (see Figure 6.22). Its name suggests that it was invented by the famous Greek mathematician, but it was probably in use as a water pump in Egypt well before his time.

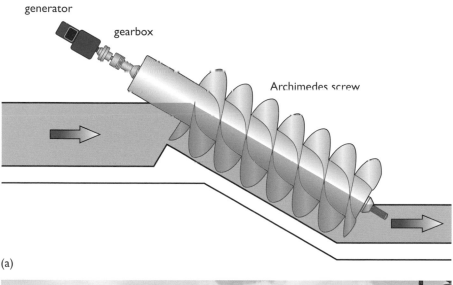

(a)

(b)

Figure 6.22 The Archimedes screw (a) schematic (b) the 12 kW turbine installed at Cragside house in Northumberland in 2014

The metal screw only half fills the trough, giving good accessibility for cleaning. The screw only turns slowly with the generator running at a higher speed, driven through a gearbox. The generator and gearbox are mounted at the top, again easily accessible for maintenance. They are used for small schemes with heads from 1 to 8 metres and need to be manufactured 'to size' with the screw occupying the full head height and with a diameter (sized to match the flow rate) of up to 5 metres. Power ratings for single screws are in the range 5–500 kW, though higher powered schemes are likely to use multiple screws.

Archimedes screws have been increasingly used in the UK since about 2007 in 'run-of river' projects. A 300 kW system using two screws each 4 metres in diameter was installed at Romney Lock on the River Thames in 2013 to provide power to Windsor Castle. The engineering work required modifications to the 200-year-old lock.

It is claimed that, given an adequate gap between the blades, this technology is 'fish friendly' allowing fish to pass right through the turbine unharmed (Kibel and Coe, 2011).

Pelton wheels

For sites of the type shown in Figure 6.13(c), with heads above 250 metres or so (or lower for small-scale systems) the **Pelton wheel** is the preferred turbine. It evolved during the gold rush days of late nineteenth century California, was patented by Lester Pelton in 1880, and is entirely different from the types described above. It is, in contrast to the reaction turbines discussed previously, an **impulse turbine**. One important difference between the turbine types is that whereas a reaction turbine runs fully submerged and with a pressure difference across the runner, impulse turbines essentially operate in air at normal atmospheric pressure.

A Pelton wheel is basically a wheel with a set of double cups or 'buckets' mounted around the rim (Figures 6.23a and 6.24). A high-speed jet of water, formed under the pressure of the high head, hits the splitting edge between each pair of cups in turn as the wheel spins. The water passes round the curved bowls, and under optimum conditions gives up almost all its kinetic energy. The power can be varied by adjusting the jet size to change the volume flow rate, or by deflecting the entire jet away from the wheel.

The efficiency of a Pelton wheel is greatest when the speed of the cups is half the speed of the water jet (Box 6.6). As the cup speed depends on the rate of rotation and the wheel diameter, and the water speed depends on the head, there is an optimum relationship between these three factors.

(a) (b)

Figure 6.23 The 16.5 MW Finlarig power station, on the shores of Loch Tay, draws its water from Loch na Lairige at a head of 415 metres. Its average annual output is 70 GWh (SSE, 2005) (a) the power station, (b) the original double twin-jet Pelton wheel and horizontal-axis 30 MW multi-pole generator.

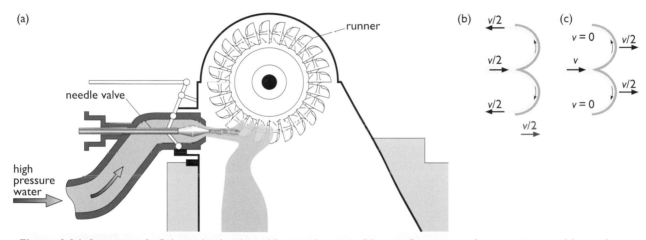

Figure 6.24 Structure of a Pelton wheel turbine: (a) vertical section, (b) water flow as seen from moving cup, (c) actual motion of water and cup

BOX 6.6 Optimum speed for a Pelton wheel

The following informal argument shows that a Pelton wheel extracts the maximum energy from the water if the cups move at half the speed of the water jet.

Consider the situation when water at speed v approaches a cup which is already moving in the same direction at half this speed ($v/2$). As seen from the cup, the water will be approaching at the difference between these two speeds, which in this case is also $v/2$.

Suppose now that the water passes smoothly round inside the curved cup until it leaves travelling in the opposite direction — as seen from the cup. You now have water moving backwards at speed $v/2$ relative to a cup that is moving forwards at just this speed. A person on the ground would see the cup moving on while the water simply falls vertically out of it. The water has given up all its kinetic energy to the wheel. 100% efficiency, in principle!

In practice this is only approximately true, and the best cup speed is a little less than $v/2$.

Input power

The power input to a Pelton wheel is determined, as usual, by the effective head and the flow rate of the water. Box 6.7 below shows that, ideally, the volume rate of flow Q in m^3 s^{-1} corresponding to an effective head H in metres is:

$$Q = A \times \sqrt{(2gH)}$$

where A is the area of the jet in m^2.

BOX 6.7 Effective head, water speed and flow rate

Although there are in practice always energy losses in forming a jet, we'll assume here that the water leaves the jet at the speed that it would have gained in 'free fall' through the effective head.

We know from Equation (1) in Section 6.5 that the potential energy in joules lost by M kg of water in falling through H metres is given by:

potential energy $= MgH$

In Chapter 1 we saw that the kinetic energy of a moving object is proportional to its mass and the square of its speed:

$$\text{kinetic energy} = \frac{1}{2}Mv^2$$

So if all the lost potential energy is converted into kinetic energy, we have:

$$\frac{1}{2}Mv^2 = MgH$$

so $v^2 = 2gH$ and

$$v = \sqrt{(2gH)} \tag{5}$$

If this water flows as a jet with a circular area of A square metres, the volume flowing out in each second Q in m^3 s^{-1} will be equal to A times v. So the volume flow rate for an effective head H is given by:

$$Q = A \times \sqrt{(2gH)} \text{ cubic metres per second}$$

Equation (2) in Section 6.5 shows that the input power to a turbine is:

$$P \text{ (watts)} = 1000 \times Q \times g \times H$$

so substituting for Q we find that:

$$P = 1000 \times A \times \sqrt{(2gH)} \times g \times H$$

Using the approximate value of g (10 m s^{-2}), the power in kilowatts becomes

$$P \text{ (kW)} = 45A\sqrt{(H^3)}$$

If adjacent cups are not to interfere with the flow, the wheel diameter needs to be about ten times the diameter of the jet. But two or even four jets can be spaced around the wheel to give greater output without increasing the size. If the number of jets is j, the power equation becomes

$$P \text{ (kW)} = 45jA \sqrt{(H^3)}$$

Turgo and cross-flow tubines

A variant on the Pelton wheel is the **Turgo turbine** (Figures 6.17 and 6.25), developed in the 1920s. The double cups are replaced by single, shallower ones, with the water entering on one side and leaving on the other. The water enters as a jet at a low angle to the plane of the turbine, striking the cups in turn, so this is still an impulse turbine (and the relationships in Boxes 6.6 and 6.7 still apply). However, its ability to handle a larger volume of water than a Pelton wheel of the same diameter gives it an advantage for power generation at medium heads.

(a) runner

(b) water flow

Figure 6.25 Flow in a Turgo turbine

The cross-flow turbine shown in Figure 6.17 (also known as the **Mitchell-Banki**, or **Ossberger turbine**) is yet another impulse type. The water enters as a flat sheet rather than a round jet. It is guided on to the blades, then travels across the turbine and meets the blades a second time as it leaves. As we will see in the next section, cross-flow turbines are often used instead of Francis turbines in small-scale plants with outputs below 100 kW or so, and some ingenious technological ideas have gone into the development of simple types of generator which can be constructed (and maintained) without sophisticated engineering facilities and are therefore suitable for remote communities.

Ranges of application

We have seen that, in general, Pelton wheels are most suitable for high heads, propellers for low heads and Francis turbines for the intermediate ranges. But the effective head is not the only factor determining the most appropriate type for a given situation. The available power also matters.

Figure 6.26 represents one way to display the ranges of application of the different turbines. It shows the ranges of head, flow rate and corresponding power which best suit each type. It should be noted however that the

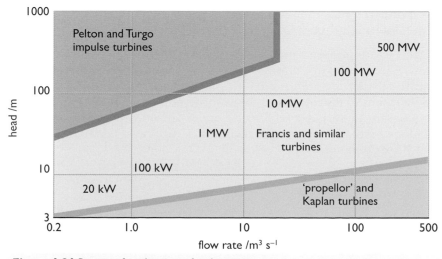

Figure 6.26 Ranges of application of turbines

boundaries are by no means clear-cut, and other technical issues, or criteria such as cost, simplicity in manufacture or ease of maintenance can lead to choices outside these ranges. Locating the site data for the Galloway plants (and others in this chapter) on Figure 6.26 and comparing the turbine types actually used with those suggested by the diagram shows that this is indeed often the case.

Small-scale installations can also be classified by the available head and flow rate, but the ranges may be very different from those of large plants. Many of these plants are run-of-river, with heads of only a few metres, and as little as 10 m could be regarded as 'high head' for a very small plant. The corresponding choices of turbine type may vary appreciably from those indicated in Figure 6.26.

6.9 Hydro as an element in a system

Even if a potential source of electric power is acceptable environmentally and financially, other factors remain that affect its viability. Few large power stations operate in isolation, and the extent to which the proposed plant can form a useful part of a supply *system* is important.

From the point of view of the operator of the system, the characteristics of the ideal power station would be:

1.   constant high availability
2.   a reserve energy store to buffer variations in input
3.   no correlation in input variations between power stations
4.   rapid response to changing demand
5.   an input which matches annual variation in demand
6.   no sudden and/or unpredictable changes in input
7.   a location which does not require long transmission lines.

Few, if any, sources meet all these criteria, but each compromise with the ideal adds to the effective cost.

Almost all hydroelectric plants score well on item 4, and, in regions with cold, dark, wet winters, on item 5 as well – unless the water is locked up as ice. Furthermore, sudden unplanned fluctuations in input (item 6) are rare, at least in large plants.

How well hydro performs on items 1, 2 and 3 depends in part on the type of plant. A high-head installation with a large reservoir will normally have little difficulty in maintaining its output over a dry period, whereas the water held behind the low dam of a run-of-river plant may not be sufficient to compensate for periods of reduced flow. A serious drought can of course affect all hydro plants over a wide region, so it cannot be said to fully satisfy the third requirement.

Figure 6.27 The Grand Coolee Dam on the Columbia River, constructed in 1942, is 170 metres high and about 1 km long. The total generating capacity of its 30 turbo-generator sets is 6.8 GW, of which 300 MW is pumped storage plant (USBR, 2011)

The final criterion is the real hurdle. Hydro locations are determined by geography, and whilst run-of-river plants may sometimes be near major centres of population, this is rare for high-head systems.

Is the case different for small-scale hydro? Smaller plants are predominantly run-of-river, or perhaps served by relatively small reservoirs, in either case a less reliable supply. On the other hand, scattered sites with different rainfall patterns could result in increased reliability. Local small-scale plants can reduce the need for long-distance transmission, reducing energy losses and costs; although the generating cost per unit of output may be greater.

Overall, hydroelectricity ranks reasonably well in terms of the above criteria. And it may also offer a bonus. A hydro plant with a large reservoir not only maintains its own reserve of energy – it might, as we have seen earlier, provide a store for the surplus output of other power stations.

6.10 **Environmental considerations**

The environmental issues associated with hydroelectricity are no less controversial than those for other energy sources. We might usefully start by briefly summarizing the environmental *benefits* of hydroelectricity compared with other types of power plant:

- in operation it releases no CO_2, and negligible quantities of the oxides of sulfur and nitrogen that lead to acid rain
- it produces no particulates or chemical compounds such as dioxins that are directly harmful to human health.
- it emits no radioactivity
- dams may collapse, but they will not cause major fires.

Moreover, a hydroelectric plant is often associated with positive environmental effects such as flood control or irrigation, and in some cases, its development leads to a valued amenity or even a visual improvement to the landscape.

However, during the twentieth century, the construction of large dams has led to the displacement of many millions of people from their homes, and dam failures have also killed many thousands. We'll consider these and other deleterious effects under three headings:

- hydrological effects – water flows, groundwater, water supply, irrigation, etc.
- other physical effects of large hydro plants
- social effects.

Hydrological effects

The three categories of effect listed above are of course not independent. Any hydrological change will certainly affect the ecology and thus the local community. A hydroelectric scheme is not basically a *consumer* of water, but the installation does 'rearrange' the resource. Diverting a river into a canal, or a mountain stream into a pipe, may not greatly change the total flow, but it can have a marked effect on the environment. Furthermore, evaporation from the exposed surface of a large reservoir may appreciably reduce the available water supply.

Such 'rearrangements' of the water supply may result in significant international disagreements. For example the Gabcikovo-Nagymaros scheme on the River Danube, first agreed in 1977, was originally intended to provide 880 MW of hydro plants on the Slovak-Hungarian border. However the diversion of the river caused a fall in the natural water table, with vegetation dying and unique forms of wildlife in danger. This resulted in cancellations and years of argument between the Slovak and Hungarian governments at the International Court of Justice at the Hague.

In September 1997, the Court ruled that *both* countries had acted illegally (ICJ, 1997), and asked that they should negotiate a new solution. The court noted that 'both Parties agree on the need to take environmental concerns seriously and to take the required precautionary measures, but they fundamentally disagree on the consequences this has for the joint Project'. A 770 MW scheme at Gabcikovo was eventually built, while the smaller scheme at Nagymaros was abandoned (Wikipedia, 2017a).

Other physical effects

Any structure on the scale of a major hydroelectric dam will affect its environment in many ways in addition to the hydrological changes discussed above. The construction process itself can cause widespread disturbance, and although the building period may be only a few years, the effect on a fragile eco-system can be long-lasting. In the longer term, a large reservoir is bound to bring other significant environmental changes.

Whether these are seen as catastrophic, beneficial or neutral will depend on the situation – in the geographical and biological sense – and certainly on the points of view and interests of those concerned.

Dam failures

The available records suggest that worldwide during the half-century since 1960 there have been about 35 major dam failures – defined as those resulting in serious material damage and/or deaths (Wikipedia, 2017b; 2017c). Many dams are of course for purposes other than hydroelectricity, such as flood control, water supply, irrigation or recreation. In the USA, for example, hydropower is the primary authorized purpose of only 2.5% of the approximately 87 000 federal dams across the country (US Department of Energy, 2016). It is not therefore surprising to find only five or six hydro plants in the world list of major dam failures.

However, the safety of hydroelectric dams is a matter of concern (and expense). A notable recent example has been the Oroville hydroelectric dam in California, which during the winter of 2016/7 suffered damage to one of its spillways as a result of heavy rainfall. This required the temporary evacuation of nearly 200 000 people while repair work took place.

The data on dam failures for China is poor. It is known that severe flooding in 1975 led to many dam failures, and estimates of the total fatalities vary between about 70 000 and over a quarter of a million (New Scientist, 2011). If the figure is indeed in this range, and hydro plants made up a significant fraction of the destroyed dams, then hydropower must rank high in any list of energy sources in terms of deaths per kWh of useful energy.

Plant failures

Hydroelectricity, with no high temperatures or combustible or radioactive fuels, has historically compared very well with other types of power plants in terms of serious damage due to plant failures. It does, however, involve high pressures, and even a head as low as 200 metres implies a pressure of 20 atmospheres. This was the head driving the ten 640 MW turbo-generators of the Sayano-Shushenskaya power station on the Yenisei River in Western Siberia, Russia, a plant whose annual output of about 25 TWh supported its claim to be the world's sixth-largest hydroelectric plant (Wikipedia, 2017d).

At 8 a.m. on Monday 17 August 2009, Turbine No.2 at Sayano-Shushenskaya disintegrated, destroying or seriously damaging the remaining nine turbines and the building, killing 75 workers, releasing 40 tonnes of transformer oil into the river and depriving the major industrial plants in the surrounding region of power (which resulted in the loss of an estimated half a million tonnes of aluminium output). Subsequent investigation found that a year or so earlier an inspection had declared all ten turbines unfit for operation due to age and poor maintenance.

Silt

Silt accumulation behind dams has been a known problem for many years – the build-up reduces the volume of stored water and consequently the hydro potential of a site. The Hoover Dam (Figure 6.14), for instance, lost about one sixth of its useful storage volume in its first 30 years, although the loss rate was reduced when the Glen Canyon dam was built 370 miles upstream.

The Aswan High Dam, built in Egypt in the 1960s, also supplies rather less electricity than originally planned, but a concern attracting greater attention than silt accumulation is that the land downstream no longer receives the soil and nutrients previously carried by the annual Nile floods. An agricultural system in existence for millennia has largely been destroyed, to be replaced by irrigation and the use of fertilizers.

Fish

The account of the Galloway Hydros in Section 6.2 mentions the effect on the salmon passing up the rivers to spawn as a major environmental issue – solved in that case by fish ladders. France had many dams constructed during the early twentieth century on rivers previously used by Atlantic fish, and as licences became due for renewal in the 1990s, stringent requirements were introduced for the construction of fish ladders or similar passages. One consequence was the *decommissioning* of dams deemed unsuitable for renewal, on environmental or economic grounds (ERN, 2017).

The decommissioning schemes implemented for a number of relatively small French dams were designed in part as pilot schemes: to identify some of the issues that might arise across Europe as more of the early and mid-twentieth century dams built for hydroelectricity become due for renewal or renovation. Recent decades have also seen calls in the USA for the decommissioning of dams on environmental grounds, leading to interesting debates with others who, also citing environmental concerns, favour hydro development.

Methane

For many years, hydroelectricity was regarded as one of the renewables that produces virtually no greenhouse gases, but this view has since come into question. It has long been known that vegetable matter that would normally decay in the air to produce carbon dioxide (CO_2) could decay anaerobically under water to produce methane (CH_4). When methane was identified as a much more potent greenhouse gas than carbon dioxide, the question arose as to whether hydro schemes that flooded land previously covered in vegetation should join the fossil fuels as significant contributors to global warming.

Following this concern, some detailed studies of individual reservoirs were carried out. One, by a team from the Swiss aquatic research institute EAWAG, was prompted by the local observation that the reservoir lake of a hydroelectric plant on the River Aare in northern Switzerland was bubbling during the summer. The detailed study estimated that each square metre of the lake surface was releasing about 0.15 grams of methane per day (EAWAG, 2010) – more than would be expected for its temperate location, and equating to a total annual methane release of about 150 tonnes. This is equivalent in global warming potential to almost 4000 tonnes of CO_2. The suggested cause was the anaerobic digestion of the large annual quantity of vegetable matter (leaves, etc.) brought down by the river (the report points out that 150 tonnes of methane is about the same amount produced by 2000 cows carrying out a similar digestion process). Such lakes can be significant emitters of greenhouse gases, although a coal-fired power plant producing the same electrical output would release approximately 40 times more greenhouse gases, expressed in CO_2 equivalent.

Currently, the issue of methane emissions seems to play a relatively small role in project assessments. The *2016 Survey of Renewable Energy Resources* of the World Energy Council (WEC, 2016) says that 'The GHG (greenhouse gas) status of freshwater reservoirs is an area of ongoing scientific research, and policy responses are still evolving as the state of knowledge progresses'.

The issue does feature in the sustainability protocol produced by the hydroelectricity industry (see below).

Social effects

The social effects of the construction of large dams can be very serious, particularly if the 'benefits' of a scheme are taken by one group of people, while another group bear the 'disbenefits'.

The Aswan and Kariba dams in Africa involved the relocation of some 80 000 and 60 000 people respectively, whilst the rising water behind the Three Gorges Dam (Box 6.8) submerged about 100 towns and displaced over a million people. It is estimated that during second half of the twentieth century, some 10 million people were displaced by reservoirs in China alone.

BOX 6.8 Three Gorges

History

China's Three Gorges plant on the Yangtze River was the largest of a series of hydro developments worldwide that attracted major opposition on environmental and social grounds. Originally proposed in 1919 by Sun Yat Sen, the project had a varied political history, culminating in its approval in 1992 against the unprecedented opposition of a third of the Congress delegates. Table 6.6 summarizes the arguments in China at the time. (The table originally appeared in ChinaOnline, in English, and the wording here is verbatim.)

Table 6.6 Summary of arguments for and against the dam

Issue	Criticism	Defence
Cost	The dam will far exceed the official cost estimate, and the investment will be unrecoverable as cheaper power sources become available and lure away ratepayers.	The dam is within budget, and updating the transmission grid will increase demand for its electricity and allow the dam to pay for itself.
Resettlement	Relocated people are worse off than before and their human rights are being violated.	15 million people downstream will be better off due to electricity and flood control.
Environment	Water pollution and deforestation will increase, the coastline will be eroded and the altered ecosystem will further endanger many species.	Hydroelectric power is cleaner than coal burning and safer than nuclear plants, and steps will be taken to protect the environment.
Local culture and natural beauty	The reservoir will flood many historical sites and ruin the legendary scenery of the gorges and the local tourism industry.	Many historical relics are being moved, and the scenery will not change that much.
Navigation	Heavy siltation will clog ports within a few years and negate improvements to navigation.	Shipping will become faster, cheaper and safer as the rapid waters are tamed and ship locks are installed.
Power generation	Technological advances have made hydrodams obsolete, and a decentralized energy market will allow ratepayers to switch to cheaper, cleaner power supplies	The alternatives are not viable yet and there is a huge potential demand for the relatively cheap hydroelectricity
Flood control	Siltation will decrease flood storage capacity, the dam will not prevent floods on tributaries, and more effective flood control measures are available	The huge flood storage capacity will lessen the frequency of major floods. The risk that the dam will increase flooding is remote.

Figure 6.28 The Three Gorges Dam

Opposition, locally and internationally, centred on the displacement of the 1.13 million people who were to lose their homes, farms and workplaces. Other serious concerns included the problem of silt that was predicted to block harbours upstream, increase flooding in some areas and ultimately reduce the plant output. Furthermore there was also the loss of one of China's most valued landscapes.

The international outcry led the World Bank and some other major financial institutions to dissociate themselves from the project, but support was found elsewhere and work started in 1993. The dam, a mile long and 181 metres high, was completed in early 2003, and the sluices were finally closed at midnight on 1 June. The river was released from its five-year diversion and by early morning the water behind the dam was over 100 m deep. Two weeks later it was almost a metre above its intended final level.

Operation of the plant

The first fourteen 700 MW Francis turbines and generators on the left bank went into operation in September 2005 and the 12 plants on the right bank in October 2008. The entire initial 26 plants operated simultaneously for the first time in June 2009. A further six 700 MW generators had been added by 2012 giving a total installed capacity of 22.5 GW.

This full power is not always available. At this location on the Yangtze, the flow rate varies in the course of a year between 5000 and 30 000 cubic metres per second. The head, in this case the height difference between the water behind the dam and the water below it, also varies depending on the season and the flood conditions.

These factors explain why Chinese claims for records by Three Gorges concentrate on 'output in a single month', whilst the Brazilians can still claim the record annual output from Itaipú despite its lower rated capacity (Box 6.10).

Sources: International Water Power & Dam Construction, 2007; International Rivers, 2017a; Hvistendahl, 2008

But even for the people immediately affected, the building of dams can have very different consequences. For those living in a valley which will become a reservoir it means the loss of your family home, possibly your livelihood, and often your entire community. In contrast, for people living on a river which periodically overflows its banks, the barrage and embankments of a hydroelectric scheme can bring freedom at last from devastating floods. On a smaller scale, the changes that mean the loss of a beloved riverside walk to some may be welcomed by others as an opportunity for new leisure activities.

Responses from the industry

Recent years have seen growing concern by the (large-scale) hydroelectricity industry over the above criticisms. The following are two responses.

Sustainable hydropower

The website Sustainable Hydropower, supported by the International Hydropower Association (IHA), an international organization representing the industry, includes the following presentation of the issues.

> After more than a century of experience, hydropower's strengths and weaknesses are equally well understood. Hydropower's negative impacts are well understood and, although not all can be eliminated, much can be done to mitigate them. These are summarized for economic, social and environmental aspects of hydropower in the following tables:

Environmental aspects

Advantages	Disadvantages
Produces no atmospheric pollutants	Inundation of terrestrial habitat
Neither consumes nor pollutes the water it uses for electricity generation purposes	Modification of hydrological regimes
	Modification of aquatic habitats
Produces no waste	Water quality needs to be monitored/managed
Avoids depleting non-renewable fuel resources (i.e. coal, gas, oil)	Greenhouse gas emissions can arise under certain conditions in tropical reservoirs
Very few greenhouse gas emissions relative to other large-scale energy options	Temporary introduction of methylmercury into the food chain needs to be monitored/managed
Can create new freshwater ecosystems with increased productivity	Species activities and populations need to be monitored/managed
Enhances knowledge and improves management of valued species due to study results	Barriers for fish migrations, fish entrainment
Can result in increased attention to existing environmental issues in the affected area	Sediment composition and transport may need to be monitored/managed
	Introduction of pest species needs to be monitored/managed

Social aspects

Advantages	Disadvantages
Leaves water available for other uses	May involve resettlement
Often provides flood protection	May restrict navigation
May enhance navigation conditions	Local land use patterns will be modified
Often enhances recreational facilities	Waterborne disease vectors may occur
Enhances accessibility of the territory and its resources (access roads and ramps, bridges)	Requires management of competing water uses
Provides opportunities for construction and operation with a high percentage of local manpower	Effects on impacted peoples' livelihoods need to be addressed, with particular attention to vulnerable social groups
Improves living conditions	Effects on cultural heritage may need to be addressed
Sustains livelihoods (fresh water, food supply)	

Economic aspects

Advantages	Disadvantages
Provides low operating and maintenance costs	High upfront investment
Provides long life span (50 to 100 years and more)	Precipitation dependent
Meets load flexibly (i.e. hydro with reservoir)	In some cases, the storage capacity of reservoirs may decrease due to sedimentation
Provides reliable service	
Includes proven technology	Requires long-term planning
Can instigate and foster regional development	Requires long-term agreements
Provides highest efficiency rate (payback ratio and conversion process)	Requires multidisciplinary involvement
	Often requires foreign contractors and funding
Can generate revenues to sustain other water uses	
Creates employment opportunities	
Saves fuel	
Can provide energy independence by exploiting national resources	
Optimizes power supply of other generating options (thermal and intermittent renewables)	

Source: Sustainable Hydropower, 2011b

The Hydropower Sustainability Assessment Protocol

The Hydropower Sustainability Assessment Protocol (HSAP) is the result of a collaboration by representatives of different sectors of the hydro industry, led by the International Hydropower Association (IHA). Essentially a list of criteria that should be satisfied by any new hydroelectricity project, it is no doubt the response of the industry to many of the problems discussed above. It was accepted by the membership of the IHA (which now administers its day-to-day operation) in November 2010.

The Protocol (HSAP, 2010) consists of an explanatory background document and four assessment sections, covering the four stages of any new project: Early Stage, Preparation, Implementation, and Operation. Collectively, these impose conditions designed to meet many of the criticisms of largescale hydro. However, by 2015 only 20 large scale hydro projects (in the range 3 MW to 3750 MW) had been assessed.

Environmental effects of small-scale systems

There is general consensus that small-scale hydro plants have fewer deleterious effects than large systems. In some respects this is evidently true – few people have been displaced from their homes by the installation of small 5 MW plants, whilst deaths from the collapse of dams across small streams seem rare.

However, not everyone agrees with the consensus. The claim, made mainly by proponents of large-scale hydro, is that a general world view – 'small is beautiful' – has been allowed to override detailed analysis. It is true that the efficiencies and the capacity factors of small-scale plants tend to be lower, and in some cases the 'reservoir area' per unit of output is greater. But as we have seen, all these factors vary significantly from site to site, and generalization is difficult.

Comparisons

It should not be forgotten that the choice may not be hydroelectricity or nothing, but hydroelectricity or some other form of power station. Despite the 'penalties' discussed above, hydroelectricity scores relatively well in terms of many other criteria. Current issues for hydropower include the question of methane emissions and the costing of long-term compensation for the people displaced by major new hydroelectric installations. Nevertheless, on the criteria used in these studies, hydro appears amongst the least harmful sources of electricity.

6.11 Economics

Generating plant can be broadly categorized either as being expensive machines for converting free or low cost energy into electrical energy or else lower cost machines for converting expensive fuels into electrical energy.

Mott MacDonald, 2010

No matter how elegant the technology, few will invest in it unless it is going to make a profit. Potential investors need to know how much each kilowatt-hour of output will cost, taking all relevant factors into consideration.

Capital costs

Hydroelectricity is well-established and most of the necessary cost information is easily available. The water-control systems, turbo-generators and output controls are standard items, covering a power range from a few hundred watts to hundreds of megawatts. The expected lifetime of the machinery is 25–50 years, and of the external structures, 50–100 years. Nevertheless, as mentioned in Section 6.3, it is difficult to generalize meaningfully about 'the cost of hydroelectric power', or to assess the economic potential for hydroelectricity in a country or a region.

The difficulty lies in the combination of the extremely site-specific construction costs and the heavy 'front-end loading' of these costs. In other words, the dominant factor in determining the cost per unit of hydro output is the initial capital cost, and a major part of this can be the civil engineering costs, which vary greatly from site to site.

Unit costs

An interesting study on hydro potential in the USA (Hall et al., 2003) assessed the costs for over two thousand sites with potential hydro capacities in the range 1–1300 MW. About half of these were green-field (blue-water?) sites, with no existing dams or hydro plants, and the estimated development costs for these, based on data for similar existing plants, fell mainly in the range US$2000 – US$4000 per kW. (At the time of writing (2017) inflation would have increased these initial costs to about US$2700 – US$5400 or approximately £2200 – £4300 per kW.)

These studies revealed the importance of the *initial costs* in determining the levelized cost of electricity (LCOE) per kilowatt-hour of output.

The civil engineering works typically accounted for 65–75% of this unit cost, whilst meeting the environmental and other criteria necessary for a licence added another 15–20%. In all, 85–95% of the capital cost was 'site' cost, with the turbo-generator and control systems accounting for only 10% or so.

With no fuel costs, and relatively low annual operation and maintenance costs, it is the interest repayments on the costs incurred in building the dams etc. that dominate the cost of energy. Box 6.9 shows the importance of the discount rate used in assessing hydroelectricity costs.

BOX 6.9 Cost comparison of hydro and CCGT plants

A combined cycle gas turbine (CCGT) plant may be regarded as the opposite extreme to a hydro plant. Built quickly, using standard components, its capital costs are relatively low; but the fuel costs are high (and unpredictable for the future) and its lifetime is likely to be much less than that of a hydro plant. Nevertheless, investors may choose to put their money into CCGT plants rather than hydro – as has indeed been the case in the UK in recent years. The following reasoning shows why.

A simple general approach to costing electricity from power plants is described in Appendix B. As described there, the key cost elements of any electrical power plant are:

- initial capital costs

- fuel costs – zero in the case of hydro – but these should include a 'carbon price' for the gas used in the CCGT

- operation and maintenance (O&M) costs

- final decommissioning costs (assumed to be zero here – although they could be significant for a hydro scheme)

Table 6.7 has relevant data for two 100 MW plants based on sample data from recent UK cost assessments (BEIS, 2016b; Arup, 2016). The main differences are obvious: the hydro plant costs much more, but the CCGT has fuel (and carbon price) costs and a much shorter expected life. In the case considered, the high CCGT capacity factor (80%) suggests that it will be in almost constant use, whilst the hydro plant (capacity factor 45%) may be used mainly as backup (or is perhaps subject to flow variations). The final column of the table is calculated from the 100 MW rated output of both plants and their respective capacity factors.

Table 6.7 Financial data and performance factors for 100 MW plants

	Capital cost / £ million	Plant lifetime / years	Average capacity factor	O&M cost per kWh of output	Fuel and carbon cost per kWh of output	Average annual plant output / million kWh
CCGT	55	30	80%	0.5 p	5.5 p [1]	701
Hydro	320	60	45%	1.5 p	-	394

[1]Assuming a bulk gas cost of 1.9 p per kWh and a fuel-to-output efficiency of 53% plus a carbon cost of £50 per tonne

The cost per kWh of electricity from each of these 100 MW plants can be calculated for different discount rates by the method described in Appendix B. Data from Table B1 in that Appendix is used to find the annual repayment for each of the above plants at four selected discount rates, shown in Table 6.8 below. Using the annual outputs in Table 6.7, these are then expressed as repayment costs per kilowatt-hour of output. Finally, adding the other unit costs from Table 6.7 leads to a *total cost per kWh* of electricity from each plant.

Table 6.8 Financial factors

Discount rate	0%	5%	10%	15%
CCGT repayment factor / £ per £1000 capital	33	65	106	152
CCGT annual repayment / £ million	1.82	3.58	5.83	8.36
Annual repayment / pence per kWh of output	0.26p	0.51p	0.83p	1.19p
Fuel + O&M cost / pence per kWh of output	6.00	6.00	6.00	6.00
CCGT total cost per kWh of output	**6.26p**	**6.51p**	**6.83p**	**7.19p**
Hydro repayment factor / £ per £1000 capital	17	53	100	150
Hydro annual repayment / £ million	5.44	16.96	32.0	48.0
Annual repayment / pence per kWh of output	1.38p	4.30p	8.12p	12.18p
O&M / pence per kWh of output	1.50	1.50	1.50	1.50
Hydro total cost per kWh of output	**2.88p**	**5.80p**	**9.62p**	**13.68p**

The clear conclusion from the results is that if discount rates are low (5% or less) then hydro is cheaper than gas generation. However if they are high (10% or more) then the reverse is true.

Quite what is a 'correct' discount rate to use is a matter of debate, particularly since 2017 bank interest rates in the UK and the USA are almost at an all-time low. The topic of discount rates and project risk is discussed in Chapter 11, but both large scale hydro power and CCGT generation are 'mature' technologies and are not seen as particularly financially risky.

It has been argued that projects with significant prospects of reducing national CO_2 emissions should be funded with (government) loans at a very low 'social' discount rate of 3.5%. This would give a hydro cost of under 5 p kWh^{-1}. Such low interest finance would, of course, also reduce the costs of rival low-carbon technologies, particularly nuclear power (CCC, 2011).

Obviously there are many other factors that can be adjusted in this calculation to tip the balance one way or the other. For example, although the CCGT is used for almost continuous generation, the hydro plant may be used to meet peak loads, possibly giving extra commercial value to its electricity. A higher carbon price for the CO_2 emissions from the CCGT would also tip the balance in favour of the hydro plant.

Table 6.9 summarizes the estimated electricity costs for new UK hydro plants of different sizes from a recent UK government study (BEIS, 2016b).

Table 6.9 UK hydroelectricity cost estimates

Hydro plant rating	Cost of electricity / p kWh^{-1}
< 100 kW	12.6
500 kW–2 MW	9.5
5–16 MW	9.7
Large storage (100 MW)	8.0

Refurbishing and upgrading

Two other options have been considered – and implemented – in recent years by countries with little remaining accessible hydro capacity:

- the installation of hydro plants at existing dams constructed for other purposes
- the refurbishing and/or upgrading of existing hydro plants.

Both approaches offer the possibility of increased hydro capacity at a lower initial cost and with fewer environmental consequences than green-field development. The US study described above noted the lower cost of these options (at least for relatively modest plants), with estimates of perhaps half the cost of green-field development when the dam already exists, and as little as a third, per 'new' kW, for refurbishment and upgrading of older plant. This option also allows the installation of extra turbines at a relatively low cost to assist with peak demands and complement other variable renewable energy sources such as wind and PV.

Many countries now have extensive refurbishment and upgrading programmes for large-scale hydro, resulting in modest increases in output from the existing dam infrastructure.

6.12 Future prospects

As shown in Figure 6.1 at the beginning of this chapter, world total electricity production in 2016 was nearly 25 000 TWh and hydro power provided 16% of this, about 4000 TWh.

The percentage contribution from hydro has been falling gently for many decades, as the building of new plants has failed to keep up with the rapid growth in total world electricity consumption. As shown in Figure 6.5 most of the increase in world hydro output since 2005 has taken place in China where in 2016 hydro provided 19% of the total electricity demand. Here, electricity consumption has been increasing at about 8% per annum, but this growth rate has been matched by hydro construction whose output has more than doubled in the last decade.

According to the estimate shown in Table 6.4, China would appear to have already developed nearly a half of its hydro technical potential. In 2016, it had 331 GW of hydro installed capacity and, according to the International Hydropower Association, the potential to develop a further 200 GW (IHA, 2017). However most of the hydro resource is located in the mountainous west of the country and will require the construction of long distance high-voltage transmission lines to distribute the power to the population centres in the east of the country.

There is also large potential for hydro development in South America, where it already supplies about a half of the electricity. This requires tapping into the flows of large rivers, which may have serious environmental consequences. Two such projects are described in Box 6.10.

BOX 6.10 Large hydroelectric projects in Brazil

Itaipú

Figure 6.29 The Itaipú dam

Construction of the 14 GW hydroelectric plant at Itaipú, on the Paraná River between Brazil and Paraguay, started in 1975, and the last two of its twenty 700 MW generators came on line in 2006–2007.

The effective head is about 200 metres and the reservoir area 1350 km^2, with an average water flow of some 10 000 tonnes a second, peaking at times to over 30 000 tonnes per second.

The plant supplies 95% of Paraguay's power and about a quarter of Brazil's, requiring a feature that must be unique in power stations: all 20 generators produce alternating current (AC) power, but half of them at a frequency of 50 Hz, and the other half at 60 Hz. Moreover, as Paraguay (50 Hz) uses only a fraction of its share, the rest is sold to Brazil, where it is rectified, transmitted as DC, and then re-converted to provide the required 60 Hz supply (Itaipú, 2011).

The total annual output obviously depends on the available flow in the Paraná and plant availability, and could of course be affected by varying demand. Over the five years 2006–2010 the average annual output was 91.1 TWh, implying an average capacity factor of slightly under 75% – significantly below the originally contracted 85%.

The overall electricity output of the dam is comparable to that of the 22 GW Three Gorges Dam in China, however the continuously flowing Paraná River gives a higher capacity factor. In 2016, Itaipú achieved its best annual output, 103 TWh (ABC TV, 2017), equivalent to nearly 30% of the UK's annual electricity demand.

Xingu river project at Belo Monte

The first suggestions for the development of hydro power on Brazil's Xingu river, a tributary of the Amazon, appeared in 1975 (the year in which Itaipú first came into operation). Initial studies led to a proposal for six large hydro plants, but the release of this plan in 1987 led to a vigorous campaign by indigenous people living in the affected areas. Their campaign attracted widespread international support, and the next two decades saw continuing opposition to a range of schemes involving between one and six dams along the Xingu. There were also objections on technical grounds – the variable flow of the river meant that the annual capacity factor of the plant could be as low as 30% (problem also experienced at Three Gorges).

In 2008, a new environmental assessment resulted in a plan involving one main power plant, the *Belo Monte*. However, in order to avoid inundating 'indigenous territory' in the bend of the river, this required the construction of the second, larger *Pimental* dam, which would create two reservoirs: one upstream along the route of the river, and a second filled by the diversion of water through two parallel 500 m wide canals on to previously dry land behind the Belo Monte dam.

In February 2010, the Brazilian government environmental agency granted a provisional environmental licence to this plan.

The following year saw a series of reversals and reinstatements of this approval, culminating in the granting on 1 June 2011 of a full licence to the Norte Energia consortium, which undertook to pay US$1.9 billion to address social and environmental problems (Barrionuevo, 2011).

The dam has now been completed and 2 GW of the total of 11 GW turbine capacity had been brought on line by the end of 2016 (IHA, 2017). Full completion is expected in 2019 and this will also require the construction of a 2550 km long transmission line to carry the electricity to Rio de Janeiro.

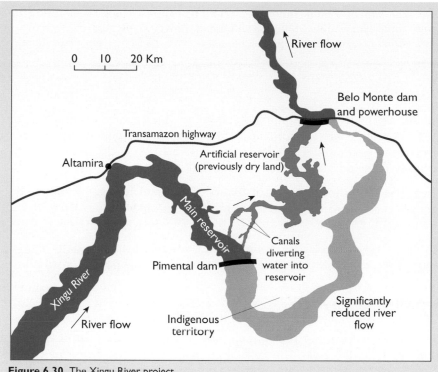

Figure 6.30 The Xingu River project

Africa, as we saw in Section 6.3, has a lower proportion of its hydro technical potential developed than any other major region of the world – a situation that has led to its place at the centre of bitter controversies, typified by the proposals for and objections to the Grand Inga project on the Congo (Box 6.11).

BOX 6.11 Grand Inga

The first Inga power plant, with a rated output of 351 MW from six turbines, was commissioned in 1972 on the Congo river, in the region of the Inga rapids in the then country of Zaïre – now the province of Bas-Congo in the Democratic Republic of the Congo. In 1982, this plant (now Inga 1) was joined by Inga 2, with its rated output of 1424 MW from eight turbines.

Unfortunately, with many financial, political and structural problems, the record of these plants has not been good. In 2008, two top directors of the company were interrogated after the disappearance of US$6.5 million intended for rehabilitation of Inga 2. Renovation work on turbines at Inga 1 and 2 is currently (2017) being carried out with financial assistance from the World Bank.

In 2008, the World Energy Council called for financing for a feasibility study of its **Grand Inga** proposal, whose main features are:

■ a third plant, Inga 3, with an output of 40 GW, requiring a 205 metre high dam forming a 15 km-long reservoir

■ a pan-African power distribution system reaching to Egypt in the north, Nigeria in the west and South Africa to the south, which would take a large proportion of the electricity.

In 2014, the World Bank announced that it had approved a US$73 million grant for the first phase of the Inga 3 project. However this funding was suspended in 2016 following disagreements with the Congolese government.

Sources: International Rivers 2016, 2017a, 2017b.

One form of hydro development that is generally expected to attract support in the coming years is *pumped storage*. World pumped storage capacity reached an estimated 150 GW at the end of 2016, with about 6.4 GW of capacity being added in that year (REN21, 2017).

China has announced a target of 40 GW of pumped storage by 2020 to help balance the output from the large increase in output from PV and wind power (IHA, 2017).

Finally, there is the hydro potential of the tens of thousands of dams in the world (possibly 30 000) that currently do not support a hydro plant.

Small-scale hydro

We saw in Section 6.4 that is not easy to assess the world output from small-scale hydro given the lack of an exact definition in terms of plant size. However, Box 6.12 summarizes the results of two detailed studies of the potential for small-scale hydro development (under 5 MW capacity) in the UK.

BOX 6.12 Small-scale hydro in the UK

The World Energy Council Resources assessment (WEC, 2010) quoted in Section 6.3 suggested that the hydro technical resource for the whole UK was 14 TWh y^{-1}, of which about 6 TWh y^{-1} has been developed. It is unlikely that any large-scale hydro projects would be given planning permission in the UK, so the future lies with small-scale schemes.

A detailed study in 2008 of the potential for small-scale hydro in Scotland found some 36 000 potential sites, with an estimated total power of 26 000 MW, giving an annual output of about 10 TWh per year. Only about 1000 sites were considered to be financially viable, but even this reduced number, with a total of 660 MW, might contribute some 2.7 TWh per year – a potential increase equal of about 40% on the present total UK hydro output (FHSG, 2008).

Two years later, a study of the potential for small-scale hydro in England and Wales was published (DECC, 2010). This differed from the Scottish study in that it was based on sites identified in an earlier analysis (reassessed in the light of technological and other changes) and also in that it excluded sites that would require the construction of new dams or weirs. Nevertheless, its estimated total potential installed capacity for England and Wales, ranged from a 'pessimistic' figure of 150 MW to an 'optimistic' one of 240 MW.

Given that, as shown earlier in Figure 6.5, world hydro generation has been growing by roughly 60 TWh y^{-1} for the last 50 years, it might be expected to increase by a further 20% by 2030. Scenarios for future world electricity generation, described in Chapter 12, suggest increases of between 15% and

35% with an annual growth rate of only 1-2% per annum. These are rather in contrast to the high continuing growth rates suggested for PV and wind power and are, no doubt, constrained by environmental concerns about new large dam projects.

6.13 **Summary**

Traditional water wheels, with their driven wheels only partially submerged in water, have been used for irrigation and driving corn mills for over 2000 years. The modern, totally submerged, water turbine was developed in the nineteenth century.

In 2016, hydro power provided one sixth of the world's electricity. The **world technical hydro resource** is about **16 000 TWh per year** – equivalent to two thirds of the world's 2016 electricity supply – however, only a **quarter** of this **technical resource** has been developed.

Most of the **world's hydro development since 2000** has taken place in **China**.

In **Norway, 95% of the country's electricity** is supplied by hydro, however, in the UK, hydro only supplies **2% of electricity** and is mainly used for supplying electricity at times of **peak electricity demand**.

Hydro **can be very flexible in its output**. It can be used to meet peak electricity demands and its generation may complement other renewable electricity technologies such as wind or PV.

Pumped storage plants are mainly used to provide **short-term backup** to the National Grid for up to a few hours. They can cover the **failure of a large fossil fuelled or nuclear plant** or **demand pickup** produced by many thousands of electric kettles being turned on at the end of a popular TV show. They require two water reservoirs, one preferably at least a hundred metres above the other. In order to store energy, water is pumped from the lower one to the upper one by a turbine driven by a large electric motor/generator. The energy can be recovered by allowing the stored water to flow back through the same turbine now driving the motor/generator to generate electricity.

The storage of **1 kilowatt-hour of electricity** requires raising **1 cubic metre of water (approximately 1 tonne)** through at least **360 metres in height**.

The **power from a hydro plant** is dependent on:

- the **efficiency of the turbine and generator (η, typically 85%)**
- the **water flow rate (Q) in cubic metres per second**
- the **effective height (H), in metres, through which the water falls** (this may be less than the 'real' height due to energy losses in flow resistance).

The key equation is:

Power (kW) = 10 × η × Q (m^3 s^{-1}) × H (m)

Hydroelectric plants use **gravitational potential energy**. The key equation for this is:

potential energy (joules) = mass (kg) × g (acceleration due to gravity) × height (metres).

In 2017, the three largest hydro plants in the world were:

- **Three Gorges in China** (22.5 GW)
- **Itaipú in Brazil** (14 GW)
- **Xiluodo in China** (13.9 GW)

Hydro schemes can range in output from **a kilowatt up to hundreds of megawatts**, with the definition of '**small scale hydro**' in the UK considered to be projects of **less than 5 MW output**.

Hydro schemes also have a very long life. Most of the Galloway scheme in Scotland was built in the 1930s and is still operational.

Hydro generation plants may be attached to **specially constructed dams** or **existing dams built for other purposes** (such as the drinking water reservoirs at the Elan Valley in Wales).

Water flows from the reservoir through a long tube called a **penstock** with the turbine situated at the lowest end. **Run-of-river plants** are situated at existing weirs or waterfalls in rivers and may only have low heads.

The **rate of rotation** of a turbine is constrained by the frequency of the electricity generated, which may be 50 Hz (cycles per second) or 60 Hz in different countries.

The **pressure of water** increases by 1 atmosphere for every 10 metres of water height.

The **Francis turbine** is the most common type used in modern hydro plants. Its moving part, the **runner**, is surrounded by **guide vanes** which deflect the water inwards towards the runner. The water then emerges from the centre of the runner. The runner is connected to a shaft that drives an electrical generator. Francis turbines are mostly used for **medium head** schemes of 50 to 200 metres head height. They are available at power ratings of up to 780 MW.

The **Pelton wheel** is commonly used for **high head** schemes of more than 100 metres height. It consists of a set of small buckets spinning in air driven by a high speed jet of water from a nozzle.

Propellor turbines consisting of a propellor in a tube are used for low head (under 15 metres height) applications. The **Kaplan** turbine is a propellor with vanes that can be adjusted with the flow rate. These may be also used in **tidal power plants**.

The **Archimedean screw** is an elongated propellor set in an open trough. These are becoming increasingly used for low head run-of-river schemes.

Hydro power does **not emit CO_2, nitrogen oxides or particulates** whilst in operation. **Fish ladders** allow fish, such as salmon, to swim up past dams to spawn in the water reservoirs.

The main disbenefits of hydro power are:

- **the likely need to dam large rivers**, possibly displacing large numbers of people. For example, the Three Gorges Dam in China displaced over one million people.
- **the risk of dam failures**, again displacing large numbers of people.

- **changes in river flows** – affecting access to inland lakes for fish – and water levels in surrounding land, which may give rise to legal arguments at a local or even international level.

- the eventual **silting up and loss of storage capacity** of reservoirs.

- **possible methane emissions** from rotting vegetation in reservoirs, a topic which still requires further research.

The **International Hydropower Association (IHA)** supervises the **Hydropower Sustainability Assessment Protocol (HSAP)** for the environmental and economic assessment of new large projects.

The key elements of the economics of hydropower are:

- **high initial capital costs**, dominated by the civil engineering works of dams.

- **an expected long operational lifetime** of 50 years or more.

- **modest operation and maintenance costs.**

- **zero fuel costs.**

The unit cost of electricity depends on the discount or interest rate used for any financial borrowing. **A low discount rate favours hydro**, a high discount rate favours fossil fuelled generation. The **refurbishment and upgrading** of existing plants may be highly economic.

Although only a quarter of the world's technical hydro resource has been exploited, scenarios of future expansion are limited by environmental considerations and only see **a growth rate of 1-2% per annum.**

The **potential for future growth in the UK is only modest** and is likely to be limited to small-scale projects.

References

ABC TV (2017) 'Brasil retiró casi 92 millones MWh de la producción récord de Itaipú', *ABC TV,* Paraguay, 2 January [Online]. Available at http://www. abc.com.py/edicion-impresa/economia/brasil-retiro-casi-92-millones-mwh-de-la-produccion-record-de-itaipu-1552332.html (Accessed 25 June 2017).

Arup (2016) *Review of Renewable Electricity Generation Cost and Technical Assumptions* [Online]. Available at https://www.gov.uk/government/publications/arup-2016-review-of-renewable-electricity-generation-cost-and-technical-assumptions (Accessed 24 June 2017).

Barrionuevo, A. (2011) 'Brazil, After a Long Battle, Approves an Amazon Dam', *The New York Times*, 1 June [Online]. Available at http://www.nytimes.com/2011/06/02/world/americas/02brazil.html?_r=2&scp=1&sq=belo%20monte&st=cse (Accessed 25 June 2017).

BEIS (2016a) *Digest of UK Energy Statistics (DUKES) 2016* [Online], London, Department for Business, Energy & Industrial Strategy. Available at https://www.gov.uk/government/collections/digest-of-uk-energy-statistics-dukes (Accessed 15 June 2017).

BEIS (2016b) *Electricity generation costs* [Online], London, Department for Business, Energy and Industrial Strategy. Available at https://www.gov.uk/government/publications/beis-electricity-generation-costs-november-2016 (Accessed 24 June 2017).

BP (2017) *Statistical Review of World Energy 2017* [Online], London, BP. Available at http://www.bp.com (Accessed 17 June 2017).

CCC (2011a) *The Renewable Energy Review* [Online], London, Committee on Climate Change. Available at https://www.theccc.org.uk/publication/the-renewable-energy-review/ (Accessed 24 June 2017).

Cruachan (2017) *Cruachan: The Hollow Mountain* [Online]. Available at https://www.visitcruachan.co.uk/pages/history.aspx (Accessed 18 June 2017).

Daily Fusion (2013) 'World's largest hydro turbine enters service in China', *Daily Fusion*, 21 August [Online]. Available at http://dailyfusion.net/2013/08/worlds-largest-hydro-turbine-enters-service-in-china-17795/ (Accessed 18 June 2017).

DECC (2010) *England and Wales Hydropower Resource Assessment* [Online] Available at https://www.gov.uk/government/publications/hydropower-resource-assessment-england-and-wales (Accessed 22 June 2017).

Dinorwig (n.d.) First Hydro Company Dinorwig Power Station [Online]. Available at http://www.fhc.co.uk/dinorwig.htm (Accessed 18 June 2017).

EAWAG (2010) 'Reservoirs: a neglected source of methane emissions', *Science Daily*, 14 October [Online]. Available at https://www.sciencedaily.com/releases/2010/10/101011090139.htm (Accessed 22 June 2017).

Elan Valley Trust (n.d.) *Elan Valley: Reservoirs and Dams* [Online]. Available at http://www.elanvalley.org.uk/discover/reservoirs-dams (Accessed 18 June 2017).

ERN (2017) *Restoring Rivers* [Online]. Available at http://www.ern.org/en/restaurer/ (Accessed 25 June 2017).

Everett, B., Boyle, G. A., Peake S. and Ramage, J. (eds) (2012) *Energy Systems and Sustainability: Power for a Sustainable Future* (2nd edn), Oxford, Oxford University Press/Milton Keynes, The Open University.

FHSG (2008) *Scottish Hydropower Resource Study Final Report* [Online], Hydro Sub Group of the Forum for Renewable Energy Development in Scotland (FHSG), Edinburgh. Available at http://www.gov.scot/Topics/Business-Industry/Energy/Energy-sources/19185/Resources/17613/FREDSHydroResStudy (Accessed 22 June 2017).

Hall, D. G., Hunt R. T., Kelly, S. R. and Greg, R. C. (2003) *Estimation of Economic Parameters of U.S. Hydropower Resources* [Online], Idaho, Idaho National Engineering and Environmental Laboratory. Available at https://www1.eere.energy.gov/water/pdfs/doewater-00662.pdf (Accessed 24 June 2017).

Hill, G. (1984) *Tunnel and Dam: the Story of the Galloway Hydros*, Glasgow, South of Scotland Electricity Board.

HSAP (2010) *Hydropower Sustainability Assessment Protocol* [Online]. Available at http://hydrosustainability.org (Accessed 22 June 2017).

Hvistendahl, M. (2008) 'China's Three Gorges Dam: An Environmental Catastrophe?', *Scientific American*, 25 March [Online]. Available at https://www.scientificamerican.com/article/chinas-three-gorges-dam-disaster/ (Accessed 22 June 2017).

ICJ (1997) *Gabcíkovo-Nagymaros Project (Hungary/Slovakia)* [Online]. Available at http://www.icj-cij.org/docket/index.php?sum=483&code=hs&p1=3&p2=3&case=92&k=8d&p3–5 (Accessed 25 June 2017).

IHA (2017) *Hydropower Status Report 2017* [Online]. Available at https://www.hydropower.org/2017-hydropower-status-report (Accessed 25 June 2017).

International Rivers (2016) *World Bank Halts Funding for Grand Inga Dam in DR Congo* [Online]. Available at https://www.internationalrivers.org/resources/world-bank-halts-funding-for-grand-inga-dam-in-dr-congo-global-construction-review-11533 (Accessed 25 June 2017).

International Rivers (2017a) *Three Gorges Dam* [Online]. Available at https://www.internationalrivers.org/campaigns/three-gorges-dam (Accessed 22 June 2017).

International Rivers (2017b) *Grand Inga Dam, DR Congo* [Online]. Available at http://www.internationalrivers.org/africa/grand-inga-dam (Accessed 25 June 2017).

International Water Power & Dam Construction (2007) 'Beyond Three Gorges in China' [Online]. Available at http://www.waterpowermagazine.com/news/newsbeyond-three-gorges-in-china (Accessed 22 June 2017).

IPCC (2012) *Renewable Energy Sources and Climate Change Mitigation* [Online]. Available at https://ipcc.ch/report/srren/ (Accessed 16 June 2017).

Itaipú (2011) *ITAIPÚ – largest power plant on Earth* [Online]. Available at www.solar.coppe.ufrj.br/itaipu_conv.html (Accessed 25 June 2017).

Kibel, P. and Coe, T. (2011) *Archimedean Screw risk assessment: strike and delay probabilities, Fishtek Consulting* [Online]. Available at http://www.fishtek.co.uk/downloads/Fishtek-screw-risk-assessment-example-report.pdf (Accessed 19 June 2017).

Mott MacDonald (2010) *UK Electricity generation costs update* [Online], Brighton, Mott MacDonald. Available at https://www.gov.uk/government/publications/uk-electricity-generation-costs-mott-macdonald-update-2010 (Accessed 18 July 2017).

National Trust (2017) *Cragside* [Online]. Available at https://www.nationaltrust.org.uk/cragside (Accessed 18 June 2017).

New Scientist (2011) Fossil fuels are far deadlier than nuclear power [Online]. Available at http://www.newscientist.com/article/mg20928053.600-fossil-fuels-are-far-deadlier-than-nuclear-power.html (Accessed 6 July 2017).

REN21 (2017) *Renewables 2017 – Global Status Report* [Online], Paris, Renewable Energy Network for the 21st Century. Available at http://www.ren21.net/gsr-2017/ (Accessed 18 June 2017).

ScottishPower (2010) *Tongland Power Station Turns 75 - But Retirement Not In Sight* [Online], Glasgow, Scottish Power. Available at https://www.scottishpower.com/news/pages/tongland_power_station_turns_75_but_retirement_not_in_sight.aspx (Accessed 16 June 2017).

ScottishPower (2011), *Galloway Hydros Technical Factsheet* [Online], Glasgow, ScottishPower. Available at https://www.scottishpower.com/userfiles/file/GallowayTechnical2011.pdf (Accessed 15 June 2017).

SSE (2005) *Power from the Glens* [Online], Perth, Scotland, SSE. Available at http://sse.com/media/87078/powerfromtheglens.pdf (Accessed 19 June 2017)

Strandh, S. (1989) *The History of the Machine*, London, Bracken Books.

Sustainable Hydropower (2011a) *About Sustainability in the Hydropower Industry: Sustainability Challenges* [Online]. Available at http://www.sustainablehydropower.org/site/info/aboutsustainability.html (Accessed 16 June 2017).

Sustainable Hydropower (2011b) *About Sustainability in the Hydropower Industry: Hydropower Strengths and Weaknesses* [Online]. Available at http://www.sustainablehydropower.org/site/info/aboutsustainability/strengthsweekness.html (Accessed 19 June 2017).

USBR (2011) *Grand Coolee Dam Statistics and Facts* [Online], Washington, D.C., U.S. Department of Interior Bureau of Reclamation. Available at www.usbr.gov/pn/grandcoulee/pubs/factsheet.pdf (Accessed 19 June 2017).

US Department of Energy (2016) Hydropower Vision: A new chapter for America's 1st renewable electricity source [Online], Oak Ridge, Tennessee, US Department of Energy, Office of Scientific and Technical Information. Available at https://energy.gov/sites/prod/files/2016/10/f33/Hydropower-Vision-10262016_0.pdf (Accessed 7 August 2017).

WEC (2010) *2010 Survey Of Energy Resources* [Online], London, World Energy Council. Available at https://www.worldenergy.org/publications/2010/survey-of-energy-resources-2010/ (Accessed 16 June 2017).

WEC (2016) *World Energy Resources 2016* [Online], London, World Energy Council. Available at https://www.worldenergy.org/publications (Accessed 4 November 2016).

Wikipedia (2017a) *Gabčíkovo-Nagymaros Dams* [Online]. Available at https://en.wikipedia.org/wiki/Gab%C4%8D%C3%ADkovo%E2%80%93Nagymaros_Dams (Accessed 25 June 2017).

Wikipedia (2017b) *Major dam failures* [Online]. Available at http://en.wikipedia.org/wiki/Dam_failure#List_of_major_dam_failures (Accessed 25 June 2017).

Wikipedia (2017c) *Hydroelectric power station failures* [Online]. Available at http://en.wikipedia.org/wiki/List_of_hydroelectric_power_station_failures (Accessed 25June 2017).

Wikipedia (2017d) 2009 *Sayano-Shushenskaya hydro accident* [Online]. Available at http://en.wikipedia.org/wiki/2009_Sayano-Shushenskaya_hydro_accident (Accessed 25 June 2017).

Chapter 7

Tidal power

By David Elliott and Mark Knös

7.1 Introduction

The rise and fall of the seas represents a vast, and as King Canute demonstrated, relentless natural phenomenon. The use of the tides to provide energy has a long history: small tidal mills on rivers were used for grinding corn in Britain and France in the Middle Ages. Subsequently, the idea of using tidal energy on a much larger scale to generate electricity emerged, using turbines mounted in large **barrages** – essentially low dams – built across suitable estuaries. More recently interest has grown in putting free-standing turbines into tidal currents.

This chapter looks at some examples of these tidal technologies and at their potential and limitations. It is, as yet, still a relatively undeveloped field, although a number of projects exist. A medium-scale 240 MW tidal barrage scheme has been built at the Rance Estuary in France (Figure 7.1) and a similar scale scheme has been built recently in South Korea. There have been proposals for much larger projects. For example, Figure 7.2 shows an artist's impression of the proposed 8.6 GW Severn Barrage, stretching 16 kilometres across the Severn Estuary in the UK.

Figure 7.1 A view of La Rance tidal power station

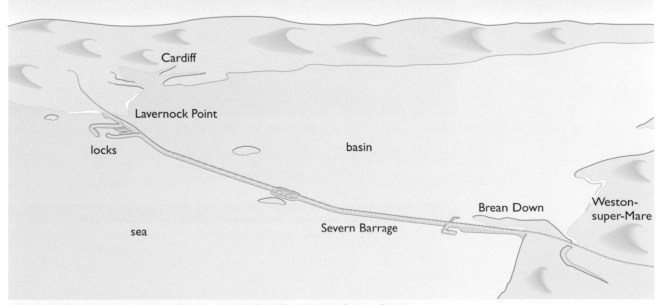

Figure 7.2 Artist's impression of the proposed Cardiff to Weston Severn Barrage

If built, this would generate 17 TWh per year, the equivalent of about 5% of the electricity generated in the UK in 2016. However, in recent years much of the emphasis has been on free-stream tidal turbines, and a range of devices are being developed and tested, with the first small arrays being deployed (Figure 7.3).

Figure 7.3 Marine Current Turbines' 1.2 MW 'SeaGen' concept

Box 7.1 outlines the early history of tidal power and describes some existing and proposed tidal projects around the world.

BOX 7.1 A brief history of tidal power

Small tidal mills, not unlike traditional watermills, were used quite widely on tidal sections of rivers and estuaries in the Middle Ages for grinding corn, but the idea of exploiting the full power of the tides in estuaries to generate electricity is relatively recent.

There have been a number of proposals for various types of barrage across the Severn, the UK's largest estuary, with the world's largest low- to high-tide range, but so far none have been taken forward. For example, a barrage concept (albeit with no provision for electricity generation) attributed to Thomas Telford was put forward in 1849. The first serious proposal involving electricity production came in 1920 from the Ministry of Transport. This was followed by a major study by the Brabazon Commission, which was set up in 1925. Its 1933 report focused on a barrage crossing the estuary along the 'English Stones line', not far from the modern Severn Bridges. It was to have 72 turbines with a total installed capacity of 804 MW and to incorporate road and rail crossings. The scheme was not taken up. It was reassessed in 1944, but, again, not implemented.

During the 1960s and 1970s a number of schemes were proposed, each crossing along different lines. These proposals culminated in another government-supported study by the Severn Barrage Committee, which was set up in 1978 under Professor Sir Hermann Bondi. The Committee reported in 1981 (Department of Energy, 1981), concluding that it was 'technically feasible to enclose the estuary by a barrage located in any position east of a line drawn from Porlock due North to the Welsh Coast'. Of all the possible crossing lines, three were favoured, the most ambitious being from Minehead to Aberthaw, which, it was estimated, could generate 20 TWh per year from 12 GW of installed capacity.

Subsequently, the less ambitious, but still very large, so-called 'inner barrage', on a line (first proposed by E. M. Wilson in 1966) from Brean Down, Weston-super-Mare to Lavernock Point near Cardiff, became the favourite, and was pursued by the Severn Tidal Power Group (STPG) industrial consortium in the 1980s. It was initially conceived as generating approximately 13 TWh per year from 7 GW installed, although this was later upgraded to 17 TWh per year from 8.64 GW installed.

Enthusiasm for tidal schemes was fuelled in part by the success of the French scheme on the Rance Estuary in Brittany, near St Malo (Figure 7.1). Construction started in 1961 and the first output from its 240 MW turbine capacity was achieved in 1966. The structure includes a road crossing. Apart from a problem with the generator mountings in 1975, it has operated very successfully. Subsequently, a much larger 15 GW scheme was proposed to enclose a vast area of sea from St Malo in the south to Cap de Carterel in the north, the so-called 'Île de Chausey Project'. This has not been implemented.

Large-scale schemes have also been proposed for the Bay of Fundy in Canada and at various sites in Northern Russia, but have not been followed up, although smaller projects have been built, including an 18 MW single unit, using a 'rim generator' (see Figure 7.15), at Annapolis Royal in Nova Scotia Canada, completed in 1984 and a 400 kW unit in the Bay of Kislaya, 100 km from Murmansk in Russia, completed in 1968.

More recently, South Korea has proposed a range of tidal projects and a 254 MW barrage was completed in 2011, making use of a causeway across the mouth of Sihwa Lake. A number of other barrage schemes have been considered around the world. Interest in the much larger Severn barrage has also continued, but, as we shall see, there are economic and environmental challenges and at present there are no plans for it to be built. However, secondary benefits, like storm and flooding protection and reduced coastal erosion from wave action, are sometimes seen as worth exploring.

The emphasis in the UK and many other countries, though, has moved onto smaller-scale in-stream tidal current turbines, operating on velocity in tidal flows rather than on (vertical) tidal height ranges. The world's first large-scale, grid-connected tidal current turbine (see Figures 7.3 and 7.35) was in service from 2008 to 2016 in the Strangford Narrows, Northern Ireland. A multitude of other tidal turbine designs are under development around the world and this area of activity seems likely to expand (see Sections 7.9 and 7.10).

The nature of the resource

It is important at the outset to distinguish *tidal* power from *hydro* power. As we saw in Chapter 5, *hydro* power is derived from the hydrological cycle (which is driven by solar energy) and is usually harnessed via hydroelectric dams. In contrast, *tidal* power is the result of the interaction of the gravitational pull of the Moon and, to a lesser extent, the Sun, on the seas. Schemes that use tidal energy rely on the twice-daily tides, and the consequent upstream flows and downstream ebbs in estuaries and the lower reaches of some rivers, as well as, in some cases, tidal movements out at sea.

Equally, we must distinguish between tidal energy and the energy in waves. Ordinary waves are caused by the action of wind over water, the wind in turn being the result of the differential solar heating of air over land and sea (see Chapter 8). If we consider wave energy, like hydroelectric energy, to be a form of solar energy, tidal energy could be called 'lunar energy'. Such distinctions are, unfortunately, not helped by the terminology which is often used — for example, the term 'tidal wave' is sometimes used to describe what are these days more usually called tsunami, the occasionally dramatic surges of water (which are neither wind-driven waves nor lunar-driven tides!) that can be produced by undersea earthquakes. There are also large climate-driven water flows in the oceans, such as the Gulf Stream, which are ultimately the result of solar heating. For the sake of completeness, we should also note that the gravitational pull of the Sun and the Moon also cause tidal phenomena in the atmosphere and in the Earth.

The energy in these various movements of water can, in principle at least, be tapped. The rise and fall of the tides can be exploited without the use of dams across estuaries, as was done in the traditional **tidal mills** on the tidal sections of rivers, as mentioned earlier. A small pond or pool is simply topped up and closed off at high tide and then, at low tide, the trapped water is used to drive a waterwheel, as with traditional watermills.

There is also the possibility of using turbines mounted independently in **tidal streams** (also called **tidal currents** — both terms are used, often interchangeably), that is *horizontal* tidal flows. Indeed, in the Middle Ages there were some floating 'undershot' waterwheels running on tidal currents. These flows of *kinetic energy* can be enhanced in some locations due to the effects of concentration in narrow channels, for example, between islands or other constrictions, or around headlands.

In addition, it may be possible to harness some of the energy in large-scale ocean streams such as the Gulf Stream. Some recent developments in the tidal current and ocean stream areas are discussed below, from Section 7.8 onwards. As noted in Box 7.1, tidal current projects are now emerging as a leading tidal option, not least since, unlike large estuary-wide barrages, they are modular and can be developed incrementally.

Before looking at tidal current systems, the initial sections of this chapter focus on tidal barrages across estuaries. The vertical difference between the water level at low tide and high tide is usually called the **tidal range**. In most *tidal range energy generation systems*, the water carried upstream by the **tidal flow** — usually called the **flood tide** — is trapped behind a barrage across the estuary. As the tide ebbs, the water level on the downstream side of the barrage reduces and a head of water develops across the barrage. The head is used to drive the water through turbine generators and the barrage scheme operates like a low-head hydro plant (see Chapter 5). Thus the barrage can be used to harness the *potential energy* provided by the *vertical* difference between tides. The main difference from hydro, apart from the salt-water environment, is that the power-generating turbines in tidal barrages have to deal with regularly varying heads of water.

A variant of the barrage concept is the tidal lagoon, which depends on a containment structure being built *within* the estuary to retain a small proportion of water in the estuary (in contrast to a barrage that acts on the whole estuary). However, the mode of operation is similar – the tidal range is used to create a head of water within the containment structure. Tidal barrages and tidal lagoons are therefore often collectively labelled **tidal range systems**, to distinguish them from systems using tidal currents.

Before looking at the details of, in turn, barrages, lagoons and tidal current schemes it is instructive to consider the physics behind tides and to appreciate what factors are involved in locating tidal energy systems.

The physics of tidal energy

The existence of tides is due primarily to gravitational interaction between the Earth and the Moon. This gravitational force, combined with the rotation of the Earth, produces, at any particular point on the globe, a twice-daily rise and fall in sea level, this being modified in height by the gravitational pull of the Sun and by the topography of land masses and ocean beds. A detailed analysis of this interaction between Earth, Moon and Sun is quite complex, but we will attempt to describe it in simple terms.

The first part of the explanation is relatively straightforward. Starting first with just the Earth and the Moon, the gravitational pull of the Moon draws the seas on the side of the Earth *nearest* to the Moon into a bulge *towards* the Moon. That gives us one tide per day at any one point, as the planet rotates through the bulge. But what about the second tide each day? This is more difficult to explain. Sometimes it is explained in simple terms by saying that the waters that make up the bulge facing the Moon are drawn from the seas at each side of the Earth, but the water at the far side is 'left behind', at its original level. However, that does not really explain the fact that the second tide is roughly the same height as the first. Neither does the fact that the water in the seas *furthest* from the Moon experiences slightly less of the lunar pull, being further away, although it may be part of the explanation.

The full explanation of the bulge that forms on the side of the Earth away from the Moon depends on a more complicated analysis, based on understanding the effect of the relative movements of the Earth and Moon (see Box 7.2).

The basic pattern described in Box 7.2 is also modified by the pull of the Sun. Although the Sun is much larger than the Moon, its distance from the Earth is much greater, and the Moon's gravitational influence on the seas is therefore approximately twice that of the Sun. The final impact depends on their relative orientation.

BOX 7.2 The Earth and the Moon

A useful mathematical analysis of the generation of tides is given in *Renewable Energy Resources* (Twidell and Weir, 2006). This identifies *two* processes at work in relation to the Earth and the Moon: a rotational effect as well as a gravitational effect.

The first process, the rotational effect, is the result of the fact that the Earth and the Moon rotate around each other, somewhat like a 'dumb-bell' being twirled. This rotation gives rise to an outward force, sometimes (rather loosely) called a centrifugal force, which acts on the water in the seas. However, this giant dumb-bell does not rotate around the halfway point between the Earth and the Moon. Since the Earth is much larger than the Moon, their common centre of rotation is close to the Earth; in fact it is just below its surface (Figure 7.4). The mutual rotation around this point produces a relatively large *outward* centrifugal force acting on the seas on the side of the Earth *furthest* from the Moon, bunching them up into a bulge. There is also a much smaller centrifugal force, directed *towards* the Moon, that acts on the seas *facing* the Moon. (This force is smaller since here the distance from the Earth's surface to the common rotation point, just below the surface, is smaller.)

The second process, the gravitational effect, is more familiar and relates to the gravitational pull of the Moon, which draws the seas on the side of the Earth *nearest* to the Moon into a bulge *towards* the Moon, whilst the seas *furthest* from the Moon, being slightly further away, experience a reduced lunar pull towards the Moon.

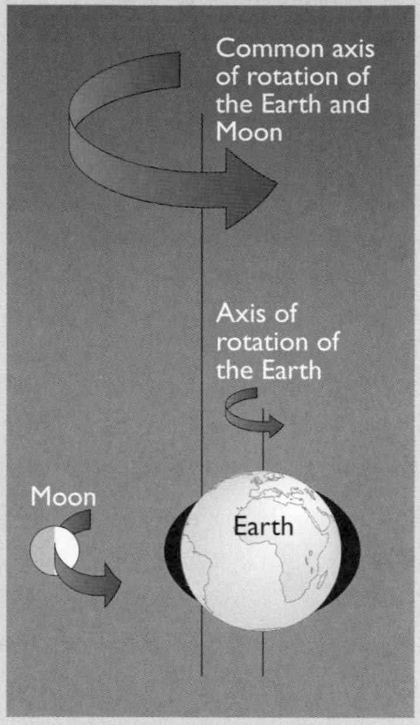

Figure 7.4 Relative rotation of the Earth and the Moon (not to scale)

There is thus, to summarize, a small centrifugal force and an increased lunar pull acting on the seas facing the Moon, and a larger centrifugal force and a decreased lunar pull acting on the seas on the other side of the Earth. The end result, on the basis of this analysis, is essentially a rough symmetry of forces, small and large, on each side of the Earth, producing tidal bulges of roughly the same size on each side of the Earth. In practice, the bulges may differ significantly, due, for example, to the tilt of the Earth's axis in relation to the orbit of the Moon and to local topographic effects.

As the Earth rotates on its axis, the lunar pull will maintain the high-tide patterns, as it were 'under' the Moon. That is, the two high-tide configurations will in effect be drawn around the globe as the Earth rotates, giving, at any particular point, *two* tides per day (or, more accurately, two tides in every 24.8 hour period), occurring approximately 12.4 hours apart. Since the Moon is also moving in orbit around the Earth, the timing of these high tides at any particular point will vary, occurring approximately 50 minutes later each day.

When the Sun and the Moon pull together (in line), whether both pulling on the same side of the Earth or each on opposite sides, the result is the very high **spring tides**; when the Sun and Moon are at 90° to each other (relative to the Earth), the result is the lower **neap tides**. The period between neap and spring tides is approximately 7 days – that is, a quarter of the 29.5-day lunar cycle (Figure 7.5). The ratio of output from a tidal range plant at the maximum spring tide to the output at a minimum neap tide can be more than two to one (Figure 7.6).

However, it is not simply a matter of the Earth's oceans rising and falling under gravitational forces. The pull of the Moon sets the whole sea into a state of oscillation, much as water might slop to and fro in a shallow dish if slightly shaken. The scale and frequency of these oscillations is influenced by the shape and size of the area of water, as defined by the surrounding land masses. The effect is somewhat like the resonance that occurs in musical instruments at various frequencies and volumes. Tidal resonances

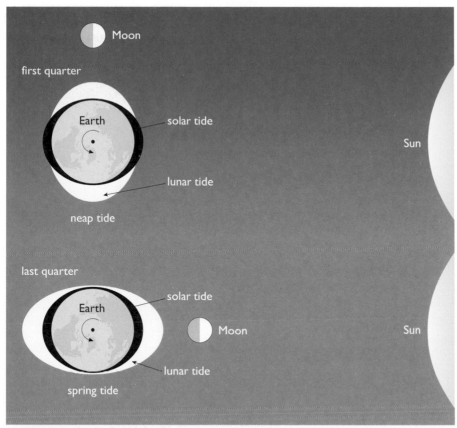

Figure 7.5 Influence of the Sun and the Moon on tidal range (not to scale)

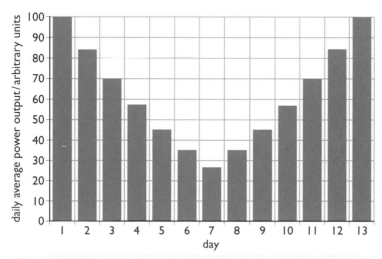

Figure 7.6 Typical variation in daily average output from a tidal range power plant over the spring–neap cycle (source: adapted from SDC, 2007)

can occur in well defined estuaries but can also be significant across entire oceans. For example, while the tidal range in open sea in mid-Atlantic is about 0.5 metres, at the coasts it can be enhanced to about 3 metres. In the case of the much wider Pacific the resonance effects are much smaller, so that the tides can be very small. In addition, the Earth's rotation causes the tidal flows to be deflected slightly, which can create a swirling pattern – in the North Atlantic there is a slow anticlockwise motion.

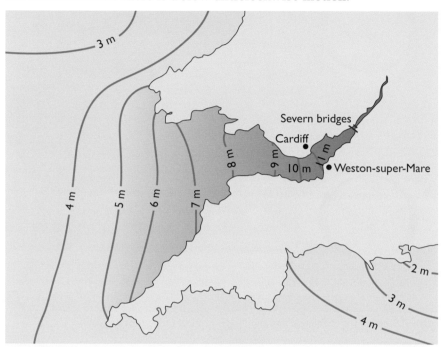

Figure 7.7 The effects of concentration of tidal flow in the Severn Estuary (tidal ranges in metres) (source: Department of Energy, 1981)

The rotation of the Earth also means that high tide occurs at each point on the Earth at a different time. Moreover, when the tide reaches the coast, the general topography of the landmass then defines the rate at which the tide moves along a coast. For example, it takes about six hours for a tide to progress down the English Channel from Land's End to Dover and there is several hours difference between high tides at other points around the east and west coast of the UK. This has important implications for the integration of the output of tidal power plants into an electricity grid.

Equally important for tidal power generation is the fact that the tidal ranges and tidal stream velocities experienced in practice at coastal sites are also sometimes significantly modified and amplified by *local* topographic variations, for example in shallow coastal waters and in estuaries. As the tide approaches the shore and the water depth decreases, the tidal flow is concentrated and the range can be increased to reach up to, typically, 3 metres. Examples of coastlines giving rise to funnelling and tidal range enhancement include the Severn Estuary in the UK (Figure 7.7), the Gulf of St Malo in France and the Bay of Fundy in Canada.

The tidal stream velocity can also be amplified if the tidal flow is constrained, for example in the Race of Alderney between the island of Alderney and the French mainland (Figure 7.8).

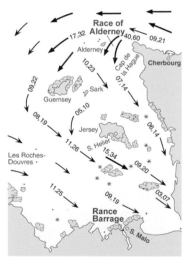

Figure 7.8 Tidal stream map for the Race of Alderney and La Rance Barrage 4 hours after high water at Dover; tidal stream rate in tenths of a knot expressed in the form: neap tide, spring tide (i.e. 10,23 means on a neap tide the stream runs at 1.0 knots and on a spring tide 2.3 knots). Note: 1 knot = 0.51 m s^{-1} (source: adapted from *Admiralty Tidal Stream Atlas: The English Channel NP250*, 1992)

However, there are also frictional effects to take into account. For example, energy is lost as the tidal flow moves over differing estuary bed materials. Thus, the extent to which the tidal range is magnified at any point depends on the balance between the energy losses and the concentration of the tidal flow by the topography. The frictional effects will usually begin to outweigh the concentration gains at some point upstream when the funnel-like layout of an estuary gives way to a more parallel-sided, flat-bottomed river configuration. On the Severn, this 'natural' optimal point normally occurs around the site of the (first) Severn Bridge, where the tidal range reaches 11 metres. The range decreases further upstream, as do flow rates.

Occasionally, in some long estuaries, dramatic tidal effects can occur upstream. For example, rather than producing a relatively slow rise, as normally happens in the main part of the Severn Estuary, the upstream tidal flow further up the Severn can be concentrated so abruptly that it rises into an almost vertical step or wave: the so-called Severn Bore. A similar effect occurs on other long estuaries, including the Humber in the UK and the Hoogly near Calcutta in India.

The water level actually experienced at a given location is also influenced by the weather – so-called **surge tides** are generated by storms rather than the Moon and Sun. Their effects can also be amplified with devastating effects by topographic features, and tidal power plants have to be built to withstand such extreme phenomena.

So, even leaving aside freak effects like the Severn Bore and occasional surge tides, there is a complex range of tidal phenomena. Fortunately for the designers of tidal barrages and other tidal devices, the end results – that is, the normal tidal patterns in estuaries – although very site specific, are predictable and reliable. The tides will continue to ebb and flow, on schedule, indefinitely.

But is the energy in the tides *really* 'renewable'? As we have seen, the primary mechanism in tide generation is the gravitational interaction between the Earth and the Moon, and the forces produced by their relative orbital movements. These forces create bulges of water. The rotation of the Earth draws these resulting tidal bulges across the seas, or, more precisely, the water in the seas rise into a bulge as the water rotates with the planet. The result is that there are horizontal tidal flows, as water is drawn into the moving bulge. The rotation of the Earth is being very gradually slowed by tidal processes (by approximately one-fiftieth of a second every 1000 years), because of frictional effects – it takes energy to drag the water along, especially through areas where there are topographical constrictions. However, the extra frictional effect that would be produced by even the widespread use of tidal barrages would be extremely small. The influence on the Moon's orbital velocity (which is also being very slowly reduced by the tidal interaction) would be even smaller. So overall, tidal energy is renewable on any reasonable interpretation of the concept.

7.2 Power generation from barrages

The basic physics and engineering of tidal barrage power generation are relatively straightforward.

Tidal barrages, built across suitable estuaries, are designed to extract energy from the rise and fall of the tides, using turbines located in water passages in the barrages. The potential energy, due to the difference in water levels across the barrage, is converted into kinetic energy in the form of fast-moving water which passes through the turbines – the spinning turbines then drive generators to produce electricity.

The *average* power output from a tidal barrage is roughly proportional to the square of the tidal range. The mathematical derivation of this is fairly simple, as is demonstrated by the analysis in Box 7.3.

BOX 7.3 **Calculation of power output from a tidal barrage**

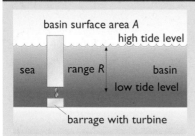

Figure 7.9 Power generation from tides (source: Twidell and Weir, 2006)

Let us assume that we have a rectangular basin with a constant surface area A, behind a barrage, and a high-to-low tidal range R (Figure 7.9). In conventional ebb generation, when the tide comes in, it is freely allowed to flow into the basin, but when the tide goes out, the water in the basin is held there, at the high-tide level. When the sea has retreated to its low-tide level, the surface of the water held behind the barrage will be at a height R above the sea.

Given a rectangular basin, the centre of gravity of the usable mass of water will be at a height $R/2$ above the low-tide level. The total volume of water in the basin will be AR and, if the density of the water is ρ it will have a mass ρAR, i.e. ρ multiplied by the volume of water (A times R). This water could all now be allowed to flow out of the barrage through a turbine to the low-tide level. The maximum potential energy E available per tide if all the water falls through a height of $R/2$ is therefore given by the mass of water (ρAR) times the height ($R/2$) times the acceleration due to gravity (g); that is, $E = \rho ARg(R/2)$. The basin could then be allowed to fill on the next incoming tide and the cycle repeated again and again. If the tidal period is T, then the average potential power that could be extracted becomes E/T or $\rho AR^2g/2T$.

Clearly, even small differences in tidal range, however caused, can make a significant difference to the viability and economics of a barrage. A mean tidal range of at least 5 metres is usually considered to be the minimum for viable power generation, depending on the economic criteria used. As the analysis in Box 7.3 indicates, the energy output is also roughly proportional to the area of the water trapped behind the barrage, so the geography of the site is very important. All of this means that the siting of barrages is a crucial element in their viability.

Many studies have been carried out on tidal power in the UK, dating from the early 1900s onwards (see Box 7.1). This is hardly surprising, as the UK holds about half the total European potential for tidal barrage energy, including one of the world's best potential sites, the Severn Estuary. There is also a range of possible medium- and small-scale sites, including locations on the Mersey and Solway Firth (further details are in Section 7.6). The total UK tidal barrage resource potential has been put at around 53 TWh per year (ETSU, 1990), which was about 14% of UK electricity generation in 2010. In practice, the contribution to electricity consumption that could be achieved in the UK and elsewhere would depend on a range of *technical, environmental* and *economic* factors. Although these factors interact, we can explore each in turn before attempting a synthesis.

Barrage designs

The input energy source for a barrage, the rise and fall of the tides, follows a roughly sinusoidal pattern (see the sea-level curves in Figures 7.10–7.12). The tides have a 12.4 hour cycle, with the actual tidal range varying from site to site as a result of complex resonance and funnelling effects as mentioned earlier.

Given the complexity of estuary configurations, the actual resonances and funnelling effects are very difficult to model accurately, with variations in depth, width and friction over differing estuary bed materials introducing many local variations.

However, it is well worth the effort required to analyse these effects when deciding on the precise siting and orientation of a barrage, since they will have a major effect on its output. Indeed, it may be possible to locate and/or operate a barrage so as to 'tune' the barrage to the estuary tidal pattern, and thus to increase energy output. Certainly, any disturbance that might reduce existing resonance effects should be avoided.

In addition to the basic issues of location and orientation, a second set of factors that influences the likely energy output of a barrage relates to its *operational pattern*.

Energy can be generated from a barrage in three main ways. The most commonly used method is **ebb generation**. Here the incoming tide is allowed to pass through the barrage sluice gates. The water is trapped behind the barrage at high-tide level by closing the sluices. The head of water then passes back through the turbines on the *outgoing* ebb tide in order to generate energy (Figure 7.10). Alternatively, **flood generation** uses the *incoming* tide to generate electricity as it passes through the turbines mounted in the barrage (Figure 7.11). In each case, two bursts of energy are produced in every 24.8 hour period. **Two-way operation**, on the ebb *and* the flood, is also possible (Figure 7.12).

The basic technology for power production is well developed, having much in common with conventional low-head hydro systems (see Chapter 6). Figure 7.13 is an artist's impression of the typical layout of a power generation scheme.

A number of different turbine configurations are possible. At La Rance, a so-called **bulb** system is used, with the turbine generator sealed in a bulb-shaped enclosure mounted in the flow (Figure 7.14). As the water has to flow around the large bulb, access (for maintenance) to the generator involves cutting off the flow of water.

These problems are reduced in the **Straflo** or **rim generator** turbine (as used at Annapolis Royal in Canada), with the generator mounted radially around the rim and only the runner (that is, the turbine blades) in the flow (Figure 7.15). Although it is more efficient than the bulb design, because the water flow is not so constricted, this design introduces extra problems with the sealing between the runner blades and the radial generator.

Alternatively, there is the **tubular** turbine configuration, with the runner set at an angle so that a long (tubular) shaft can take rotational energy out to an external generator (Figure 7.16). This design also avoids constricting

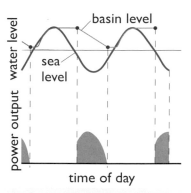

Figure 7.10 Schematic diagram of water levels and power outputs for an ebb generation scheme (source: Department of Energy, 1981)

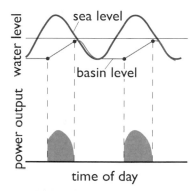

Figure 7.11 Schematic diagram of water levels and power outputs for a flood generation scheme (source: Department of Energy, 1981)

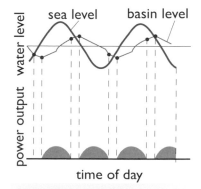

Figure 7.12 Schematic diagram of water levels and power outputs for a two-way generation scheme (source: Department of Energy, 1981)

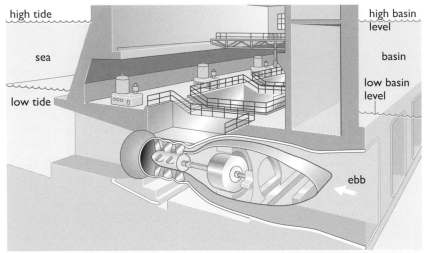

Figure 7.13 Artist's impression of the typical layout of a power generation scheme

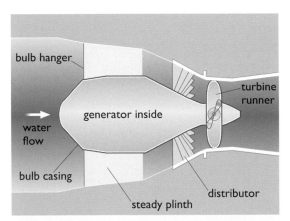

Figure 7.14 Bulb turbine as used at La Rance

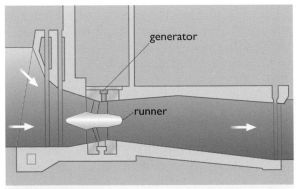

Figure 7.15 Straflo or rim generator turbine as used at Annapolis Royal

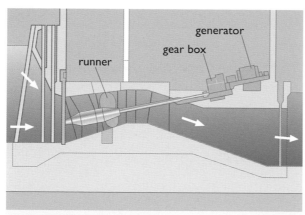

Figure 7.16 Tubular turbine

the flow of the water and, since the generator is not in a confined space, there is room for a gearbox, which can allow for efficient matching to the higher speed generators usually used with hydro plants. Several such units have been used in hydro power plants in the USA, the largest being rated at 25 MW. However, there have been vibration problems in the long drive shaft and, so far, bulb turbines have proved to be the most popular with barrage designers.

As mentioned above, the rotational speeds of the turbines in tidal barrages tend to be lower than those for turbines in hydro plants (50–100 revolutions per minute, in comparison to 200–450 revolutions per minute for hydro generators) and therefore wear is also reduced. Since large volumes of water have to pass through the barrage in a relatively short time, large numbers of turbines are required in a large-scale barrage (see Boxes 7.4 and 7.5 for details of La Rance and the proposed Severn Barrage).

In simple ebb or flood generation, this large installed capacity is used only for a relatively short period (three to six hours at most) in each tidal cycle, producing a large but short burst of power, which may not match the demand for electricity. However, using reversible-pitch turbines it is possible to operate on both the ebb and the flood in a two-way operation. Reversible-pitch turbines are more complex and costly than standard turbines and, although the output will be more evenly distributed over time, there will be a net decrease in electricity output for each phase compared with a simple ebb generation scheme. This is because, in order to be ready for the next cycle, neither the ebb nor the flow generation phases can be taken to completion: it is necessary to open the sluices and reduce water levels ready for the next flood cycle, and vice versa for ebb generation (see Figure 7.12).

Flood pumping is another option for electricity generation. Here, the turbine generators are run in reverse and act as motor-pump sets, powered by electricity from the grid. Additional water is pumped behind the barrage into the basin at around high tide, when there is a low head difference across the barrage. This provides extra water for the subsequent ebb generation phase when there is a high head difference. This is, in effect, a way to store excess off-peak power from the grid, but with a net energy gain.

BOX 7.4 **La Rance**

The 740-metre long Rance Barrage was constructed
between 1961 and 1967. It has a road crossing
and a ship lock (Figure 7.17) and was designed
for maximum operational flexibility. It contains

24 reversible (that is, two-way) turbines, each of
10 MW capacity, operating in a tidal range of up
to 12 metres, with a typical head of approximately
5 metres.

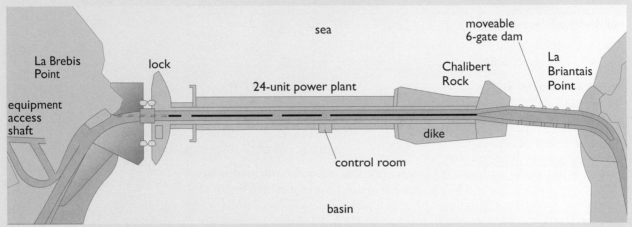

Figure 7.17 Layout of the Rance Barrage (source: Department of Energy, 1981)

The operational pattern initially adopted at La Rance
was to optimize the uniformity of the power output
by using a combination of two-way generation
(which meant running the turbines at less than the
maximum possible head of water) and incorporating
an element of pumped storage. For spring tides,
two-way generation was favoured; for neap tides and
some intermediate tides, direct pumping from sea
to basin was sometimes carried out to supplement
generation on the ebb.

Although some mechanical problems were
encountered in 1975, which subsequently led to
two-way operation mostly being avoided, overall
the barrage has been very successful. Typically
the plant has been functional and available for
use more than 90% of the time, and net output
has been approximately 480 GWh y^{-1} with,
in some years, significant energy gains from
pumping.

The construction of the barrage involved building
two temporary coffer-dams, with the water then
being pumped out of the space in between, to
allow work to be carried out in dry conditions
(see Figure 7.18). River water was allowed to

Figure 7.18 La Rance coffer-dam during construction of the
barrage

pass via sluices, but the reduced ebb and flow
resulted in effective stagnation of the estuary and
the subsequent partial collapse of the ecosystem
within it. Since construction, exchange of water
between the open sea and the estuary has restored
the estuarine ecosystem, but because there was no
monitoring it is difficult to establish what changes
the barrage caused to the original environment in
the estuary.

BOX 7.5 The proposed Severn Barrage

Basic data for the Cardiff to Weston Barrage is given below (Department of Energy, 1989). The same design specifications were used for the assessment carried out by the Sustainable Development Commission in 2007 (SDC, 2007) and in the Department of Energy and Climate Change/Welsh Assembly Government/SWRDA study in 2010 (DECC, 2010a).

Number of turbine generators	216
Diameter of turbines	9.0 metres
Operating speed of turbines	50 rpm
Turbine generator rating	40 MW
Installed capacity	8640 MW
Number of sluices, various sizes	166
Total clear area of sluice passages	35 000 m^2
Average annual energy output	17 TWh
Operational mode	ebb generation with flood pumping
Length of barrage:	
total	15.9 km
including: powerhouse caissons	4.3 km
sluice caissons	4.1 km
other caissons	3.9 km
embankments	3.6 km
Area of enclosed basin at mean sea level	480 km^2

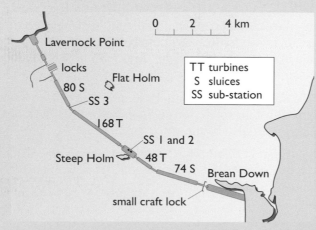

Figure 7.19 Layout of the proposed Cardiff to Weston Severn Barrage (source: adapted from Department of Energy, 1989)

In addition, as will be discussed in more detail later, many different types of **double-basin** system have been proposed (see, for example, Figure 7.20), often using pumping between the basins. During periods of low demand, excess electricity generated by the turbines of the first basin can be used to pump water into the second basin, ready for the latter to use for generation when power is required.

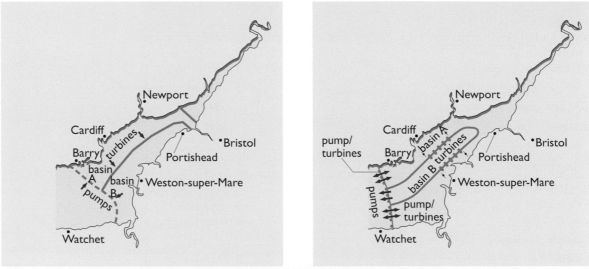

Figure 7.20 Severn Estuary with possible double-basin schemes (source: adapted from Department of Energy, 1981)

Whatever the precise configuration chosen for a barrage, the basic components are the same: *turbines*, *sluice gates* and, usually, *ship locks*, to allow passage of ships, all linked to the shore by *embankments*.

The turbines are usually located in large concrete units. For the Severn Barrage, the use of large concrete, or steel, caissons (containment and support structures) to house the turbines has been proposed. These could be constructed on shore, in dry dock facilities, and then floated onto site and sunk into place. It has been claimed such techniques could cut civil engineering costs by up to 30% as compared to the temporary coffer-dam approach taken at La Rance (although there could be additional difficulties in placing caissons in strong currents).

Sluice gates are another essential operational feature of a barrage, to allow the tide to flow through ready for ebb generation, or back out after flow generation. These can also be mounted in caissons.

The rest of a barrage is relatively straightforward to construct. La Rance, for example, has a rock-filled embankment, whilst one design proposed for the Severn used sand-filled embankments faced with suitable concrete or rock protection.

7.3 **Environmental considerations for tidal barrages**

The construction of a large barrier across an estuary will clearly have a significant effect on the local ecosystem. Some of the effects will be negative, and some will be positive. The negative local impacts have to be weighed up against the role that barrages could play in helping to resolve some global environmental and energy problems (such as global warming caused by CO_2 emissions from the burning of fossil fuels), and in offering improved energy security through decreased reliance on imported fuel.

In the UK, much research has gone into trying to ascertain the probable final balance of positive and negative impacts, and the overall cost effectiveness, focusing mainly on the proposed Severn Tidal Barrage (Department of Energy, 1989). The most recent review was carried out as part of the UK government's Severn Tidal Power Feasibility Study (DECC, 2010a). This looked at a number of tidal range projects for the Severn, including the Cardiff–Weston Barrage and some smaller barrages and tidal lagoons, and concluded that of all the schemes studied, the Cardiff–Weston Barrage would 'have the greatest impact on habitats and bird populations and the estuary ports', as well as very high capital costs.

Certainly the most obvious potential impact of any barrage would be on local wildlife, that is, fish and birds, many of the latter being migratory. The UK's estuaries play host to approximately 28% of European swans and ducks and to 47% of European geese. There are also large populations of fish: the Severn, for example, is well known for its salmon and eels (elvers). Many of these species rely on the estuaries for food, and access to that supply might be affected by a tidal barrage (Department of Energy, 1989).

The proposed Severn Barrage would decrease a large area (200 km² or more) of mud flats exposed each day, since the water level variations behind the barrage would be significantly reduced (Figure 7.21). Some species (for example, mud-wading birds) feed on worms and other invertebrates from the exposed mud flats, and could be adversely affected. Similar issues would apply for salt marshes that might be exposed daily by the tides at other potential barrage sites.

However, the barrage could have a compensating impact on the level of silt and sediment suspended in the water – the action of the tide in churning up silt makes the water in the Severn Estuary impenetrable to sunlight. With the barrage in place and the tidal ebbs and flows reduced, some of this silt would drop out, making the water clearer. Given this change in **turbidity** sunlight would penetrate further down, increasing the biological productivity of the water and therefore increasing the potential food supplies for fish and birds. The net impact is likely to be mixed: some species might not find a niche in the new ecological balance, whilst others previously excluded from the estuary might become established.

This rather simplified example illustrates the general point that there are complex interactions at work making it difficult to predict the outcome.

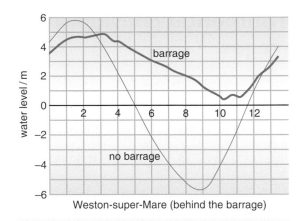

Figure 7.21 Variation in tide levels with and without the Severn Barrage over a 12 hour period (source: Department of Energy, 1989)

Similar interactions and trade-offs occur in relation to other ways in which barrages can impact on their surroundings. Clearly, the construction of a barrage across an estuary will impede any shipping, even though ship locks are likely to be included. The fact that the sea level behind the barrage would, on average, be higher could improve navigational access to ports, the net effect depending on tidal cycles and the precise location of the barrage and of any ports.

Visually, barrages present fewer problems than comparable hydro schemes. Even at low tide, the flank exposed would not be much higher than the maximum tidal range. From a distance, all that would be seen would be a line on the water.

Barrages could also play a useful role in providing protection against floods and storm damage, since they could be operated to control very high tidal surges and limit local wave generation. Conversely, for some sites, due to the change in tidal patterns (with the tide upstream staying above mid-level for longer periods), there might be a need for improved land drainage upstream.

A barrage would have some effect on the local economy both during the construction phase and subsequently, in terms of employment generation and local spending, tourism and, in particular, enhanced opportunities for water sports. Depending on the scale and the site, there could also be the option of providing a new road or rail crossing, as with the Rance Barrage. The incorporation of a public road was part of the plans proposed for the Severn Barrage.

Whether these local infrastructural improvement options represent environmental benefits or costs depends on your views on industrial and commercial development (some conservation and wildlife groups, for example, baulk at the prospect of increased tourism), but many people would be likely to welcome local economic growth. Indeed, that was the message from local populations faced with barrage proposals. Local commercial and civic interests and the wider public have on the whole been supportive of such plans while other special interest groups have opposed barrages. For example, the Royal Society for the Protection of Birds (RSPB) sees barrages as inherently damaging, reducing habitats for key species, particularly migrant birds. This problem could clearly be compounded if several barrages were to be built. In 2008, when the Severn Barrage idea was moving up the UK political agenda, the National Trust, RSPB and World Wide Fund for Nature (WWF), in a coalition with other groups, came out strongly against the barrage.

7.4 Integration of electrical power from tidal barrages

The electricity produced by barrages must usually be integrated with the electricity produced by the other power plants that feed into a national grid power transmission network.

The key problem in feeding power from a tidal barrage into national grid networks is that with conventional ebb or flood generation schemes the tidal energy inputs come in relatively short bursts at approximately twelve-hour intervals. Typically, power can be produced for five to six hours during spring tides and three hours during neap tides, within a tidal cycle lasting 12.4 hours (Figure 7.22).

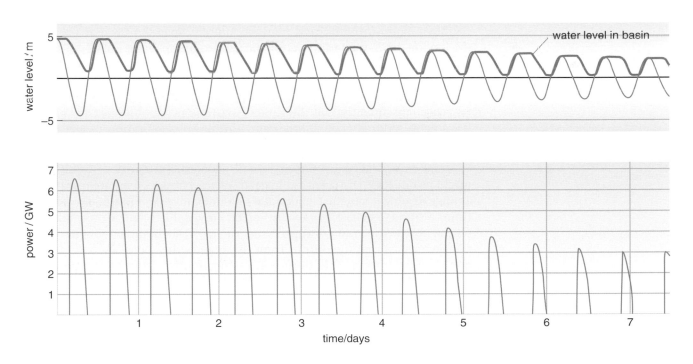

Figure 7.22 Water level and power output of the proposed Severn Barrage over a spring–neap tide cycle (source: Laughton, 1990)

Clearly, such availability of power from a barrage will not always match the pattern of demand for power on the grid. With a large, well-developed grid, as in the UK, with a large load and many other types of power plant connected into it, this problem might not be too severe, depending on the size of the barrage and the flexibility of the output. The availability of large amounts of electricity storage capacity could also help. Even so, with a barrage the size of that proposed for the Severn, absorbing all the power would clearly represent a significant task. At its peak, it would be generating over 8 GW of power, which represents a sizeable proportion of the UK's peak electricity demand of under 60 GW (in 2016), and an even larger proportion of the minimum demand level during summer nights (around 20 GW – see Figure 11.13).

Two-way generation, on the incoming flood as well as the ebb, might offer some advantages since that would give *four* bursts of power for each 24 hour cycle, but as noted earlier the more complex turbines required add to the cost. There is also the option of using modified turbines as pumps, running on grid power (e.g. during off-peak periods), to pump extra water behind a barrage during the flood phase. Typically, between 5% and 15% extra output can be gained, with little additional capital cost and with no loss of generating efficiency. This has been the favoured option for all the barrage schemes proposed in the UK. That said, the overall *economic* advantages of pumping may be fairly small (the electricity used has to be paid for), and they depend crucially on tidal timing, which will not necessarily coincide with off-peak grid power availability.

Another option we noted earlier for providing power output more evenly over time is to construct two (or more) basins. For example, the first basin could operate in the normal manner, but any excess power it produced during low-demand periods, could be used to pump water out of the second basin to keep it below tide level, so that power could be generated from the second basin when needed by filling it through its turbines. As these configurations involve building what amounts to two or more barrages, costs increase significantly – at the time of writing, no double-barrage systems have been seriously considered, and simple ebb generation is seen as the most economic option.

Whichever system is used, the overall economic viability of tidal power might be enhanced, in theory at least, if several barrages were in operation in a range of sites, given the fact that the tidal maxima occur at slightly different times around the coast. For example, the Solway Firth and Morecambe Bay on the northwest coast are approximately five to six hours out of phase with the Severn, so that the output from these and other possible barrage sites could be fed into the grid to provide a contribution from tidal power over a longer period of time, although at neap tides this input would be low.

It is possible that there might be synergies between tidal and other renewable energy systems, for example via the installation of wind turbines along barrages in the same way as some wind farms have been constructed on causeways in harbours. The wind plant might even be used at times for pumping water behind the barrage, although the energy contribution that could be made, even if wind turbines were located regularly along the entire length of a barrage, would be relatively small compared with the output of a large barrage. For example, if, say, thirty 2 MW wind turbines were installed along parts of the 16 km long proposed Severn Barrage, their total annual electricity output would be just over 1% of that from the barrage.

Finally, the linking of power from barrages to the National Grid could present some practical problems. As with many other types of large, new, power plants, extra grid connections would have to be made, and in some circumstances existing local grid lines strengthened to carry the extra power. Fortunately, most potential barrage sites in the UK are reasonably near to existing power lines, so the problems such as those of harnessing deep-sea wave power (much of which would have to be transmitted from the north of Scotland; see Chapter 9) would be avoided. In the case of the proposed Severn Barrage new power lines, perhaps stretching to the major loads in the Midlands, would probably be required. It has been estimated that these grid connections might add 10% to the capital costs of the barrage.

In summary, the key issue to integration is cost, whether this is for additional grid linkages or for systems that allow power to be produced on a more nearly continuous basis. We now move on to look at economic factors more generally.

7.5 The economics of tidal barrages

The overall economics of tidal barrages depends both on their operational performance and their initial capital costs. The latter are high – it was estimated in the late 2000s that the 10 mile Cardiff (Lavernock Point) – Weston-super-Mare (Brean Down) barrage would cost over £30 billion (DECC, 2010a). The civil engineering works are the single largest element in the total cost, closely followed by turbine manufacture and installation. In addition to the construction and equipment costs, there is also the cost of borrowing the money, with interest having to be paid during the course of construction, when of course there is no income since the barrage is not yet operating. This raises a key issue for the economics of large capital-intensive projects like this with long construction times.

Although there are running costs (approximately 1% of the total capital cost per year), tidal barrages, like all renewable energy systems based on natural flows, have no fuel cost. After initial construction, they can generate power for many years without major civil engineering effort with the low-speed turbines needing replacement perhaps every 30 years.

Furthermore, in addition to the significant initial capital costs, the period during which power can be produced, at least for simple single-basin ebb or flood systems, is clearly less than for a conventional power plant. For example, because it would only operate during tidal cycles, the 8.6 GW turbine capacity of the Severn Barrage could only offer the same output, averaged out over a year, as a conventional plant with around 2 GW of generating capacity. In other words, the barrage requires a large investment in expensive capacity which is only used intermittently and can therefore only replace a limited amount of conventional plant output. The value of a barrage as a replacement for conventional plant(s) will depend in practice on the scale and timing of the outputs of the plants they can replace, not all of which will be able to generate continuously either.

As explained in Chapters 1 and 6, a convenient way to compare systems is in terms of their capacity factors (the ratio of a plant's actual energy output over a given period to the theoretical output if the plant operated at its full-rated capacity). The annual capacity factor for the proposed Severn Barrage is estimated at around 23%. By comparison, in typical UK conditions the average annual plant capacity factor for nuclear stations has in recent years been around 77%, and for combined cycle gas turbine power plants around 84%. Thus, compared with most other types of power plant, tidal projects have a relatively high capital cost in relation to the energy output with, consequently, long capital payback times and low rates of return on the capital invested, the precise figures depending on the price that can be charged for the electricity.

In the 1980s, when the Severn Tidal Power Group analysed the economics of a Severn Barrage, the UK electricity industry was state-owned. As such it was expected to make a return on investment of about 5% per annum. However, the policy of the Conservative government of the time was to privatize the industry (which eventually happened in 1989). Any barrage project would thus be expected to be competitive with existing coal-fired and nuclear plants. It would also be privately financed and required to make commercial rates of return (10–15%). This would require a high electricity price to cover the interest on the large capital outlay. The Severn Barrage was not considered to be economic under these conditions (STPG, 1986). That was one reason why the project was not supported at that particular time.

Although fuel prices have increased dramatically and concerns about climate change have grown, the basic economic assessment has not changed significantly since then. It has, however, been argued that, given wider strategic concerns, such as climate change and security of energy supplies, it might be justifiable to accept a lower rate of return, similar to that used for public projects.

In its 2003 Green Book, the UK Treasury suggests that public sector organizations should make a real return of 6% on investments, though this point is qualified with a long discussion on the proper evaluation of risk and 'optimism bias' (HM Treasury, 2003).

In its 2007 study of tidal options for the Severn, the UK government's Sustainable Development Commission (SDC) noted that, 'The high capital cost of a barrage project leads to a very high sensitivity to the discount rate used' (a truism for all projects with significant initial capital expenditure, as exemplified in Chapter 6 where a cost comparison of combine cycle gas turbine (CCGT) and hydro plants was made). At a low discount rate of 2%, which it argued could be justified for a climate change mitigation project, the cost of electricity output 'is highly competitive with other forms of generation'. On that basis they felt that the project should be financed as a public project (SDC, 2007).

In contrast they noted that, at a more 'commercial' discount rate of 8% or more, costs escalated significantly, making private sector investment unlikely without significant market intervention by government. For example they calculated that at a 2% discount rate the Cardiff–Weston Barrage would generate at between 2.27 and 2.31p per kWh depending on how long it took to build, whereas at 10% discount rate the cost would rise to 11.18–12.37p per kWh. These costs were expressed in the money of the day, i.e. £(2006).

In its 2010 study of Severn tidal schemes the UK government compared the impact of 3.5% public sector 'social' discount rates with 10% commercial 'investor' discount rates (see Chapter 11 and Appendix B for a further discussion of discount rates). Table 7.1 illustrates the dramatic difference in energy costs and also the different levels of environmental impact of different options. Table 7.1 also compares the costs of large and smaller barrage schemes on the Severn. It might be expected that the smaller schemes further upstream would be more attractive financially, because they can be built more rapidly – but this is evidently not the case, in part due to the lower energy output from the smaller tidal ranges at these sites. So whether under 'investor' or 'social' financial frameworks, smaller barrages on the Severn are less attractive in terms of the cost of electricity produced. Although in the case of the Severn this seems clear-cut, small barrages in other estuaries, in locations with higher tidal ranges, might do better.

Table 7.1 Comparison of costs and impacts of large and small barrages on the Severn

Scheme	Installed capacity /MW	Annual energy generated /TWh y^{-1}	Levelized energy cost/£ MWh^{-1}		Inter-tidal habitat loss/km^2
			Investor (10% discount rate)	Social (3.5% discount rate)	
Cardiff–Weston Barrage	8640	15.6	312	108	160
Shoots Barrage	1050	2.7	335	121	33
Beachley Barrage	625	1.2	419	151	27

Source: DECC, 2010a

Given the lifetime of the low-speed turbines (replacement needed perhaps only every 30 years), low running costs (approximately 1% of the total capital cost per year) and zero fuel cost, it has been argued that, like hydro projects, barrages make a very good long-term investment. This is unfortunately not reflected in contemporary financial approaches, with the emphasis usually being on short-term returns.

Quite apart from the problems of funding and the vagaries of finance capital, interest rates, etc., on the 'supply' side, there are technical and environmental uncertainties, with trade-offs between operational efficiency and likely impact. To these uncertainties must be added the 'demand side' uncertainties, with the price that can be charged for tidal electricity having to be estimated over the very long term and compared with conventional fuel prices.

At £30 billion or so for cyclically varying power, the Severn barrage might not be the best use of any money available. Indeed a study by Frontier Economics for the National Trust, RSPB, WWF and other environmental groups opposed to the Severn barrage, suggested that large barrages were significantly more expensive on a £ per MWh basis than any other major energy generation option (Frontier Economics, 2008; see Figure 7.23). On this basis, it would seem that other options are likely to be more attractive, a view that now seems to be shared by the government, which in its 2010 study of the Severn tidal options commented: 'The Government believes that other options, such as the expansion of wind energy, carbon capture and storage (CCS) and nuclear power without public subsidy, represent a better deal for taxpayers and consumers at this time' (DECC, 2010a).

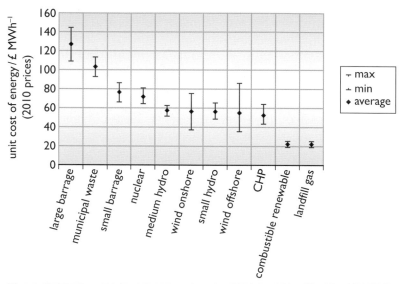

Figure 7.23 Comparison of costs of energy from tidal and other renewable energy projects, at 7% discount rate (source: adapted from Frontier Economics, 2008)

Although it seems likely that fossil fuel (and possibly nuclear) electricity prices will increase over time, so that tidal barrage projects will become more attractive, under the conditions present at the time of writing tidal barrages appear to be relatively unattractive commercial investment options. Nevertheless, with concerns about climate change and energy security, interest in the concept might revive, in part because of the large amount of renewable energy that could be generated.

In carbon balance terms, barrages do not generate CO_2 during their operation, but building them inevitably does. The UK SDC estimated that the carbon dioxide that would be produced by making the materials for and constructing the Cardiff–Weston barrage effectively amounted to 2.42 gCO_2 per kWh generated. This translates into a carbon payback time (based on how long it would take to recoup those emissions by the use of the barrage rather than a fossil-fuelled plant) of around 5–8 months (SDC, 2007).

Once built, a barrage could be seen as either displacing the output of existing fossil plant, or displacing the need for some other form of new capacity, and the associated emissions. The SDC decided to focus on the latter and assume that a Severn barrage would displace the need for new gas-fired combined cycle gas turbine (CCGT) plant, since 'a Severn barrage is unlikely to be operational for at least ten years, during which time much of the UK's coal capacity will be taken out of service' (SDC, 2007).

On this basis, SDC calculated that the Cardiff–Weston Barrage would avoid the emission of 5.5 MtCO_2 per year compared to alternative CCGT gas-fuelled electricity generation, which represents a 0.92% reduction in UK carbon emissions compared to the 1990 baseline level used by the government. That might be seen as a relatively small saving for an estimated outlay of £30 billion or more.

A 2010 UK government review made it clear that there was little chance of the scheme obtaining public funding support, given the high construction cost, long lead times and also the potentially large environmental impacts (DECC, 2010a). But it did say that private developers could come up with proposals.

In 2012, Hafren proposed a privately funded two-way ebb and flow barrage design for the Severn, operating on a lower head with a larger number of lower speed turbines. It claimed that the environmental impact would be lower (Hafren, 2013). However, that too was turned down, on the basis that the economics were still uncertain. Smaller barrage proposals are nevertheless still under consideration in the UK and also elsewhere, and might prove to be easier to finance, depending in part on the tidal range available at the chosen sites.

7.6 Tidal barrages: potential projects

United Kingdom

The UK has some of the largest, and most studied, tidal energy resources in the world. In general, the best potential tidal barrage sites in the UK are on the west coasts of England and Wales, where the highest tidal ranges are to be found. Despite its indented coastline, the tidal energy potential of Scotland is very small (1–2 TWh y^{-1}) due to its generally low tidal range.

As we have seen, the practical potential of tidal power depends crucially on economics, as well as on environmental factors. In theory, the exploitable potential, assuming that every practical UK scheme was developed, could rise to approximately 53 TWh per year, or around 14% of UK electricity generation in 2010. Approximately 90% of this potential (48 TWh y^{-1}) lies in eight large sites, each offering between 1 TWh y^{-1} and 17 TWh y^{-1}, while the remaining 10% relates to 34 small sites, each providing somewhere in the range 20–150 GWh y^{-1} (ETSU, 1990).

The Cardiff–Weston Barrage, if built, would make the largest contribution, approximately 17 TWh y^{-1}, but initial estimates have suggested that the Wash, the Mersey, the Solway Firth, Morecambe Bay and possibly the Humber, amongst others, could also make significant contributions (Figure 7.24).

In addition to these larger sites, there are many smaller estuaries and rivers that could be used. Feasibility studies were carried out in the 1980s on the Loughor Estuary (8 MW) and Conwy Estuary (33 MW) in Wales, the Wyre (64 MW) in Lancashire, and the Duddon (100 MW) in Cumbria. Overall, the total UK potential for small tidal schemes (that is, schemes of up to 300 MW capacity), has been estimated at nearly 2% of UK electricity requirements.

Studies have continued on some of these smaller sites, for example, on the Duddon, and there have also been proposals for larger schemes, including a 1 GW rated £2 billion barrage 11 miles across the Wash, and a 4.75 km² tidal lagoon in the Thames estuary, coupled, in a £2–£4 billion project, with a tunnel crossing and tidal surge barrier between Medway and Canvey Island.

In addition there is continuing interest in other locations, where some even larger schemes might be possible. For example, a study led by the North

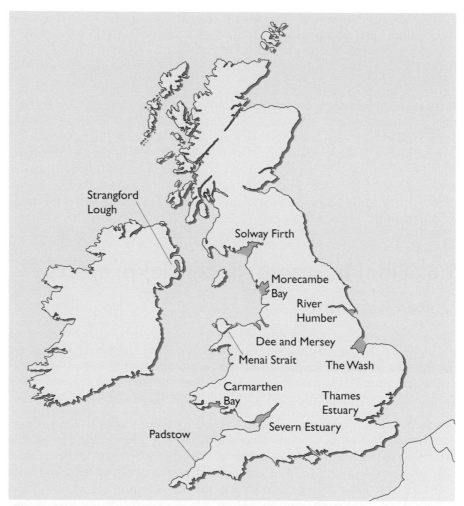

Figure 7.24 Some potential locations for tidal barrages in the UK

West Regional Development Agency looked at tidal energy options for the Solway Firth, the UK's third largest estuary, on the border of England and Scotland. It identified nine main options including four tidal barrages, the largest scheme running from Workington to Abbey Head.

The schemes were compared on the basis of their likely environmental impact and the levelized cost of electricity (LCOE) generated. The conclusions for the barrage projects were as follows:

- Workington to Abbey Head: £16 billion cost and large environmental impact; 5.9 GW installed, delivering 11.5 GWh y^{-1} – this scheme had a cost of energy (LCOE) of £183 per MWh (18.3p/kWh).

- Southerness Point to Beckfoot: £6.1 billion cost and substantial environmental impact; 2.7 GW installed – this had the lowest (LCOE) of all the Solway schemes studied at £175 per MWh (17.5 p/kWh).

- Bowness to Annan: £1.2 billion cost; 316 MW installed delivering 320 GWh y^{-1} – (LCOE) of £389 per MWh (38.9p/kWh).

- Morecambe Bay barrage (located out of the main estuary to reduce ecological impact): 113 MW installed delivering 120 GWh y^{-1} – (LCOE) of £553 per MWh (55.3p/kWh).

■ For comparison, the 8.6 GW Cardiff–Weston Barrage (using an 8% rate of return, not the 10% discount rate in Table 7.1) was estimated to give a CoE of £160 per MWh (16.0 p/kWh).

The non-barrage tidal projects examined (including lagoons and tidal flow turbine systems) all had similarly high CoEs, but lower environmental impacts (Halcrow Group Ltd et al., 2009).

A 2010 study by the North West Regional Development Agency and renewable energy developer Peel Energy looked at a range of tidal options for the Mersey, including barrages of various sizes but also tidal lagoons and a variety of tidal current turbine concepts. The study reviewed an earlier 700 MW barrage proposal by the Mersey Barrage Company, but suggested that, although viable, the impacts of a barrage on shipping, sedimentation, water quality and the local ecology would need to be very carefully assessed. A smaller 500 MW barrage was also seen as viable, delivering 900 GWh y^{-1}. Moreover, it also found that some of the other tidal options, although producing less energy, might have significantly lower environmental impacts (Peel Energy/NWRDA, 2010).

World

The total power in the tides globally is very large – of the order of 3000 GW.

In practice, since there are geographical access and location constraints on siting barrages or other means of extracting tidal energy, and they also have to be reasonably near to major lines of a national grid, the realistically available resource is much smaller. Jackson (1992) has put the realistically recoverable resource at 100 GW. Although that is under 12% of the existing global hydroelectric capacity, it still represents a significant resource. As Table 7.2 indicates, there are a number of potential large-scale sites for tidal barrages around the world, in Russia, Canada, the USA, Argentina, Korea, Australia, France, China and India, with an estimated total potential of perhaps as much as 300 TWh y^{-1}. There are some locations where some very large projects might be possible (e.g. the potential 87 GW scheme at Penzhinsk in Russia) as well as many smaller ones.

So far, only a few of these potential sites have been exploited, notably, some small projects in Canada, Russia and the medium-scale project of La Rance in France. However, South Korea has a significant tidal barrage programme, which is developing some sites in addition to those identified by the World Energy Council in 2001 (see Table 7.2).

A 2006 study (Lee, 2006) estimated that South Korea had around 2400 MW of barrage potential, including possibly 700–1000 MW at Incheon, 600–800 MW at Cheonsu, and 480–520 MW at Garolim (this estimate being an increase on the 400 MW shown in Table 7.2). A 2008 study reported that a new 430 MW barrage project was planned near a dam at Saemangeum, along with a 520 MW project at Garolim and an 812 MW Gangwha/Incheon project (Jo, 2008). While these projects are still on the agenda, a 254 MW tidal range project has been built at Sihwa, in Gyeonggi province, bordering the West Sea. This generates power from a tidal rise of 5.64 metres and is currently the world's largest tidal plant.

There have also been some very ambitious (if speculative) proposals for very large barrages, such as huge barrages across the Bering Straits and the Irish Sea; although this latter proposal would be considered by many engineers to be technologically unrealistic, as well as economically and politically challenging.

Table 7.2 Potential non-UK tidal power sites

Country	Mean tidal range/m	Basin area/km²	Installed capacity/MW	Approx annual output/TWh	Annual plant capacity factor/%
Argentina					
San José	5.8	778	5040	9.4	21
Golfo Nuevo	3.7	2376	6570	16.8	29
Rio Deseado	3.6	73	180	0.45	28
Santa Cruz	7.5	222	2420	6.1	29
Rio Gallegos	7.5	177	1900	4.8	29
Australia					
Secure Bay	7.0	140	1480	2.9	22
Walcott Inlet	7.0	260	2800	5.4	22
Canada					
Cobequid	12.4	240	5338	14.0	30
Cumberland	10.9	90	1400	3.4	28
Shepody	10.0	115	1800	4.8	30
India					
Gulf of Kutch	5.0	170	900	1.6	22
Gulf of Khambat	7.0	1970	7000	15.0	24
South Korea[1]					
Garolim	4.7	100	400	0.836	24
Cheonsu	4.5	–	–	1.2	–
Mexico					
Rio Colorado	6–7	–	–	5.4	–
USA					
Pasamquoddy	5.5	–	–	–	–
Knik Arm	7.5	–	2900	7.4	29
Turnagain Arm	7.5	–	6500	16.6	29
Russian Federation					
Mezen	6.7	2640	15 000	45	34
Tugur[2]	6.8	1080	7800	16.2	24
Penzhinsk	11.4	20 530	87 400	190	25

[1] See text for more recent additions/updates
[2] 7000 MW variant also studied
Source: WEC, 2001

7.7 **Tidal lagoons**

The major environmental changes associated with the building of a barrage across an estuary has led to interest in potentially less invasive offshore 'bounded reservoir' or 'tidal lagoon' systems, consisting of circular low-head dams in open water, trapping water at high tide, with the water then being released to drive turbines in the usual way (Figure 7.25).

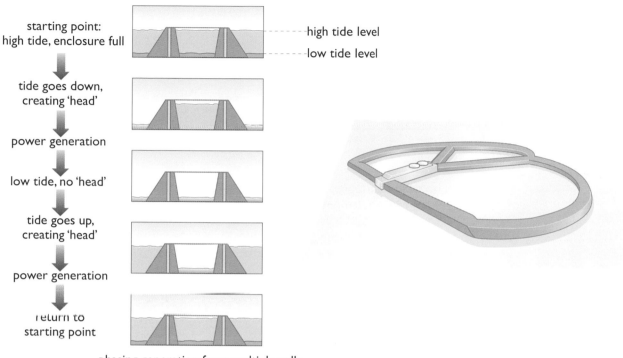

starting point:
high tide, enclosure full

tide goes down,
creating 'head'

power generation

low tide, no 'head'

tide goes up,
creating 'head'

power generation

return to
starting point

high tide level

low tide level

phasing generation from multiple cells

Figure 7.25 Tidal Electric's 'bounded' reservoir proposal

Lagoons would be in relatively shallow water and would be constructed like causeways with rock infill. Lagoons have the key advantage that although, as with conventional barrages, they would involve the creation of a head of water, they would not involve blocking off an estuary and so should have a lower environmental impact and avoid interfering with the passage of ships.

As well as generating power directly for the grid, lagoons might also be used as a low-head offshore pumped storage facility (see Chapter 6). Moreover, as with the double barrage idea, segmented lagoons could enable phased operation and pumping between segments.

The US company Tidal Electric has been investigating potential sites for tidal lagoons in Alaska, Africa, Mexico, India and China. At the time of writing a 1000 MW project is being considered in India and the company has also been in discussion with the Chinese government, which, it says, has expressed interest in a 300 MW offshore tidal lagoon to be built near the mouth of the Yalu River. Three sites are also under consideration in the UK, off the coast of Wales, including a 60 MW lagoon off Swansea. As illustrated in Figure 7.26 many other locations in that area have also been seen as possible lagoon sites.

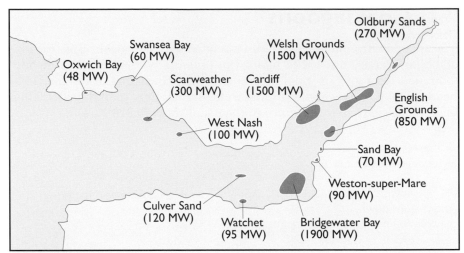

Figure 7.26 Possible tidal lagoon sites in and around the Severn estuary (source: Salter and Walker, 2010)

The disadvantage of offshore lagoons is that the entire containment wall has to be constructed, therefore increasing the cost. This is in contrast to barrages where the estuary shore provides free containment for the bulk of the water held behind it at high tide. Even so, given that lagoon construction is expected to be easier and quicker than for barrages, it is claimed that the cost of power from offshore lagoons can be competitive (Atkins Ltd, 2004).

While Tidal Electric have focused on fully offshore lagoons, a compromise option, to reduce cost further, would be to site the lagoon near to the shore, with part of the containment being the shore line. However, inshore lagoons of this type are more likely to impact on mud flats, and the wildlife that uses them, than fully offshore schemes.

A 2010 DECC review of Severn tidal options (DECC, 2010a) concluded that although the cost of energy was relatively high, a near-shore lagoon at Bridgwater Bay on the English side could be feasible, with 'lower environmental impacts than barrage options', but that the Welsh Grounds near-shore lagoon was not viable (see Table 7.3). Tidal Electric's fully offshore scheme for Swansea Bay was not reviewed, being deemed to be outside of the geographical area of the study.

However, a 320 MW tidal lagoon partly linked to and enclosing some of the shoreline has been proposed for the Bay of Swansea in Wales. The promoters claimed that it would be able to generate at around £150/MWh, similar to the then current figure for offshore wind. Moreover, they added that subsequent projects, at sites with higher tidal ranges, could reduce that to give prices comparable to those for on-land wind projects. A 1.8–2.8 GW tidal lagoon off Cardiff was proposed as a follow-up (TLP, 2014). At the time of writing, the Swansea Tidal Lagoon project is one of the subjects of a new government-backed review of tidal options, with the economics being a key focus. The initial review supported the idea strongly as a pilot 'pathway' project, but the government has yet to respond (Hendry, 2016).

Lagoon proposals have also emerged for other locations around the UK, notably the Solway Firth and the Mersey estuary (see Section 7.6 above), but, so far, none have gone ahead.

Table 7.3 Comparison of costs and impacts of lagoons in the Severn (see Table 7.1 for the equivalent information for barrages)

| Scheme | Installed capacity /MW | Annual energy generated /TWh y⁻¹ | Levelized energy cost/£ MWh⁻¹ | | Inter-tidal habitat loss/km² |
			Investor (10% discount rate)	Social (3.5% discount rate)	
Welsh Grounds Lagoon	1000	2.1	515	169	73
Bridgwater Bay Lagoon	3600	6.2	349	126	25

Source: DECC, 2010a

No doubt the debate over the relative merits of barrages, large and small, and lagoons near shore and offshore, will continue. The combined resource is certainly not insignificant – an assessment of the energy resource offered by tidal range projects in the UK carried out in 2010 suggested that, assuming a maximum credible expansion programme, 20 GW of tidal range capacity might be expected by 2050 delivering 40 TWh per year. That included contributions from tidal lagoons, as well as barrages on the Severn, Mersey and Solway (DECC, 2010b). However, the 2010 DECC Severn Tidal Feasibility Study concluded that, not only was the Cardiff–Weston Barrage high cost and high risk, the smaller barrages and lagoons were even less attractive economically (Figure 7.27). The DECC report did say that the results for other locations around the UK might be different, but it is hard to see how they could do better than the Severn – the best tidal range site. However, in the long-term they might be seen as good investments for the future, since barrages and lagoons will last up to a century or more without the need for many upgrades apart from the turbines. That was the view taken on lagoons in Hendry's 2016 review. It suggested that some lagoons might ultimately run with contracts as low as £66/MWh (Hendry, 2016).

Globally, barrages and lagoons are still on the agenda. A review by the International Renewable Energy Agency said there were over 100 GW of tidal range projects under consideration around the world (IRENA, 2014). However, although some small- to medium-scale tidal range projects may go ahead, for example, in South Korea and possibly the UK (lagoons especially), it is not likely to be a major area of expansion globally in the near term.

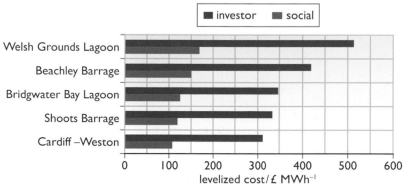

Figure 7.27 DECC estimates of Severn barrage and lagoon electricity prices at 'investor' (10%) and 'social' (3.5%) discount rates (source: DECC, 2010a)

7.8 Tidal streams/currents

As noted in the introduction to this chapter, the use of tidal streams or currents is being followed up with some enthusiasm. Instead of using costly and invasive barrages in estuaries to create *potential* energy heads with trapped water, it is possible to harness the energy in the *horizontal* movement of the tides – that is, the *kinetic energy* of the tidal ebbs and flows.

In open sea, the speed of the ebb and flow of the tide is relatively low, but constrictions caused by islands, channels and headlands can significantly accelerate the streams. The enhanced velocity of the tidal streams in these locations concentrates the energy and provides the most cost effective site to harness this resource. There are many technologies designed for this, but turbines closely resembling underwater 'wind turbines' have proven most popular with developers. These 'conventional' horizontal turbines can be either mounted on the sea bed (Figure 7.3) or tethered mid-stream on a buoyant hull (Figure 7.28).

Figure 7.28 Sustainable Marine Energy's floating tidal current system (source: SME 2015)

As was noted in Section 7.1, the terms 'tidal current' and 'tidal stream' are often used interchangeably, and, to confuse matters more, not all the water flows in oceans are actually 'tidal' in the sense of being driven by the Moon's gravity. Some flows are the result of complex interactions between warm equator layers and cooler polar layers of water around the world, and the associated effects of varying salinity and densities. These are in effect solar-driven and form large ocean-spanning currents, like the Gulf Stream, which in turn contribute to global weather patterns. Strictly speaking the correct terms to use for such non-tidal flows are **ocean currents** or **ocean streams**. From a technology perspective they share many of the same challenges as tidal systems, with some developers targeting both markets. We will discuss these developers and their challenges later in this section.

Whatever type of flow is used, the basic physics of power extraction via turbines is similar to that for wind turbines. As Chapter 8 will explain, the power P in the wind is given by the formula $P = 0.5 \, \rho a v^3$, where ρ is the density of air, v the wind speed, and the area of the circle swept by the rotor (which is πr^2, where r is the rotor blade length). Thus the energy available is proportional to the square of the rotor size and, in the case of a turbine mounted in water, the cube of the water velocity.

The most obvious difference between wind turbines and tidal devices is the concentration of energy due to water being much denser than air. Less obvious is the resulting load fluctuations on the drive train and blades from turbulence generated from the tidal flows interacting with the seabed and other topographical features.

These effects are much more significant (given the density of water and the narrow and shallow channels involved) than in the case of wind flows. A wind turbine would never be built in the bottom of a deep valley, but this is the ideal location for a tidal turbine, as the seabed valley offers the flow acceleration needed for commercial extraction of the power. Accelerating the flow in a smooth walled channel is used in hydrodynamics to reduce turbulence. However, in nature there are no smooth topographies, with the more complex channels having high degrees of turbulence with fixed and shed vortices. The resonance effects of flows in estuaries have been discussed in Section 7.1 but a more relevant phenomenon to tidal streams are fixed vortices, which sometimes form dramatic whirlpools such as the one in the Gulf of Corryvreckan.

Tidal flows will also be affected by atmospheric conditions, the atmospheric pressure, as well as the interaction between air and water at the sea/river surface. Estimating the energy resource available is therefore complex and site specific, requiring detailed local modelling (see Hardisty, 2007; Bryden et al., 2007). So, while areas where there are high tidal ranges are important (see Figure 7.24), the local topography is crucial in determining whether tidal currents/streams can be utilized effectively – flow rates can sometimes be low at sites with a high tidal range and vice versa.

The resource and its location

Although tidal current resources are site specific, there is a lot of energy in the enhanced tidal currents in appropriate locations. For example, it has been estimated that the power flowing through the north channel of the Irish Sea is equivalent to 3.6 GW, while the flow through Pentland Firth in Scotland is the equivalent of 6.1 GW.

Mean peak spring tidal currents of at least 2–2.5 m s^{-1} (~4–5 knots) and a depth of between 20 and 35 metres are seen as necessary for economic exploitation. On this basis, the UK has some of the world's best sites for producing energy from tidal streams, as shown (see Figure 7.29). Table 7.4 gives details of the expected energy outputs in some of these sites.

Figure 7.29 Possible tidal current sites around and near the British Isles

Table 7.4 Some major tidal current sites in the vicinity of the British Isles and their associated resource potential

Site name	Location	TWh y^{-1}
Pentland Skerries	Pentland Firth	3.9
Strøm	Pentland Firth	2.8
Duncansby Head	Pentland Firth	2.0
Casquets	Alderney	1.7
South Ronaldsay	Pentland Firth	1.5
Hoy	Pentland Firth	1.4
Race of Alderney	Alderney	1.4
South Ronaldsay	Pentland Firth	1.1
Rathlin Island	North Channel	0.9
Mull of Galloway	North Channel	0.8

Sources: SDC, 2007

The 2050 Pathways report, produced by the Department of Energy and Climate Change in 2010, saw UK tidal current projects as potentially delivering a total of up to 69 TWh per year by 2050 from 21.3 GW of installed capacity, and when assuming maximum possible development (DECC, 2010b). Moreover, a study by an industry group, led by the Public Information Research Centre (PIRC), put the practical resource potential for tidal currents much higher, at 116 TWh per year, from 33 GW installed capacity (OVG, 2010).

The UK clearly has some of the best sites, but there are also good resources elsewhere (Hardisty, 2009). In its reviews of tidal energy potential, IRENA claims that the long-term total global tidal stream resource is much larger than that for tidal range, although no figure was given, only a combined total estimate, including tidal range, of 1000 GW (IRENA 2014).

Tidal current turbine design constraints and options

There are topographical and other local constraints on how much energy can be extracted from this resource. When considering device design and location it is crucial to note that friction with the seabed produces a vertical 'shear' effect in the current: surface water moves fastest, with the flow being slower lower down. As a result, around 75% of the energy available in the flow is contained in the top half of the water (Figure 7.30). Therefore, there is a desire to locate the turbine in this zone, near to the surface, while maintaining full submersion for optimal turbine efficiency. Given that most early sites under development are relatively shallow (below 40–50 metres), current tidal devices have a size constraint that does not apply to wind turbines. However, this size limit can be avoided by moving to mid-depth locations. From Figure 7.31 it can be seen that there is a large potential for extracting energy from deeper water sites.

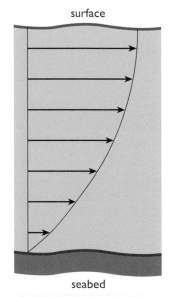

Figure 7.30 The shear effect – water speed (as indicated by arrow length) reduces with depth

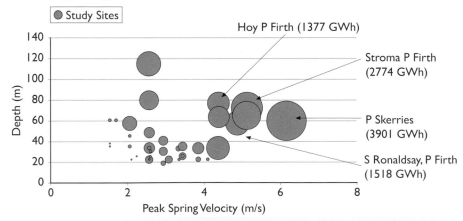

Figure 7.31 Relationship of water depth to available resources (source: Oceanflow Energy Ltd)

Since the available energy is proportional to the square of the rotor blade length and the cube of the water velocity, even a small increase in blade size or water speed will yield a significantly increased amount of energy production. Figure 7.32 shows how the relationship works out in practice – the faster the flow and the larger the rotor (subject to practical considerations) the better.

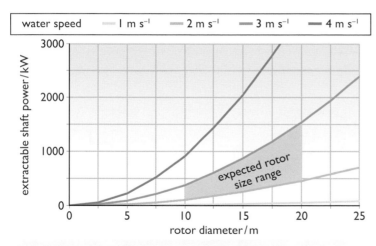

Figure 7.32 Relationship between rotor diameter, water speed and extractable power (source: adapted from Marine Current Turbines, 2010)

The similarities between wind turbines and tidal turbines mean that they both operate at the same tip speed ratio. However, the increased fluid density more than compensates for the lower water velocities and smaller turbine diameters, resulting in lower turbine rpms than equivalent rated wind turbines.

The output from a tidal stream turbine is likely to be lower than that from an equivalent-sized conventional turbine in a barrage, which has the advantage of harnessing the resources of a large body of water behind a barrage and funnelling the concentrated energy through turbines. However, free-standing tidal current systems tend to have a lower power rating with larger rotors and do not need a change in water head to function. Indeed, large tidal ranges can be detrimental to stream turbines as they limit the turbine size to a safety margin below low water.

An advantage of tidal current devices over simple ebb or flood tidal barrages is that they can operate on both ebb and flood tides and thereby attain a much higher capacity factor. Other advantages are that individual units can be constructed on a modular basis and installed incrementally in gradually expanding arrays giving both economies of scale and a much shorter lead time between investment and gaining revenue.

As with wave energy systems (see Chapter 9) there may be problems of fouling (for example, by seaweed, lost fishing nets, etc.), tethering and power take-off to overcome, but on the other hand, expensive and environmentally intrusive barrages are not required. There will be little visual impact, as in most designs all of the device would be under water.

As noted earlier, the most common type of tidal turbine is the horizontal axis turbine, similar to wind turbines. The components can be broken down as shown in Figure 7.33.

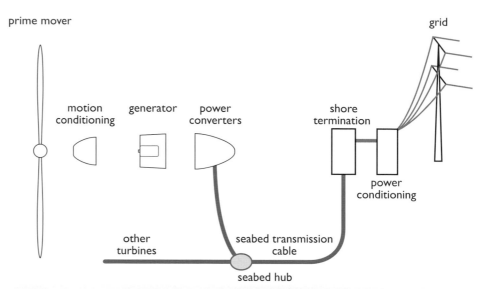

Figure 7.33 Schematic layout of a typical tidal turbine system (source: Oceanflow Energy Ltd)

Figure 7.33 shows a conventional turbine split up into key component parts. The prime mover is the turbine; the motion conditioning is done in a gear box, electricity is produced in the generator, while the power converter controls the generator load and produces grid compliant electricity. The seabed transmission cable then gets the power ashore.

As we shall see, there are many turbine designs, some of them quite novel. While horizontal axis turbines, as in Figure 7.33, dominate, vertical axis designs have also been used. In some cases the turbines are supported under floating pontoons tethered to the seabed, in others they are fixed directly on the seabed, or mounted on towers. Ducts can also be used to accelerate the water flow through the turbine. Alternatively, use can be made of hydroplane arrangements, involving seabed mounted hydrofoils oscillating up and down in the tidal flows.

Whatever the design, arrangements have to be made to enable access for repair and maintenance, which is not always easy at sea. Sending repair boats out is expensive and sometimes risky, so low-maintenance designs are at a premium, but, as will be later outlined, a variety of configurations have been developed to give access to the turbines and components, for repair either on site, or, if necessary, on shore.

The following sections look at some examples of approaches to turbine design and at the status, at the time of writing, of development in the UK and elsewhere.

Some of the projects reviewed below are well established, having been fully tested at sea, and some have been deployed at full scale. However, it should be stressed that many of the projects being discussed are at a relatively early stage of development, with the claims about potential energy outputs and generation capacities still unproven – indeed some descriptions are of currently speculative design concepts and proposals. When examining novel proposals, care has to be taken to assess the credibility of the claims being made. For example, generation capacities are sometimes claimed which could only be achieved, given the size of the device, at very high water speeds. While many of the devices and proposals described in the sections below may not eventually prove successful, a wide sample of projects has been included here, to give a feel for the range and vitality of the innovative and experimental processes underway.

7.9 An overview of projects and tidal stream concepts in UK waters

Tidal current/stream development has only progressed on a significant scale in the UK relatively recently, partly because it was previously thought that the cost of generating electricity via such technologies was likely to be high. Such cost considerations meant that work on tidal stream technology was given a low priority in the renewable energy development strategy adopted in the early 1990s by the (then) Department of Trade and Industry (ETSU, 1993).

Subsequently, following a review by the Marine Foresight panel set up by the Office of Science and Technology, support for tidal stream projects has grown, partly because the report argued that economies of scale via the production of large numbers of turbines (albeit only after a substantial development programme) could reduce these costs (OST, 1999) and also because novel devices have begun to emerge.

In 1994, in a pioneering project, a prototype two-bladed 10 kW tidal current turbine was tested in the Corran Narrows in Loch Linnhe, near Fort William in Scotland, by a consortium involving IT Power, the National Engineering Laboratory and Scottish Nuclear. This was the world's first tidal current turbine. The tests involved a rotor of 3.9 metres in diameter submerged under a small catamaran pontoon in the sea at a depth of 5 metres. In a fully operational system it was proposed that the rotor unit would be supported by a cable attached to a floating buoy and also tethered to the seabed. It could therefore swivel around on the change of the tide to absorb power from tidal currents in either direction (Figure 7.34).

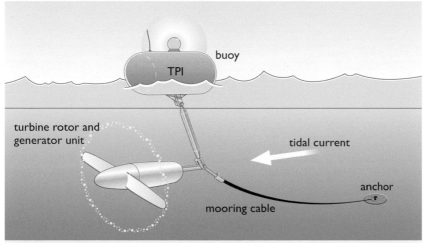

Figure 7.34 The IT Power tidal current turbine concept, developed and tested at Loch Linnhe, Scotland (source: adapted from Marine Current Turbines, 2010)

This free-floating, mid-depth swivelling system concept might in some locations have cost advantages over fixed bottom-mounted devices, e.g. in places where there is sufficient room for the structure to swing around. However, the need for flexible marine cable power take-off arrangements could add to the overall cost, and Marine Current Turbines (MCT) (an offshoot of IT Power) went on to develop a system with a rotor mounted on a fixed steel pile driven into the seabed.

Subsequently, MCT implemented this approach in a two-bladed 300 kW experimental prototype named Seaflow, which was tested 1 km off north Devon near Lynmouth, between 2003 and 2006, before being decommissioned in 2009 after testing had finished.

MCT then developed a larger two-bladed twin-rotor design of 1.2 MW capacity (at 2.4 m s^{-1}), called SeaGen, which they installed in Strangford Narrows, Northern Ireland in 2008 (Figure 7.35). The rotors of the first prototype are 16 metres in diameter and their pitch can be rotated through 180 degrees, to re-orient them to the tide when it changes direction, thus allowing efficient operation on both the ebb and the flood. They can be raised out of the water for easy access. Although the prototype is rated at a relatively low power level, its successors are expected to be capable of up to 2 MW, using rotors of up to 20 m in diameter.

Figure 7.35 MCT 1.2 MW SeaGen in Strangford Narrows with the rotor blades raised out of the water for access

SeaGen has proved very successful, delivering power to the grid and, from 2009 onwards, earning Renewable Obligation Certificates under the UK Renewable Obligation scheme for the electricity supplied, and achieving a very high load/capacity factor of up to 66%. Following this project, the aim was to build a 10 MW 'tidal farm' array using several machines off the coast of Wales near Anglesey, with a similar project planned for Kyle Rhea off the Isle of Skye in Scotland. In addition, MCT and ESB International looked at a 50–100 MW project off the Antrim coast in Northern Ireland. MCT also had plans for a 100 MW project off the Orkney Islands.

Some of these plans were continued when, in 2012, MCT was taken over by Siemens, but funding proved to be difficult and in 2015 Siemens sold MCT off, with Atlantis Resources Corporation, which has developed and tested its own propeller-type tidal turbine design, then taking over some of the projects and MCT's technology.

Atlantis is to include MCT designs in the second phase of the 398 GW MeyGen tidal current project in Pentland Firth. The first 6 MW phase involves one Atlantis machine and three of the turbines initially developed by Norwegian company Hammerfest Strom, now part of Andritz Hydro Hammerfest (Atlantis, 2016).

In terms of technology, conventional axial flow or 'propeller-type' designs, like MCT's SeaGen and the Atlantis and Hammerfest devices, seem the most obvious way forward: they mimic the successful approach used by the wind industry. Although not yet fully developed, tested or deployed like MCT's systems, there have been several similar designs, like Swanturbine's seabed mounted Cignet, developed at Swansea University, and Tidal Energy's DeltaStream triangular frame seabed-mounted system, being tested off Wales – although both of these use three-bladed turbines, rather than two-bladed ones as in SeaGen. A double rotor system, with two triple-bladed contra-rotating rotors mounted on a common shaft, has also been developed by Strathclyde University and a 500 kW version has been tested by spin-off company Nautricity. The 1 MW AK 1000 device initially developed by Atlantis also had two triple-bladed rotors on one shaft, but in this case they were run independently, one rotor running on the ebb tide and the other on the flood tide, with, in each tide, the other rotor being feathered. Subsequently, however, for its larger 1.5 MW AR 1500 system, as used in Pentland Firth, Atlantis switched to a single rotor with variable pitched blades.

Novel designs

The turbines we have looked at so far may look like aircraft or ship propellers, but of course they do not propel anything – they absorb energy from the tidal flow. Propeller-type configurations have certainly been popular, but they are not the only option and the tidal turbine field is currently at the innovative stage where many different concepts are being proposed and tested. For example, the Engineering Business group developed a novel design – the **Stingray** – with a set of totally submerged hydroplanes, oscillating up and down in the tidal flows and driving a generator mounted on the seabed (Figure 7.36).

A 150 kW prototype was tested at sea in 2002 and again 2003. However it proved to be very inefficient and was subsequently abandoned. It seems the energy conversion efficiency was low (around 9.5%) due in part to the relatively slow oscillation rate, but a

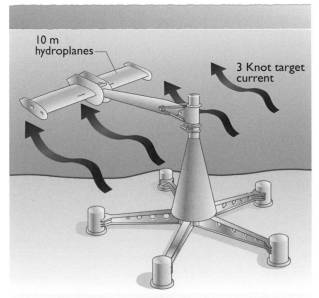

Figure 7.36 Stingray tidal generator

derivative, with two sets of hydrofoils on a 'see-saw' arrangement oscillating much more rapidly, was subsequently developed by Pulse Tidal. In 2009, a 100 kW test rig was tested in the Humber estuary, with power being delivered to shore. The company has been looking in particular at Kyle Rhea as a possible site for a 1.2 MW unit, possibly to be followed by linking eight tidal power devices together in a 9.6 MW tidal fence (Figure 7.37).

Given the relatively shallow depth of most tidal channels, hydroplane type designs are claimed to have an advantage over propeller-like designs in that they can operate in shallower water, with the blades able to traverse a larger cross-sectional area of tidal flow than the circular area defined by the rotating blades of propeller-like devices. This comparison does need to be treated with some caution, since reciprocating devices operate most efficiently in mid-stroke and lose energy as they slow at the end of each stroke.

Many other designs have been proposed. One approach has been to enhance/accelerate the tidal flow using an annular duct surrounding a conventional horizontal-axis rotor (Figure 7.38). A very different approach to ducting has been adopted by Neptune Renewable Energy. Their Proteus system has a vertical-axis rotor mounted in a duct system submerged under a floating pontoon (Figure 7.39). The system was tested at the University of Hull and a 150 tonne prototype was then built and installed in the Humber Estuary, the aim being to use the electricity generated to power The Deep, an aquarium centre in Hull.

An issue with ducting is whether the enhancement of the flow justifies the extra cost of the duct. With wind turbines, this does not seem to be the case, especially since wind directions vary, but tidal flows always follow fixed paths, back and forth, so fixed ducting might be economically viable. However, after tests it was found that the Proteus system did not live up to expectation and the project was abandoned. The combination of vertical axis turbines and ducting was perhaps too ambitious (Steiner-Dicks, 2013).

One of the most novel designs is the 'open-centred' turbine developed by Irish company OpenHydro, which has rotors mounted around the inner rim of a sealed circular generator system, which also acts as a duct (Figure 7.40). Following tests of a 250 kW rated prototype at the European Marine Energy

Figure 7.37 Pulse Tidal's two hydrofoils mounted on a see-saw concept

Figure 7.38 Example of a ducted rotor being developed by Rotech

Figure 7. 39 Neptune's Proteus – a vertical-axis ducted rotor

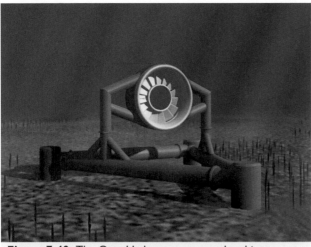

Figure 7.40 The OpenHydro open-centred turbine

Centre (EMEC), a 1 MW version was developed and installed for testing in the Bay of Fundy, Canada in 2008, and, in 2010, a 200 MW project was given a site in the Pentland Firth off the northern coast of Scotland. In addition there are plans for a 284 MW project off Alderney and 1 MW project backed by EDF in France.

BOX 7.6 A European test centre

The European Marine Energy Centre (EMEC) in Stromness in the Orkney Islands, provides support for development in tidal and wave power with a site at the Fall of Warness off the island of Eday supporting tidal power research and a site at Billia Croo (Mainland) where wave power devices may be tested.

Apart from offering sites that allow developmental systems to be deployed in the open sea and connected to a grid, EMEC has expertise in device monitoring and with its links to different developers can support best practice throughout this part of the renewables industry. The site is not just restricted to UK devices as the centre is being used by developers from across the world.

Perhaps the most original design so far, however, is Minesto's tidal kite. This Swedish concept has an aerofoil wing, with a rotor and generator mounted on it, tethered to the seabed or supported from a float, but free to move in the tidal flow, under rudder control, at enhanced speed in a figure of eight pattern. It is being tested off Northern Ireland. Cable stresses are an obvious issue, since the cable also has to carry the electrical connection. However, the developers claim that, since the flow over the rotor blades is speeded up by perhaps 10 times, the system could extract power economically from relatively low tidal flow rates, thus, in effect, expanding the potential tidal resource (Minesto, 2015).

One of the key factors in tidal turbine design, as with wave energy devices, is accessibility for maintenance and repair, since this is likely to be a dominant factor in the running costs. Many devices are designed so that the rotor unit and/or generator can be winched to the surface by a boat.

Alternatively some have in-built mechanisms for raising the turbine assembly out of the water, as shown in Figure 7.35 in the case of MCT's SeaGen. In a similar vein, TidalStream's Triton design has a series of propeller-type units on a hinged structure fixed to the seabed which can be tilted up so that the turbines are on the surface (Figure 7.41). A one-tenth scale prototype has been tested in the Thames. MCT have also proposed something similar for the follow-up SeaGen deployments – a hinged frame fixed to the seabed (Figure 7.42).

Figure 7.41 TidalStream's Triton

Figure 7.42 MCT's next generation SeaGen – the seabed-mounted array can be raised to the surface for access

Most of the tidal current schemes and designs discussed above involve single machines or groups of machines in free-standing 'tidal farm' arrays, but there are also proposals for more integrated schemes, with devices mounted in structures such as permeable 'tidal fences' across estuaries (see Box 7.7).

BOX 7.7 Integrated tidal current turbine structures for the Severn

Some integrated schemes, along with barrages and lagoons, were looked at in the first round of the UK government's Severn Tidal Feasibility Study. Although they were not included on the final shortlist, an ancillary DECC-supported assessment programme, the Severn Embryonic Technology Scheme (SETS), which started in 2008, followed up some of the ideas described here.

Tidal Fence

As an alternative to the Severn Tidal Barrage, the Severn Tidal Fence group (STF) proposed a 1.3 GW permeable tidal fence in the same location, with a row of large diameter ducted tidal current turbines, which STF said would be less environmentally invasive than a barrage – since it was not trying to create a head of water, it would not have to block the entire estuary (Figure 7.43).

Figure 7.43 The STF Tidal Fence proposal

Tidal Reef

The Tidal Reef concept, developed by Evans Engineering, is something of a compromise between a full barrage and the tidal fence, designed to minimize environmental impact. It would be a 15 mile long causeway from Minehead to Aberthaw in Wales, with, in one version, vertical-axis tidal turbines running on both ebb and flood tides, mounted in floating caissons, which would rise, from the supporting causeway, with the tides. The project would comprise over 1000 5 MW turbines of 10 m diameter with an annual generation of around 20 TWh. This compares to the 17 GWh expected from the Cardiff–Weston Barrage.

The basic concept was for the reef to be very permeable, so, unlike the Cardiff–Weston barrage it would not hold back the full height of the tide or create a large head of water – the low-speed turbines would run on only about a 2 m head, so that the environmental impact should be lower. Atkins and Rolls Royce subsequently developed a new version of the idea (the Rolls Royce/Atkins Tidal Bar) which was chosen, along with STF's Tidal Fence, for further assessment funded by the government's SETS scheme.

Spectral Marine Energy Converter (SMEC)

A third project looked at under SETS was the Spectral Marine Energy Converter (SMEC), a novel device being developed by VerdErg using the venturi effect – the reduction in fluid pressure that results when a fluid flows out through a constriction. In this system the tidal flow passes through vanes, creating a pressure drop which is used to drive a turbine and generator via a secondary water flow. The secondary flow will have a higher velocity than the primary flow. Furthermore, this secondary flow can power a turbine and generator located more conveniently out of the water.

This concept could be used at a variety of scales – in free-standing mid-stream locations in rivers, or across whole estuaries in a permeable fence-like structure. Vertical vanes would be used in the latter case to produce the pressure drop, horizontal vanes for shallow mid-channel use in rivers. An estuary-wide tidal fence version would, VerdErg claimed, produce nearly as much electrical power as a full conventional barrage, but at two thirds of the cost and with much less environmental impact. Subsequently VerdErg dropped the 'SMEC' label and now calls it Venturi-Enhanced Turbine Technology (VerdErg, 2016).

Assessment

Reporting on the outcome of the SETS project in 2010, DECC's Severn Tidal Feasibility Study noted that the tidal bar/reef and the Spectral Marine Energy Converter 'showed promise for future deployment within the Severn estuary – with potentially lower costs and environmental impacts than either lagoons or barrages'. However, they added, 'these proposals are a long way from technical maturity and have much higher risks than the more conventional schemes the study has considered. Much more work would be required to develop them to the point where they could be properly assessed' (DECC, 2010a).

7.10 Tidal current projects and concepts around the world

The idea of using tidal currents is also being explored around the world, although, since UK tidal current resources are amongst the best globally, some of the devices being developed by non-UK companies are being tested in UK waters, with some subsequently being designated for full scale deployment in UK waters, sometimes as joint projects with UK companies. Examples include the devices developed by Norway-based Hammerfest Strøm and Singapore-based Atlantis Resources Corporation, both of which were mentioned earlier (see Figures 7.44 and 7.45).

Figure 7.44 The Hammerfest Strøm tidal turbine

Figure 7.45 Atlantis AK-1000 1MW double rotor tidal turbine

Tidal current projects – a world overview

A wide variety of other concepts are being developed and tested around the world, in Europe (including most notably Norway, but also France, the Netherlands, Germany and Italy), and in the USA, Canada, New Zealand and Australia. This section looks at a sample.

South Korea

Following tests with a 100 kW tidal current power facility installed in a narrow channel with a 5.5 m s^{-1} current speed in 2003 in Uldolmog on the south coast, a 1 MW rated facility is being built and there are plans for a larger project in the Daebang Strait of up to 20 MW. There are also plans for an ocean current power farm in the Sihwa area where there is already a tidal barrage. The largest tidal stream project proposal so far is at Wando, with plans to have 300 MW of tidal current turbines installed, possibly with Voith Hydro 1 MW Seaturtle turbines.

A unique feature of Voith's design is that it has unsealed sea-water compatible bearings, which, it is claimed, reduce maintenance requirements (Figure 7.46).

Figure 7.46 Voith 1 MW turbine

Norway

In addition to the Hammerfest Strøm project already mentioned, Hydra Tidal are developing a multi-rotor floating propeller-type device with laminated wood composite blades. A 1.5 MW four-rotor two-bladed version, Morild II, with 23 metre diameter blades, was officially opened offshore from Lofoten, northern Norway, in 2010, with grid links planned.

USA

Six of Verdant Power's small (35 kW) three bladed propeller-type tidal turbines have been tested in New York City's East River near Roosevelt Island. By 2008 they claimed to have clocked up 9000 turbine-hours of operation (Figure 7.47). There are plans for expanding the project to 30 turbines, possibly feeding power to the United Nations building nearby. In addition, the Massachusetts Tidal Energy Company has been looking at the possibility of installing up to 150 2 MW devices at Vineyard Sound on the New England coast.

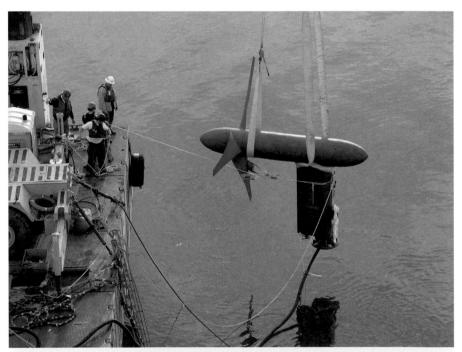

Figure 7.47 Verdant Power's turbine being lowered into East River, New York City in 2007

Meanwhile, following tests with a prototype, the UEK Corporation, based in Annapolis, has plans for a project at the mouth of the river Delaware, using its two-way 45 kW Underwater Electric Kite ducted turbines.

In addition, the Ocean Renewable Power Company (ORPC) is deploying its cross-flow tidal turbine in Eastport, Maine and in 2010 reportedly landed power from its 60 kW prototype (Figure 7.48). The device has a helical turbine design similar to that developed by Gorlov (see next section), but mounted horizontally. The company is also looking at sites in Alaska, which has some of the USA's largest tidal resources.

On the US West coast, the Pacific Gas & Electric Company has signed an agreement with the City and County of San Francisco, as well as the Golden Gate Energy Company, to assess possibilities for harnessing the tides in San Francisco Bay.

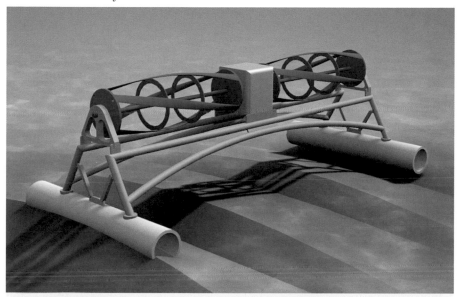

Figure 7.48 ORPC cross-flow turbine: a 250 kW version is planned

Canada

Canada has had a long involvement with tidal barrage technology and is now also looking at tidal current systems. For example, the Atlantic Tidal Energy Consortium has outlined plans to install up to 600 MW of tidal systems in the Bay of Fundy. Canadian company Clean Current has developed a version of the ducted rotor design and plans to test it in the Bay of Fundy. In addition, Canada's Minas Basin Pulp and Power Company has an agreement with the UK's MCT to install a 2 MW SeaGen device in the Bay of Fundy and the OpenHydro turbine has also been tested there (a 1 MW device was installed in 2008, although it suffered blade failure).

Australia and New Zealand

In Australia, Tenax Energy has plans for a 200 MW tidal project in fast-flowing waters in the Clarence Strait between northern Australia and the Tiwi Islands; while in New Zealand Neptune Power is planning an array of 1 MW floating sub-sea turbines in the tidal currents off the Cook Strait between the North and South Islands, and CREST Energy have plans for a major $400m 300 MW project at Kaipara Harbour near Auckland.

India

The Atlantis Resources Corporation is planning to install a 50 MW tidal farm in the Gulf of Kutch in Gujarat, western India, to be followed up by a 250 MW project if all goes well.

Other novel projects and devices

An EU-funded study has been made of the potential for tidal current generation in the Strait of Messina, between Sicily and mainland Italy. This involves installing 100 vertical-axis turbines on the seabed at a depth of 100 metres (Figure 7.49). In preparation for this, the EU-backed Enemar project tested a novel vertical-axis turbine in the Strait of Messina. Interest has been shown in this system by China, for a possible project in the Jintang Strait in the Zhoushan Archipelago.

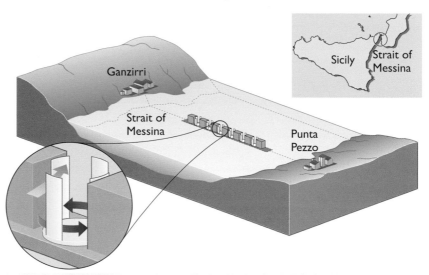

Figure 7.49 The Strait of Messina project

Figure 7.50 Gorlov's helical turbine

There are many other novel projects under development around the world, some of which are quite exotic, including tidal sails – current driven 'sails' mounted on an underwater 'ski-lift' like arrangement.

Some devices are designed for slower water speeds and for use in rivers with smaller rotors, although the basic physics is against them (to get the same output as from fast-moving water, you would need much larger rotors). Although in general big is best, there may be locations where useful and commercially viable energy can be obtained from such systems.

Large tidal projects and ocean current schemes

While small projects may have a place, there are also proposals for some very large projects. For example, Dr Alexander Gorlov, Professor of Mechanical Engineering at Northeastern University in Boston, has been developing ideas for turbine devices for use with *ocean* currents further out to sea.

Dr Gorlov's patented Helical Turbine (Figure 7.50) is a variant of the Darrieus vertical-axis wind turbine design (see Chapter 6).

Small prototypes have been tested and Gorlov has proposed large-scale applications, in particular using the Gulf Stream. He noted that the total power of the kinetic energy of the Gulf Stream near Florida is equivalent to approximately 65 000 MW. Gorlov's aim is to construct a very large tidal farm of one hundred power modules each with hundreds of triple-helix Gorlov turbines mounted in a lattice array.

Although Gorlov's scheme is still highly-speculative, interest in ocean stream power is growing in the USA. Florida Atlantic University's Center for Ocean Energy Technology is looking at the potential of the Gulf Stream more generally, and has suggested that up to 10 GW of capacity could ultimately be installed.

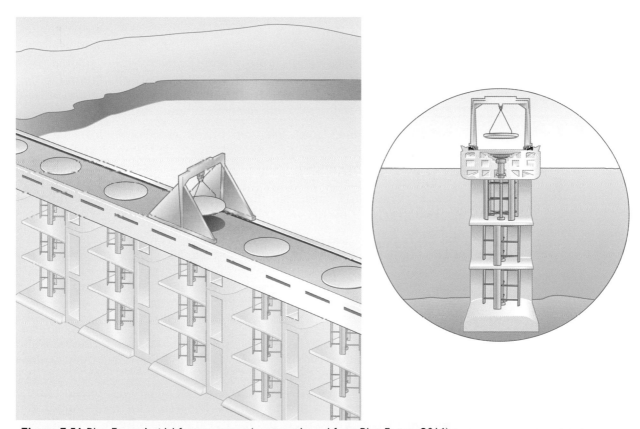

Figure 7.51 Blue Energy's tidal fence concept (source: adapted from Blue Energy, 2016)

Equally ambitious in conception, the Canadian company Blue Energy has developed a tidal fence concept in which H-shaped vertical-axis turbines are mounted in a modular framework structure (Figure 7.51). Blue Energy has been investigating the idea of a 50 MW rated demonstration tidal power plant in the Philippines, possibly to be followed by a 1000 MW plant, with arrays of vertical-axis turbines mounted in a permeable causeway, or tidal fence, between two islands, extracting power from tidal current flows.

Finally, as something of a hybrid of tidal current and tidal barrage ideas, mention should be made of the Dynamic Tidal Power concept, developed by Dutch companies and now being taken up by China. It has tidal turbines mounted in a causeway angled out to sea in such a way that the tidal flows parallel to the shore cause a localised head of water to be created on one side. There is no full barrage enclosing the water, just localized 'piling up' against the causeway wall, making it less invasive and cheaper than a barrage and also, it is claimed, suitable for areas where the tidal range is low. An 8 GW design is being looked at (DTP, 2014).

7.11 Tidal current assessment

Although tidal barrages offer larger amounts of energy per project, and tidal current technology is in its infancy, at the time of writing, the prospects for the latter look more promising. Tidal current systems may also be easier to develop than the wave energy systems described in Chapter 9. Whereas wave energy systems have to operate in a chaotic interface between air and water and have to try to capture energy from multiple-vectored wave motions using complex technology, tidal currents are smoother, laminar flows, in a more stable, although still turbulent, undersea environment, and the energy can be captured using relatively simple turbines.

Not surprisingly, then, given the relatively easier technological challenge, tidal current projects have proliferated – there are around 100 different devices under development around the world at various scales. Few are likely to be commercially successful, but those that are may reap large rewards. One study suggested that the annual world market for tidal technology might be £155–444 billion (Westwood, 2005).

The UK is well placed to exploit tidal flows in geographical terms (Atlas of UK Marine Renewable Energy Resources, 2011). A study by ABPmer has suggested that ultimately about 36 GW of tidal current device capacity might in theory be installed in the UK. It estimated that devices in under 40 m depth might generate around a total of up to 94 TWh per year – about a quarter of the 2010 UK electricity requirement (ABPmer, 2007).

However, it is unlikely to be possible to harness all of that resource, and certainly it could take time to do so. The DTI/Carbon Trust Renewables Innovation Review in 2004 put the UK's total practical tidal current resource at around 31 TWh per year, about 10% of UK electricity demand at the time, while a subsequent study by the Carbon Trust, taking economic and other constraints into account, put the tidal figure at only 18 TWh per year (Carbon Trust, 2006).

These estimates may be pessimistic and depend on the assumptions used. For example, Prof. Stephen Salter has argued that existing estimates of the tidal stream resource ignore changes in the resource that might be caused by the physical installation of turbines lowering the current velocity (and thus lowering the energy 'lost' to seabed friction). Looking at the Pentland Firth, he claimed that, 'Any small reduction in velocity caused by turbine installations will release large amounts of energy. About one third of the present total friction loss could be extracted, giving a possible resource of 10–20 GW, much higher than previous estimates' (Salter, 2008).

On a similar basis, Prof. David Mackay at Cambridge University claimed that the UK tidal current resource has been seriously underestimated, by perhaps a factor of 10 or more (MacKay, 2007). Similarly, DECC's 2050 Pathways report noted the following:

> It has been widely quoted that the total UK tidal stream potential is of the order of 17 TWh/year. This is derived from a method that provides the most conservative estimate. ... However, academic research has highlighted uncertainties surrounding the calculation of practical resource and other methods of estimating the tidal stream resource have resulted in higher technical potentials of up to 197 TWh/year. ...

> Industry and academics across a range of disciplines, including oceanography, turbulence, marine energy and physics, need to collaborate to come to a consensus on the appropriate methods for estimating resource and the subsequent predictions that result.

> (DECC, 2010b)

Whether or not these views are confirmed, the resource is far from small and significant resources have also been identified elsewhere in the world (Hardisty 2007, 2009).

Perhaps inevitably, estimates of what might actually be obtained in practice are fluid when a technology is at a relatively early stage of development. It is not clear which systems will be successful in the long run, much less where they can best be sited, how much of the resource they can harvest, and at what cost.

It is reasonable to draw parallels with wind power (see Chapter 8): significant cost-reducing improvements can be expected when and if capacity is established and as economies of volume production are achieved. A 2011 Carbon Trust report claimed that, given targeted and accelerated development, by 2025 electricity costs could be comparable to those from offshore wind projects, and, for the best sites, possibly even to those from land-based wind projects and nuclear plants (Carbon Trust, 2011).

As it stands, the UK may have up to 1 GW of tidal stream projects in the early 2020s, by which time costs should be clearer, and multi-GW deployment in the UK and elsewhere may then follow. IRENA has suggested that, on the basis of some European studies, by 2020 tidal projects might be generating between € 0.17/kWh and € 0.23/kWh, although that may be optimistic: current demonstration projects suggested the levelized cost of energy (LCOE) would be in the range of € 0.25-0.47/kWh (IRENA, 2014).

Environmental impact and integration issues

In addition to operational and economic issues, if tidal current devices are to be used on a wide scale then a key issue will be their environmental impact. As indicated earlier, this is a major concern for tidal barrages and may also prove to be of some significance for tidal lagoons. In contrast, most studies of tidal current turbines so far have suggested that impacts will be low. Even large arrays will not impede flows significantly and as the rotor blades will turn slowly (slower than wind turbines, and much slower than the turbines in barrages and lagoons) they should not present a hazard to marine life – fish will be unaffected. Certainly, experience with MCT's SeaGen has not indicated any problems. At the time of writing, no seal deaths have been attributed to the turbines since their installation in 2008. A sonar system has been used to detect the approach of any marine mammals and shut the turbines down.

All structures put in the sea will have some impact, and this needs to be, and already is, carefully assessed when considering possible locations. The installation process for seabed mounted systems can be disruptive and there is the issue of the impact of noise on sea creatures. There have also been studies of their sensitivity to electrical fields from undersea cables, but these impacts are not unique to tidal, or indeed offshore wind and wave, systems.

Comparing various tidal schemes and their impact is difficult. Small barrages may have less impact/MW than large barrages, and lagoons may be better than both, but the total impact will depend on the total number of projects and their locations. The output/MW installed may be less from small projects in sites with low tidal range, but the total capacity installed would have to be larger than that of a large barrage to get the same output. That may push up the impacts. Similarly, although tidal current turbines may have less impact than barrages or lagoons, many 2 MW tidal turbines would be required to equal the output of, say, an 8 GW barrage. However, in this case, given that tidal current turbines have higher load factors than barrages, the total capacity needed to get the same output could be less.

The point here is that comparisons of impacts, and indeed of overall costs and performance, have to be made at the integrated energy *systems* level. Tidal projects differ from systems using intermittent wind, wave or solar energy since tidal energy is highly predicable and, storm surges apart, unrelated to weather. It does vary with the lunar cycle and will be low at neap tides. Moreover, there are locational variations: high tides occur at different times around the coast. However, that means that a geographically distributed network of small to medium scale tidal plants can deliver more continuous output than a single large plant.

Figure 7.52 illustrates the difference by comparing a single large barrage as on the Severn estuary and a hypothetical two tidal current turbine system. The location of the turbine projects and the separation between them would be all important to ensure that the local high tide peaks for each one filled in the lulls for the other. Since peak tides occur every 12 hours or so, albeit at different times at each site, and the turbines can generate on the ebbs and the flows every six hours or so, locations with a three-hour high tide separation is what is needed to fill in the gap. For example, sites off Portland

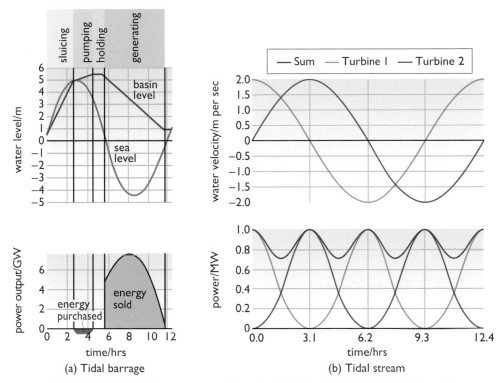

Figure 7.52 (a) A tidal power station such as the Severn Barrage would produce enormous bursts of power of up to 8 GW in magnitude and six hours long; (b) theoretically, two tidal stream turbines at two locations where the time of high tide differs by about 3 hours could, in combination, produce an almost continuous supply of electricity

Bill and the Isle of Wight would be suitable in the UK context. As can be seen from Figure 7.52, in theory that separation would give a reasonable combined output, which could be further improved with more projects at other suitably chosen sites around the coast.

On this basis it has been claimed that a series of tidal current turbines with a total capacity of 200 MW sited around the UK might provide a net continuous supply equivalent to that from a 'firm' (i.e. always available) capacity of about 45 MW (Hardisty, 2008). However, as with all tidal systems, during the low part of the two-week neap-spring tidal cycle, all the projects, wherever sited, would be producing less output, so the overall system would not provide unvarying output. Moreover, in practice it might be hard to find sites that fit exactly in phase all around the coast, so there might be gaps in coverage.

That limitation might be even more of an issue in a possible extension of the network to include small to medium scale tidal barrages and lagoons along with the distributed network of tidal current turbines. Sites for lagoons and barrages are even more geographically defined; however, it is possible that, if some could be included at the right locations, the overall balancing effect could be increased further, especially if these tidal range plants could also offer pumped storage facilities (Yates et al., 2013).

It is worth noting that, at slack water times (when the tidal direction is changing), the individual tidal current projects at specific sites in the network would, briefly, not be producing any output. The barrages and lagoons would also have periods of low or no output, but the network as a whole would be able to deliver: it has to be viewed in system terms.

Furthermore, the presence of a large barrage like that proposed for the Severn, with its large cyclic output, would not help too much with overall grid balancing in such a network, unless there was a large storage capacity available. Its output would at times swamp the system, and the smaller tidal projects at other sites could not collectively equal its output at other times when it was not generating.

However, as can be seen, a range of suitably located small to medium scale tidal projects might make a useful contribution to grid balancing and system integration. Moreover, such a network may be able to deliver more continuous power than wind, wave and solar based systems, helping to balance the overall grid system (Elliott, 2016). We will be returning to the system integration issue in a later chapter.

7.12 Summary

The **tides are driven by the gravitational pull of the Moon, modified by that of the Sun, and are very predictable.** Although the energy available from them does vary over time, it is a **carbon-free source of energy** and it can be collected in two main ways:

- By building **hydro-like dams or barrages across an estuary**, trapping high tides behind them for power generation via hydro-type turbines.
- By installing **free-standing tidal current turbines** in tidal flows, extracting energy from the horizontal movement of tidal waters.

A 250 MW tidal barrage has been operating in France since the 1960s and a similar sized project is now running in South Korea. Larger projects have been proposed, but have not yet been developed.

Large estuary-wide tidal barrages are major civil engineering constructions, which are **expensive to build** and are also likely to be **environmentally intrusive, blocking major natural flows and impacting on wildlife and the local ecosystem.** However, a variant is the **tidal lagoon**: an impoundment structure built in the estuary but not blocking it. That means its environmental impact should be less. Part of the impoundment may include some of the coast line.

Since they use the vertical rise and fall of tides, **barrages and lagoons are termed 'tidal range' systems**, to differentiate them from systems using 'tidal flows'. The latter are, in effect, **underwater wind turbines, running on water flows which have much higher energy densities than wind**. They can be installed with much shorter lead-times than barrages or lagoons, on a modular basis, as funding allows, and should have low environmental impacts. They do not block the tidal flows significantly and rotor speeds are relatively low, much lower than for barrage or lagoon turbines.

The basic physics of barrages means that the **power output is proportional to the square of the tidal range**, so sites with **large tidal ranges are best**. That limits them to specific estuary locations, but there are many good sites around the UK and around the world.

For **tidal turbines**, the power available follows the same law as for wind turbines: it is **proportional to the square of the rotor blade radius and the cube of the flow speed**. Sites with high flow rates, therefore, are very advantageous.

High tidal flows are usually found in areas where there are natural constrictions, for example, in channels between landmasses or islands, of which the UK has many.

Barrages, lagoons and tidal current turbines can all in theory be run on both the ebb and flow of the tides, with there being four tidal movements in and out every 24 hours or so. However, this means having two-way generation turbines, which can be expensive for the high-speed turbines embedded in barrages and lagoons, but less so for free-standing lower-speed tidal current turbines since the whole unit can be turned to face the changing tidal direction, or its rotors can be made two-way relatively easily.

There is continuing interest in tidal barrages around the world and there are some significant barrage projects in Korea, but at the time of writing tidal current turbines are attracting the most attention, with the UK in a leading position, given that it has some of the best resources and also well-established test sites. There are many projects under development in the UK and elsewhere following the UK's early pioneering 1.2 MW SeaGen project in Northern Ireland, which demonstrated technical viability.

Tidal current turbines are still a new technology, with relatively high costs, however they are moving down their cost and performance learning curve, with the expectation being that, as with wind power, they will reach economic viability in due course. The largest project currently planned in the UK is the 398 MW Meygen scheme in Pentland Firth in Scotland. All being well, multi GW deployment may then follow.

Although no tidal lagoons have been built anywhere as yet, a small one, with 320 MW generation capacity, has been proposed for Swansea Bay in Wales as a pathway project, with similar learning curve expectations for subsequent larger ones. Several have been proposed around the UK. Some small barrages have also been proposed in the UK, but, although there can be local non-energy benefits (e.g. new river crossings), **given the relatively high capital costs and impacts of barrages, tidal current turbines seem likely to be favoured**. However, it is the case that barrages and lagoons, once built, can run for a century or more, without major upgrading, apart from the turbines. On that basis, despite the high construction costs, they may be seen as **good long-term investments**.

The balance between the options could shift with new technology or new system-level approaches. **Barrage/lagoon technology is essentially mature**, being similar to hydro, so, new impoundment construction methods apart, the opportunity for significant cost reductions is limited. However, there may be some **operational gains to be made by using barrages or lagoons in pumped storage mode**.

In addition, a network of barrages, lagoons and tidal current turbines suitably dispersed at sites around the coast of the UK (or in other similar locations) could provide more **continuously available energy output, since high tides occur at different times at each geographical point**.

Otherwise, with just individual devices, there would be periods of low or zero output, with peak output not necessarily being well matched to energy demand. In the case of large barrages, such as the 8.2 GW barrage, at one time proposed for the Severn estuary in the UK, the **large bursts of lunar cycle-driven output could be problematic unless major electricity storage facilities were developed.**

The output may be cyclic and at times low (at neap tides especially), but the **overall tidal energy resource is large**.

For the UK, in terms of practically viable projects, it has been put conservatively at 20 GW for barrages and lagoons potentially delivering 40 TWh p.a., and at 21.3 GW for tidal current turbines, delivering 69 TWh p.a. (DECC 2010b). The larger output in the latter case is because the load factors are higher for tidal current turbines. Some have achieved over 50% compared to the 25% expected for tidal range systems.

Obtaining these high outputs in practice would depend on **costs and securing site permissions**, but the potential is clearly there and in theory even more could be obtained.

Globally, the **total tidal resource has been put at 1000 GW**, and, although not all of that could be used effectively, it is of the same order as the world's current hydro capacity.

The large potential of tidal energy has been talked about for many years, but the economics have often been seen as challenging. Now, with increasing concerns about climate change and energy security, it seems that priorities may be changing and more effort is being put into developing new technologies. While interest in barrages remains, the rapid development of tidal current technology, and also the growing interest in lagoons, could well mean that tidal energy will become a significant reality.

References

ABPmer (2007) *Quantification of Exploitable Tidal Energy Resources in UK Waters*, Hampshire, ABP Marine Environmental Research Ltd [Online]. Available at https://www.iwight.com/azservices/documents/2782-FF5-Quantification-of-Exploitable-Tidal-Energy-Resources-in-UK-Waters.pdf (Accessed 20 March 2017).

Admiralty Charts and Publications (1992) *Admiralty Tidal Stream Atlas: the English Channel NP250*, Taunton, UK Hydrographic Office.

Atkins Consultants Ltd (2004) *WS Atkins' Feasibility Study* [Online], Surrey, Atkins Consultants Ltd. Available at http://www.tidalelectric.com/ws-atkins-feasibility-study (Accessed 22 March 2017).

Atlantis (2016) *MeyGen* [Online], Edinburgh, Atlantis Resource Ltd. Available at http://atlantisresourcesltd.com/projects/meygen-scotland.html (Accessed 20 March 2017).

Atlas of UK Marine Renewable Energy Resources (2011) *Atlas of UK Marine Renewable Energy Resources* [Online]. Available at http://www.renewables-atlas.info/ (Accessed 20 March 2017).

Blue Energy (2016) *Blue Energy* [Online]. Available at http://www.bluenergy.com (Accessed 20 March 2017).

Bryden, I. G., Couch, S. J., Owen, A. and Melville, G. (2007) 'Tidal current resource assessment', *Proc IMcchE, Part A: Journal of Power and Energy*, vol. 221, no. 2, pp. 125–135.

Carbon Trust (2006) *Future Marine Energy: Results of the Marine Energy Challenge: Cost competitiveness and growth of wave and tidal stream energy* [Online], London, Carbon Trust. Available at http://ec.europa.eu/ourcoast/download.cfm?fileID=967 (Accessed 20 March 2017).

Carbon Trust (2011) *Accelerating Marine Energy* [Online], London, Carbon Trust. Available at https://www.carbontrust.com/media/5675/ctc797.pdf (Accessed 20 March 2017).

DECC (2010a) *Severn Tidal Power: Feasibility Study Conclusions and Summary Report* [Online], London, Department of Energy and Climate Change. Available at https://www.gov.uk/government/uploads/system/uploads/attachment_data/file/50064/1._Feasibility_Study_Conclusions_and_Summary_Report_-_15_Oct.pdf (Accessed 20 March 2017).

DECC (2010b) *2050 Pathways Analysis* [Online], London, Department of Energy and Climate Change. Available at https://www.gov.uk/government/uploads/system/uploads/attachment_data/file/68816/216-2050-pathways-analysis-report.pdf (Accessed 20 March 2017).

Department of Energy (1981) *Tidal Power from the Severn Estuary, Vol. I*, London, HMSO, Energy Paper 46.

Department of Energy (1989) The Severn Barrage Project: General Report, London, HMSO, Energy Paper 57.

ETSU (1990) *Renewable Energy Research and Development Programme*, Oxfordshire, Energy Technology Support Unit, R56.

ETSU (1993) Tidal Stream Energy Review, Oxfordshire, Energy Technology Support Unit, ETSU-T-05/00155/REP.

Elliott, D. (2016) *Balancing Green Power*, Bristol, Institute of Physics Publications.

Frontier Economics (2008) *Analysis of a Severn Barrage, A report prepared for the NGO steering group*, London, Frontier Economics [Online]. Available at http://assets.wwf.org.uk/downloads/frontier_economics_barrage_repo.pdf (Accessed 20 March 2017).

Fundy (2016) Fundy Ocean Research Center for Energy [Online]. Available at http://fundyforce.ca/ (Accessed 20 March 2017).

Hafren Power (2013) *The Severn Barrage* [Online]. Available at https://web.archive.org/web/20130303010720/http:/www.hafrenpower.com/severn-barrage/index.html (Accessed 22 March 2017).

Halcrow Group Ltd, Mott MacDonald and RSK Group plc (2009) *Solway Energy Gateway Feasibility Study* [Online]. Available at http://enviroeconomynorthwest.com/download/energy_and_environmental_technologies_sector/projects/Solway%20Energy%20Gateway%20feasibility%202009.PDF (Accessed 21 March 2017).

Hardisty, J. (2007) 'Assessment of tidal current resources: case studies of estuarine and coastal sites', *Energy & Environment*, vol. 18, no. 2, pp. 233–49.

Hardisty, J. (2008) 'Power intermittency, redundancy and tidal phasing around the United Kingdom', *The Geographical Journal*, vol. 174, no. 1, pp. 76–84 [Online]. DOI 10.1111/j.1475-4959.2007.00263.x (Accessed 22 March 2017).

Hardisty, J. (2009) *The Analysis of Tidal Stream Power*, Chichester, Wiley.

Hendry, C. (2016) *The Role of Tidal Lagoons* [Online]. Available at https://hendryreview.files.wordpress.com/2016/08/hendry-review-final-report-english-version.pdf (Accessed 20 March 2017).

HM Treasury (2003) *The Green Book: Appraisal and Evaluation in Central Government* [Online], London, TSO. Available at https://www.gov.uk/government/uploads/system/uploads/attachment_data/file/220541/green_book_complete.pdf (Accessed 20 March 2017).

IRENA (2014) *Ocean Energy: Technology Readiness, Patents, Deployment Status and Outlook* [Online], Abu Dhabi, United Arab Emirates, International Renewable Energy Agency. Available at http://www.irena.org/DocumentDownloads/Publications/IRENA_Ocean_Energy_report_2014.pdf (Accessed 20 March 2017).

Jackson, T. (1992) 'Renewable energy: Summary paper for the renewables series', *Energy Policy*, vol. 20, no. 9, pp. 861–83.

Jo, C. H. (2008) 'Recent Development of Ocean Energy in Korea', paper presented at *British Wind Energy Association Marine '08*. Edinburgh, 28 February.

Laughton, M. A. (1990) *Renewable Energy Sources, Report 22*, London, Watt Committee on Energy/Elsevier Applied Science.

Lee, K-S. (2006) *Tidal and Tidal Current Power Study in Korea* [Online], Ansan, Korea Ocean Research & Development Institute. Available at http://www.energybc.ca/cache/tidal/KOREA.pdf (Accessed 22 March 2017).

MacKay, D. J. C. (2007) *Under-estimation of the UK Tidal Resource* [Online], Cavendish Laboratory, University of Cambridge. Available at http://www.inference.phy.cam.ac.uk/sustainable/book/tex/TideEstimate.pdf (Accessed 20 March 2017).

Marine Current Turbines (2010) 'Presentation to 4th International Tidal Energy Summit', paper presented at *4th International tidal Energy Summit*, London, 23–24 November.

Minesto (2015) *Minesto* [Online]. Available at http://www.minesto.com (Accessed 21 March 2017).

OST (1999) *Energies From the Sea: Towards 2020*, London, Marine Foresight Panel.

OVG (2010) *The Offshore Valuation – A valuation of the UK's offshore renewable energy resource* [Online], Machynlleth, Public Interest Research Centre [Online]. Available at http://www.ppaenergy.co.uk/web-resources/resources/467ac5b8919.pdf (Accessed 20 March 2017).

Peel Energy/NWRDA (2010) *Mersey Tidal Power* [Online]. Available at http://www.merseytidalpower.co.uk/content/project (Accessed 20 March 2017).

Salter, S. (2008) *Renewable electricity-generation technologies, Fifth Report of Session 2007–08, Volume II*, London, House of Commons: Innovation, Universities, Science and Skills Committee/TSO.

Salter, M. and Walker, C. (2010) *The Severn Barrage Digest* [Online], London, All Party Parliamentary Group on Angling. Available at http://www.martinsalter.com/wp-content/uploads/2010/04/severn-barrage-digest-chapter-6-post-short-list-conclusions.pdf (Accessed 21 March 2017).

SDC (2007) *Turning the Tide: Tidal Power in the UK* [Online], London, Sustainable Development Commission. Available at http://www.sd-commission.org.uk/data/files/publications/Tidal_Power_in_the_UK_Oct07.pdf (Accessed 21 March 2017).

SME (2015) PLAT-O: *A Commercially Viable Tidal Energy Platform* [Online], Edinburgh, Sustainable Marine Energy Ltd. Available at http://www.all-energy.co.uk/__novadocuments/86819?v=635675468470900000 (Accessed 21 March 2017).

Steiner-Dicks, K. (2013) 'Neptune Renewable Energy: chosen approach technically flawed; plans for liquidation', *Tidal Today*, 13 February.

STPG (1986) *Tidal Power from the Severn*, Hertfordshire, Severn Tidal Power Group.

TLP (2014) *Harnessing the power of our tides*, Swansea, Tidal Lagoon Power [Online]. Available at http://www.tidallagoonswanseabay.com/ (Accessed 20 March 2017).

Twidell, J. W. and Weir, A. J. (2006) *Renewable Energy Resources* (2nd edition), Oxford, Taylor & Francis.

VerdErg (2016) VerdErg [Online]. Available at http://www. verdergrenewableenergy.com (Accessed 21 March 2017).

WEC (2001) 2001 *Survey of Energy Resources*, London, World Energy Council [Online]. Available at https://www.worldenergy.org/wp-content/uploads/2012/10/PUB_Survey-of-Energy-Resources_2001_WEC.pdf (Accessed 21 March 2017).

Westwood, A. (2005) *Refocus Marine Renewable Energy Report: Global Markets, Forecasts and Analysis 2005–2009*, London, Elsevier Science.

Yates, N., Walkington, I., Burrows, R. and Wolf, J. (2013) 'Appraising the extractable tidal energy resource of the UK's western coastal waters', *Philosophical Transactions Of the Royal Society A*, vol. 371 [Online]. DOI: http://dx.doi.org/10.1098/rsta.2012.0181 (Accessed 20 March 2017).

Chapter 8

Wind energy

By Derek Taylor

8.1 Introduction

Wind energy has been used for thousands of years for milling grain, pumping water and other mechanical power applications. Today, there are many thousands of windmills in operation around the world, a proportion of which are used for water pumping. But it is the use of wind energy as a pollution-free means of generating electricity on a significant scale that is attracting most current interest in the subject. Strictly speaking, a wind*mill* is used for milling grain, so modern technology for electricity generation is generally differentiated by use of the term **wind turbines**, partly because of their functional similarity to the steam and gas turbines that are used to generate electricity, and partly to distinguish them from their traditional forebears. Wind turbines are also sometimes referred to as **wind energy conversion systems (WECS)** and sometimes described as **wind generators** or **aerogenerators**.

Attempts to generate electricity from wind energy have been made (with various degrees of success) since the late nineteenth century when Professor James Blyth of the Royal College of Science and Technology, now Strathclyde University, built a range of wind energy devices to generate electricity, his first being in 1887. A later design built at Marykirk in Scotland continued to generate electricity for over 20 years. In 1888 Professor Charles Brush built a 12 kW wind generator, based on a 50 m diameter, 19th century multiblade style, horizontal axis wind turbine that ran for 20 years. In 1891 Professor Poul la Cour began experimenting with an electricity generating windmill to produce and store hydrogen. This was to be used for lighting at Askov Folk High School in Denmark. He later initiated the Danish Wind Electricity Society (DVES) in 1903 and his experiments became the foundation of the Danish wind energy industry.

For many years, small-scale wind turbines have been manufactured to provide electricity for remote houses, farms and remote communities, and for charging batteries on boats, caravans and holiday cabins (thousands of small turbines similar to that shown in Figure 8.1 are in use worldwide). More recently they have been used to provide electricity for cellular telephone masts, traffic signs, street lighting and remote telephone boxes.

However, it is only since the 1980s that the technology has become sufficiently mature to enable rapid growth of the sector. Between the early 1980s and the late 2000s the cost of wind turbines fell steadily and the rated capacity of typical machines increased significantly. Now, on reasonably windy and accessible sites, wind turbines are one of the most cost-effective methods of electricity generation. Given continuing improvements in cost,

Figure 8.1 The Marlec Rutland 1200 Windcharger which can generate up to 500 watts (source: Marlec, 2016)

capacity and reliability, it can be expected that wind energy will become even more economically competitive over the coming decades. Moreover, as wind turbines are increasingly deployed offshore, where wind speeds are generally higher and planning constraints perhaps less demanding, the technically accessible wind resource is massively increased. Of course, as will be seen later, there are significant additional technical challenges associated with offshore wind and the cost of generation is inevitably higher.

The improvements in wind power technology have made it one of the fastest growing renewable energy technologies worldwide in terms of installed rated capacity. A total of over 432 GW of wind generating capacity had been installed by the end of 2015, with almost 36 GW added in that year. This is about 24 times the capacity that had been installed by the end of 2000, and, at the time of writing, the current average growth rate is around 17% cumulatively and 22% per annum. In June 2017, global wind power capacity went past half a terawatt (500 GW).

To understand the machines and systems that extract energy from the wind involves an appreciation of many fields of knowledge, from meteorology, aerodynamics and planning to electrical, structural, civil and mechanical engineering. Hence, this chapter begins with a description of the atmospheric processes that give rise to wind energy. Wind turbines and their aerodynamics are then described, together with various ways of calculating their power and energy production. This is followed by discussions of the environmental impact and economics of wind energy, together with an examination of recent commercial developments and a discussion of its future potential. The final section looks at offshore wind power, which seems likely to be one of the most important areas of wind energy development in coming decades, especially for the UK and northern Europe.

8.2 The wind

As mentioned in Chapter 3, one square metre of the Earth's surface on or near the equator receives more solar radiation per year than one square metre at higher latitudes. The curvature of the Earth means that its surface becomes more oblique to the Sun's rays with increasing latitude. In addition, the Sun's rays have further to travel through the atmosphere as latitude increases, so more of the Sun's energy is absorbed *en route* before it reaches the surface. As a result of these effects, the tropics are considerably warmer than higher latitude regions.

This differential solar heating of the Earth's surface causes variations in atmospheric pressure, which in turn give rise to the movement of atmospheric air masses which are the principal cause of the Earth's wind systems (see Box 8.1 for more details).

BOX 8.1 The Earth's wind systems

Like all gases, air expands when heated, and contracts when cooled. Thus warm air is less dense than cold air and will rise to high altitudes when strongly heated by solar radiation.

A low pressure belt (with cloudy and rainy weather patterns) is created at the equator due to warm humid air rising in the atmosphere until it reaches the tropopause (the top of the troposphere). At the surface the equatorial region is called the 'doldrums' (from an old English word meaning dull) by early sailors who were fearful about becoming becalmed.

At the tropopause in the northern hemisphere the air moves northwards and in the southern Hemisphere it moves southwards. This air gradually cools until it reaches latitudes of about 30 degrees, where it sinks back to the surface, creating a belt of high pressure at these latitudes (with dry clear weather patterns). The majority of the world's deserts also occur in these high-pressure regions.

Some of the air that reaches the surface at these latitudes is forced back towards the low-pressure zone at the equator. These air movements are known as the '**trade winds**'. On reaching the equator these air movements complete the circulation of what is known as the **Hadley cell** – named after the scientist (George Hadley) who first described them in 1753.

However, not all of the air that sinks at the 30 degree latitudes moves toward the equator. Some of it moves poleward until it reaches the 60 degree latitudes, where it meets cold air coming from the poles at what are known as the 'polar fronts'. The interaction of the two bodies of air causes the warmer air to rise and most of this air cycles back to the 30-degree latitude regions where it sinks to the surface, contributing to the high-pressure belt. This completes the circulation of what is known as the **Ferrel cell** (named after William Ferrel who first identified it in 1856).

The remaining air that rises at the polar fronts moves poleward and sinks to the surface at the poles as it cools. It then returns to the 60-degree latitude region completing the circulation of what is known as the **polar Hadley cell** or **polar cell**.

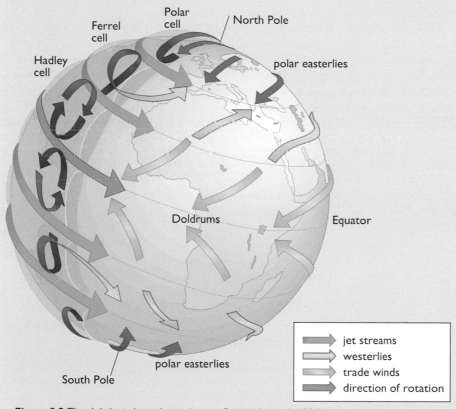

There is a further complication in that, because the Earth itself rotates, winds moving across the Earth's surface are subject to a phenomenon known as the Coriolis Effect. The net result of this effect, given the Earth rotates in an eastwards direction, is that in the northern hemisphere 'north bound' winds are caused to veer 'right'. Such winds are known as 'westerlies' as, whilst they are veering toward an easterly direction, it is the convention when referring to wind direction to use the direction from which winds blow. In the southern hemisphere 'north bound' winds veer to the 'left' ('trade winds'). Likewise, 'south bound' winds veer 'right' ('trade winds') in the northern hemisphere and 'left' ('westerlies') in the southern hemisphere. Figure 8.2 shows the overall pattern of global wind circulation.

Figure 8.2 The global wind circulation (source: Burroughs et al., 1996)

Atmospheric pressure is the pressure resulting from the weight of the column of air above a specified surface area, with the unit of atmospheric pressure being known as the bar. Atmospheric pressure is measured by means of a barometer (Figure 8.3). These devices are usually calibrated in millibars (mbar), that is, thousandths of a bar. The average atmospheric pressure at sea level is about 1013.2 mbar (approximately 1 bar). The SI unit of pressure, the pascal (Pa), is defined as one newton per square metre and 1 bar is equivalent to 100 kPa.

On the weather maps featured in television weather forecasts or in newspapers, there are regions marked 'high' and 'low', surrounded by contours (Figure 8.4). The regions marked 'high' and 'low' relate to the atmospheric

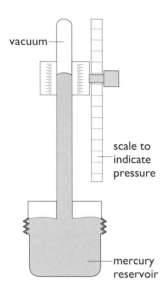

Figure 8.3 A Fortin barometer, an example of a barometer used to measure atmospheric pressure. Variations in atmospheric pressure acting on the mercury in the reservoir cause the mercury in the column to rise or fall

pressure and the contours represent lines of equal pressure called **isobars**. The high-pressure regions tend to indicate fine weather with little wind, whereas the low-pressure regions tend to indicate changeable windy weather and precipitation.

In addition to the main global wind systems shown in Box 8.1 there are also local wind patterns, such as sea breezes (Figure 8.5) and mountain valley winds (Figure 8.6).

Energy and power in the wind

The energy contained in the wind is its kinetic energy, and as we saw in Chapter 1 the kinetic energy of any particular moving mass (moving air in this case) is equal to half the mass, m, (of the air) times the square of its velocity, V:

$$\text{kinetic energy} = \text{half mass} \times \text{velocity squared} = \tfrac{1}{2}mV^2 \qquad (1)$$

where m is in kilograms and V is in metres per second (m s^{-1}).

We can calculate the kinetic energy in the wind if, first, we imagine air passing through a circular ring or hoop enclosing an area A (say 100 m^2) at a velocity V (say 10 m s^{-1}) (see Figure 8.7). As the air is moving at a velocity of 10 m s^{-1}, a cylinder of air with a length of 10 m will pass through the ring each second. Therefore, a volume of air equal to $10 \times 100 = 1000$ cubic metres (m^3) will pass through the ring each second. By multiplying this volume by the density of air, ρ (which at sea level is 1.2256 kg m^{-3}), we obtain the mass of the air moving through the ring each second. In other words:

mass (m) of air per second = air density × volume of air passing per second
$$\qquad\qquad = \text{air density} \times \text{area} \times \text{length of cylinder of air}$$
$$\qquad\qquad\qquad \text{passing per second}$$
$$\qquad\qquad = \text{air density} \times \text{area} \times \text{velocity}$$

that is:

$$m = \rho\, AV$$

Substituting for m in (1) above gives:

$$\text{kinetic energy per second} = 0.5\, \rho\, AV^3 \text{ (joules per second)}$$

where ρ is in kilograms per cubic metre (kg m^{-3}), A is in square metres (m^2) and V is in metres per second (m s^{-1}).

If we recall that energy per unit of time is equal to power, then the power in the wind is P (watts) = kinetic energy in the wind traversing the circular ring per second (joules per second), that is:

$$P = 0.5\, \rho\, AV^3 \qquad (2)$$

The main relationships that are apparent from the above calculations are that the power in the wind is proportional to:

■ the density of the air

■ the area through which the wind is passing (i.e. through a wind turbine rotor), and

■ the cube of the wind velocity.

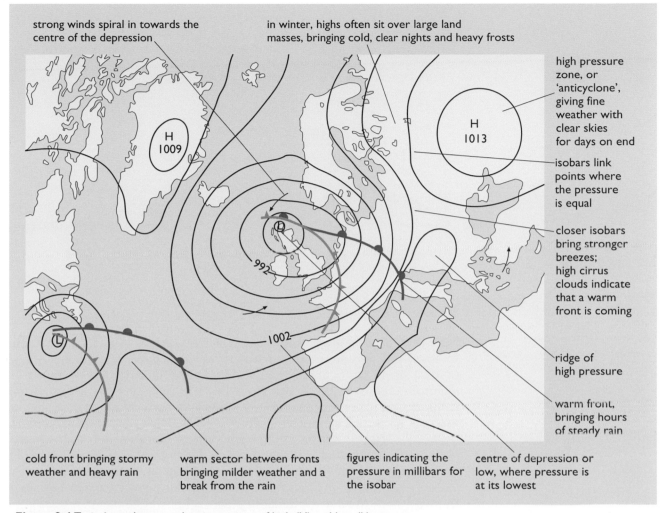

strong winds spiral in towards the centre of the depression

in winter, highs often sit over large land masses, bringing cold, clear nights and heavy frosts

high pressure zone, or 'anticyclone', giving fine weather with clear skies for days on end

isobars link points where the pressure is equal

closer isobars bring stronger breezes; high cirrus clouds indicate that a warm front is coming

ridge of high pressure

warm front, bringing hours of steady rain

cold front bringing stormy weather and heavy rain

warm sector between fronts bringing milder weather and a break from the rain

figures indicating the pressure in millibars for the isobar

centre of depression or low, where pressure is at its lowest

Figure 8.4 Typical weather map showing regions of high (H) and low (L) pressure

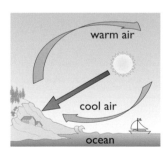

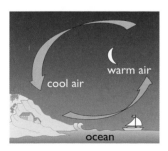

Figure 8.5 Sea breezes are generated in coastal areas as a result of the different heat capacities of sea and land, which give rise to different rates of heating and cooling. The land has a lower heat capacity than the sea and heats up quickly during the day, but at night it cools more quickly than the sea. During the day, the sea is therefore cooler than the land and this causes the cooler air to flow shoreward to replace the rising warm air on the land. During the night the direction of air flow is reversed

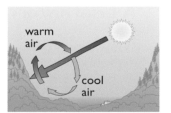

Figure 8.6 Mountain valley winds are created when cool mountain air warms up in the morning and, as it becomes lighter, begins to rise: cool air from the valley below then moves up the slope to replace it. During the night the flow reverses, with cool mountain air sinking into the valley

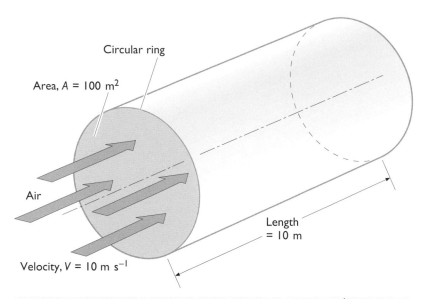

Circular ring

Area, A = 100 m²

Air

Velocity, V = 10 m s⁻¹

Length = 10 m

Figure 8.7 Cylindrical volume of air passing at velocity V (10 m s⁻¹) through a ring enclosing an area, A (100 m⁻²), each second

Note that the air density is lower at higher elevations (e.g. in mountainous regions) and, perhaps more importantly, average densities in cold climates may be significantly higher than in hot regions (more than 10% higher for example than in tropical regions). Also, wind velocity has a very strong influence on power output because of the 'cube law'. For example, a wind velocity increase from 6 m s⁻¹ to 8 m s⁻¹ will increase the power in the wind by a factor of more than two. It is also important to appreciate that the power contained in the wind is not in practice the amount of power that can be extracted by a wind turbine. This is because losses are incurred in the energy extraction/conversion process (see Section 8.4 on aerodynamics). Moreover there are additional mechanical-to-electrical power conversion losses.

8.3 Wind turbines

A brief history of wind energy

Wind energy was one of the first non-animal sources of energy to be exploited by early civilizations. It is thought that wind was first used to propel sailing boats, but the static exploitation of wind energy by means of windmills is believed to have been taking place for about 4000 years.

Windmills have traditionally been used for milling grain, grinding spices, dyes and paint stuffs, making paper and sawing wood. Traditional wind pumps were used for pumping water in Holland and East Anglia in the UK, and, because they often used identical forms of sails and support structures, they were (and are) often also referred to as windmills.

Many early windmills were of the *vertical-axis* type and, unlike modern wind turbines which are driven by lift forces (see below), these were drag-driven devices and relied on differences in drag on either side of the vertical shaft in order to function.

Screen wind machines

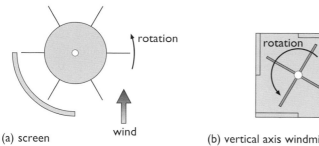

(a) screen   (b) vertical axis windmill screened by walls

Clapper-type wind machines   **Wind machine with cyclic pitch variation**

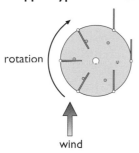 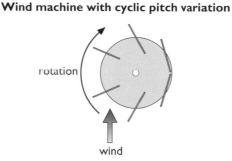

Cup-type wind machines

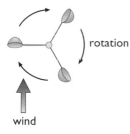 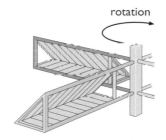

(a) cup anemometer   (b) 'streamlined anemometer sail windmill' invented by Faustus Verantius, a seventeenth century bishop and engineer (Needham, 1965)

Figure 8.8 Some examples of traditional vertical-axis windmills

Some examples are shown in Figure 8.8 and include the following.

Screened windmills. These windmills employ screens or partial walls around the windmill, which are positioned to screen the windmill sails from the wind during the 'backward' part of the cycle, when the sails are moving towards the wind.

'Clapper' windmills. These windmills are so called because the moveable sails 'clap' against stops as the rotor turns with the wind (forwards), maximizing their air resistance, but align themselves with the wind (like a weather vane) when on the part of their cycle in which they are moving into the wind (backwards), so reducing their air resistance.

Cyclically pivoting sail windmills. These windmills are similar to the 'clapper' windmills, but use a more complex mechanism to achieve progressive changes in sail orientation. The pitch angle of each sail is cyclically adjusted according to its position during its rotation cycle and to the direction of the wind. This gives a difference in resistance on either side of the windmill's rotation axis, causing it to rotate when exposed to a wind stream.

Differential resistance or cup type windmills. In these windmills, the blades are shaped to offer greater resistance to the wind on one surface compared with the other. This results in a difference in wind resistance on either side of the windmill axis, so allowing the windmill to turn. The first electricity producing wind generator, invented by Professor James Blythe, mentioned above, was a 10 m diameter vertical axis cup type device. A modern example of this type of wind-driven device is the cup anemometer, an instrument used for measuring wind speed. The simple 'S' type and multi-bladed S-type windmills are also examples of this type, as is the 'Savonius rotor' (a 'split-S' shaped rotor as shown in Figure 8.9). Savonius rotors are used for powering fans in trucks and vans and have been used for simple do-it yourself windmills. They are produced as micro wind generators, including variants with helically twisted semi-cylindrical 'cups'.

The more familiar *horizontal-axis* windmills are thought to have appeared in Europe in the twelfth century. These traditional machines consisted of radial arms supporting sails that rotated about a horizontal axis, in a plane that faced into the direction from which the wind was blowing. The sails or blades themselves were set at a small oblique angle to the wind and moved in a plane at right angles to the wind direction. Another characteristic of these windmills is that their rotation axes were usually manually or automatically aligned with the wind direction.

In the Mediterranean regions of Europe, the traditional windmills took the form of triangular canvas sails attached to radial arms. In northern Europe, such windmills were characterized by long rectangular sails consisting of either canvas sheets on lattice frameworks, so-called 'common sails', or 'shutter-type sails', which resembled venetian blinds. The shuttering arrangement gave a degree of control over starting, regulating and stopping the windmill according to wind strength.

In northern Europe there were two main forms of windmill. One was the less common 'post mill', in which the whole windmill was moved about a large upright post when the wind direction changed; the other was the more common 'tower mill' (Figure 8.10), in which the rotor and cap were supported by a relatively tall tower, usually of masonry. In the tower mill, only the cap (in combination with the rotor and its shaft) were moved in response to changes in wind direction. The sails turned fairly slowly and provided mechanical power.

At their zenith, before the Industrial Revolution, it is estimated that there were some 10 000 of these windmills in Britain (Golding, 1955) and they formed a familiar feature of the countryside.

As an indication of how numerous windmills were in the English countryside, during the Eastern Tour of his Rural Rides in the 1830s William Cobbett wrote:

> The windmills in the vicinage are so numerous that I counted, whilst standing in one place, no less than seventeen. They are all painted or washed white; the sails are black; it was a fine morning, the wind was brisk, and their twirling altogether added greatly to the beauty of the scene, which having the broad and beautiful arm of the sea on the one hand, and the fields and meadows, studded with farm-houses, on the other, appeared to me to be the most beautiful sight of the kind that I had ever behold.

> (Cobbett, 1830)

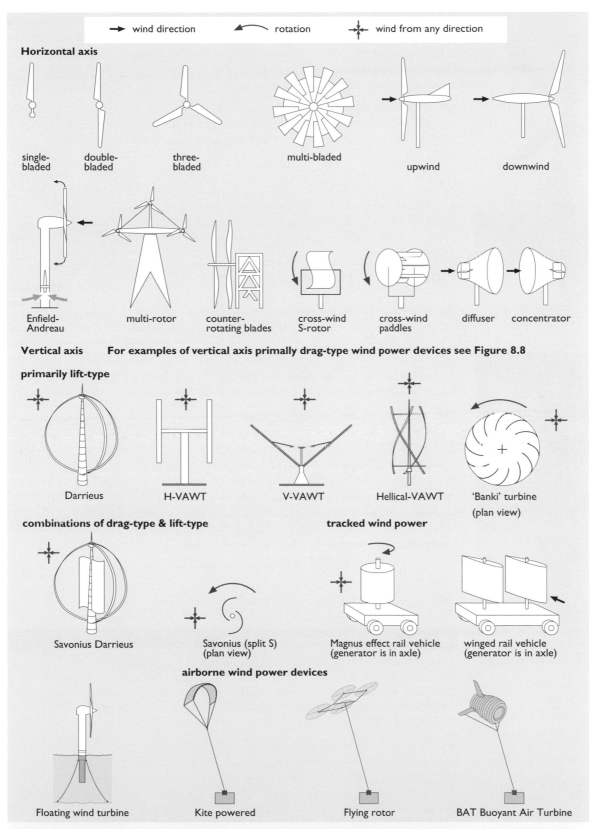

Figure 8.9 Some examples of the machines that have been proposed for wind energy conversion. (Source: partly based on Eldridge, 1975. For further information on these machines see Eldridge, 1975 and Golding, 1955). The figure includes mainly Horizontal axis and Vertical axis machines, but also includes a number of other types.

Figure 8.10 Traditional north European tower windmill

Figure 8.11 Multi-bladed wind pump

Wind turbine types

The variety of machines that has been devised or proposed to harness wind energy is considerable and includes many unusual devices. Figure 8.9 shows a small selection of the various types of machines that have been proposed over the years.

Most modern wind turbines come in one of two basic configurations: horizontal axis and vertical axis. Horizontal axis turbines are predominantly of the 'axial flow' type (i.e., the rotation axis is in line with the wind direction), whereas vertical axis turbines are generally of the 'cross flow' type (i.e., the rotation axis is perpendicular to the wind direction). They range in size from very small machines that produce a few tens or hundreds of watts to very large turbines producing as much as 8 MW. Larger turbines rated at 10 to 15 MW are now being considered and 20 MW designs and even 50 MW designs are being investigated.

Horizontal axis wind turbines

Horizontal axis wind turbines (HAWTs) generally have either two or three blades, but can have many more. Multi-bladed wind turbines have what appears to be virtually a solid disc covered by many solid blades (usually of slightly cambered sheet metal construction). They have been used since the nineteenth century for water pumping on farms (Figure 8.11). Appropriately for their application they produce high torque at low rotor speeds.

The term '**solidity**' is used to describe the fraction of the swept area that is solid. Wind turbines with large numbers of blades, such as these multi-bladed devices, have highly solid swept areas and are referred to as **high-solidity** wind turbines. Wind turbines with small numbers of narrow blades have a swept area that is largely void: only a very small fraction of the area appears to be 'solid' – such devices are referred to as **low-solidity** wind turbines. Multi-blade wind pumps have **high-solidity rotors** and modern electricity-generating wind turbines (with one, two or three blades) have **low-solidity rotors**.

Low-solidity devices work effectively at much higher rotational speeds making them attractive for electricity generation (Box 8.2 discusses the effect of blade number on turbine characteristics).

BOX 8.2 **Effect of the number of blades**

The speed of rotation of a wind turbine is usually measured in either revolutions per minute (rpm) or radians per second (rad s^{-1}). The **rotation speed** in revolutions per minute (rpm) is usually symbolized by N and the **angular velocity** in radians per second is usually symbolized by Ω. The relationship between the two is given by:

$$1 \text{ rpm} = \frac{2\pi}{60} \text{ rad s}^{-1} = 0.10472 \text{ rad s}^{-1}$$

A useful alternative measure of wind turbine rotor speed is **tip speed**, U, which is the **tangential velocity** of the rotor at the tip of the blades, measured in metres per second. It is the product of the angular velocity, Ω, of the rotor and the **tip radius**, R (in metres):

$$U = \Omega R$$

Alternatively, U can be defined as:

$$U = \frac{2\pi RN}{60}$$

By dividing the tip speed, U, by the **undisturbed wind velocity**, V_0, upstream of the rotor, we obtain a non-dimensional ratio known as the **tip speed ratio**, usually symbolized by λ. This ratio provides a useful measure against which aerodynamic efficiency can be plotted. The aerodynamic efficiency of a wind turbine is usually described as its **power coefficient** (effectively the ratio of power output from the turbine to the theoretical power in the wind). This quantity is symbolized by C_P and given as a fraction, such that 1 equates to 100% efficiency. When the power coefficient is plotted against tip speed ratio, such $C_P - \lambda$ curves provide an effective way to present the performance of a rotor and to compare wind turbines with differing characteristics.

A wind turbine of a particular design can operate over a range of tip speed ratios, but will usually operate with its best (maximum) efficiency at a particular tip speed ratio, i.e. when the velocity of its blade tips is a particular multiple of the wind velocity. This optimum tip speed ratio (λ_{opt}) is also commonly denoted as λ_{max} with the corresponding efficiency (i.e. power coefficient) being C_{Pmax}. The optimum tip speed ratio for a given wind turbine rotor will depend upon both the number of blades and the width of each blade.

In order to extract energy as efficiently as possible, the blades have to interact with as much as possible of the wind passing through the rotor's **swept area**. The blades of a high-solidity, multi-blade wind turbine interact with all the wind at very low tip speed ratios, whereas the blades of a low-solidity turbine have to travel much faster to 'virtually fill up' the swept area, in order to interact with all the wind passing through. If the tip speed ratio is too low, some of the wind travels through the rotor swept area without interacting with the blades; whereas if the tip speed ratio is too high, the turbine offers too much resistance to the wind, so that some of the wind goes around it. A two-bladed wind turbine rotor with each blade the same width as those of a three-bladed rotor will have an optimum tip speed ratio *one-third higher* than that of a three-bladed rotor. Optimum tip speed ratios for modern low-solidity wind turbines range between about 6 and 20.

In theory, the more blades a wind turbine rotor has, the more efficient it is. However, when there are large numbers of blades in a rotor, the flow becomes more disturbed, so that they aerodynamically interfere with each other. Thus high-solidity wind turbines tend to be less efficient overall than low-solidity turbines. Of low-solidity machines, three-bladed rotors tend to be the most energy efficient; two-bladed rotors are slightly less efficient and one-bladed rotors slightly less efficient still. Wind turbines with more blades can be generally expected to generate less aerodynamic noise as they operate at lower tip speeds (see Section 8.6) than wind turbines with fewer blades.

The mechanical power that a wind turbine extracts from the wind is the product of its angular velocity and the torque imparted by the wind. **Torque** is the moment about the centre of rotation due to the driving force imparted by the wind to the rotor blades. Torque is usually measured in newton metres (N m) (see Box 8.4). For a given amount of power, the *lower* the angular velocity the *higher* the torque; and conversely, the *higher* the angular velocity the *lower* the torque.

The pumps that are used with water pumping wind turbines require a high starting torque to function. Multi-bladed turbines are therefore generally

Figure 8.12 Two-bladed HAWT (WEG MS400 turbine)

Figure 8.13 Three-bladed HAWT (Vestas V52 850 kW turbine)

Figure 8.14 Single bladed HAWT (MBB 600 kW turbine)

used here because of their low tip speed ratios and resulting high torque characteristics.

Conventional electrical generators run at speeds many times greater than most wind turbine rotors so they generally require some form of gearing when used with wind turbines. Low-solidity wind turbines are better suited to electricity generation because they operate at high tip speed ratios and therefore do not require as high a gear ratio to match the speed of the rotor to that of the generator. In addition, many low-solidity small wind turbines (and even certain very large wind turbines) have avoided using gearboxes by using directly coupled low-speed multi-pole generators.

Modern *low-solidity* HAWT rotors evolved from traditional windmills and superficially resemble aircraft propellers. Wind turbines with such rotors are by far the most common design manufactured today. They have a clean streamlined appearance, due in part to their design being driven by aerodynamic considerations derived largely from developments in aircraft wing and propeller design. HAWT rotors generally have two or three wing-like blades (Figures 8.12 and 8.13). They are almost universally employed to generate electricity. Some experimental single bladed HAWTs have also been produced (Figure 8.14) and continue to be researched. Three blades are the most common, though two and single bladed rotors may be beneficial in very large rotors being research for offshore HAWTs.

Vertical axis wind turbines

Vertical axis wind turbines (VAWTs) that employ aerofoil type blades, unlike their horizontal axis counterparts, can harness winds from any direction without the need to reposition the rotor when the wind direction changes. However, despite this advantage, they have found little commercial success to date, in part because – unlike HAWTs – they have not been able to benefit from the experience curve, and in part it may be due to issues with power quality, cyclic loads on the tower systems and the lower efficiency of some VAWT designs. At the micro/small scale, they tend to have lower aerodynamic efficiency and tend to require more expensive rotors compared to micro/small HAWTs. A technical description of how aerofoil-based VAWTs operate is given in Section 8.4.

The modern aerofoil-based VAWT evolved from the ideas of the French engineer, Georges Darrieus, whose name is used to describe one of the VAWTs that he invented in 1925 – such devices were independently reinvented in Canada by South and Rangi at the National Aeronautical Establishment of the National Research Council in the 1960s (South and Rangi, 1972). This device, which resembles a large eggbeater, has curved blades (each with a symmetrical aerofoil cross-section) the ends of which are attached to the top and bottom of a vertical shaft (see Figure 8.15). Several hundred were manufactured in the USA and installed in wind farms in California in the 1980s. A small number were produced in Canada, including 'Eole', the largest VAWT yet built, a 100 m tall 60 m diameter turbine. Eole operated for six years from 1988 in Quebec and achieved a 94% availability (see Figure 8.15(b)).

These Darrieus VAWTs were guyed structures, which added complexity and limited their height, though a Dutch 15 m diameter 100 kW Darrieus

VAWT on a cantilevered tower was built in the 1980s and a new design of Darrieus VAWT, mounted on a free-standing cantilevered tower that avoids the difficulties of guys, is under development in New Mexico (VPM, 2017). There was also a floating enhanced Darrieus VAWT design under development (FWC, 2017). Sandia Laboratories carried out a retrospective review of VAWTs and are also researching the potential for large Darrieus VAWTs for offshore applications.

The blades of a Darrieus VAWT take the form of a 'troposkein' (the curved, arch-like shape taken by a spinning skipping rope). This shape is a structurally efficient one, well suited to coping with the relatively high centrifugal forces acting on VAWT blades. However, they can be difficult to manufacture, transport and install, though the advent of modern composites and manufacturing methods may help to address some of the difficulties. In order to overcome these problems, straight-bladed VAWTs have been developed: these include the 'H'-type vertical axis wind turbine (H-VAWT) and the 'V'-type vertical axis wind turbine (V-VAWT). Interestingly Sandia Laboratories in their retrospective review of VAWTs concluded 'that VAWTs do have significant advantages over HAWTs in offshore applications. … Their primary disadvantage remains the longer blade length required by the full Darrieus VAWT configurations. … Both the 'H' and the 'Y' (e.g. the 'V') configurations offer potential for variable, cost effective designs.' (Sutherland et al., 2012).

The H-VAWT (Figure 8.16) consists of a tower (which may house a vertical shaft), capped by a hub to which is attached two or more horizontal cross arms that support the straight, upright, aerofoil blades. In the UK, this type of turbine was developed by VAWT Ltd which built 125 kW and 500 kW prototypes at Carmarthen Bay and a 100 kW turbine on the Isles of Scilly in the 1980s. There continues to be interest in H-VAWTs both at small scale and for large offshore applications (VertAx, 2017).

The V-VAWT consists of straight aerofoil blades attached at one end to a hub on a vertical shaft and inclined in the form of a letter 'V'. Its main features include a shorter tower, shorter shaft compared to the Darrieus VAWT, ground/water level-mounted generator options, ground/water level blade installation via hinged blades and the ability to self start without needing complex variable pitch blades or the electrical starting required by other types of VAWTs. Experimental prototypes were tested at the Open University (Figure 8.17(a)) as was the *Sycamore Rotor*, a single bladed version. New generation *V2 turbine* variants (Figure 8.17(b)) and other novel derivative configurations suited to very large scale and offshore fixed/ floating applications are being researched by the author.

At the present time, VAWTs are not generally economically competitive with HAWTs. However, they continue to attract research as they should, in principle, offer significant advantages over HAWTs in terms of blade loading and fatigue, if they can be built in very large sizes (such as are becoming desired for use in offshore applications). Whilst VAWTs are subject to wind-induced cyclic loads (which do not progressively increase with increasing size of turbine) they are not subject to the major gravitational cyclic loadings (which do progressively increase with rotor diameter) that large diameter HAWTs experience. As the Canadian Eole demonstrated in the 1980s/90s, large VAWTs can be operated with very high reliability.

(a)

(b)

Figure 8.15 (a) Seventeen metre diameter Darrieus-type VAWT at Sandia National Laboratories, New Mexico.
(b) Sixty metre diameter Eole VAWT in Quebec, Canada

Figure 8.16 500 kW 'h'-type VAWT at Carmarthen Bay, Wales

More recent variants in aerofoil-based vertical axis wind turbines are VAWTs with helically shaped blades. These were first advocated in the 1990s by the Swedish engineer Olle Ljungstrom and additionally by US engineer Alexander Gorlov. Gorlov also suggested such turbines could be used in hydro and, as discussed in Chapter 7, tidal current applications (Gorlov, 1998). Stimulated by perceived concerns about cyclic torque (due to the cyclic variation in the position of a VAWTs blade, relative to the wind direction – see Section 8.4), in the 1990s Ljungstrom produced designs of Darrieus VAWTs that employed helically shaped blades. A number of more recent designs have employed such blades. In practice, however, simply employing three blades should usually be sufficient to even out the torque variation for VAWTs, without the extra complexity and high cost of manufacturing helical blades (Musgrove, 1990, cited in Freris, 1990).

As was mentioned above, most types of aerofoil-based VAWTs are not able to self-start without some extra mechanism (as they are generally unable to produce sufficient aerodynamic starting torque). Examples of such additions are drag-driven 'vertical axis starter rotors' (which can reduce aerodynamic efficiency), complex variable pitch blades, or some form of electrical starting mechanism. Electrically assisted starting is not a major issue for the medium/large scale VAWTs employed in wind farms, but it is a major shortcoming for small/micro scale VAWTs, especially for off-grid applications or on relatively low wind speed sites, when electrically started VAWTs can consume large amounts of electricity, thus greatly reducing their net productivity (Day et al., 2010).

8.4 Aerodynamics of wind turbines

Aerodynamic forces

When a force is transferred by a moving solid object to another solid object, the second object will generally move in either the same direction or in a direction at a small angle (less than 90 degrees) to the direction of motion of the first object, unless subjected to another force. However, the method by which forces are transferred from a fluid to a solid object is very different.

Wind turbines are operating in an unconstrained fluid, in this case air. To understand how they work, two terms from the field of aerodynamics will be introduced. These are 'drag' and 'lift'.

An object in an air stream experiences a force that is imparted from the air stream to that object (Figure 8.18). We can consider this force to be equivalent to two component forces acting in perpendicular directions, known as the *drag force* and the *lift force*. The magnitude of these drag and lift forces depends on the shape of the object, its orientation to the direction of the air stream, and the velocity of the air stream.

The **drag force** is the component that is in line with the direction of the air stream. A flat plate in an air stream, for example, experiences maximum drag forces when the direction of the air flow is perpendicular (that is, at right angles) to the flat side of the plate; when the direction of the air stream is in line with the flat side of the plate, the drag forces are at a minimum.

(a)

(b)

Figure 8.17 (a) V-VAWT prototype developed and tested at the Open University in Milton Keynes in the 1980s (b) multi megawatt scale V-Turbine concept in offshore configuration

Traditional vertical axis windmills and undershot water wheels (see Chapter 6) are driven largely by drag forces.

Objects designed to minimize the drag forces experienced in an air stream are described as streamlined, because the lines of flow around them follow smooth, stream-like lines. Examples of streamlined shapes are teardrops, the shapes of fish such as sharks and trout, and aeroplane wing sections (aerofoils) (Figure 8.19).

The **lift force** is the component that is at right angles to the direction of the air stream. It is termed 'lift' force because it is the force that enables aeroplanes to *lift* off the ground and fly, though in other applications it may induce a *sideward* (as in a sailboat) or *downward* force (as in the downforce aerofoil used in some racing cars). Lift forces acting on a flat plate are smallest when the direction of the air stream is at a zero angle to the flat surface of the plate. At small angles relative to the direction of the air stream – that is, when the so-called *angle of attack* (see below for more detail) is small – a low pressure region is created on the 'downstream' (or 'leeward') side of the plate as a result of an increase in the air velocity on that side (Figures 8.20 and 8.21 show this effect on aerofoil sections).

In this situation, there is a direct relationship between air speed and pressure: the faster the airflow, the lower the pressure (i.e. the greater the 'suction effect'). This phenomenon is known as the **Bernoulli effect** after Daniel Bernoulli, the Swiss mathematician who first explained it. The lift force thus acts as a 'suction' or 'pulling' force on the object, in a direction at right angles to the airflow.

As well as enabling aeroplanes and gliders to fly, it is the lift force that propels modern sailing yachts, and supports and propels helicopters. Lift is also the principal force that drives a modern wind turbine rotor and thus allows it to produce power.

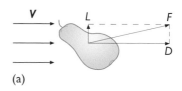
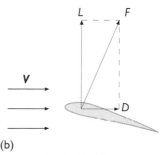

(a)

(b)

Figure 8.18 (a) and (b) An object in an air stream is subjected to a force, F, from the air stream. This is composed of two component forces: the drag force, D, acting in line with the direction of air flow and the lift force, L, acting at 90° to the direction of air flow

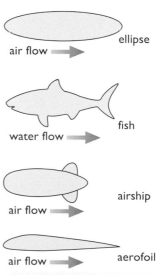

Figure 8.19 Some examples of streamlined shapes

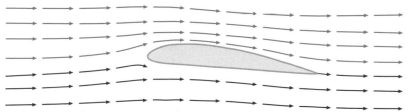

Figure 8.20 Streamlined flow around an aerofoil section

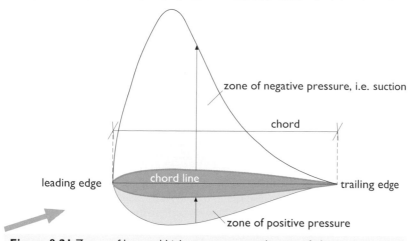

Figure 8.21 Zones of low and high pressure around an aerofoil section in an air

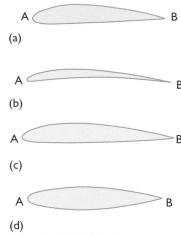

Figure 8.22 Types of aerofoil section: (a), (b) and (c) are various forms of asymmetrical aerofoil section and (d) is a symmetrical aerofoil section

Aerofoils

Arching or cambering a flat plate will cause it to induce higher lift forces for a given angle of attack, but the use of so-called **aerofoil sections** is even more effective. There are two main types of aerofoil section that are conventionally distinguished: asymmetrical and symmetrical (Figure 8.22). Both have a markedly convex upper surface, a rounded end called the 'leading edge' (which faces the direction from which the air stream is coming), and a pointed or sharp end called the 'trailing edge'. It is the shape of the 'under surface' or high pressure side of the sections that identifies the type. Asymmetrical aerofoils are optimized to produce most lift when the underside of the aerofoil is closest to the direction from which the air is flowing. Symmetrical aerofoils are able to induce lift equally well (although in opposite directions) when the air flow is approaching from either side of the **chord line** (the 'length', from the tip of its leading edge to the tip of its trailing edge, of an aerofoil section).

The angle which an aerofoil (or flat or cambered plate profile) makes with the direction of an airflow, measured against a reference line (usually the chord line of the aerofoil), is called the **angle of attack** α (alpha) (Figure 8.23). When airflow is directed towards the underside of the aerofoil, the angle of attack is usually referred to as positive.

When employed as a wing profile, asymmetrical aerofoil sections will, subject to a net incident airflow velocity (in aircraft this is due to forward flight), tend to accelerate the airflow over the more convex 'upper' surface. The high air speed thus induced results in a large reduction in pressure over

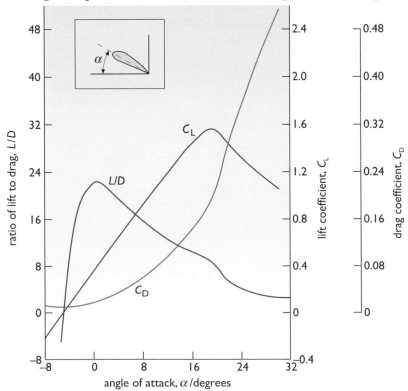

Figure 8.23 Lift coefficient, C_L, drag coefficient, C_d, and lift to drag ratio (L/D) versus angle of attack, α (shown inset), for a Clark Y aerofoil section. The region just to the right of the peak in the C_L curve corresponds to the angle of attack at which stall occurs

the upper surface relative to the lower surface. This results in a 'suction' effect which 'lifts' the aerofoil-shaped wing, although it should be noted that this lift can only be sustained if the airflow leaves the aerofoil at the downstream edge (known as the trailing edge) in a smooth manner that prevents the high pressure air recirculating around the trailing edge and cancelling out the reduced pressure. The strength of the lift force induced by an aerofoil section is well demonstrated by its ability to support the entire mass of a large aircraft such as the Airbus A380.

The lift and drag characteristics of many different aerofoil shapes, for a range of angles of attack, have been determined by measurements taken in wind tunnel tests, and catalogued (e.g. in Abbott and von Doenhoff, 1958). The lift and drag characteristics measured at each angle of attack can be described using non-dimensional **lift** and **drag coefficients** (C_L and C_D) or as **lift to drag ratios** (C_L/C_D). These are defined in Box 8.3.

BOX 8.3 Aerofoil sections and lift and drag coefficients

Note that the chord of an aerofoil section is also the same as the *width* of the blade in a wind turbine at a given position along the blade.

Drag coefficient (C_D)

The drag coefficient of an aerofoil is given by the following expression:

$$C_D = \frac{D}{0.5pV^2A_b}$$

where:

D is the drag force in newtons (N)

ρ is the air density in kilograms per cubic metre (kg m^{-3})

V is the velocity of the air approaching the aerofoil in metres per second (m s^{-1})

A_b is the blade area (i.e. chord × length) in square metres (m^2).

In the case of a blade element, the area is equal to the mean chord × length of the blade element.

Lift coefficient (C_L)

The lift coefficient of an aerofoil is given by the following expression:

$$C_L = \frac{L}{0.5pV^2A_b}$$

where L is the lift force in newtons.

The lift and drag coefficients of an aerofoil can be measured in a wind tunnel at different angles of attack and wind velocities. The results of such measurements can be presented in either tabular or graphical form as in Figure 8.23.

Each aerofoil has an angle of attack at which the lift to drag ratio (C_L/C_D) is at a maximum. This angle of attack results in the maximum force and is thus the most efficient setting of the blades of a HAWT. Consequently, plots of this ratio against angle of attack can be useful to turbine designers (Figure 8.23).

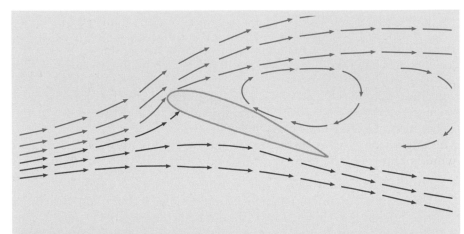

Figure 8.24 Aerofoil section in stall

Another important characteristic relationship of an aerofoil is its **stall angle**. This is the angle of attack at which the aerofoil exhibits stall behaviour. Stall occurs when the flow suddenly leaves the suction side of the aerofoil (when the angle of attack becomes too large), resulting in a dramatic loss in lift and an increase in drag (Figures 8.23 and 8.24). When this happens during the flight of an aeroplane, it can be extremely dangerous unless the pilot can make the plane recover. One of the methods used by wind turbines to limit the power extracted by the rotor in high winds takes advantage of this phenomenon; such turbines are known as stall regulated (see below for more details).

Aerofoils can also now be designed with the aid of specially developed software, and new aerofoils are being designed and optimized to be more efficient in the aerodynamic conditions experienced by wind turbines. Examples of wind turbine specific aerofoils include the DU airfoils (Rooij and Timmer, 2004) and NREL airfoils (Tangler and Somers, 1995). Figure 8.23 shows typical lift and drag coefficients, and lift to drag ratios, for one aerofoil section. Knowledge of these coefficients is essential when selecting appropriate aerofoil sections in wind turbine blade design. The lift and drag forces experienced are both proportional to the energy in the wind.

Relative wind velocity

When a wind turbine is stationary, the direction of the wind as 'seen' from a wind turbine blade is the same as the undisturbed wind direction. However, once the blade is moving, the direction from which it 'sees' the wind approaching effectively changes in proportion to the blade's velocity. (In the case of a moving vertical axis wind turbine blade, the direction from which the blade 'sees' the wind is additionally affected by its position during its rotation cycle – see Figure 8.27). Two-dimensional **vectors** are used to represent this effect graphically. A two-dimensional vector is a quantity that has both magnitude and direction. A velocity vector can be represented graphically in the form of an arrow, the length of which is proportional to speed, and the angular position of which indicates the direction of flow.

The wind as seen from a point on a moving blade is known as the **relative wind**, and its velocity is known as the **relative wind velocity** (usually symbolized by W). This is a vector which is the resultant (i.e. the vector sum) of the **wind velocity at the rotor**, V_1 (i.e. the **undisturbed wind velocity vector**, V_0, reduced by a factor known as the axial interference factor) and the tangential velocity vector of the blade at that point on the blade, u (see Box 8.4). Note that the tangential velocity, measured in metres per second (m s^{-1}), is distinct from the angular velocity, which is measured in radians per second or in revolutions per minute (Box 8.2).

The angle from which the point on the moving blade sees the relative wind is known as the **relative wind angle** (usually symbolized by ϕ) and is measured from the tangential velocity vector, u. The **blade pitch angle**, β, at this point on the blade is the relative wind angle minus the angle of attack, α, at that point on the blade (see Box 8.4).

Harnessing aerodynamic forces

Modern horizontal and vertical axis wind turbines make use of the aerodynamic forces generated by aerofoils in order to extract power from the wind, but each harnesses these forces in a different way.

In the case of a HAWT with fixed-pitch blades with its rotor axis assumed to be in constant alignment with the undisturbed wind direction, for a given wind speed and constant rotation speed, the angle of attack at a given position on the rotor blade *stays constant throughout its rotation cycle* (Box 8.4).

BOX 8.4 **HAWT rotor blades wind forces and velocities**

Figure 8.25 shows a section through a moving rotor blade of a HAWT. Also shown is a vector diagram of the forces and velocities at a position along the blade at an instant in time.

Because the blade is in motion, the direction from which the blade 'sees' the relative wind velocity, W, is the resultant of the tangential velocity, u, of the blade at that position and the wind velocity at the rotor, V_1. Note that in this diagram, the direction of u is shown in the opposite direction to the direction of motion of the rotating blade, in the same manner that a flag on a motor boat moving in calm weather points to the stern of the boat, showing the air to be 'flowing' from the opposite direction to the boat's forward motion.

The wind velocity at the rotor, V_1, is the undisturbed wind velocity upstream of the rotor, V_0, reduced by a factor that takes account of the wind being slowed down as a result of power extraction. This factor is often referred to as the **axial interference factor**, and is represented by a.

The tangential velocity, u, (in metres per second) at a point along the blade is the product of the angular velocity, Ω (in radians per second) of the rotor and the local radius, r, (in metres), at that point, that is:

$$u = \Omega r \tag{3}$$

Albert Betz showed in 1928 that the maximum fraction of the power in the wind that can theoretically be extracted is 16/27 (59.3%). This occurs when the undisturbed wind velocity is reduced by one-third, in other words, when the axial interference factor, a, is equal to one-third. The value of 59.3% is often referred to as the **Betz limit**.

The relative wind angle, ϕ, is the angle that the relative wind makes with the blade (at a particular point with local radius, r, along the blade) and is measured from the plane of rotation. (Note: if it were not for the fact that the wind is slowed down as a result of the wind turbine extracting energy – in other words if V_0 was not reduced to V_1 at the rotor – the tip speed ratio would be equal to the reciprocal of the tangent of the relative wind angle at the blade.) The angle of attack, α, at this point on the blade can be measured against the relative wind angle, ϕ. The blade pitch angle (usually represented by β) is then *equal* to the relative wind angle *minus* the angle of attack. Since the rotor is constrained

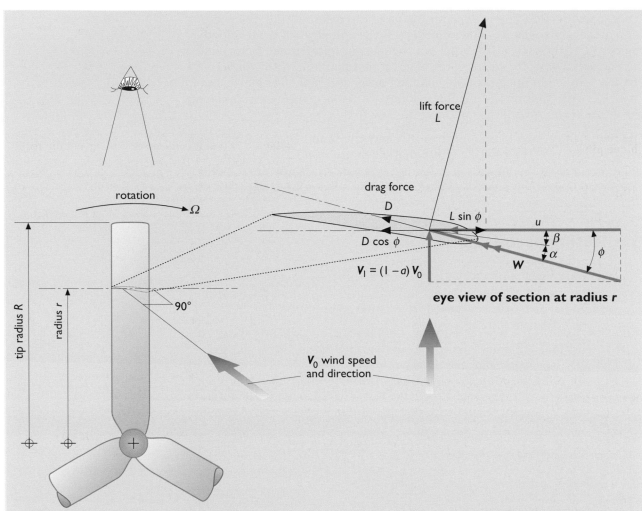

Figure 8.25 Vector diagram showing a section through a moving HAWT rotor blade. Notice that the drag force, D, at the point shown is acting in line with the direction of the relative wind, W, and the lift force, L, is acting at 90° to it

to rotate in a plane at right angles to the undisturbed wind, the driving force at a given point on the blade is that component of the aerofoil lift force that *acts in the plane of rotation*. This is given by the product of the lift force, L, and the sine of the relative wind angle, ϕ (that is, $L \sin \phi$). The component of the drag force in the rotor plane at this point is the product of the drag force, D, and the cosine of the relative wind angle, ϕ (that is, $D \cos \phi$).

The torque, q (that is, the moment about the centre of rotation of the rotor in the plane of the rotor), in newton metres (N m) at this point on the blade is equal to the *product* of the *net driving force in the plane of rotation* (that is, the component of lift force in the plane of rotation minus the component of the drag force in the rotor plane) and the local radius, r. The total torque, Q, acting on the rotor can be calculated by summing the torque at all points along the length of the blade and multiplying by the *number* of blades. The power from

the rotor is the *product* of the total torque, Q and the rotor's angular velocity, Ω.

Why are rotor blades twisted?

The magnitude and direction of the relative wind angle, ϕ, varies along the length of the blade according to the local radius, r. Equation (3) shows that the tangential velocity varies with radius, so as the tangential speed *decreases* towards the hub, the relative wind angle, ϕ, *progressively increases* (see Figure 8.26). A HAWT rotor designed for optimum performance will have a tapered blade, and to have a constant angle of attack along its length (assuming the same aerofoil section is used throughout its length), it will have to have a built-in twist. The amount of twist will vary (as the relative wind angle varies) progressively from tip to root. Figure 8.26 demonstrates the progressive twist of such a HAWT rotor blade. Most manufacturers of HAWT blades use tapered and twisted blades, although it is possible to build functional HAWT rotor blades that are not twisted. These are cheaper, but less efficient and how well they function depends in part on both the aerofoil characteristics and the overall blade pitch angle.

Blade pitch

As well as the blade pitch angle defined above, the term **blade pitch** also refers to the whole blade's angular position about the blade's longitudinal axis (also known as the pitch axis) such that in the case of a **variable pitch** rotor blade, the whole blade is able to be rotated about its pitch axis. In most cases all of the blades of a variable pitch HAWT rotor change pitch at the same time in order to:

optimise the turbine's power production across a range of wind speeds (in order to maintain the angle

of attack at or near to the optimal angle across a range of wind speeds)

reduce its output at high wind speeds

to stop the rotor during very high wind speeds or to 'park' the rotor (such that the blade pitch is in its 'feathered' position, e.g. each blade pitch is at or near 90 degrees, relative to the plane of rotation), when it is necessary to prevent the rotor from operating for any reason.

In the case of a **fixed pitch** rotor the blade pitch angle remains unchanged, which makes them less productive and less controllable. Most large wind turbines employ variable pitch blades, but most micro wind turbines and small wind turbines and some medium-scale wind turbines use fixed pitch blades.

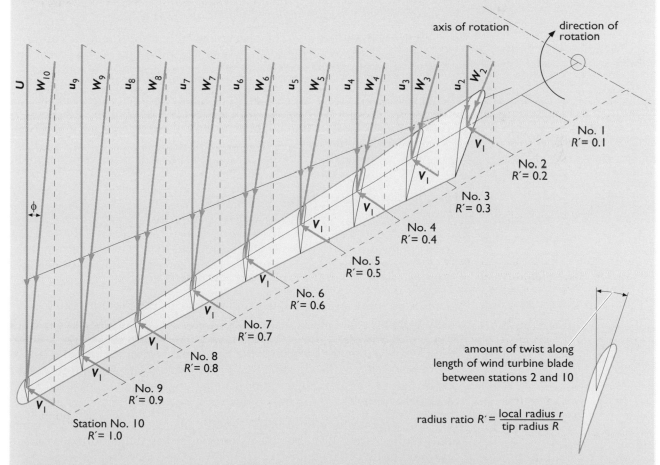

Figure 8.26 Three-dimensional view of an optimally tapered and twisted HAWT rotor blade design (the blade is shown in a horizontal position and moving through the upward part of its cycle about its axis of rotation). The figure shows how the relative wind angle, ϕ, changes along the blade span (length). Note that the blade aerofoil section and the angle of attack are assumed to be constant along the length of the blade. The diagram (lower right) of the view along the blade indicating the amount of built-in twist along the length of the blade shows the blade cross-section at station 2: the cross section of the blade at station 10 has been omitted for clarity

Stall control of wind turbines

Let us assume that a wind turbine is rotating at a constant rotation speed, regardless of wind speed, and that the blade pitch angle is fixed. As the wind speed *increases* the tip speed ratio *decreases*. At the same time, the relative wind angle *increases*, causing an increase in the angle of attack.

It is possible to take advantage of this characteristic to control a turbine in high winds, if the rotor blades are designed so that above the rated wind speed they become less efficient because the angle of attack approaches the stall angle. This results in a loss of lift, and thus torque, on the regions of the blade that are in 'stall'.

This method of so-called **stall regulation** has been employed successfully on numerous fixed-pitch HAWT rotors and has also been employed on most modern lift-driven aerofoil-based VAWT rotors. However, as turbines have increased in size this approach to power regulation has generally been discontinued in HAWTs in favour of variable speed turbines with variable pitch regulation. The main reasons for this are that it is very hard for designers to predict exactly when a given rotor will stall and that variable pitch rotors are more efficient over a range of wind speeds.

By contrast, in an aerofoil-based VAWT with fixed pitch blades, under the same conditions, the angle of attack at a given position on the rotor blade is *constantly varying throughout its rotation cycle.*

During the normal operation of a horizontal axis rotor, the direction from which the aerofoil 'sees' the wind is such that the angle of attack remains positive throughout.

In the case of an aerofoil-based vertical axis rotor, however, the angle of attack changes from positive to negative and back again over each rotation cycle. This means that the 'suction' side reverses during each cycle, so a symmetrical aerofoil has to be employed to ensure that power can be produced irrespective of whether the angle of attack is positive or negative.

Horizontal axis wind turbines

Most horizontal axis wind turbines are axial flow devices – the rotation axis is maintained in line with wind direction by a 'yawing' mechanism, which constantly realigns the wind turbine rotor in response to changes in wind direction.

In addition to its swept area and rotor diameter, the performance (power output, torque and rotation speed) of an axial flow horizontal axis wind turbine rotor is dependent on numerous other factors. These include the number and shape of the blades, the choice of aerofoil section, the length of the blade chord, the tip speed ratio, the blade pitch angle, the relative wind angle and angle of attack at positions along the blade, and the amount of blade twist between the hub and tip (see Box 8.4).

Box 8.4 explains how the relative wind velocity, W, and relative wind angle, ϕ, both vary along the blade, and describes their influence on the optimum blade pitch angle.

A few examples of cross-flow horizontal axis windmills have also been proposed historically (see Figure 8.9) and cross-flow horizontal axis turbines are under development for use in building integrated wind energy systems (see Section 8.8 and Taylor, 1998). In this context, cross-flow horizontal axis turbines function in a similar manner to the vertical axis wind turbines described below.

Vertical axis wind turbines

Modern VAWTs, unlike HAWTs, are 'cross-flow devices'. This means that the direction from which the undisturbed wind flow comes is at right angles to the axis of rotation; that is, the wind flows across the axis. As the rotor blades turn, they sweep a three-dimensional surface (an ovoid-like surface in the case of the Darrieus VAWT, a cylindrical surface in the case of an H-VAWT or a conical surface in the case of a V-VAWT), as distinct from the single circular plane swept by a HAWT's rotor blades.

In contrast to traditional vertical-axis windmills, the blades of modern aerofoil-based vertical axis wind turbines extract most of the power from the wind as they pass across the front and rear – as distinct from one side (relative to the undisturbed wind direction) of the swept surface.

A vertical axis wind turbine will function with the wind blowing from any compass direction, but let us assume initially that it is blowing from one particular direction and also that the setting angle of the blade is such that its chord is in line with a tangent to the circular path of rotation (that is, it has 'zero set pitch'). Clearly, the angle of the blade, in relation to the direction of the undisturbed wind, changes from zero to 360 degrees over each cycle of rotation. It might appear that the angle of attack of the wind to the blade would vary by the same amount, and so it might seem impossible for an aerofoil-based VAWT to operate at all. However, we have to take into account the fact that when the blade is moving, the relative wind angle 'seen' by the blade is the resultant (W) of the wind velocity V_1 at the rotor and the blade velocity u (see Box 8.4). Provided that the blade is moving sufficiently fast relative to the wind velocity (in practice, this means at a tip speed ratio of three or more), the angle of attack that the blade makes with the relative wind velocity W will only vary within a small range (see Figure 8.27).

V_1 = wind velocity at rotor

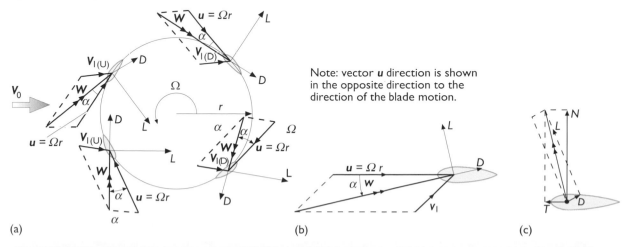

Note: vector u direction is shown in the opposite direction to the direction of the blade motion.

(a) (b) (c)

Figure 8.27 The lift and drag forces acting on VAWT rotor blades can be resolved into two components: 'normal', N, (that is, in line with the radius) and 'tangential', T, (that is perpendicular to the radius). The magnitude of both components varies as the angle of attack varies: (a) blade forces and relative velocities for a VAWT, showing angles of attack at different positions; (b) detail of aerodynamic forces on a blade element of a VAWT rotor blade; (c) normal (radial) and tangential (chord-wise) components of force on a VAWT blade. Note $V_{1(U)}$ is the wind velocity at the rotor on the upwind side; $V_{1(d)}$ is the wind velocity at the rotor on the downwind side

8.5 **Power and energy from wind turbines**

How much power does a wind turbine produce?

The power output of a wind turbine varies with wind speed: every turbine has a characteristic wind speed–power curve, often simply called the **power curve**. The shape of a wind speed–power curve is influenced by the:

- rotor swept area
- choice of aerofoil
- number of blades
- blade shape
- optimum tip speed ratio
- speed of rotation
- cut-in wind speed (the wind speed at which a turbine begins to generate power)
- rated wind speed (the wind speed at which a turbine generates its rated power)
- shut-down or cut-out wind speed (the wind speed at which a turbine is shut down and stops generating – also known as the furling wind speed)
- aerodynamic efficiency (power coefficient)
- gearing efficiency, and
- generator efficiency.

An example of such a curve is shown in Figure 8.28.

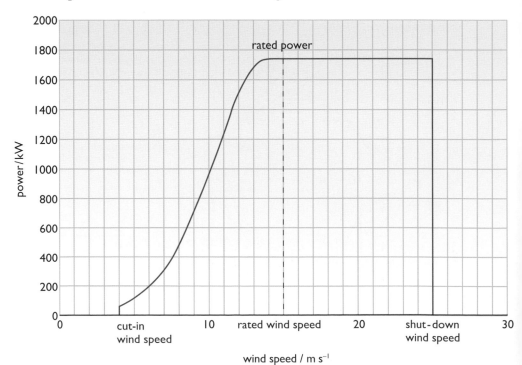

Figure 8.28 Typical wind turbine wind speed–power curve

How much energy will a wind turbine produce?

The energy that a wind turbine will produce depends on both its wind speed–power curve and the **wind speed frequency distribution** at the site. The latter is essentially a graph or histogram showing the number of hours for which the wind blows at different wind speeds during a given period of time. Figure 8.29 shows a typical wind speed frequency distribution.

For each incremental wind speed within the operating range of the turbine (that is, between the cut-in wind speed and the shut-down wind speed), the energy produced at that wind speed can be obtained by multiplying the number of hours of its duration by the corresponding turbine power at this wind speed (given by the turbine's wind speed–power curve). This data can then be used to plot a **wind energy distribution** such as that shown in Figure 8.30. The total energy produced in a given period is then calculated by summing the energy produced at all the wind speeds within the operating range of the turbine.

The best way to determine the wind speed distribution at a site is to carry out wind speed measurements with equipment that records the number of hours for which the wind speed lies within each given 1 m s^{-1} wide speed band, e.g. 0–1 m s^{-1}, 1–2 m s^{-1}, 2–3 m s^{-1}, etc.

The longer the period over which measurements are taken, the more accurate is the estimate of the wind speed frequency distribution. As the power in the wind is proportional to the cube of the wind velocity – see Equation (2) – a small error in estimating the wind speeds can produce a large error in the estimate of the energy yield.

Additional factors that affect the total energy generated include transmission losses and the availability of the turbine. **Availability** is an indication of the reliability of the turbine installation and is the fraction or percentage

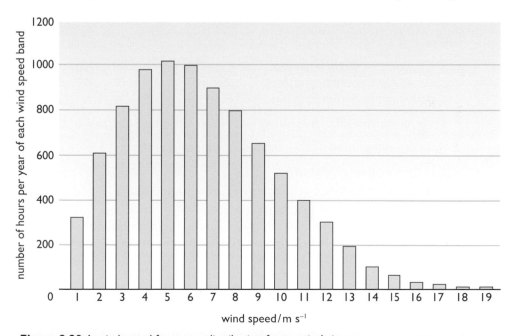

Figure 8.29 A wind speed frequency distribution for a typical site

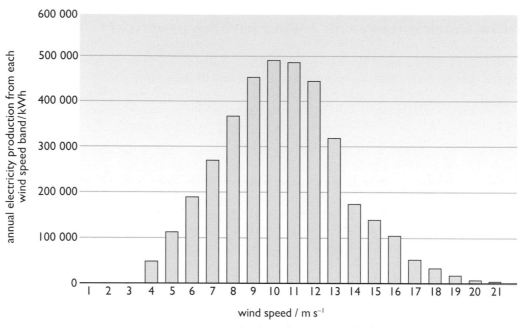

Figure 8.30 Wind energy distribution for the same site as in Figure 8.29, showing energy produced at this site by a wind turbine with the wind speed–power curve shown in Figure 8.28

of a given period of time for which a wind turbine is available to generate, when the wind is blowing within the turbine's operating range. Current commercial wind turbines typically have annual availabilities in excess of 90%, many have operated at over 95% and some are achieving 98%.

If the mean annual wind speed at a site is known, or can be estimated, the following formula (Beurskens and Jensen, 2001) can be used to make a rough *initial estimate* of the electricity production (in kilowatt-hours per year) from a number of wind turbines:

Annual electricity production = $K\,V_m^3\;A_t\,T$

where:

   $K = 3.2$ and is a factor based on typical turbine performance characteristics and an approximate relationship between mean wind speed and wind speed frequency distribution (see below)

   V_m is the annual mean wind speed at the site in metres per second

   A_t is the swept area of the turbine in square metres

   T is the number of turbines.

This formula should be used with great caution, however, because it is based on an average of the characteristics of the medium- to large-scale wind turbines currently available and assumes an approximate relationship between annual mean wind speed and the frequency distribution of wind speeds that may not be accurate for an individual site. It also does not allow for the different power curves of wind turbines that have been optimized either for low or high wind speed sites. The K factor of 3.2 given above assumes a well designed turbine suited to its site (Beurskens and Jensen, 2001), but it should not be used with small-scale wind turbines as their performance varies greatly and they are more likely to be located at sites with lower annual mean wind speeds and potentially in suburban and urban areas.

For a small wind turbine (less than 200 m^2 swept area) estimates of potential electricity production can be derived if the supplier provides a British Wind Energy Association Reference Annual Energy (RAE) value (BWEA, 2008) or an American Wind Energy Association Rated Annual Energy (RAE) value together with an AEP (Annual Energy Production) curve (AWEA, 2009). However there will still be a high level of uncertainty in urban areas. Knowing the IEC 61400 wind speed class that the turbine has been tested for also helps to estimate energy production.

Estimating the wind speed characteristics of a site

It is expensive to carry out detailed measurements at a site and wind speed measurements are often not carried out for small wind turbine installations. However the use of remote sensing methods such as SODAR (SOnic Detection And Ranging) and Doppler LIDAR (LIght Detection And Ranging) makes it feasible to monitor wind speeds without the need for tall towers. In addition, lower cost instrumentation is becoming available for monitoring small wind turbine sites.

It is preferable to record the wind speed and direction as close as possible to the proposed site for at least a year. However this will only give information for a particular time period, and weather patterns change. In order to ascertain the longer term wind speed characteristics, it is useful to correlate the measured data with data measured at one or more nearby meteorological stations or other wind recording sites. Then by statistical analysis of the two data sets, and extrapolating over the long-term data from the meteorological station, an estimate of the longer term wind speed characteristics at the site can be made. This technique is referred to as the **Measure–Correlate–Predict** or **MCP** method.

There are a number of different ways of implementing the MCP methodology all based on different statistical analysis techniques (Rogers et al, 2005, 2006a). These methods are embedded into different software packages, but their application requires careful judgement – consistency in the use of the methods is important and more than one algorithm should be employed in order to avoid bias.

If it is not possible to carry out wind speed and direction measurements at a proposed site, or where a preliminary analysis is required prior to installing instrumentation, there are a number of techniques that can be employed to give an approximate estimate of the wind speed characteristics of a site.

Using wind speed measurements from a nearby location

This involves making use of existing wind speed measurements from one or more locations nearby and deriving the data for the proposed site by interpolation or extrapolation, taking into account differences between the proposed site and the sites for which measurements are available.

Wind resources at 50 m above ground level for five different topographic conditions										
		Sheltered terrain		Open plain		At a sea coast		Open sea		Hills and ridges	
		m s⁻¹	W m⁻²	m s⁻¹	W m⁻²	m s⁻¹	W m⁻²	m s⁻¹	W m⁻²	m s⁻¹	W m⁻²
Wind class	5	>6.0	>250	>7.5	>500	>8.5	>700	>9.0	>800	>11.5	>1800
	4	5.0–6.0	150–250	6.5–7.5	300–500	7.0–8.5	400–700	8.0–9.0	600–800	10.0–11.5	1200–1800
	3	4.5–5.0	100–150	5.5–6.5	200–300	6.0–7.0	250–400	7.0–8.0	400–600	8.5–10.0	700–1200
	2	3.5–4.5	50–100	4.5–5.5	100–200	5.0–6.0	150–250	5.5–7.0	200–400	7.0–8.5	400–700
	1	<3.5	<50	<4.5	<100	<5.0	<150	<5.5	<200	<7.0	<400

Figure 8.31 Annual mean wind speeds and wind energy resources over Europe (EU countries) combining land-based and offshore wind atlases (source: Troen and Petersen, 1989)

Using wind speed maps and atlases

Maps are available that give estimates of the mean wind speeds over the UK and many other countries. However, most of these maps were made using data from meteorological stations, which tend to be located in places that are often not appropriate for wind energy, so wind speed maps and atlases specifically for wind energy purposes have also been developed for many countries.

Using long-term wind measurements and the WAsP model mentioned below, a *European Wind Atlas* (Troen and Petersen, 1989) has been produced by the Risø Laboratory in Denmark for the European Commission. This document includes maps of various areas within the European Union (for example, Figure 8.31), which show the annual mean wind speed at 50 m above ground level for five different topographic conditions: sheltered terrain, open plain, sea coast, open sea, hills and ridges. The atlas includes a series of procedures for taking account of site characteristics to estimate the wind energy likely to be available. These procedures work quite well on sites with a gentle topography but are not so good for very hilly terrain or urban areas. A similar atlas (included in Figure 8.31) has also been produced to cover the *offshore* wind energy resource in the European Union (Risø, 2009). Similar wind atlases based on the same approach have also been produced for Russia, South Africa and parts of North Africa as well as a Global Wind Atlas. Also wind speed/energy atlases have been produced for Ireland, USA and Canada. There is a new higher resolution European Wind Atlas currently under development.

The Energy Technology Support Unit (ETSU) also prepared a wind atlas and database of the UK. Using wind speed data from meteorological stations, a digital terrain model of the UK and a wind speed prediction computer model known as NOABL (Numerical Objective Analysis of Boundary Layer), ETSU estimated an annual mean wind speed (AMWS) value for each 1 km × 1 km Ordnance Survey grid square in the UK (Burch and Ravenscroft, 1992). Whilst this is a useful atlas/database for rural areas, it consistently over predicts the AMWS in urban and suburban areas and (like most wind atlases developed for wind energy development in windy areas) it should not be used for that purpose. The UK Microgeneration Installation Standard, MIS 3003, (MIS, 2015 and MCS, 2015), include some adjustment factors that try to take account of this for small or micro wind turbines, but it is still not very reliable when used in those situations. This atlas is no longer being updated, but, at the time of writing, can still be accessed via the RenewableUK and Department for Climate Change (DECC) websites (RenewableUK, 2017; DECC, 2017).

Because of the unreliability of the NOABL database, the UK Energy Savings Trust has made available the wind speed data for Scotland that it accumulated whilst carrying out field trials of domestic small wind turbines. This database of AMWS can be accessed by entering a Scottish postcode together with the rural, suburban or urban site classification (EST, 2016).

Wind flow simulation computer models

A number of computer models have been developed that aim to predict the effects of topography on wind speed. Data from the nearest wind speed measurement station, together with a description of its site, is required and local effects are taken into account to arrive at estimated wind data for the proposed wind turbine site. Examples include NOABL and WAsP. As described above, NOABL was used in the development of the UK wind atlas/database and WAsP was used in the development of the *European Wind Atlas* (Figure 8.31). WAsP also forms the basis of at least two proprietary wind speed assessment computer software models. There are also improved versions of WAsP, for example for offshore use, and some that are better suited to modelling wind speeds in complex terrain.

CFD (computational fluid dynamics) is increasingly used to model wind flows in complex terrain, over and around forests and for designing wind farms. Berge et al., (2006) compares two CFD models with WAsP in modelling complex terrain.

Used with care, CFD models can be useful for carrying out initial assessments though they are complex, computer intensive, require training and need to be well understood. They tend to be used for wind farm projects in difficult terrain. There is also a need to have access to high quality digital terrain maps (special types of files containing three-dimensional map data), although some models are able to utilize/synchronize with Google Earth.

8.6 Environmental impact

Wind energy development has both positive and negative environmental impacts. The scale of its future implementation will rely on successfully maximizing the positive impacts whilst keeping the negative impacts to the minimum (see for example the US National Research Council's report for Congress, (NRC, 2007) which gives a comprehensive overview of the positive and negative environmental impacts of wind energy).

The generation of electricity by wind turbines does *not* involve the release of carbon dioxide or pollutants that cause acid rain or smog or respiratory/cardiovascular/dementia, diseases/deaths, that are radioactive, or that contaminate land, sea or water courses. Large-scale implementation of wind energy within the UK would probably be one of the most economic and rapid means of reducing carbon dioxide emissions and air pollution from fossil fuels - which is a dominant cause of ill health/death. According to WHO (WHO, 2014 and WHO, 2016), 7 million premature deaths per year globally are attributable to air pollution, principally from combustion and according to Kendrick (2017) air pollution may well be as important, and likely a more important link to heart disease than dietary factors. It can therefore be inferred that a major benefit from wider deployment of wind power generation will be helping to improve the health of large numbers of people from the resulting cleaner air quality. Over its working lifetime, a wind turbine can generate approximately 40 to 80 times the energy required to produce it (see Everett et al., 2012).

Some 44% of the water used in the EU is used to cool thermal or nuclear power plants, but wind turbines do not require the consumption of water and their use saved 387 million cubic metres of water in the EU in 2012 (EWEA, 2014). This benefit could be of growing importance if water shortages occur with increasing frequency in the future. (One idea, when there is excess wind power available than is needed by the electricity grid, is to use the wind power to purify water or to pump water around the network, or between reservoirs, or possibly as a flood amelioration strategy.)

Of course wind power is not without certain negative (or perceived negative) repercussions and the following subsections will look into the following issues:

- noise
- electromagnetic interference
- aviation related issues
- wildlife
- public attitudes and planning.

Wind turbine noise

Whilst wind turbines are often described as noisy by opponents of wind energy, in general they are not especially noisy compared with other machines of similar power rating (see Table 8.1 and Figure 8.32). However, there have been incidents where wind turbine noise has been cited as a nuisance. Currently available modern wind turbines are generally much quieter than their predecessors and conform to noise immission level requirements (see below). **Noise immission** is a measure of the cumulative noise energy to which an individual is exposed over time. It is equal to the average noise level to which the person has been exposed, in decibels, plus 10 times the logarithm (log10) of the number of years for which the individual is exposed, (Note: Figure 8.32 shows indicative sound levels interpolated from the pictorial positions along the horizontal line. For more precise values see Table 8.1 and its source reference.)

Table 8.1 Noise of different activities compared with wind turbines

Source/activity	Noise level in dB(A)*
Threshold of pain	140
Jet aircraft at 250 m	105
Pneumatic drill at 7 m	95
Truck at 48 km h⁻¹ (30 mph) at 100 m	65
Busy general office	60
Car at 64 km h⁻¹ (40 mph) at 100 m	55
Wind farm at 350 m	35–45
Quiet bedroom	20
Rural night-time background	20–40
Threshold of hearing	0

*dB(A): decibels (acoustically weighted to take into account that the human ear is not equally sensitive to all frequencies)
Source: ODPM, 2004b and SDC, 2005

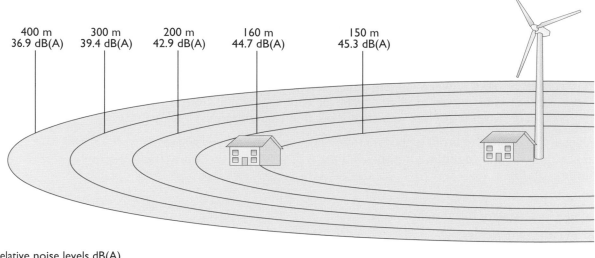

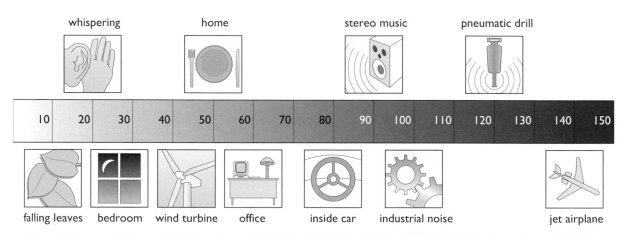

Figure 8.32 Wind turbine noise pattern from a typical wind turbine (source: EWEA, 1991)

There are two main sources of wind turbine noise. One is that produced by mechanical or electrical equipment, such as the gearbox and the generator, known as **mechanical noise**; the other is due to the interaction of the air flow with the blade, referred to as **aerodynamic noise**.

If noise does occur, then mechanical noise is usually the main problem, but it can be remedied fairly easily by the use of quieter gears, mounting equipment on resilient mounts, and using acoustic enclosures – or eliminating the gearbox altogether by opting for a direct drive low speed generator, with more and more wind turbines utilizing this option with the added benefit of improved efficiency.

Aerodynamic noise

The aerodynamic noise produced by wind turbines can perhaps best be described as a 'swishing' sound. It is affected by: the shape of the blades; the interaction of the airflow with the blades and the tower; the shape of the blade trailing edge; the tip shape; whether or not the blade is operating in stall conditions; and turbulent wind conditions, which can cause unsteady forces on the blades, causing them to radiate noise.

Aerodynamic noise will tend to increase with the speed of rotation. For this reason, some turbines are designed to be operated at lower rotation speeds during periods of low wind. Noise nuisance is usually more of a potential risk in light winds: at higher wind speeds, background wind noise tends to mask wind turbine noise. Increasing the numbers of blades is also likely to reduce aerodynamic noise.

Noise regulations, standards, controls and reduction

Most commercial wind turbines undergo noise measurement tests in accordance with one of the following:

■ the recommended procedure developed by the International Energy Agency (Ljunggren and Gustafsson, 1994; Ljunggren, 1997). This procedure lies behind an online computer model provided by the National Physical Laboratory (NPL, 2016)

■ a procedure conforming to the Danish noise regulations (see below)

■ the method documented in the IEC (International Electrotechnical Commission) international standard 61400-11 (IEC, 2012).

The measured noise level values from such tests provide information that enables the turbines to be sited at a sufficient distance from habitations to minimize (or avoid) the risks of noise nuisance. This standard procedure also allows manufacturers to identify any noise problem and take remedial action before the commercial launch of the machine.

In Denmark, in order to control the effects of noise from wind turbines, there is a standard that specifies that the maximum wind turbine noise level permitted at the nearest dwelling in open countryside should be 44 dB(A) at 8 m/s wind velocity and 42 dB(A) at 6 m/s. At habitations in residential areas a noise level of only 39 dB(A) and dB(A) at 8 and 6 m/s respectively is permitted. For low frequency noise heard indoors there is a limit of 20dB at 6 and 8 m/s (LCTU, 2013). These noise limits have been demonstrated to be achievable with commercially available turbines.

In the UK, the current guidance (principally for medium- and large-scale turbines) is that noise limits should be set relative to background noise with different limits for daytime and night-time (35 – 40 dB(A) for daytime and 43 dB(A) for night-time). The decision of which daytime limit to use – either 35 or 40 dB(A) – depends on the following:

■ the number of dwellings in the neighbourhood of the wind farm

■ the effect of noise limits on the number of kWh generated

■ the duration and level of exposure.

For micro/small wind turbines the noise limit is slightly higher at 45dB LAEQ 5 min (i.e. the equivalent continuous A-weighted sound level over a 5 minute period) at 1 metre from the window of a habitable room in the façade of any neighbouring residential property (DCLG, 2007 and 2009).

Turbines that comply with the MCS Planning Standard (MCS, 2015) must provide a noise immission map – drawn up using specified methods

(BSI, 2013) – that shows the noise at different distances from the hub for different wind speeds across the range of operation of the turbine. At the time of writing, further information on wind turbine siting in the UK is provided in planning guidance on renewable and low carbon energy (Planning Guidance, 2015) that accompanies the UK National Planning Policy Framework (DCLG, 2012). Fiumicelli and Triner (2011) provide extensive information together with a wind farm noise complaint methodology. The Microgeneration Certification Scheme (MCS) discussed in the section on small-scale wind turbines (Section 8.8) also addresses this topic. See also the RenewableUK planning guidance for small wind turbines (RenewableUK, 2011a and 2011b) and the Sustainable Development Commission's wind power guide (SDC, 2005).

Noise is a sensitive issue. Unless it is given careful consideration at both the turbine design and the project planning stages, taking into account the concerns of people who may be affected, opposition to wind energy development is likely.

By eliminating the gearbox, mechanical turbine noise can be considerably reduced – as mentioned previously, some manufacturers have developed gearbox free turbines with low speed generators directly coupled to the rotor. This makes the turbines very quiet and able to be more comfortably sited close to buildings.

There is much ongoing research to reduce wind turbine noise. A useful overview of this work is included in Legerton (1992) and refinements in methods of measurement and noise propagation modelling are described in Kragh et al. (1999), Rogers et al. (2006b) and IEC (2012). Further useful work is being carried out by the Institute of Acoustics who have produced a good practice guide on the application of ETSU-R-97 (IOA, 2013) plus a method for rating amplitude modulation in wind turbine noise, (IOA, 2016). The EWEA/WindEurope also provides extensive information with regard to reducing wind turbine noise (EWEA, 2012). There have been concerns raised periodically about possible health effects from wind turbine noise, but a scientific investigation by Health Canada concluded that there was no evidence for a link between exposure to wind turbine noise and negative health effects in people and that wind turbine noise did not have any measurable effect on illness and chronic disease, stress and sleep quality (Health Canada, 2014).

Electromagnetic interference

When a wind turbine is positioned between a radio, television or microwave transmitter and receiver (Figure 8.33) it can sometimes reflect some of the electromagnetic radiation in such a way that the reflected wave interferes with the original signal as it arrives at the receiver. This can cause the received signal to be distorted significantly.

The extent of electromagnetic interference caused by a wind turbine depends mainly on the materials used to make the blades and on the surface shape of the tower. If the turbine has metal blades (or glass-reinforced plastic blades containing metal components), electromagnetic interference may occur if it is located close to a radio communications service. The laminated timber blades used in some turbines absorb rather than reflect radio waves so do not generally present a problem. There has also been research into applying radar absorbing materials (RAMs) into blades. Faceted towers reflect more than smooth rounded towers, due to their flat surfaces.

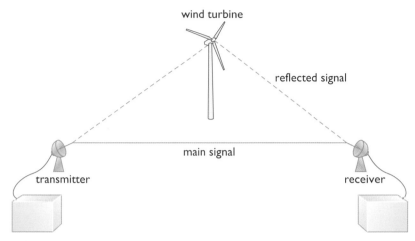

Figure 8.33 Scattering of radio signals by a wind turbine

Chignell (1987) and ODPM (2004b) give some simplified guidance about how to prevent/minimize electromagnetic interference, and Ofcom have examined the impact of tall structures, including wind turbines, on broadcasting and other wireless services (Ofcom, 2009).

The most likely form of electromagnetic interference is to television reception. This is relatively easily dealt with by the installation of relay transmitters or by connecting cable television services to the affected viewers. In the UK, the BBC used to provide an online assessment tool to evaluate the impact of wind turbines on TV broadcast services, but this is not currently available, though on its 'Help Receiving TV and Radio' website it states that 'We are not aware of any problems caused to TV reception by small domestic wind turbine installations.' (BBC, 2016).

Microwave links, very high frequency (VHF) omni-directional ranging (VOR) and instrument landing systems (ILS) can also be affected by wind turbines. A method of determining an acceptable exclusion zone around radio transmission links has been developed which takes account of the characteristics of antennae (Bacon, 2002). In the UK, Ofcom commissioned detailed research into potential wind farm interference to fixed link and scanning telemetry devices (Randhawa and Rudd, 2009); at the time of writing Ofcom maintains a web page that lists necessary procedures for developers and links to certain publications that provide guidance on wind farms and electromagnetic interference (Ofcom, 2017). The Joint Radio Company has also produced further guidance information (JRC 2014a, 2014b and 2016). See also the RenewableUK planning guidance for small wind turbines (RenewableUK, 2011a and 2011b) and the Sustainable Development Commission's wind power guide (SDC, 2005).

Wind turbines and aviation

According to a RenewableUK members survey cited on the BEIS web site 'approximately 3 GW of wind farm applications submitted for planning approval in 2013 were subject to objections or conditions from the aviation sector. The industry estimates that radar issues account for approximately 12 GW of objections in the planning system as a whole.' (BEIS, 2013).

The UK Ministry of Defence (MOD) has voiced concern about the interference with military radar that could be caused by wind turbines. In addition, the MOD is concerned that wind turbines (particularly those with large diameters and tall towers), when located in certain areas, will penetrate the lower portion of the low flying zones used by military aircraft. These various concerns have led to the MOD's intervention which has impeded the development of a number of wind farms in the UK. NATS (National Air Traffic Services Ltd) has produced a series of maps that indicate the areas around the UK where radar interference may be considered a potential hazard to aviation. The maps cover different wind turbine tip heights (20 m to 200 m).

NATS maintains a website (NATS, 2016) giving information about wind turbines and aviation, including a series of maps from NATS and MOD that show the consultation zone areas in the UK for which NATS requires notification of wind turbine planning applications. The Civil Aviation Authority (CAA) has produced a document that provides guidance and a range of mitigation techniques (CAA, 2016), see also CAA (2015) and CAA (2010). The Airport Operators Association has also produced an advice note for safeguarding of aerodromes (AOA, 2016) and DECC produced an Aviation Plan that aims to enable the implementation of mitigation measures to reduce the impact of wind turbines on radar (DECC, 2015). The Defence Infrastructure Organisation maintains a website for guidance with regard to wind farms and MOD safeguarding, which also facilitates a pre-application consultation process (DIO, 2014).

There have been a number of studies (for example Jago and Taylor, 2002) aimed at clarifying the precise nature of the effects of wind turbines on radar. It seems the experience of a number of European countries is that the effect of wind turbines on military aviation was not a major problem, though it does seem to be becoming an issue in parts of the USA and this had led to a series of publications to address/mitigate the issues. These include Barrett *et al.*, (2014) and Barrett and DeVita, (2011).

The UK Government and the wind industry have funded a number of projects to address these issues and these seem to be arriving at solutions acceptable to the MOD and CAA. One involves adapting the design of wind turbine blades to include RAMs (radar absorbing materials). A joint project between QinetiQ and Vestas (Appleton, 2010 and Marsh, 2010) has tested a turbine equipped with a set of RAM blades in Norfolk and this 'stealthy' turbine has demonstrated a substantially reduced impact on radar. Vestas has now included turbines with stealthy blades in its product line to enable installations in radar sensitive locations, including a 96 MW project in France. Another approach is the development of filtering systems that can reliably filter out the interference from wind turbines. One such system is BAE's ADT (Advanced Digital Tracking) system (Butler, 2007) and this appears to be another viable solution which may be able to be deployed at affected radar sites around the UK.

The Airspace & Safety Initiative produced an overview of some of the issues with regard to aviation planning considerations for wind turbines (ASI, 2013), and RenewableUK has also produced a document that provides guidance with regard to offshore renewables and aviation (RenewableUK, 2016a).

Impact on wildlife

In the UK, a collaboration between English Nature, the Royal Society for the Protection of Birds (RSPB), the World Wide Fund for Nature (WWF) and the British Wind Energy Association (BWEA) yielded a guidance document (English Nature et al., 2001) for nature conservation organizations and developers when consulting over wind farm proposals in England. This covers nature conservation, environmental impact assessments, the planning process and a checklist of possible impacts of relevance to nature conservation (both flora and fauna). Since this report was published Natural England, the Countryside Council for Wales and Scottish Natural Heritage have all developed extensive guidance on wind energy and wildlife.

In the case of offshore wind turbines, there are concerns about the possible impact on fish, crustaceans, marine mammals, marine birds and migratory birds and these are the subject of ongoing research by a number of organizations including Natural England, Scottish Natural Heritage, COWRIE (Collaborative Offshore Wind Research Into the Environment) and CEFAS (Centre for Environment, Fisheries and Aquaculture Science) amongst others.

Elsewhere, the US Fish and Wildlife Service has produced guidelines (USFWS, 2012) providing advice and recommendations for land-based wind energy projects to follow to avoid or minimise impacts on wildlife and habitats.

The impact of wind turbines on bats has been of particular concern and is discussed further below.

Wind turbines and birds

In addition to potential disturbance, barriers and potential habitat loss, the main potential hazard to birds presented by wind turbines is that they could be killed by flying into the rotating blades (Drewitt and Langston, 2006).

So far the worst location for bird strikes has been the Altamont Pass in California, where raptor species have been killed. However, apart from bird strikes on wind turbines at Tarifa and Navarra in Spain, this raptor mortality does not seem to have been duplicated elsewhere, so it may be due to special circumstances.

Erickson compared avian mortality sources and reports that 200 000–350 000 bird collisions occur per year with wind turbines, 6.5 million with towers, 12–64 million with power lines, 89–380 million with roads/vehicles, and 365–988 million with buildings and windows. 1.4–3.7 billion are estimated to be killed by cats (Erickson, 2015).

The US FWS guidelines (USFWS, 2012), mentioned above, includes guidance to avoid and minimize the impacts of wind turbines on birds. In addition, the US National Wind Coordinating Collaborative has developed a guide for studying wind turbines and wildlife interactions (NWCC, 2011), a mitigation tool box (NWCC, 2007) and a birds and bat factsheet has been produced by the American Wind & Wildlife Institute (AWWI, 2016).

According to Natural England (2010), there is little evidence that wind farms in England have a significant impact on birds, but nonetheless Natural England and Scottish Natural Heritage provide guidance about wind turbines and birds, and post-construction monitoring of bird impacts.

Studies were carried out by Denmark's National Environmental Research Institute on the offshore wind farm at Tunø Knob (which was deliberately located where there was a large marine bird population, in order to monitor the interaction of birds – mainly eiders – and wind turbines). The institute's conclusions were that the eiders keep a safe distance from the turbines but are not scared away from their foraging areas – it was felt that the offshore wind turbines had no significant impact on water birds (NERI, 1998).

Nonetheless there are concerns that a substantial increase in the number of wind turbines could result in an increase in bird strikes, so projects will need to take account of bird sensitivity areas and to take particular care when locating turbines in bird migratory routes. The RSPB has produced bird sensitivity maps and guidance for England (Bright, Langston and Anthony, 2009) and Scotland (Bright et al., 2006), and an accompanying UK bird sensitivity map (RSPB, 2015) to inform developers about the appropriate siting of wind turbines. It is also now possible to install bird control radar systems which automatically detect approaching birds and, if there is a likelihood of collisions, bird deterrent devices can be activated, or the turbines slowed down or shut down until after the birds have passed.

In general, RSPB has been very supportive of wind projects and while it has opposed 6% of wind energy projects because they threaten bird populations, the RSPB installed an 800 kW wind turbine adjacent to its headquarters in Bedfordshire in order to generate an equivalent of 50% of the organisation's electricity needs at 127 sites in the UK. RSPB stated: 'We hope that by siting a wind turbine at our UK headquarters, we will demonstrate to others that with a thorough environmental assessment and the right planning and design, renewable energy and a healthy thriving environment can go hand in hand' (RSPB, 2016a), indicating that, if sensitively planned, wind turbines can coexist with wildlife.

The RSPB has some concerns with certain shallow sea locations, but it is also supportive of offshore wind farms – particularly in deeper water – and of the use of floating wind turbines so long as they are not negatively impacting seabirds or other marine wildlife (RSPB, 2016b).

To put the wind turbine impact on birds in context, a major impact on farmland species has been the detrimental effect of changes in agricultural practices, particularly the progressive change (since WW2) from pasture/ meadow-rich mixed farming to large-scale carbon-intensive/fossil-fuel-intensive mechanized arable agriculture. The 2016 State of Nature report assessed the risk of extinction for 1118 farmland species. Of 26 bird species, almost half (46%) are in danger of going extinct, including the corn bunting and the turtle dove and their numbers are still declining. Climate change has had a significant impact too, although its impact has been mixed, with both beneficial and detrimental effects on species. 'Nevertheless, we know that climate change is one of the greatest long-term threats to nature globally.' (Hayhow et al., 2016). Mark Eaton, a lead author of the report stated: 'We now know that farming practices over recent decades have had the single largest impact on the UK's wildlife.' (Marshall, 2016).

Wind turbines and bats

There is growing concern that wind turbines may have an impact on bats – particularly along migration routes.

Natural England produced interim guidance (Natural England, 2009a and 2009b) to help planners and wind turbine operators take account of potential impacts to bats when developing or assessing wind turbine developments.

The UK Bat Conservation Trust (BCT) produced a scoping report (BCT, 2009) with regard to bats and wind farms, but has also raised specific concerns about bats and micro wind turbines. BCT has released a Position Statement (BCT, 2017) to raise awareness of the potential hazards.

EUROBATS have also produced guidelines for consideration of bats in wind farm projects which were revised in 2014 (Rodrigues et al., 2014).

The UK Department of Agriculture, Food and Rural Affairs (DEFRA) funded Exeter University to research such impacts. This was called the National Bats and Wind Turbines Project and was completed in 2016. It focused on large projects and surveyed some 46 commercial wind turbine sites equally distributed (approximately) over England and Wales. This found there is still much uncertainty and variability and there is a need to improve the prediction of risk with further research using daily sampling of turbines, but there is one important finding:

> It is possible to be confident that most nights with wind speeds >5 ms[-1] will have no bat casualties ... A simple strategy that should be considered at all sites where technically feasible is to restrict the rotation of wind turbines below the cut-in speed (e.g. by feathering the blades). This will have a positive outcome for bats, as the amount of time the blades are turning at low wind speeds will be reduced, whilst also involving no loss of energy generation.
>
> (Mathews et al., 2016)

This would only need to be at night time so such a strategy where bats are at risk would seem sensible, though turbines designed for low mean wind speed sites may be impacted (as they may well have cut-in wind speeds below 5 ms[-1]) though these would be more likely to be single turbines. It seems the bats may be more impacted when there are larger numbers of turbines on a site.

Public attitudes to wind power/planning considerations

Additional environmental factors that should be considered in assessing the impact of a wind turbine installation include safety and shadow flicker (ODPM, 2004b), however there are also concerns about the visual aspect of turbines.

Visual impact and attitudes

The visual perception of a wind turbine or a wind farm is determined by a variety of factors. These will include physical parameters such as turbine size, turbine design, number of blades, colour, the number of turbines in a wind farm, the layout of the wind farm and the extent to which moving rotor blades attract attention. Figure 8.34 compares wind turbines and other large constructions in the UK.

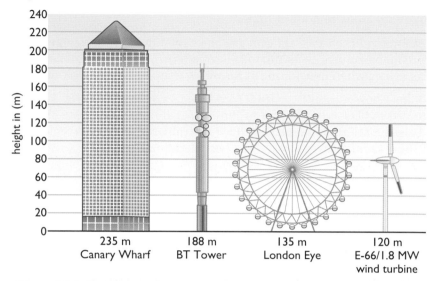

Figure 8.34 Comparison of wind turbines and other structures in the landscape

'Flicker' may be caused by sunlight interacting with rotating blades on sunny days. A comprehensive peer-reviewed study commissioned by DECC into wind-turbine related flicker showed that 'the frequency of flickering caused by wind turbines is such that it should not cause a significant risk to health' (PB, 2011).

An individual's overall perception of a wind energy project will also depend on a variety of less easily defined psychological and sociological parameters. These may include the individual's level of understanding of the technology, opinions on what sources of energy are desirable, and his or her level of involvement with the project. Newspaper and television reports are often the only source of information to which many people have access about wind energy, and these may well influence their opinions on the subject.

Much of the controversy about wind energy development has been due to opposition to changes to the *visual* appearance of the landscape. However, whether this is due to a visual dislike of wind turbines specifically, or simply to a general dislike of changes in the appearance of the landscape is often unclear. Resistance to the visual appearance of new structures or buildings is not a new phenomenon and opinions often change once structures become familiar.

Since the 1990s, over 60 independent survey projects have been carried out to monitor public attitudes and they have consistently shown that on average 70% to 80% support the development of wind farms in the UK (see for example NOP, 2005 and YouGov, 2010). Since July 2012, DECC, now BEIS, have been carrying out quarterly Public Attitude Tracker (PAT) surveys and the most recent in October 2016 in Wave 19 (BEIS, 2016) has shown a 'Total Support' score of 71% but a combined score of 92% (including 'Total support' + 'neither support or oppose' + 'Don't knows') for onshore wind (the highest such scores since the PAT survey commenced) with just 2% 'strongly opposed' and 6% 'opposed' to onshore wind. In the case of offshore wind, the 'Total support' score was 75% with a combined score of 95%. These were the highest such scores for offshore wind since the PAT survey commenced. In the Wave 17 (PAT) survey there was 56% agreement in favour of 'having a large scale renewable energy project' in the respondents' area, with an 82% combined score (including 'Total agree' + 'Neither agree or disagree' + 'Don't knows'), the second highest such scores since the PAT surveys commenced, whilst 9% 'slightly disagreed' and 9% 'strongly disagreed' (DECC, 2016).

However, there is still opposition to change and it is important for projects to be well designed and planned, and also to engage with local communities to provide trusted and reliable information together with meaningful community benefits.

Planning

Planning controls have a major influence on the deployment of wind turbines and some local authorities have developed policy guidelines about the planning aspects of wind energy. However, the planning aspects of wind energy have been treated differently in different locations.

The UK Government previously included planning guidance for wind energy in Planning Policy Statement 22 (PPS 22) (ODPM, 2004a), however, with the implementation of National Planning Policy Framework, (DCLG, 2012), PPS 22 was superseded, though it and its companion guide (ODPM, 2004b) do provide useful guidance. It has been replaced by Planning practice guidance for renewable and low carbon energy (DCLG, 2013).

Guidelines for developers and planners have also been prepared by Natural England, Scottish Natural Heritage and the Countryside Council for Wales.

A useful wind energy planning conditions guidance note was produced for the UK Department of Business, Enterprise and Regulatory Reform (BERR) by TNEI Services (TNEI, 2007). This summarizes the various factors and conditions that are likely to affect the granting of planning permission for a wind energy project.

However, the most major changes to planning policy for onshore wind (particularly projects greater than 50 MW) were announced by a Written Ministerial Instrument (WMS) in June 2015 (Clark, 2015). This states:

> When determining planning applications for wind energy development involving one or more wind turbines, local authorities should only grant planning permission if:
>
> • the development site is in an area identified as suitable for wind energy development in a Local or Neighbourhood Plan; and
>
> • following consultation, it can be demonstrated that the planning

impacts identified by affected local communities have been fully addressed and therefore the proposal has their backing.

In applying these new considerations, suitable areas for wind energy development will need to have been allocated clearly in a Local or Neighbourhood Plan. Maps showing the wind resource as favourable to wind turbines, or similar, will not be sufficient. Whether a proposal has the backing of the affected local community is a planning judgement for the local planning authority.

(Clark, 2015)

This has led to local planning authorities developing individual Supplementary Planning Guidance and whilst the (DCLG, 2013) advised that local planning authorities should not rule out otherwise acceptable renewable energy projects through inflexible rules on buffer zones or separation distances, many local planning authorities appear to have included very stringent separation distances such that future wind energy developments in such locations may now be difficult to achieve.

Small and micro wind turbines can be installed under 'permitted development' regulations (see MCS, 2015), though there have been a number of conditions imposed. Notably, the number of turbines allowed under Permitted Development is just one with a maximum swept area of 3.8 m^2 (i.e. in the case of HAWT, a maximum diameter of around 2.2 m). The maximum height if building–mounted is 15 m and 11.1 m if it is pole mounted (DCLG, 2015).

8.7 Economics

Calculating the costs of wind energy

The economic appraisal of wind energy involves a number of specific factors. These include:

- the annual energy production from the wind turbine installation;
- the capital cost of the installation;
- the discount rate being applied to the capital cost of the project (see Chapter 10 and Appendix B);
- the length of the contract with the purchaser of the electricity being produced;
- the number of years over which the investment in the project is to be recovered (or any loan repaid), which may be the same as the length of the contract;
- the operation and maintenance costs, including maintenance of the wind turbines, insurance, land leasing, offshore leasing etc.

A simple procedure for calculating the cost of wind energy is given in Appendix B. More information on costing and investing in energy projects can be found in Everett et al. (2012).

The estimates for land-based operation and maintenance costs are quite varied but seem to be equivalent to 5 to 8% of capital cost per year, though increasingly sophisticated condition monitoring systems are anticipated to improve reliability and reduce operations and maintenance (O&M) costs (Milborrow, 2013). It is extremely difficult to predict the O&M costs for offshore wind energy projects although one source (McMillan and Ault, 2007) estimates that they could be of the order of 3 to 5 times land-based costs. However there are a number of initiatives underway that are focused on reducing the cost of offshore wind energy O&M (and other) costs, which include Crown Estate (2012), UK Offshore Wind Cost Reduction Task Force (Jamieson, 2012), Carbon Trust (2008), Mathiesen (2014), ORE-Catapult (2016), ORE-Catapult, (2017) and Carbon Trust, (2016).

Greenacre et al. (2010) gives some indication of the difficulty in predicting offshore O&M costs as these may be influenced by a range of factors that include location, distance offshore, water depth, turbine redundancy and reliability, remote condition monitoring, uncertain 'weather windows' (especially in winter), material supply chains, currency exchange rates and vessel availability.

As we have seen, the annual energy produced by a wind turbine installation depends principally on the wind speed–power curve of the turbine, the (hub height), wind speed frequency distribution at the site, and the availability of the turbine.

As already noted in earlier chapters *capacity factor* is widely used to describe the productivity of a power plant over a given period of time.

If a wind turbine were able to operate at full rated power throughout the year, it would have an annual capacity factor of 100%. However, in reality, the wind does not blow constantly at the full-rated wind speed throughout the year, so in practice a wind turbine will have a much lower capacity factor. On moderate land-based wind sites in the UK, with annual mean wind speeds equivalent to half the rated wind speed, a turbine capacity factor of 0.25 (i.e. 25%) is typical. However, on better land-based wind sites, such as Carmarthen Bay in Wales, St Austell in Cornwall or the Orkney Islands, capacity factors of 0.35–0.40, are achievable, and 0.35–0.5 or more for offshore wind.

The capital cost of large and medium scaled land-based wind turbines currently ranges from approximately US\$950 to US\$1240 per kilowatt of output (WEC, 2016). With 15 to 20 year contracts and on sufficiently windy sites, wind energy is likely to be competitive with conventional forms of electricity generation, if the costs of the latter are calculated on a comparable basis. The cost of wind-generated electricity is very dependent on the way the plant is financed and this can strongly affect the price of the electricity produced.

The World Economic Forum (WEF) estimates that the Levelised Cost of Electricity (LCOE) of wind generated electricity is around \$50/MWh (\$0.05/kWh) and has reached grid parity in a number of countries. Indeed in some countries 'it has become more economical to install wind capacity than coal capacity' (WEF, 2017). WEF estimates:

> more than 30 countries have reached grid parity without subsidies, and around two thirds of the global market should reach grid parity in the next couple of years. If electricity costs were to rise by 3%

annually, 80% of the global market would reach parity in the next couple of years according to Deutsche Bank.

Usually associated with intermittent generation, wind energy (and solar PV) attractiveness has been additionally boosted by technical advances in battery and storage. Cost–efficient storage can overcome the seasonality issue of wind (and solar) power generation, and battery costs have fallen dramatically in recent years. Boosted by the growth of electrical vehicle markets, the average price of battery packs has fallen from $1000/kWh in 2010 to $350/kWh in 2015, according to the United Nations Environment Programme (UNEP). Further price compression is expected as electric vehicles become more widespread and battery production increases.

(WEF, 2017)

As the cost of wind energy does not include the cost of fuel, it is relatively straightforward to determine, compared with the cost of energy from fuel-consuming power plants which are dependent on estimates of future fuel costs. High or escalating fuel prices tend to favour zero (or low) fuel cost systems such as wind energy, but steady or falling fuel prices are less favourable to them.

Wind turbines (on land) are very quick to install, so they can be generating before they incur significant levels of interest on the capital expended during construction – in contrast to many other highly capital-intensive electricity generating plant (e.g. large hydro stations, tidal barrages and nuclear power stations).

8.8 Commercial development and wind energy potential

Wind energy developments worldwide

The present healthy state of the wind energy industry is due largely to developments in Denmark and California in the 1970s and 1980s, and Germany in the 1990s and 2000s.

In Denmark, unlike most other European countries that historically employed traditional windmills, the use of wind energy never ceased completely, largely because of the country's lack of fossil fuel reserves and because windmills for electricity generation were researched and manufactured from the nineteenth century until the late 1960s. Interest in wind energy took on a new impetus in the 1970s, as a result of the 1973 'oil crisis'. Small Danish agricultural engineering companies then undertook the development of a new generation of wind turbines for farm-scale operation.

It was California, however, that gave wind energy the push needed to take it from a small, relatively insignificant industry to one with the potential for generating significant amounts of electricity. A rapid flowering of wind energy development took place there in the mid-1980s, when wind farms began to be installed in large numbers. As a result of generous environmental tax credits, an environment was created in which it was possible for companies to earn revenue both from the sale of wind-generated electricity to Californian utilities and from the manufacture of wind turbines. The new Californian market gave Danish manufacturers an opportunity to develop a successful export industry, taking advantage of the experience acquired within their home market.

Since the 1980s, Europe had taken the lead in wind energy, with over 147 700 MW of wind generating capacity (over 34% of the world total) installed by the end of 2015. Germany in particular has been in the vanguard of deployment in Europe to date and by the end of 2015 had installed almost 45 000 MW. However by the end of 2015, Asia had become the leading region with over 175 800 MW installed (40% of world total) and China had achieved the world's largest wind energy capacity, with over 145 000 MW installed. The USA has the next largest with over 74 000 MW installed (GWEC, 2016a).

World wind power capacity reached over 432 GW by the end of 2015, around 7% of total global power capacity. A record 63 GW was added in 2015 with a global growth rate of 17.2%. Global wind power generation amounted to 950 TWh in 2015, nearly 4% of total global power generation, while Denmark produced 42% of its electricity from wind turbines and Germany's wind power contributed 13% of its electricity consumption (WEC, 2016).

In the UK, over 6600 grid connected wind turbines had been installed by the end of 2015, representing a combined capacity of over 13 600 MW, generating 11% of UK electricity demand - enough electricity for 8.25 million homes (over 30% of UK households) and offsetting some 15.5 million tonnes of CO_2 per year (GWEC, 2016a and RenewableUK, 2016b). Wind energy also provided 17% of electricity demand in the month of December 2015 and supplied 20% of demand in the last week of 2015 (RenewableUK, 2016b). Figure 8.35 shows the location of onshore wind energy projects in the UK at the time of writing.

Details of levels of wind generating capacity installed in various countries and continents in 2015 are given in Table 8.2.

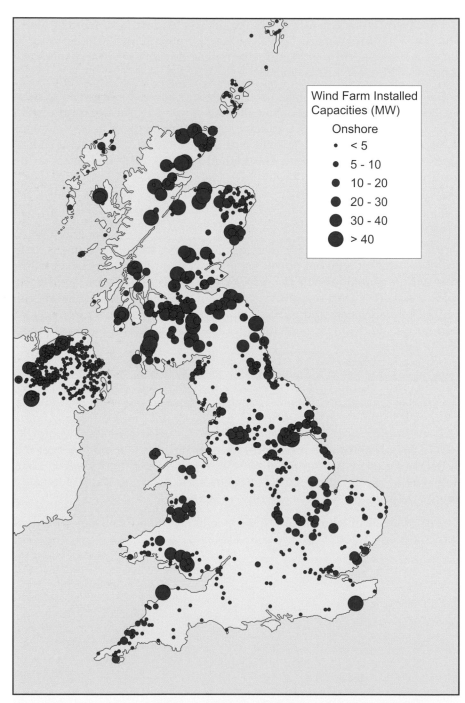

Figure 8.35 Locations of onshore wind energy projects in the UK (sources: DUKES, 2016)

Table 8.2 Global installed wind power capacity (MW) – regional distribution

	Start 2015	End 2015
Africa & Middle East		
South Africa	570	1053
Morocco	787	787
Egypt	610	810
Tunisia	245	245
Ethiopia	171	324
Jordan	2	119
Other	151	151
Total	**2 536**	**3 489**
Asia		
PR China	114 609	145 362
India	22 465	25 088
Japan	2 794	3 038
South Korea	610	835
Taiwan	633	647
Pakistan	256	256
Thailand	223	223
Phillippines	216	216
Other	167	167
Total	**141 973**	**175 831**
Europe		
Germany	39 128	44 947
Spain	23 025	23 025
UK	12 633	13 603
France	9 285	10 358
Italy	8 663	8 958
Sweden	5 425	6 025
Poland	3 834	5 100
Portugal	4 947	5 079
Denmark	4 881	5 063
Turkey	3 738	4 694
Netherlands	2 865	3 431
Romania	2 953	2 976
Ireland	2 262	2 486
Austria	2 089	2 411
Belgium	1 959	2 229
Rest of Europe	6 564	7 387
Total Europe	**134 251**	**147 771**
of which EU-28	129 060	141 578

Latin America & Caribbean

Brazil	5 962	8 715
Chile	764	933
Uruguay	529	845
Argentina	271	279
Panama	35	270
Costa Rica	198	268
Honduras	126	176
Peru	148	148
Guatemala	-	50
Caribbean	250	250
Others	285	285
Total	**8 568**	**12 220**
North America		
USA	65 877	74 471
Canada	9 699	11 205
Mexico	2 359	3 073
Total	**77 935**	**88 749**
Pacific Region		
Australia	3 087	4 187
New Zealand	623	623
Pacific Islands	12	13
Total	**4 442**	**4 823**
World Total	**369 705**	**432 883**

Source: GWEC, 2016a

Small-scale wind turbines

Small-scale wind turbines are more expensive per kilowatt of capacity than medium-scale and large-scale wind turbines. In most cases, the cost of the power they produce is not competitive with mains electricity (except in remote areas or rural windy locations) without support schemes. The need for batteries also tends to greatly increase the cost of such systems, though this could be changing as the costs of battery storage systems are falling rapidly, partly as a result of the innovations from the electrical vehicle sector. However, when wider deployment of feed-in tariffs occurs, such as those in place in certain European countries and were introduced for a while in the UK in 2010, there can be fresh opportunities for carefully sited small wind systems that are grid-linked. Grid-linking of small wind turbines initially permitted the omission of batteries, though with the falling prices of batteries, coupled with reducing feed-in tariff rates, batteries are now being included in such systems and allow time-shifting of electrical usage.

The steady demand from people interested in obtaining electricity from pollution-free sources, or who are in locations where conventional supplies are not available, already provides enough support to sustain a significant number of manufacturers of small-scale wind turbines throughout the world (Figures 8.1 and 8.36). However, whilst there are still a few UK small wind turbine manufacturers, many have ceased operations in recent years unless they have been able to focus on niche markets.

Figure 8.36 Small-scale wind turbine

Planning controls were relaxed in the UK for so-called micro wind turbines, subject to a number of constraints in terms of numbers of turbines, maximum size, maximum height and maximum noise etc., these micro wind turbines are to be considered 'permitted development' (that is they do not in general require planning permission). However the maximum heights permitted are unfortunately too low for satisfactory output from small wind turbines, as these turbines would invariably require relatively tall towers to be productive – also to be effective in lower wind speed areas small wind turbines would require larger diameter rotors for a given power rating, which would not be possible in the current permitted development requirements. There is therefore a likelihood that these constraints could well reduce small/micro wind turbine installations and there is a risk that inappropriate installations of micro/small wind turbines will occur as a result of the current permitted development requirements.

Recent experience of ill-conceived micro and small wind turbines (both horizontal axis and vertical axis designs) and inappropriate use of the NOABL wind speed database (principally in urban and suburban locations) and inappropriate siting has, perhaps not surprisingly, damaged the reputation for micro/small scale turbines in the UK.

To improve the situation RenewableUK has introduced a small wind turbine (SWT) standard (RenewableUK, 2014), as has AWEA (AWEA, 2009). These standards require manufacturers of micro/small wind turbines with swept areas of 200 m^2 or less to comply with the IEC 61400 series of wind turbine standards, especially IEC 61400-2 (SWTs), but also IEC 61400-1 and be acoustically compliant with IEC-61400-11 and performance tested to IEC-61400-12. The AWEA SWT standard requires SWT manufacturers to provide consistent, independently verified data including the AWEA Rated Power output at 11 m s^{-1} and a standard AWEA RAE (discussed in Section 8.5). In addition there is also a UK Microgeneration Scheme standard (and website (MSC, 2017)) for micro and small wind turbines (MCS 006, see MSC 2009) and a Microgeneration Installation Standard (MIS 3003, see MIS 2015) for installing wind turbines that utilize the UK feed-in tariff. In addition, there is an MCS planning standard for SWTs (MCS, 2015) and RenewableUK has also produced planning and safety guidance for SWTs (RenewableUK, 2011a, 2011b and 2015).

There is also a need for special care when assessing the wind characteristics of the site for a micro/small wind turbine. If possible, wind speed measurements should be carried out, together with a survey and site visit, to filter out obviously inappropriate sites, e.g. those surrounded by trees, large numbers of buildings or other obstacles to wind flow.

Local community and co-operatively owned wind turbines

Another type of wind energy development that is gaining support is that of local community wind turbines. This can take a variety of forms. In Denmark it usually involves a group of people from a local community buying a turbine or group of turbines. The local community benefits from the sale of the electricity produced, or makes use of it for its own purposes.

This approach can encourage a positive attitude towards wind energy in communities that might be opposed to commercial wind energy developers from outside the area. A number of organizations have attempted to develop such projects in the UK, and there are now a number of innovative projects and several more are being planned. The first community-owned wind farm (Baywind) has been operating successfully in Cumbria since the 1990s. There are a number of other community wind farms in the UK, with most in Scotland – including some which have brought useful income to Scottish island communities such as those on the Isle of Skye and the Isle of Gigha.

One innovative project is the Westmill Wind Farm Cooperative (Westmill Wind Farm Cooperative, 2017) in Oxfordshire that has five 1.3 MW wind turbines (Figure 8.37). In 2016 this was the UK's largest community-owned wind farm.

It is being increasingly recognized that local people are likely to be more supportive of community wind turbines, and interest in the concept of the 'village wind turbine' or 'town wind farm' appears to be increasing.

Numerous single medium- and large-scale wind turbines are currently operating in the UK, either as community wind turbines, or in supplying factories, hospitals, supermarkets or housing projects.

Whilst it is still a protracted process, community wind turbine projects may become more popular as a means of supplying lower cost electricity and developments in smart metering, energy storage and associated technologies may further help its deployment. The transition towns movement and social media is raising awareness and enthusiasm about low carbon energy sources and there are a number of organisations around that are able to facilitate community groups interested in such projects. There are a range of funding options available, particularly as appropriately and properly installed wind energy is seen as good investment. Also properly developed community wind turbines/farms are more likely to win support from the local public and planners.

(a)   (b)

Figure 8.37 (a) Westmill Wind Farm Co-operative, (b) opening ceremony

Wind energy and buildings

The established wisdom of experienced designers and installers of small wind turbines for many years has been that 'wind turbines and buildings do not generally mix'. A small wind turbine should be located at a distance from the nearest building equivalent to around ten times the height of the building, and supported on a tower with a hub height equivalent to at least twice the height of the building. This is still good advice, but unfortunately it is not consistent with the UK conditions of permitted development for micro/small wind turbines. An example of what happens if this good advice is not followed is demonstrated by the poor performance of the building-mounted wind turbines monitored in the Energy Savings Trust's field trial of domestic wind turbines (EST, 2009).

In spite of the established wisdom, the UK government promoted the installation of building-mounted micro/small wind turbines, in part because of over-optimistic estimates from new manufacturers of building-mounted micro wind turbines, over-predicted wind speeds from NOABL in urban and suburban areas, and because a number of studies and reports made predictions of the likely electricity production from building-mounted wind turbines in these areas that were very optimistic. Whilst estimates for tower mounted wind turbines in these areas were also over optimistic, building mounted wind turbines (on most typical traditional buildings) experience higher levels of turbulence which further degrades productivity. Some building mounted-wind turbines are attached to the wall of a gable end of a building, which is similar to mounting a wind turbine near a cliff edge: a type of location discouraged by established manufacturers of small turbines. This raises structural integrity issues, and damage to the house is quite possible depending on the size of turbine. Other mounting approaches are possible, and some manufacturers give guidance (Udell et al., 2010) and suggest low cost mounting arrangements, but great care is needed.

There is a great deal of technical naivety on the part of architects and buyers of micro/small wind turbines (who often unadvisedly treat them as just another electrical appliance) and often have a lack of understanding about appropriate siting or know how much electricity a turbine is really likely to generate. That coupled with the fact that the internet has many dubious 'revolutionary' designs of micro/wind turbines available may have damaged perceptions about wind turbines in general as the public is often not able to differentiate between these (often motionless) devices and the proven technology used in the commercial wind industry. The US National Renewable Energy Laboratory (NREL) has developed guidance on developing wind turbines in the built environment which covers some of the issues (Fields et al., 2016).

If the building is designed to integrate the wind turbines, the turbines may be less affected by turbulence and damaging flows, but will still be dependent on the local wind speed conditions – although taller buildings may be able to take advantage of higher wind speeds. Such approaches are likely to be expensive, bespoke and not easily replicated on other buildings.

However there may be opportunities for using buildings (in appropriate locations) as a means of enhancing wind energy production by carefully designing the building form to accelerate wind velocities. In doing so it may be possible to reduce the size of wind turbine required for a given power output and to offset the topographical roughness effects which slow down local winds in urban and suburban environments.

The author of this chapter has patented an approach using one or more wing-like 'planar concentrators'. These can be used as free-standing systems, or as building-augmented wind energy systems when used in combination with a building surface such as a roof (as in the 'Aeolian Roof', Figure 8.38). Axial flow or cross-flow wind turbines are located between the planes, or between the planes and the building surface (e.g. roof or wall). For a given wind speed, this approach has demonstrated a significant increase in power output compared to the same wind turbine in normal non-augmented operation. Such systems can be deployed on both new and existing buildings and could generate a high proportion of the electricity requirements of appropriately oriented energy-efficient buildings.

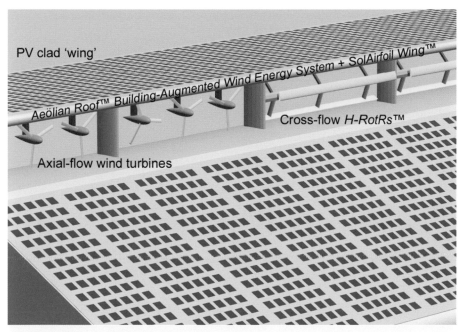

PV clad 'wing'

Aeölian Roof™ Building-Augmented Wind Energy System + SolAirfoil Wing™

Cross-flow H-RotRs™

Axial-flow wind turbines

Figure 8.38 Aeölian Roof building-augmented wind energy systems

Wind energy potential

In an extensive study, Greenblatt (2005) scaled a previous estimate of the world land-based wind energy potential (Grubb and Meyer, 1993) at Class 4+ sites, to take account of taller towers and increasing hub heights (Class 4+ meaning Class 4 sites and above – see Figure 8.31). They arrived at a figure of 185 000 TWh per year.

With the advent of turbines capable of electricity generation from lower wind speed sites, Greenblatt (2005) also explored the wind energy potential if Class 3+ (i.e. including Class 3 sites and above) were included in the analysis, concluding that the global annual wind generated electricity resource could be around 335 000 TWh per year (Table 8.3).

The Global Climate and Energy Project at Stanford University carried out an evaluation of global potential of wind energy using five years of data from the US National Climatic Data Center and the Forecasts Systems Laboratory (Archer and Jacobson, 2005). This study estimated the global wind speeds at 80 metres above ground level (based on wind speed data for the year 2000) and found that using only 20% (e.g. 14.4 TW or 10 800 Mtoe) of the potential viable land-based resource wind energy could satisfy the global electricity demand (given at that time as 1.6 to 1.8 TW) *seven times over*. Jacobson and Delucci and their team subsequently built on this research to develop a roadmap to 2050 for 139 countries to achieve 100% clean renewable energy provision from WWS (wind, water and solar energy technologies) to supply all energy needs (Jacobson and Delucci, 2016). In this scenario they envisage that the total demand for 2050 to be 11.840 TW and of this 23.52% to be supplied by 1582 345 onshore 5 MW wind turbines (plus 5.04 GW from existing turbines) and 13.62% from 935 150 offshore 5 MW turbines (plus 0.26 GW from existing turbines).

Table 8.3 Available world 'land-based' wind resources and future electricity demand

Region of the world	Electricity demand by 2025/TWh y^{-1}	Installed capacity /GW	Wind resource /TWh y^{-1} (Class 4+ sites)	Wind resource /TWh y^{-1} (Class 3+ sites)
North America	6700	18 700	62 400	93 500
Latin America	1800	6100	20 400	36 300
Europe	6200	15 200	50 500	92 500
Western	3100	4400	14 700	21 000
Eastern and Former Soviet Union	3100	10 800	35 800	71 300
Africa/Middle East	2200	10 400	34 700	71 300
Asia	8700	1900	6400	21 500
India	1300			
China	4300			
Other Asia	3100			
Australia/Oceania	400	3200	10 700	20 200
World Total	**26 000**	**70 400**	**185 000**	**335 400**

Note: Class 4+ = Class 4 and above; Class 3+ = Class 3 and above
Source: Greenblatt, 2005

In 2016, the GWEC (Global Wind Energy Council) produced a series of global wind energy outlook scenarios (GWEC, 2016b) to examine the future potential for wind energy up to 2020, 2030 and 2050. These were based on four scenario assumptions:

(1) The IEA's New Policies Scenario (NPS) based on the projections in the International Energy Agency 2016 World Energy Outlook (IEA, 2016).

(2) The 450 Scenario (450) sets out an energy pathway which requires the concentration of greenhouse gases in the atmosphere to be limited to around 450 parts per million of CO_2 equivalent (ppm CO_2-eq).

(3) A moderate scenario (MS) which takes into account policy measures to support renewable energy and targets either enacted or in the planning stages around the world but it also assumes that commitments for emission reductions agreed by governments at COP21 (Paris, 2015) will be implemented – though on the modest side.

(4) An advanced scenario (AS) which has more ambitious assumptions based on an estimate of the extent to which the wind industry could grow in a best case 'wind energy vision'.

Figure 8.39 shows the predicted increases in global cumulative wind power capacity based on these scenarios up to 2050.

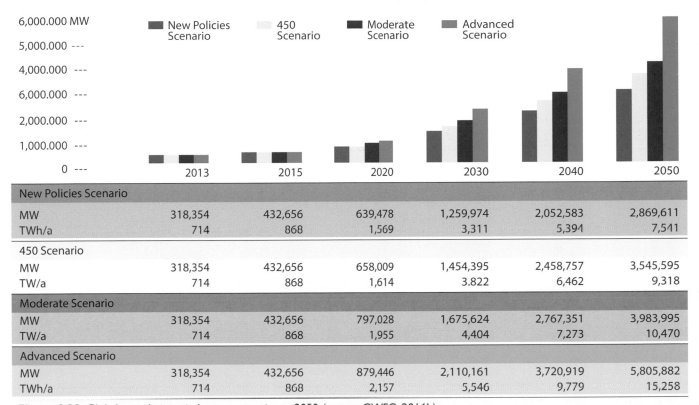

Global Cumulative Wind Power Capacity

New Policies Scenario						
	2013	2015	2020	2030	2040	2050
MW	318,354	432,656	639,478	1,259,974	2,052,583	2,869,611
TWh/a	714	868	1,569	3,311	5,394	7,541
450 Scenario						
MW	318,354	432,656	658,009	1,454,395	2,458,757	3,545,595
TW/a	714	868	1,614	3.822	6,462	9,318
Moderate Scenario						
MW	318,354	432,656	797,028	1,675,624	2,767,351	3,983,995
TW/a	714	868	1,955	4,404	7,273	10,470
Advanced Scenario						
MW	318,354	432,656	879,446	2,110,161	3,720,919	5,805,882
TWh/a	714	868	2,157	5,546	9,779	15,258

Figure 8.39 Global cumulative wind power capacity to 2050 (source GWEC, 2016b)

European potential

The EEA (European Environment Agency) carried out a detailed land/sea use analysis of the land-based and offshore wind energy potential within Europe (EEA, 2009). This analysis estimated the technical potential (TP) that could be generated in 2020 and 2030 assuming 80 m hub height on land and 120 m hub height offshore. This was then filtered to exclude environmentally sensitive areas plus a number of offshore constraints, such as shipping lanes etc, and zoning to yield the 'constrained potential' (CP). The 'economically competitive potential' (ECP) was calculated on the basis of projected costs of developing and running wind farms in 2020 and 2030 derived from EC, 2009. The 2009 EEA report estimates the EPC for 2030 as 27 000 TWh per year from land-based and 3 400 TWh per year from offshore projects. Table 8.4 gives a summary of the TP, CP and ECP estimated on this basis for 2020 and 2030.

In 2015, the EWEA (European Wind Energy Association – now Wind Europe) produced a series of wind energy scenarios for 2030 (Corbetta et al., 2015) to establish the European wind energy industry's vision to 2030. The Central Scenario expects 320 GW of wind energy capacity to be installed in the EU in 2030: 254 GW as onshore and 66 GW as offshore. The 96 000 wind envisaged turbines estimated to produce 778 TWh of electricity (24.4% of the EU's demand) and avoid the emission of 436 million tonnes (Mt) of CO_2. EWEA's Low Scenario only foresees 251 GW of wind energy installations, 22% lower than in the Central

Scenario, equal to 19% of EU electricity demand and avoids 339 Mt of CO_2 emissions from 76 000 wind turbines. EWEA's High Scenario expects 392 GW to be installed by 2030, 23% higher than in the Central Scenario, equal to meet 31% of EU electricity demand and avoid 554 Mt of CO_2 emissions from 114 000 wind turbines.

Table 8.4 Projected technical, constrained and economically competitive potential for European wind energy development in 2020 and 2030

		Year	TWh	% share of 2020 and 2030 demand (*)
Technical potential	Onshore	2020	45 000	11–13
		2030	45 000	10–11
	Offshore	2020	25 000	6–7
		2030	30 000	7
	Total	2020	70 000	17–20
		2030	75 000	17–18
Constrained potential	Onshore	2020	39 000	10–11
		2030	39 000	9
	Offshore	2020	2 800	0.7–0.8
		2030	3 500	0.8
	Total	2020	41 800	10–12
		2030	42 500	10
Economically competitive potential	Onshore (*)	2020	9 600	2–3
		2030	27 000	6
	Offshore	2020	2600	0.6–0.7
		2030	3400	0.8
	Total	2020	12 200	3
		2030	30 400	7

Note: (*) European Commission projections for energy demand in 2020 and 2030 (EC2008 a, b) are based on two scenarios: 'business as usual' (4 078 TWh in 2020 – 4 408 TWh in 2030) and 'EC Proposal with RES trading' (3 537 TWh in 2020 – 4 279 TWh in 2030). The figures here represent the wind capacity relative to these two scenarios, e.g. onshore capacity of 45 000 TWh in 2020 is 11–12.7 times the size of projected demand.
(*) These figures do not exclude Natura 2000 areas (which form an EU-wide network of nature protection areas).
Source: EEA, 2009

A European Commission report (EC, 2009) which refers to investing in technologies for the Strategic Energy Technologies plan (SET-Plan (EU, 2010)) states that 'With additional research efforts, and crucially, significant progress in building the necessary grid structure over the next ten years, wind energy could meet one fifth of the EU's electricity demand in 2020, one third in 2030 and half by 2050', (see also Zervos and Kjaer, 2009). Achieving this would require achieving 400 GW wind energy capacity in 2030 and 600 GW capacity in 2050 with the majority (350 GW) of the 2050 capacity coming from offshore turbines.

Table 8.5 gives a summary of the wind energy capacity needed to meet the European Commission's SET-Plan targets.

Table 8.5 Wind energy capacity needed to meet the European Commission's SET-plan targets

	Onshore wind (GW)	Offshore wind (GW)	Total wind energy capacity (GW)	Average capacity factor onshore	Average capacity factor offshore	TWh onshore	TWh offshore	TWh Total	EU-27 gross electricity consumption*	Wind power's share of electricity demand
2020**	210	55	265	26.0%	42.3%	479	204	683	3494	20%
2030	250	150	400	27.0%	42.8%	592	563	1155	3368	34%
2050	250	350	600	29.0%	45.0%	635	1380	2015	4000	50%

* Electricity demand assumes the European Commission's New Energy Policy $100 oil/barrel scenario until 2020 and High Renewables/Energy Efficiency scenario for 2030. Demand in 2050 is assumed to be 4000 TWh.
** Assuming 265 GW by 2020 in accordance with EWEA's 'high' scenario combined with the European Commission's 'New Energy Policy' Assumption for demand.
Source: Zervos and Kjaer, 2009

8.9 Offshore wind energy

The capital costs of energy from offshore wind farms are generally higher than those of onshore installations because of the extra costs of civil engineering for substructure, higher electrical connection costs and the higher specification materials needed to resist the corrosive marine environment.

However, offshore wind speeds are generally higher and more consistent than on land (apart from certain mountain and hill tops) and test results from the Tunø Knob offshore wind farm in Denmark indicate that actual output is 20–30% higher than estimated from wind speed prediction models. Availability was also higher than expected with an average of 98% being achieved, though this may not necessarily be typical. These wind energy characteristics, together with likely reductions in offshore costs as experience is gained in this environment, are expected to make offshore wind energy costs competitive in the medium to long term (although it should also be noted that capital costs doubled in the five years up to 2009 (Willow and Valpy, 2011)). In deeper water, further offshore, capital and operational costs will be higher, but it is anticipated that the increased energy yield, particularly from larger rotors (offshore, it is more feasible to utilize very large-scale wind turbines than it is on land, see Box 8.5) will more than compensate for these additional costs (Willow and Valpy, 2011) though this may depend on the proportion of novel and untested engineering approaches that get deployed – especially in deeper more turbulent waters.

Europe is the world's leader in offshore wind, having installed 3230 offshore turbines (over 11027.3 MW in 84 wind farms) with grid connections to 11 European countries by the end of 2015, in shallow waters – depths mainly up to 30–45 m (EWEA, 2016). EWEA forecasts that once the six offshore projects under construction are completed a further 1.9 GW of capacity will be added, bringing the cumulative installed capacity in Europe to 12.9 GW. EWEA has identified 26.4 GW of consented offshore projects that could be constructed over the next decade. A total of 63.5 GW of projects are understood to be in the planning phase (EWEA, 2016).

BOX 8.5 Very large turbines

Figure 8.40 (a) and (b) The 164 m diameter MHI-Vestas V164-8 MW wind turbine designed for offshore operation which employs British made blades produced on the Isle of Wight. These turbines are currently being installed in large numbers on various UK offshore wind farms. (Source: MHI-Vestas, 2016). (c) The 8 MW 154 m diameter Siemens SWT-8-154 prototype direct drive turbine being tested at the Østerild test site, Denmark, installed 30 January 2017 (source: Siemens, 2017). (d) The 7MW 167 m diameter MHI MT7167/7.0 Hydraulic drive wind turbine (or Digital Displacement Transmission (DDT) turbine) being tested at the Hunterston test site in Scotland. Currently this has the world's largest rotor diameter employing 81.6 m blades (source: MHI, 2015)

It is more feasible to utilize very large-scale wind turbines offshore than on land. This may improve economic viability, as more energy can be captured from a single platform (also known as 'power per tower') and this can have benefits in terms of reduced maintenance costs. However, working against these benefits is the tendency to increasing capital costs associated with turbines above approximately 2 MW rated capacity. The latter effect is due to the scaling laws for strength and weight of materials with increasing rotor swept area, together with the much higher torque loading experienced by gearboxes.

Very large turbines include the 150 m diameter GE (formerly Alstom) Haliade 150–6 MW, 154 m diameter Siemens 154–7 MW/8 MW (Figure 8.40(c)), 164 m diameter MHI-Vestas V164–8 MW (Figure 8.40 (a) and (b)) employing 80 m long blades developed on the Isle of Wight and manufactured on the island. During its testing phase the V164–8 MW turbine generated 192 MWh in a 24 hour period (October 2014) and as of December 2016, 32 MHI-Vestas V164–8 MW turbines were installed in 3 months on the Burbo Bank Extension project off the coast of Liverpool.

At the time of writing the turbine with the largest diameter is the 167 m diameter Mitsubishi 7 MW Hydraulic Drive Turbine (HDT) or Digital Displacement Transmission (DDT) Turbine (Figure 8.40(d)) – formerly known as the SeaAngel turbine – being tested at Hunterston in Scotland and also on a floating platform in Japan. In contrast to the latest Haliade and Siemens turbines that employ direct drive generators, the Mitsubishi 7 MW HDT utilizes a novel high efficiency hydraulic transmission system developed in the UK by Artemis Intelligent Power and based on the ideas of Professor Stephen Salter. The drive avoids the difficulties experienced by gearboxes and reduces the tower top mass which may be a critical feature in the viability of very large scale floating wind turbine systems. Subject to test results Mitsubishi (MHI) also plans to incorporate the hydraulic drive into future MHI-Vestas V164-8MW turbines.

UK based Blade Dynamics (now part of GE) developed a modular 78 m blade which was successfully tested at the ORE-Catapult test centre at Blyth, HAWT. LM Wind Power (now also part of GE) produced an 88.4 m blade in June 2016 for a 180 m diameter 8 MW wind turbine for Adwen (now part of a merged Siemens-Gamesa company).

Hendrik Stiedal (former CTO of Siemens) revealed that Siemens has produced a design for a 10 MW turbine which would have a 210 m diameter rotor, 140 m tall tower, innovative direct drive generator and predicted to generate 50 000 MWh per year (Stiedal, 2014).

AMSC Wintec Solutions has developed a design for a 190 m diameter 10 MW HAWT called SeaTitan based on an innovative super-conductor-based direct drive generator which is said to substantially reduce the tower top mass compared to other systems. At the time of writing a SeaTitan has yet to be installed.

Enercon has been manufacturing 126 m diameter 7.5 MW direct drive turbines for land-based applications.

Designs for 15 MW turbines are being considered for offshore applications by both GE and Gamesa.

The EU funded a five year long wide-ranging UpWind project to investigate and evaluate the practicality of developing 20 MW HAWTs for offshore operation. The project included 40 partners from the wind turbine industry, universities and others fields, and in the final report they concluded that such turbines (which would have rotor diameters of over 200 m) are feasible, though the blades would have to be lighter and would be more flexible than today's largest turbine blades.

Sandia Laboratories in the USA has been designing 100 m long blades for a 13 MW HAWT and are also leading a project to develop a 50 MW 'exoscale' offshore wind turbines known as 'Segmented Ultralight Morphing Rotor' (SUMR) funded by the US Department of Energy. The 200+ m long blades would morph (flex inward) in high winds.

It is uncertain how much larger horizontal axis turbines can be scaled. This will involve developing blades which will have to be lighter and have improved fatigue resistance, because when HAWTs are built to the sizes necessary to achieve very high power ratings, the reversing gravity loads on the rotor (as the blades move up and downwards during their rotation cycles) become a significant structural limitation.

Similarly, there may be an increase in the cyclic impact on large HAWT wind turbine blades due to higher levels of 'wind shear' that occur with large rotor diameters. As wind speed increases with height, the blade tip of a large diameter HAWT will move through wind velocities that vary in magnitude between the upper and lower parts of the rotation cycle and the larger the diameter the greater is the difference in wind speed experienced. Design feasibility studies are underway to explore whether larger HAWTs up to 20 MW are possible and to explore the benefits of material substitution together with improvements to the structural design of blades.

An alternative strategy is the use of multi-rotor turbines. These have been attempted periodically in the past but have not been successful commercially to date, though they may be a potential way to maximize the 'power per tower' and avoid some of the scaling issues of very large turbines mentioned above, such as high gravitational cyclic loading and high torque etc. In spite of this there are several challenges related to yawing, rotor interactions, balancing the rotor loads etc., but there are a number of projects underway to explore this approach.

Vertical axis wind turbines do not experience the gravity-driven fatigue loading encountered in large HAWTs. VAWT blades are also not as greatly affected by wind shear generated cyclic loading, and towers for some large swept area VAWT designs do not have to be built as tall in order to provide sufficient clearance above the sea level – thus reducing overturning moments. So in principle, VAWTs could be scaled-up to potentially very large sizes.

The 2009 EEA study (EEA, 2009) mentioned previously estimates that the 'economically competitive' European offshore wind energy potential 'in 2020 is 2600 TWh y^{-1}, equal to between 60% and 70% of projected electricity demand, rising to 3400 TWh y^{-1} in 2030 equal to 80% of the projected EU electricity demand.' EEA, 2009, also estimates the technical potential of offshore wind energy to be 'seven times greater than the projected EU electricity demand in 2030'.

When realized, the ambitious programmes for substantial offshore projects established by several European countries will mean that wind power will ultimately become a major provider of electricity in those countries.

One interesting development which has taken place in Norway was an innovative floating 2.3 MW wind turbine (Figure 8.42(b)), known as the Hywind project, that was successfully operated in a sea water depth of 220 m. This has significant implications for expanding the potential offshore wind energy beyond the shallow continental shelf sites so far developed. Hywind 2, a 30 MW floating pilot wind farm – at a depth of 95–120 m – is currently being developed near Peterhead using a scaled-up Hywind employing 6 MW Siemens wind turbines (Figure 8.43). See Box 8.6, which discusses floating wind turbines in more detail.

Several Asian countries, including China, Japan, Taiwan and South Korea are stimulating offshore wind energy development in their territorial waters (see Figure 8.44 and Table 8.7). China is currently the leading market for offshore wind energy in Asia with 3% of global offshore wind energy installed, it has plans to install 10 GW by 2020. Prompted by the Fukushima nuclear power plant meltdown, the Japanese government has committed to actively develop offshore wind energy, but as the coastal waters are deep the main focus has been on floating turbines. South Korea installed its first offshore wind turbines in 2012 and the government plans a phased development of a 2500 MW offshore project and 500 MW of projects are in the pipeline. Taiwan has also announced plans to deploy 600 MW of capacity by 2020 and 4 GW by 2030. India is reportedly planning some pilot offshore projects that could be operational by 2019 (Smith et al., 2015).

Whilst the USA has the world's second largest installed land-based wind energy capacity, the first US offshore wind farm – 30 MW Block Island Wind Farm off the coast of New England – began operating in December 2016. A major study of the offshore wind energy potential around the USA was completed by the National Renewable Energy Laboratory (NREL). The US offshore wind energy resource study assumed 3 MW/km² spacing and was based on 6 MW turbines and a 100 m hub height. Taking into account a range of constraints including water depth and exclusion zones, the US Technical offshore wind energy capacity was estimated at 2058 GW and calculated to generate some 7203 TWh/year - which is roughly double the 3863 TWh of electricity used in the USA 2014 (Musial et al., 2016).

BOX 8.6 **Floating wind turbines**

Floating wind turbines present a new way to access a previously unattainable clean renewable resource, namely wind energy resources at sea at water depths too deep for fixed bottom-mounted offshore wind turbines.

The potential wind energy resources at deep water locations is truly enormous so consequently there is considerable activity ongoing to develop viable floating wind energy systems. There over 30 concepts currently being developed, principally in Europe, Japan and the USA based largely on conventional horizontal axis wind turbines but a few have proposed vertical axis designs. Most of the projects are designed around a single wind turbine, though a number have proposed multi-wind turbine platforms. The concepts are generally based on a

variation of three main types of floating platforms developed by the offshore oil and gas industries known as 'semi-submersible', 'spar-buoy' and 'tension leg platform' or TLP (Figure 8.41).

Prototype or demonstration floating wind turbine units based on semi-submersible and spar-type have been successfully tested in Europe, Japan and the USA (Figure 8.42). These demonstrations have helped to grow confidence in the concept and more ambitious experimental floating wind projects are in the process of being developed and deployed.

Figure 8.41 Different types of floating wind turbine platforms. Spar (left), Semi-submersible (centre) and Tension leg platform of TLP (right). (Source: DNV-GL 2016)

The initial prototype projects were utilising turbines rated at under 2 MW, but the current round of experimental floating wind turbine projects are utilizing much larger wind turbines such as the five Siemens 6 MW turbines on spar-buoys employed on the Statoil Hywind-Scotland project (Figure 8.43) and the 167 m diameter Mitsubishi 7 MW Hydraulic Drive Turbine – mentioned in Box 8.5 – being deployed on a V shaped semi-submersible floating platform in Japan (Figure 8.42(d)).

There are some major floating wind projects being planned around the world.

Figure 8.42 Selection of various floating wind turbine concepts being investigated and under development. (a) Principle Power WindFloat 8 MW (b) Statoil Hywind 2 MW scaling to 6 MW Siemens for Scotland (c) DCNS SeaReed 6 MW GE Haliade turbine (d) Mitsubishi Hydraulic Drive 7 MW on Fukishima Forward II Semi-sub, Japan (e) GustoMSC Tri-Floater with 5 MW wind turbine, Netherlands/France (f) DBD Systems Eco TLP (g) PelaStar TLP (Glosten) GE Haliade 6 MW turbine (h) Ideol FloatGen 2 MW, France (i) Toda Consortium-Hitachi 5 MW on hybrid spar, Japan (j) Aqua Ventus/VolturnUS University of Maine, USA

Figure 8.43 Statoil Hywind – Scotland floating wind farm planned off Peterhead. 5 × 6 MW Siemens wind turbine on spar buoy floating platforms. (Source: Statoil)

Floating VAWTs

Whilst floating VAWT concepts are being researched in various locations, the most comprehensive research is being carried out by Sandia Laboratories in the USA. They are investigating the technical and economic feasibility of floating VAWTs and have carried out detailed aerodynamic and structural rotor design studies and are progressing VAWT-specific floating platform designs, system-level studies and VAWT LCOE modelling. The main focus has been on the use of the curved blade Darrieus VAWTs, though they are also investigating other VAWT configurations and recently concluded: 'Despite technical challenges associated with floating VAWTs, the systems have inherent advantages over floating HAWTs for reducing capital costs and reducing life-cycle O&M costs.' (Griffith et al., 2016).

Floating Wind Potential

Table 8.6 Offshore wind resource potential floating wind capacity in Europe, USA, and Japan (US NREL, 2013; EWEA, 2013; Marine International Consulting, 2013).

Country/Region	Share of off-shore wind resource in deep water locations (>60m depth)	Potential floating wind capacity
Europe	80%	4000 GW
USA	60%	2450 GW
Japan	80%	500 GW

(Source: James and Costa Ros, 2015).

About 210 US offshore wind energy projects (with a combined capacity of over 15 GW) were in the planning and permitting process as of June 2015, with the majority of these being planned for the north east and mid-Atlantic regions of the US coastline, with some being considered at the Great Lakes, the Gulf of Mexico and along the Pacific Coast (Smith et al., 2015). Beyond 2020, the US Department of Energy's Wind Vision scenario (US DOE, 2015) includes the deployment of 22 GW by 2030 and 86 GW by 2050, with future deployment occurring in all major US coastal regions (Smith et al., 2015).

At the time of writing the total current global offshore wind energy capacity is 12 107 MW, installed with 2739 turbines across 73 offshore wind farms in 15 countries, with 92% in European waters. Current policy plans global offshore wind capacity to reach 72 GW in 2030 (WEC, 2016), with another 460 MW installed (not yet commissioned) by end of 2015, a further 2.56 GW under construction and 8.13 in pre-construction stages. According to the Global Offshore Wind Farm Database (4C Offshore, 2016), over 900 offshore wind farms are being planned in 36 countries around the world. Table 8.7 and Figure 8.44 summarize the global offshore wind projects as of June 2015. Ernst & Young have projected that by 2020, offshore wind will be about 10% of global installed capacity (Ernst & Young, 2015).

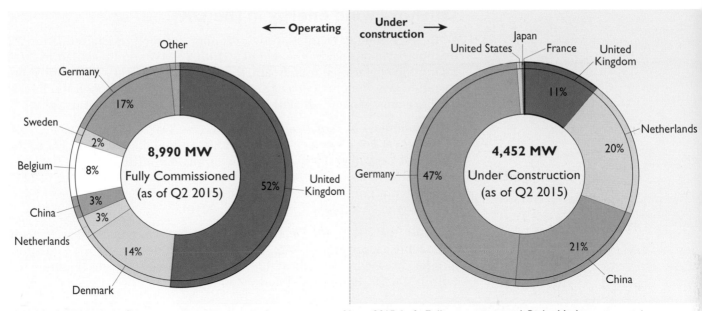

Figure 8.44 Global offshore wind power capacity by country as of June 2015. Left: Fully commissioned. Right: Under construction (source: Smith et al., 2015)

Table 8.7 Summary of Operating and Under Construction Offshore Wind Projects by Country (as of 30 June 2015)

	Operating (MW)	Under Construction (MW)	Total (MW)	Rank
United Kingdom	4625	503	5128	1
Germany	1505	2108	3613	2
Denmark	1271	0	1271	3
China	310	918	1228	4
Netherlands	247	873	1120	5
Belgium	712	0	712	6
Sweden	202	0	202	7
Japan	52	13	64	8
Finland	32	0	30	9
United States	0	30	30	10
Ireland	25	0	25	11
France	0	8	8	12
South Korea	5	0	5	13
Norway	2	0	2	14
Portugal	2	0	2	15
Total	8990	4452	13442	

Note: Totals may not sum due to rounding.

Offshore wind energy in the UK

With over 5066 MW offshore wind energy capacity installed (from 22 offshore wind farms) by the end of 2015 (572 MW were installed in 2015 taking the operational total to 29), the UK has the world's largest offshore installed wind energy capacity at the time of writing (though the largest offshore wind capacity installed in 2015 was 2.28 GW in Germany).

UK offshore wind farms generated over 17.4 TWh in 2015 – an increase of 30% over 2014 – which is 5.3% of UK annual electricity (7.3% in the month of December 2015), enough to supply 4.2 million homes (i.e. over 16% of UK households) and avoided the emission of 7.4 tonnes of CO_2. Another nine projects are under construction that account for another 4.5 GW when complete (Crown Estate, 2016).

In 2008, the UK BERR Department published an update to its marine renewable energy atlas (BERR, 2008). This includes a map of UK offshore annual mean wind speeds at 100 m height above sea level (Figure 8.46) within the 'UK Renewable Energy Zone'.

BOX 8.7   London array – world's largest offshore wind farm

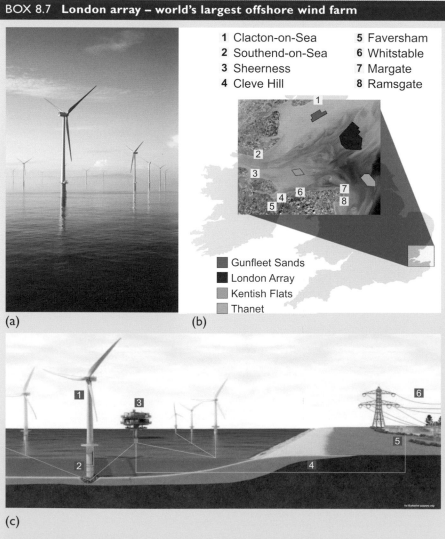

1 Clacton-on-Sea   5 Faversham
2 Southend-on-Sea   6 Whitstable
3 Sheerness   7 Margate
4 Cleve Hill   8 Ramsgate

Gunfleet Sands
London Array
Kentish Flats
Thanet

(a)   (b)   (c)

Figure 8.45 The London Array (a) view of wind farm (b) location (c) schematic of operation (1) Wind turns the blades on each individual wind turbine to generate electricity. (2) Array cables buried in the seabed take the electricity from the wind turbines to the offshore substations. (3) The offshore substations boost the voltage to reduce transmission losses. (4) The export cables from the offshore substations bring the electricity to shore. (5) The cables run underground to the onshore substation at Cleve Hill, where the electricity feeds straight into the high voltage network. (6) The electricity is carried through the local network and the National Grid into homes, businesses and industries.

Project ownership: Dong Energy (50%) Eon (30%) Mazdar (20%).

The world's largest single offshore wind farm at the time of writing (Figure 8.45) began officially operating in July 2013. It started in 2010 at a site 20 km off the Kent and Essex coasts, in the mouth of the Thames estuary, located between two sandbanks – Long Sand and Deep Knock. It is a Round 2 project that forms the fourth wind farm (of potentially a total of five) in the Thames Estuary after Kentish Flats, Gunfleet Sands and Thanet Wind Farms. The London Array wind farm cost £1.95 billion, took two years to construct, was completed in 2012 and has been operational since 2013.

The project consists of 175 Siemens SWT-3.6–120 turbines each rated at 3.6 MW, giving a total installed capacity of 630 MW. The 120 m diameter turbines are spaced 650 m apart in one direction and 1200 m apart in the other, over an area of 100 km2, in a water depth of up to 25 m. Each turbine has a hub height above sea level of approximately 87 m. The wind farm is estimated to generate an electricity output equivalent to the consumption of more than 500 000 homes. The array is intended to reduce annual CO_2 emissions by about 900 000 tonnes per year, said to be 'equivalent to eliminating the emissions of 300 000 passenger cars' and had a capacity factor of 45.3% in 2015.

London Array also set a new record for the amount of clean electricity produced by an offshore wind farm in a single calendar month (London Array, 2016).

December 2015 saw its 175 turbines generate 369 000 MWh of electricity – considerably above target and well above the previous best of 317 000 MWh set in the previous November. The capacity factor for the month, which saw average wind speeds of 11.9 m/s (27 mph), was 78.9%.

The two successive months of production brought net overall output for the year to some 2500 000 MWh, enough to meet the needs of more than 600 000 UK households (based on an average household consumption of 4,115 kWh per year).

(Source: London Array (2012))

The atlas resulted in a range of more detailed assessments of the UK offshore wind energy potential, including one from the Offshore Valuation Group (OVG, 2010). This evaluated all of the marine renewable energy resources available in the UK (including tidal and wave energy as well as offshore wind). It estimated that the total practical resource from offshore fixed wind turbines (i.e. turbines fixed to the seabed) was some 406 TWh per year (including those already installed). It also estimated the total practical resource from offshore floating wind turbines that could be deployed in deeper water to be 1533 TWh per year indicating a total combined output (fixed and floating wind turbines) of over 1900 TWh per year from offshore wind energy.

Since the first UK offshore wind energy project was installed at Blyth in 2000, many more offshore wind farms have been installed both in the UK (Figure 8.47) and in other parts of Europe and much has been learnt, important experience and confidence gained to encourage a substantial growth both in successfully operating offshore turbines and attracting investment into substantial planned projects.

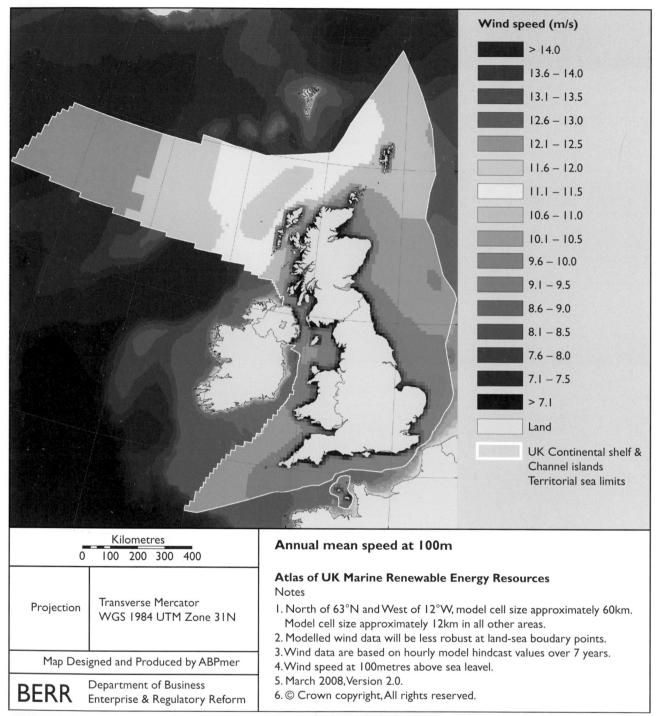

	Wind speed (m/s)
	> 14.0
	13.6 – 14.0
	13.1 – 13.5
	12.6 – 13.0
	12.1 – 12.5
	11.6 – 12.0
	11.1 – 11.5
	10.6 – 11.0
	10.1 – 10.5
	9.6 – 10.0
	9.1 – 9.5
	8.6 – 9.0
	8.1 – 8.5
	7.6 – 8.0
	7.1 – 7.5
	> 7.1
	Land
	UK Continental shelf & Channel islands Territorial sea limits

Annual mean speed at 100m

	Kilometres
	0 100 200 300 400

Projection	Transverse Mercator WGS 1984 UTM Zone 31N

Map Designed and Produced by ABPmer

BERR Department of Business Enterprise & Regulatory Reform

Atlas of UK Marine Renewable Energy Resources
Notes
1. North of 63°N and West of 12°W, model cell size approximately 60km. Model cell size approximately 12km in all other areas.
2. Modelled wind data will be less robust at land-sea boudary points.
3. Wind data are based on hourly model hindcast values over 7 years.
4. Wind speed at 100metres above sea leavel.
5. March 2008, Version 2.0.
6. © Crown copyright, All rights reserved.

Figure 8.46 Annual mean wind speeds offshore at 100 m above sea level from the Atlas of UK Marine Renewable Energy (source: BERR, 2008)

Operational: Total capacity of offshore wind farms that have been fully commissioned.

Capacity MW	
01 Barrow	90
02 Blyth	4
03 Burbo Bank	90
04 Greater Gabbard	504
05 Gunfleet Sands Demonstration	12
06 Gunfleet Sands I	108
07 Gunfleet Sands II	65
08 Gwynt y Mör	576
09 Humber Gateway	219
10 Inner Dowsing	97
11 Kentish Flats	90
12 Kentish Flats extension	50
13 Lincs	270
14 London Array	630
15 Lynn	97
16 Levenmouth Demonstration Turbine	7
17 North Hoyle	60
18 Ormonde	150
19 Rhyl Flats	90
20 Robin Rigg East	90
21 Robin Rigg West	90
22 Scroby Sands	60
23 Sheringham Shoal	317
24 Teeside	62
25 Thanet	300
26 Walney (Phase 1)	184
27 Walney (Phase 2)	1834
28 Westermost Rough	210
29 West od Duddon Sands	389
Total	**5,094**

Under construction: Total capacity of offshore wind farms that are under construction or where the developer has confirmed a final investment decision, but are not yet fully operational.

Up to capacity MW	
30 Burbo Bank extension	258
31 Dudgeon	402
32 East Anglia ONE	714
33 Galloper[1]	336
34 Hornsea project 1	1200
35 Hywind 2 Demonstration (Buchan Deep)	30
36 Race Bank[1]	546
37 Rampion[1] (Southern Array)	400
38 Walney extension	660
Total	**4,546**

Government support on offer: Total capacity of offshore wind farms that have secured a Contract for Difference or whose publicly stated timescales are consistent with accessing the Renewables Obligation.

Capacity MW	
39 Aberdeen Demonstration	66
40 Beatrice	664
41 Blyth Demonstration	99
42 Neart na Gaoithe (NNG)	448
Total	**1,211**

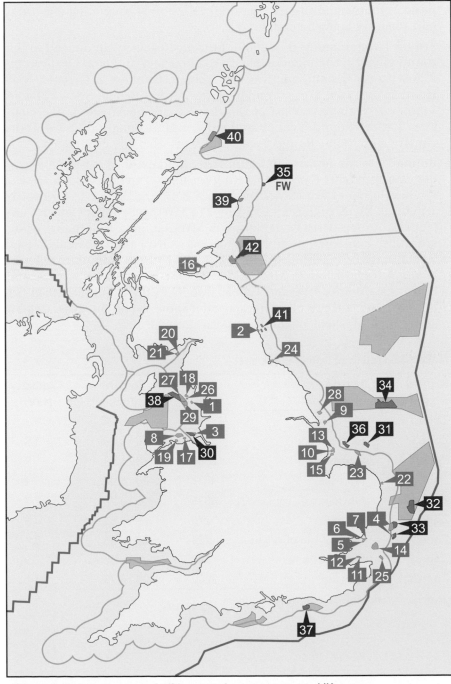

Territorial Water's Limit
UK Continental Shelf
Round 3 offshore projects
FW Floating wind turbines

Figure 8.47 Location map of offshore wind energy project in UK waters

The UK government supported the initial offshore wind farms with capital grants and, through the Crown Estates Office (which is responsible for the area of seabed that surrounds the UK), has allocated licences for three rounds of offshore wind energy development (see Figure 8.47).

- Round One (announced in 2001). This was effectively a demonstration phase to allow offshore developers and wind turbine manufacturers to gain experience and to transfer expertise from the UK's offshore oil and gas industries. Round One sites had water depths less than 20 metres and were situated within 12 km of the coast. Of the seventeen Round One projects that were initially awarded, 15 are now generating electricity and some are being extended.

- Round Two (announced in 2003). Fifteen sites were announced in this round giving a combined capacity of 7000 MW in 2003. Of these the following are operational: Greater Gabbard, Gunfleet 2, Gwynt y Mor, Humber Gateway, Lincs, London Array 1, Sheringham Shoal, Thanet, Walney 1, Walney 2, West of Duddon Sands and Westermost Rough. Two further Round Two sites (Dudgeon and Race Bank) are under construction. Some Round Two projects are also being extended.

- Round Three (announced in 2010). This round is aimed at facilitating the development of 33 GW of generating capacity. The nine areas allocated for development are shown in Figure 8.47 and listed as follows in Table 8.8. Many of these, including 714 MW East Anglia One, 1.2 GW Hornsea Project 1 (HOW01) and 400 MW Rampion (Southern Array), are under construction and several are well in the Consented stage: Bravo (Firth of Forth), Creyke Beck A and B (Dogger Bank), Lackenby A and B (Dogger Bank), MacColl (Moray Firth), Southern Array Zone, Stevensen (Moray Firth) and Telford (Moray Firth). A number are also at the planning and pre-planning stage but some in certain areas, notably at the West of Isle of Wight, have been stopped.

Table 8.8 Details of Round Three sites

Round 3 zone	Name	Area /km²	Depth /m	Capacity /GW
Zone 1	Moray Firth	520	30 to 57	1.3
Zone 2	Firth of Forth	2852	30 to 80	3.5
Zone 3	Dogger Bank	8660	18 to 63	9
Zone 4	Hornsea	4735	30 to 40	4
Zone 5	Norfolk	6036	5 to 70	7.2
Zone 6	Hastings	270	19 to 62	0.6
Zone 7	West of Isle of Wight	723	22 to 56	0.9
Zone 8	Bristol Channel	949	19.5 to 60.9	1.5
Zone 9	Irish Sea	2200	28 to 78	4.2

Source: Derived from Crown Estate, 2008a

The Crown Estate has further awarded exclusivity agreements in Scottish Territorial Waters for ten sites.

In 2009, the *UK Renewable Energy Strategy* (DECC, 2009), was published. This included the target of obtaining more than 30% of electricity from renewables by 2020. It projected that most of this would come from a combination of land-based and offshore wind energy projects.

This has the potential to be a step change in serious support for wind energy development in the UK. If and when these projects are successfully delivered, the UK would then have started to become a major generator of electricity from wind energy. However, as a result of changes in UK government policy in 2016, with regard to support for land based wind farms, future development on land is becoming uncertain even though there is considerable technical potential.

The UK Government 2050 Pathways Analysis (DECC, 2010) mentioned above, assumed a number of scenario trajectories and four levels of offshore wind energy deployment. Level four represents the most ambitious of these. As a result of evidence submitted in response to consultation (DECC, 2011a), DECC doubled its estimates, increasing the potential contribution from offshore wind from 430 TWh y^{-1} to 929 TWh y^{-1} of electricity by 2050 (assuming a 45% annual capacity factor). It also included the possibility of including floating wind turbines. In the December 2011 update of its Carbon Plan, DECC also expanded the proposed contribution from wind energy – both from land and offshore – envisaged in 2050. In this plan, it is not only envisaged that wind energy would be making a major contribution to electricity generation by 2050, but also contributing to low/zero carbon heating of buildings (via heat pumps) and also providing a major proportion of the electricity to power vehicles (in parallel to electrifying the car fleet) in 2050 (DECC, 2011b).

The size of the turbines continues to increase. Those being utilized in the offshore wind farms now under construction are deploying 7 and 8 MW wind turbines from Siemens and MHI-Vestas, which are helping to reduce the cost of offshore wind energy, particularly at the scale of some of these newer projects – some of which exceed a GW in capacity.

8.10 **Summary**

At the global scale, **winds are produced as a result of the transfer of solar heat from the warm tropical regions towards the colder regions** at the North and South poles. Complex swirling weather systems are produced by the effect of the Earth's rotation on the moving air.

Wind turbines extract the **kinetic energy of the moving air. The power in the wind is proportional to the cube of the wind velocity**.

Before the Industrial Revolution, **Britain may have had over 10 000 windmills**.

Modern wind turbines may be **vertical axis** or, more commonly, **horizontal axis**. Commercial horizontal axis turbines are available in ratings from a **few tens of watts** electrical output **up to 8 MW**. An **8 MW turbine** may have **blades 80 m long**. Designs for 10–15 MW output are being considered.

Turbines use **blades with aerofoil sections** (like aircraft wings). As they move through the air it produces **lift** and **drag forces**. The turbine blades **are set at an angle** or **pitch** so that **the lift force drives the blade forwards**, producing rotation of the turbine and extracting power from the air. Often the blade pitch can be varied allowing the turbine power to be varied or to stop rotation completely.

The **maximum theoretical proportion of energy that can be extracted from the wind is 59%**, the **Betz limit**.

The **speed-power curve** of a wind turbine **describes a turbine's output at different wind speeds**. It has several parts. As the wind speed increases there is:

- a **cut-in wind speed** below which it does not produce any power at all
- **a range over which the power increases roughly in proportion to the cube of the wind speed** until it reaches
- **a range over which it produces its full rated power**
- a **shut-down wind** speed above which the turbine's power may be reduced or completely shut down to prevent damage.

The **annual electrical output of a turbine** can be estimated using the **speed-power curve**, as supplied by the manufacturer, and **the wind speed distribution** for a particular location, which can be estimated from the annual average wind speed.

Average wind speeds over a year can vary from under 4 m s^{-1} in sheltered areas on land, up to over 10 m s^{-1} on hills and ridges on land and in the open sea. **The best wind resources** in Europe **on land are in Scotland and the western coasts of Ireland, Denmark and Norway. Offshore, the Irish Sea, the North Sea and the English Channel have high average wind speeds**.

The main environmental problems of wind power are: **visual intrusion, low frequency noise, electromagnetic interference with TV and radar** due to signal reflections from the blades, and **bird collisions**.

The cost of electricity from a wind turbine is dominated by its **capital costs** and **installation costs. Operation and maintenance costs are low, typically 5-8% of the capital cost per year**. Wind turbines do not generate continuously. On land **capacity factors** may vary from **25% in sheltered sites up to 35-40% in windy coastal sites. Offshore** they range from **35% to 50%**. The World Economic Forum (WEF) estimate that the **cost of wind power** has reached **grid parity with conventional generation in 30 countries**.

In 2015 wind power produced 11% of the UK's electricity.

In 2017 the world total installed capacity of wind turbines exceeded 500 GW.

Globally, **the wind power resource** may be **more than 10 times the projected 2025 world electricity consumption**.

Offshore wind speeds are higher and more consistent than on land. In 2015 the UK had the world's largest total capacity of offshore turbines, over 4600 MW. UK offshore locations may be classified as Round 1, Round 2 and Round 3. Round 1 locations are early demonstration sites close inshore; Round 2 sites are further offshore and are now well developed. **Round 3 sites are still further offshore and are currently being developed with large wind farms.**

Floating wind turbines are now being developed that can be **located in deep water.**

References

4C Offshore (2010) *4C Offshore* [Online]. Available at http://www.4coffshore. com (Accessed 8 January 2010).

4C Offshore (2016) *Offshore Wind Overview Report*, [Online], Lowestoft, Suffolk, 4C Offshore Ltd. Available at http://www.4coffshore.com/ windfarms/downloads/samples/20160404_OverviewReportSample.pdf (Accessed 24 February 2017).

Abbott, I. H. and von Doenhoff, A. E. (1958) *Theory of Wing Sections*, New York, Dover Publications Inc.

ABC (2011) *American Bird Conservancy (ABC) policy on bird collisions*, [Online]. Available at http://www.abcbirds.org/abcprograms/psolicy/ collisions/index.html (Accessed 8 November 2011).

AOA (2016) *Safeguarding of Aerodromes: Advice Note 5 Renewable Energy and Impact on Aviation*, [Online], London, Airport Operators Association. Available at http://www.aoa.org.uk/wp-content/uploads/2016/09/Advice-Note-5-Renewable-Energy-2016.pdf (Accessed 24 February 2017).

Appleton, S. (2010) *Stealth blades – a progress report*, [Online], Glasgow, All-Energy Available at http://onlinelibrary.wiley.com/ doi/10.1029/2004JD005462/full (Accessed 20 March 2017).

Archer, C. L. and Jacobson, M. Z. (2005) 'Evaluation of global wind power', *J. Geophys Res.*, 110, D12110, [Online]. Available at http://www.agu.org/ journals/jd/ jd0512/2004JD005402/ (Accessed 8 November 2011).

ASI (2013) *Aviation Planning Considerations for Wind Turbines: Managing the Impact of Wind Turbines on Aviation*, [Online], London, Airspace & safety initiative. Available at http://airspacesafety.com/wp-content/upl oads/2013/09/20130701ManagingTheImpactOfWindTurbinesOnAviati on_Script_FINAL_V1.pdf (Accessed 24 February 2017).

AWEA (2009) *AWEA Small Wind Turbine Performance and Safety Standard – AWEA 9.1* [Online], Washington, American Wind Energy Association. Available at smallwindcertification.org/wp-content/uploads/2011/05/ AWEA_2009-Small_Turbine_Standard1.pdf (Accessed 9 January 2017).

AWWI (2016) *Wind Turbine Interactions with Wildlife and their Habitats: A Summary of Research Results and Priority Questions*, [Online], Washington, American Wind & Wildlife Institute. Available at https://awwi.org/wp-content/uploads/2016/07/AWWI-Wind-Wildlife-Interactions-Summary-June-2016.pdf (Accessed 21 December 2016).

Bacon, D. F. (2002) *Fixed-link wind turbine exclusion zone method, A proposed method for establishing an exclusion zone around a terrestrial fixed radio link outside of which a wind turbine will cause negligible degradation of the radio link performance*, [Online], London, Ofcom. Available at https://www.ofcom.org.uk/__data/assets/pdf_file/0031/68827/ windfarmdavidbacon.pdf (Accessed 1 February 2017).

Barrett, S. B. and DeVita, P. M (2011) *Investigating Safety Impacts of Energy Technologies on Airports and Aviation – A Synthesis of Airport Practice*,

ACRP Synthesis 28, [Online], Washington, Transport Research Board. Available at www.TRB.org (Accessed 24 February 2017).

Barrett, S. B., DeVita, P. M. and Lambert, J. R. (2014) *Guidebook for Energy Facilities Compatibility with Airports and Airspace*, ACRP Report 108, [Online], Washington, Transportation Research Board. Available at www. TRB.org (Accessed 24 February 2017).

BBC (2016) *Windfarm Assessment Tool* [Online]. Available at http://www. bbc.co.uk/reception/info/windfarm_tool.html (Accessed 12 December 2016).

BCT (2009) *Determining the potential ecological impact of wind turbines on bat populations in Britain, Phase 1 report Final Report*, [Online], Scoping & Method Development Report, May. Available at www.bats.org.uk/pages/ wind_turbines.html and http://www.bats.org.uk/data/files/determining_ the_impact_of_wind_turbines_on_british___bats_final_report_29.5.09_ website.pdf (Accessed16 March 2017).

BCT (2010) *Position Statement: Microgeneration Schemes: Risks, Evidence and Recommendations* [Online], London, Bat Conservation Trust. Available at http://www.bats.org.uk/publicatio ns.php?keyword=microgeneration+s chemes&month=&year=&category=&search=Search (Accessed 9 July 2017).

BCT (2017) *Microgeneration Schemes*: A Position Statement from BCT, [Online], Bat Conservation Trust. http://www.bats.org.uk/pages/ microgeneration_issues.html (Accessed 16 March 2017).

BEIS (2013) *Wind turbines, aviation and radar, Onshore wind: part of the UK's energy mix* [Online]. Available at https://www.gov.uk/guidance/ onshore-wind-part-of-the-uks-energy-mix (Accessed on 19 December 2016).

BEIS (2016) *Energy and Climate Change Public Attitude Tracker – Wave 19*, [Online]. London, UK Department for Business, Energy & Industrial Strategy (BEIS), Available at https://www.gov.uk/government/uploads/system/ uploads/attachment_data/file/563236/Summary_of_key_findings_BEIS_ Public_Attitudes_Tracker_-_wave_19.pdf, and accompanying *Wave 19 Summary Tables* spreadsheet, https://www.gov.uk/government/statistics/ public-attitudes-tracking-survey-wave-19, (Accessed 3 January 2017).

Berge, E., Gravdahl, A. R., Schelling, J., Tallhaug, L. and Undeheim, O (2006) *A comparison of WAsP and two CFD-models*, [Online]. Presentation for EWEC 2006 Available at http://www.windchina.info/ cwpp/files/1613/9821/8733/WAsPCFD_WAsP_WindSim_ewec_berge.pdf (Accessed February 2017).

BERR (2008) *Atlas of UK Marine Renewable Energy Resources*, [Online]. Southampton, Department for Business & Regulatory Reform, Available at http://www.renewables-atlas.info (Accessed 17 March 2017).

Beurskens, J. and Jensen, P. H. (2001) 'Economics of wind energy – Prospects and directions', *Renewable Energy World*, July–Aug.

Bright, J. A., Langston, R. H. W., Bullman, R., Evans, R. J., Gardner, S., Pearce-Higgins, J. and Wilson, E. (2006) *Bird Sensitivity Map to provide locational guidance for onshore wind farms in Scotland, RSPB Research Report No 20* [Online], Sandy, Bedfordshire, Royal Society for the Protection of Birds.

Available at https://www.rspb.org.uk/Images/sensitivitymapreport_tcm9-157990.pdf (Accessed 24 February 2017).

Bright, J. A., Langston, R. H. W. and Anthony, S. (2009) *Mapped and written guidance in relation to birds and onshore wind energy development in England, RSPB Research Report No 35* [Online], Sandy, Bedfordshire, Royal Society for the Protection of Birds. Available at http://www.rspb.org.uk/Images/EnglishSensitivityMap_tcm9-237359.pdf (Accessed 24 February 2017).

BSI (2013) Wind turbines: Acoustic noise measurement techniques, (BS) EN 64100-11:2013, London, British Standards Institution.

Burch, S. F. and Ravenscroft, F. (1992) *Computer Modelling of the UK Wind Energy Resource: Final Overview Report*, ETSU WN7055, ETSU Oxon, Crown copyright.

Burroughs, W. J., Crowder, B., Robertson, E., Vallier-Talbot, E. and Whitaker, R. (1996) *Weather – the ultimate guide to the elements*, London, HarperCollins.

Butler, G. (2007) 'Advanced Digital Tracker – Update' at All Energy '07 Aviation and technical Workshop, May http://www.all-energy.co.uk/UserFiles/File/2007GeoffButler.pdf (Accessed 9 December 2011).

BWEA (2008) BWEA Small Wind Turbine Performance and Safety Standard, 29 February. Available at https://www.solarcollect.co.uk/downloads/bwestandard2008.pdf (Accessed 17 March 2017).

CAA (2010) *Lighting of En-Route Obstacles and Onshore Wind Turbines*, Policy Statement, [Online], London, Directorate of Airspace Policy, Civil Aviation Authority. Available at https://publicapps.caa.co.uk/docs/33/DAP_LightingEnRouteObstaclesAndWindTurbines.pdf (Accessed 24 February 2017).

CAA (2015) *ACN-2015-00-0460: Liverpool Windfarm Trial (AL1)*, [Online], London, Safety & Airspace Regulations Group, Civil Aviation Authority. Available at https://publicapps.caa.co.uk/docs/33/2015000460ACN11LverpoolWindfarmTrial.pdf (Accessed 24 February 2017).

CAA (2016) *CAP 764 - CAA Policy and Guidelines on Wind Turbines*, Safety & Airspace Regulations Group, [Online], London, Civil Aviation Authority. Available at https://publicapps.caa.co.uk/docs/33/CAP764%20Issue6%20FINAL%20Feb.pdf (Accessed 24 February 2017).

Carbon Trust (2008) *Offshore wind power: big challenge, big opportunity*, [Online], London, Carbon Trust. Available at https://www.carbontrust.com/resources/reports/technology/offshore-wind-power/ (Accessed 24 February 2017).

Carbon Trust (2016) *European offshore wind developers join forces with the Carbon Trust to slash costs of offshore wind*, Press Release 11 July 2016, https://www.carbontrust.com/about-us/press/2016/07/european-offshore-wind-developers-join-forces-with-the-carbon-trust-to-slash-costs-of-offshore-wind (Accessed 17 March 2017).

Chignell, R. J. (1987) *Electromagnetic Interference from Wind turbines – A Simplified Guide to Avoiding Problems*, East Kilbride, National Wind Turbine Centre, National Engineering Laboratory.

Clark, G (2015) *House of Commons: Written Statement (HCWS42) - Local Planning*, [Online], London, Department for Communities and Local Government. Available at http://www.parliament.uk/documents/commons-vote-office/June%202015/18%20June/1-DCLG-Planning.pdf (Accessed 3 January 2017).

Cobbett, W (1830) *Rural rides in the southern, western and eastern counties of England*, London, Penguin Classics.

Corbetta, G., Ho, A. and Pineda, I. (2015) *Wind energy scenarios for 2030*, [Online], Brussels, Belgium, WindEurope. Available at https://www.ewea.org/fileadmin/files/library/publications/reports/EWEA-Wind-energy-scenarios-2030.pdf (Accessed 24 February 2017).

Crown Estate (2008a) *Round 3 offshore wind farms table*, The Crown Estate, June.

Crown Estate (2008b) *Round 3 Offshore wind farms map*, The Crown Estate, http://www.thecrownestate.co.uk/media/158181/round3_map.pdf (Accessed 8 November 2011).

Crown Estate (2012) *Offshore Wind Cost Reduction Pathways Study*, [Online], London, The Crown Estate. Available at https://www.thecrownestate.co.uk/media/5493/ei-offshore-wind-cost-reduction-pathways-study.pdf (Accessed 9 Jan 2017).

Crown Estate (2016) *Offshore wind operational report: January – December 2015*, [Online], London, Crown Estate. Available at https://www.thecrownestate.co.uk/media/5462/ei-offshore-wind-operational-report-2016.pdf (Accessed 24 February 2017).

Day, A., Dance, S., Moseley, T. and Dunlop, B. (2010) *Ashenden Wind turbine trial: Phase II Progress Report*, [Online], London, South Bank University. Available at https://www.wind-power-program.com/Library/Performance%20of%20individual%20wind%20turbine%20installations/ashenden_turbine_trial_end_of_phase_II(QR5).pdf (Accessed 24 February 2017).

DCLG (2007) *Domestic Installation of Microgeneration Equipment – Final report from a Review of the related Permitted Development Regulations*, [Online], London, Department for Communities and Local Government. Available at https://planningjungle.com/wp-content/uploads/Domestic-Installation-of-Microgeneration-Equipment-Final-report-from-a-Review-of-the-related-Permitted-Development-Regulations-April-2007-Part-1.pdf (Accessed 24 February 2017).

DCLG (2009) *Permitted development rights for small scale renewable and low carbon energy technologies, and electric vehicle charging infrastructure – Consultation*, [Online], London, Department for Communities and Local Government. Available at https://planningjungle.com/wp-content/uploads/Permitted-development-rights-for-small-scale-renewable-and-low-carbon-energy-technologies-and-electric-vehicle-charging-infrastructure-Consultation-November-2009.pdf (Accessed 17 March 2017).

DCLG (2011) *The National Planning Policy Framework*, [Online], London, Department for Communities and Local Government. Available at https://www.publications.parliament.uk/pa/cm201012/cmselect/cmcomloc/1526/1526.pdf (Accessed 24 February 2017).

DCLG (2012) *National Planning Policy Framework*, [Online], London DCLG. Available at https://www.gov.uk/government/uploads/system/uploads/attachment_data/file/6077/2116950.pdf (Accessed 17 March 2017).

DCLG (2013) *Planning practice guidance for renewable and low carbon energy*, [Online], London, Department for Communities and Local Government. Available at https://www.gov.uk/government/uploads/system/uploads/attachment_data/file/225689/Planning_Practice_Guidance_for_Renewable_and_Low_Carbon_Energy.pdf (Accessed 24 February 2016).

DCLG (2015) *The Town and Country Planning Order 2015, No. 596, Part 14* [Online], Surrey, The National Archives. Available at http://www.legislation.gov.uk/uksi/2015/596/pdfs/uksi_20150596_en.pdf (Accessed 24 February 2017).

DECC (2009) *the UK Renewable Energy Strategy*, [Online], London, Department of Energy and Climate Change, HMSO. Available at https://www.gov.uk/government/uploads/system/uploads/attachment_data/file/228866/7686.pdf (Accessed 24 February 2017).

DECC (2010) *2050 Pathways Analysis*, [Online], London, Department of Energy and Climate Change. Available at https://www.gov.uk/government/uploads/system/uploads/attachment_data/file/42562/216-2050-pathways-analysis-report.pdf (Accessed 24 February 2017).

DECC (2011a) *2050 Pathways Analysis – Response to Call for Evidence*, [Online], London, Department of Energy and Climate Change. Available at https://www.gov.uk/government/uploads/system/uploads/attachment_data/ile/68821/2050-pathways-analysis-response-pt1.pdf (Accessed 24 February 2017).

DECC (2011b) *The Carbon Plan: Delivering our low carbon future*, London, Department of Energy and Climate Change. Available at https://www.gov.uk/government/uploads/system/uploads/attachment_data/file/47613/3702-the-carbon-plan-delivering-our-low-carbon-future.pdf (Accessed 24 February 2017).

DECC (2015) *The Aviation Plan – In respect of the interaction of wind turbines and aviation interests*, [Online], London, Department of Energy and Climate Change. Available at https://www.gov.uk/government/uploads/system/uploads/attachment_data/file/397208/Aviation_Plan_Update_2015_FINAL.pdf (Accessed 24 February 2017).

DECC (2016) *Energy and Climate Change Public Attitude Tracker – Wave 17*, London, Department of Energy & Climate Change. Available at https://www.gov.uk/government/statistics/public-attitudes-tracking-survey-wave-17 (Accessed 24 February 2017).

DECC (2017) Department of Energy and Climate Change's wind speed database website URL: http://webarchive.nationalarchives.gov.uk/20121217150421/http:/decc.gov.uk/en/content/cms/meeting_energy/wind/onshore/deploy_data/windsp_databas/windsp_databas.aspx (Accessed 17 March 2017) or http://www.bwea.org/noabl/index.html (Accessed 17 March 2017).

DIO (2014) *Guidance – Wind farms: MOD safeguarding*, [Online], Sutton Coldfield, West Midlands, Defence Infrastructure Organisation, Ministry of Defence. Available at https://www.gov.uk/government/publications/wind-farms-ministry-of-defence-safeguarding/wind-farms-mod-safeguarding (Accessed 24 February 2017).

DNV-GL (2014) *Electrifying the Future*, Oslo, DNV-GL

Drewitt, A. L. and Langston, R. H. W. (2006) 'Assessing the impacts of wind farms on birds', *British Ornithologists' Union*, vol. 148, pp. 29–42.

EC (2009) *Investing in the Development of low Carbon technologies (SEt-Plan)*, [Online], Brussels, European Commission. Available at http://www.biofuelstp.eu/downloads/2009_comm_investing_development_low_carbon_technologies.pdf (Accessed 24 February 2017).

EEA (2009) *Europe's on shore and offshore wind energy potential – An assessment of environmental and economic constraints, EEA Technical Report No. 6/2009,* [Online], Copenhagen, European Environment Agency. Available at http://www.eea.europa.eu/publications/europes-onshore-and-offshore-wind-energy-potential (Accessed 24 February 2017).

Eldridge, F. R. (1975) *Wind Machines*, (Report to the NSF, Mitre Corporation, USA, 1975).

English Nature, RSPB, WWF-UK, BWEA (2001) *Wind farm development and nature conservation*, Sandy, Bedfordshire, RSPB.

Erickson, W., Rabie, P., Taylor, K. and Bay, K. (2015) *Comparison of avian mortality sources and evaluation and development of compensatory mitigation options for birds, National Wind Coordinating Collaborative 2014 Washington*, American Wind Wildlife Institute.

Ernst & Young (2015) *Offshore wind in Europe: Walking the tightrope to success*, [Online], France, Ernst & Young. Available at http://www.ewea.org/fileadmin/files/library/publications/reports/EY-Offshore-Wind-in-Europe.pdf (Accessed 24 February 2017).

EST (2009) *Location, location, location: Domestic small-scale wind field trial*, [Online], London, Energy Savings Trust. Available at http://www.solacity.com/docs/BEST%20turbine%20trial%20report.pdf (Accessed 24 February 2017).

EST (2016) *Wind Speed Predictor*, [Online]. http://www.windspeed.energysavingtrust.org.uk/ (Accessed 8 December 2016).

EU (2010) *The European Strategic Energy technology Plan – SET-Plan* [Online]. Available at https://ec.europa.eu/energy/en/topics/technology-and-innovation/strategic-energy-technology-plan (Accessed 24 February 2017).

Everett, B., Boyle, G. A., Peake S. and Ramage, J. (2012) *Energy Systems and Sustainability: Power for a Sustainable Future* 2nd edn, Oxford, Oxford University Press/Milton Keynes, The Open University.

EWEA (1991) T*ime for Action: Wind Energy in Europe,* Brussels, European Wind Energy Association.

EWEA (2010) *2050: Facilitating 50% Wind Energy – Recommendations on transmission infrastructure, system operation and electricity market integration*, [Online], Brussels, European Wind Energy Association. Available at http://www.ewea.org/fileadmin/files/library/publications/position-papers/EWEA_2050_50_wind_energy.pdf (Accessed 27 February 2017).

EWEA (2012) *Workshop on Wind Turbine Noise: From Source to Receiver,* [Online]. Available at http://www.ewea.org/events/workshops/past-workshops/wind-turbine-noise/ (Accessed 12 December 2016).

EWEA (2014) *Saving water with wind energy* [Online], Brussels, European Wind Energy Association. Available at https://windeurope.org/fileadmin/files/library/publications/reports/Saving_water_with_wind_energy.pdf (Accessed 27 February 2017).

EWEA (2016) *The European offshore wind industry – key trends and statistics 2015* [Online], Brussels, European Wind Energy Association. Available at https://www.ewea.org/fileadmin/files/library/publications/statistics/EWEA-European-Offshore-Statistics-2015.pdf (Accessed 27 February 2017).

Fields, J., Oteri, F., Preus, R. and Baring-Gould, I. (2016) *Deployment of Wind Turbines in the Built Environment: Risks, Lessons, and Recommended Practices* [Online], Washington, National Renewable Energy Laboratory. Available at http://www.nrel.gov/docs/ty16osti/65622.pdf

Fiumicelli, D. and Triner, N. (2011) *Wind Farm Noise Statutory Nuisance Complaint Methodology* [Online], Beckenham, AECOM for DEFRA. Available at https://www.gov.uk/government/uploads/system/uploads/attachment_data/file/69222/pb-13584-windfarm-noise-statutory-nuisance.pdf (Accessed 27 February 2017).

FWC (2017) *Floating windfarms corporation* [Online]. Available at http://floatingwindfarms.com/index.html (Accessed 31 January 2017).

Golding, E. W. (1955) 'Generation of Electricity by Wind Power', London, E. and F. N. Spon.

Gorlov, A. (1998) *Development of the helical reaction hydraulic turbine, Final Technical Report (DE-FG01-96EE 15669)*, [Online], Washington, Northeastern University for US Department of Energy. Available at https://www.osti.gov/scitech/servlets/purl/666280/ (Accessed 27 February 2017).

Greenacre, P., Gross, R. and Heptonstall, P. (2010) *Great Expectations: the cost of offshore wind in UK waters – understanding the past and projecting the future*, London, UK Energy Research Centre.

Greenblatt, J. B. (2005) *Wind as a source of energy, now and in the future*, Amsterdam, Environment Defense for InterAcademy Council.

Griffiths, T., Paquette, J., Barone, M., Owens, B. and Bull, D. (2016) 'Floating Offshore Vertical Axis Wind Turbine Project', Poster presented at *AWEA 2016*, SAD2016-4691C, Sandia National Laboratories.

Grubb, M. and Meyer, N. (1993) *Wind energy: resources, systems and regional strategies*, in Johansson, T.B. et.al (eds) Renewable Energy – Sources for Fuels and Electricity, Washington, Island Press.

GWEC (2016a) *Global Wind Report – Annual Market Update for 2015*, [Online]. Available at http://www.gwec.net/wp-content/uploads/vip/GWEC-Global-Wind-2015-Report_April-2016_19_04.pdf (Accessed 19 March 2017).

GWEC (2016b) *Global Wind Energy Outlook – 2016*, [Online]. Available at http://www.gwec.net/publications/global-wind-energy-outlook/global-wind-energy-outlook-2016/ (Accessed 19 March 2017). (Accessed 27 February 2017).

Hayhow, D. B., Burns, F., Eaton, M. A., Al Fulaij, N., August, T. A., Babey, L., Bacon, L., Bingham, C., Boswell, J., Boughey, K. L., Brereton, T., Brookman, E., Brooks, D. R., Bullock, D. J., Burke, O., Collis, M., Corbet, L., Cornish, N., De Massimi, S., Densham, J., Dunn, E., Elliott, S., Gent, T., Godber, J., Hamilton, S., Havery, S., Hawkins, S., Henney, J., Holmes, K., Hutchinson, N., Isaac, N. J. B., Johns, D., Macadam, C. R., Matthews, F., Nicolet, P., Noble, D. G., Outhwaite, C. L., Powney, G. D., Richardson, P., Roy, D. B., Sims, D., Smart, S., Stevenson, K., Stroud, R. A., Walker, K. J., Webb, J. R., Webb, T. J., Wynde, R. and Gregory, R. D. (2016) *State of Nature 2016*, [Online], Newark, State of Nature partnership, http://www.wildlifetrusts.org/sites/default/files/state_of_nature_uk_report_pages_1_sept.pdf (Accessed 28 February 2017).

Health Canada (2014) *Wind Turbine Noise and Health Study* [Online]. Available at http://www.hc-sc.gc.ca/ewh-semt/noise-bruit/turbine-eoliennes/summary-resume-eng.php (Accessed on 28 February 2017).

IEA (2016) *World Energy Outlook* [Online]. Available at https://www.iea.org/newsroom/news/2016/november/world-energy-outlook-2016.html (Accessed 28 February 2017).

IEC (2012) *Wind turbine generator systems – Part 11: Acoustic noise measurement techniques, IEC Standard 61400-11,* Geneva, International Electrotechnical Commission.

IOA, 2013 *A good practice guide to the application of ETSU-R-97 for the assessment and rating of wind turbine noise*, [Online], St. Albans, Hertfordshire, The Institute of Acoustics. Available at http://www.ioa.org.uk/sites/default/files/IOA%20Good%20Practice%20Guide%20on%20Wind%20Turbine%20Noise%20-%20May%202013.pdf (Accessed 28 February 2017).

IOA (2016) *A Method for Rating Amplitude Modulation in Wind Turbine Noise*, [Online], St. Albans, Hertfordshire, The Institute of Acoustics. Available at http://ioa.org.uk/sites/default/files/AMWG%20Final%20Report-09-08-2016_0.pdf (Accessed 28 February 2017).

IPCC (2011) *Special Report of Renewable Energy Sources and Climate Mitigation (SEREN) - Summary Report*, Abu Dhabi, Inter-government Panel on Climate Change.

Jacobson, M. Z., Delucchi, M. A., Bauer, Z. A. F., Goodman, S. C., Chapman, W. E., Cameron, M. A., Bozonnat, C., Chobadi, L., Clonts, H. A., Enevoldsen, C. P., Erwin, J. R., Fobi, S. N., Goldstrom, O. K., Hennessy, E. M., Liu, J., Lo, J. Meyer, C. B., Morris, S. B., Moy, K. R., O'Neill, P. L., Petkov, I., Redfern, S., Schucker, R., Sontag, M. A., Wang, J., Weiner, E. and Yachanin, A. S. (2016) *100% Clean and Renewable Wind, Water, and Sunlight (WWS) All Sector Energy Roadmaps for 139 Countries of the World*, [Online], California, Stanford University. Available at https://web.stanford.edu/group/efmh/jacobson/Articles/I/CountriesWWS.pdf (Accessed 1 March 2017).

Jago, P. and Taylor, N. (2002) *Wind turbines and aviation interests – European Experience and Practice*, STAYSIS Ltd for the DTI, ETSU W/14/00624/REP (DTI PUB URN No 03/5151). [Online], Brussels, Association pour la Promotion des Energies Renouvelables. Available at http://www.apere.org/manager/docnum/doc/doc1288_aviation.fiche115 (Accessed 1 March 2017).

James, R & Costa Ros, M (2015) *Floating Offshore Wind: Market & Technology Review* [Online], London, Carbon Trust. Available at https://www.carbontrust.com/media/670664/floating-offshore-wind-market-technology-review.pdf (Accessed 1 March 2017).

Jamieson, A. (2012) *Offshore Wind Cost Reduction Task Force Report, June 2012*, [Online], London, RenewableUK. Available at https://www.gov.uk/government/uploads/system/uploads/attachment_data/file/66776/5584-offshore-wind-cost-reduction-task-force-report.pdf (Accessed 1 March 2017).

JRC (2014a) *Calculation of Wind Turbine clearance zones for JRC managed fixed services with particular reference to UHF (460MHz) Telemetry Systems when turbine sizes and locations are accurately known*, Issue 4.2, Joint Radio Company, December 2014. Available at http://www.jrc.co.uk/sites/default/files/Calculation%20of%20the%20Clearance%20Zone%204.2.1_0.pdf (Accessed 20 March 2017).

JRC (2014b) *Procedure for coordination with wind energy developments*, Issue 2.1, Joint Radio Company, December 2014. Available at http://www.jrc.co.uk/sites/default/files/JRC_Procedure_for_co-ordination_with_wind_energy_developments-2.1.0.pdf (Accessed 20 March 2017).

JRC (2016) *Radio Link Coordination with Wind Farms*, [Online]. Available at http://www.jrc.co.uk/wind-farms (Accessed 12 December 2016).

Kendrick, M. (2017) 'Does diet play a role in cardiovascular disease' presented at 2017 Public Health Collaboration Conference, Manchester, 17-18 June.

Kragh, J. et al. (1999) *Noise immission from wind turbines*, National Engineering Laboratory for the Energy Technology Support Unit (ETSU), ETSU W/13/00503/REP.

LCTU (2013) *Energy Policy Toolkit on Physical Planning of Wind Power: Experiences from Denmark*, [Online], Copenhagen, Danish Energy Agency.

Available at https://ens.dk/sites/ens.dk/files/Globalcooperation/physical_ planning_of_wind_power.pdf (Accessed 1 March 2017).

Legerton, M. (ed.) (1992) *Wind turbine Noise Workshop Proceedings*, ETSU-N-123, Kirkstile, Department of Trade and Industry/British Wind Energy Association.

Ljunggren, S. and Gustafsson, A. (1994) *Recommended practices for Wind turbine testing. 4. Acoustics Measurement of Noise Emission from Wind turbines* (3rd edn), submitted to the Executive Committee of the International Energy Agency Programme for Research and Development on Wind Energy Conversion Systems.

Ljunggren, S. (1997) Recommended practices for Wind turbine testing. 10. Measurement of Noise Emission from Wind turbines, submitted to the Executive Committee of the International Energy Agency Programme for Research and Development on Wind Energy Conversion Systems.

London Array (2012) The world's largest offshore wind farm, [Online], London, London Array. Available at https://web.archive.org/ web/20110920071946/http://www.londonarray.com/downloads/london_ array_brochure.pdf. (Accessed 1 March 2017).

London Array (2016) *Renewable Energy Record Achieved at London Array*, [Online], London, London Array. Available at http://www.londonarray. com/project/renewable-energy-record-achieved-at-london-array/ (Accessed 1 March 2017).

Marsh, G (2010) 'Going stealthy with composites', Reinforced Plastics Magazine, November/December 2010 issue. Available at url http://www. materialstoday.com/composite-applications/features/going-stealthy-with-composites/ (accessed 1 February 2017).

Marshall, C (2016) 'Nature loss linked to farming intensity', BBC News, 14 September [Online]. Available at www.bbc.co.uk/news/science-environment-37298485 (Accessed 1 March 2017).

Mathews, F., Richardson, S., Lintott, P. and Hosken, D. (2016) *Understanding the Risk to European Protected Species (bats) at Onshore Wind Turbines to inform Risk Management*, Exeter, University of Exeter.

Mathiesen, J, (2014) 'Carbon Trust Offshore Wind Accelerator: Driving down the cost of offshore wind,' Presentation at *EERA DeepWind 2014*, http://www.sintef.no/globalassets/project/deepwind2014/presentations/ closing/j-matthiesen_carbon-trust.pdf (Accessed 9 January 2017).

McMillan, D. and Ault, G. (2007) 'Quantification of Condition Monitoring Benefit for Offshore Wind Turbines', *Wind Engineering*, vol. 31, no. 4, pp. 267–285.

MCS (2009) *MCS 006 Issue 1.5 – Product certification scheme requirements: Micro and Small Wind turbines*, MCS Working Group 3 'Micro and Wind Systems' for DECC, http://www.microgenerationcertification.org/images/ MCS%20006%20-%20Issue%202.1%20Product%20Certification%20 Scheme%20Requirements%2015%20Jan%202014.pdf (Accessed 9 December 2011).

MCS (2015) *MCS 020 Planning standards for permitted development installations of wind turbines and air source heat pumps on domestic premises, Issue 1.2*, [Online], London, Department of Energy and Climate Change. Available at http://regulations.completepicture.co.uk/pdf/Mcs/MCS%20020%20Planning%20Standards%20Issue%201.2.pdf (Accessed 8 July 2017).

MCS (2017) *Microgeneration Certification Scheme*, [Online]. Available at www.microgenerationcertification.org (Accessed 1 March 2017).

Milborrow, D., (2013) 'Turbine advances cut O&M costs', *WindPower Monthly* magazine, 1 June [Online]. Available at www.windpowermonthly.com/article/1183992/turbine-advances-cut-o-m-costs (Accessed 9 January 2017).

MIS (2015) *MIS 3003: Issue 3.4 Microgeneration Installation Standard: Requirements for MCS contractors undertaking the supply, design. Installation, set to work commissioning and handover of micro and small wind turbine systems* [Online], London, Microgeneration Certification Scheme. Available at http://www.microgenerationcertification.org/images/MIS_3003_Issue_3.4_Micro_Wind.pdf (Accessed 20 March 2017).

Musgrove, P. J. (1990) 'Vertical axis WECS design' in Freris, L. L. (1990) *Wind Energy Conversion Systems*, Prentice Hall.

Musial, W., Heimiller, D., Beiter, P., Scott, G. and Draxi, C. (2016) *2016 Offshore Wind Energy Resource for the United States*, Colorado, National Renewable Energy Laboratory.

NATS (2016) *Self-assessment maps*, [Online]. Available at http://www.nats.aero/services/information/wind-farms/self-assessment-maps/ (Accessed 1 March 2017).

Natural England (2009a) Technical Information Note TIN051: *Bats and onshore wind turbines – Interim guidance*, Worcester, Natural England.

Natural England (2009b) Technical Information Note TIN059: *Bats and single large wind turbines: Joint Agencies interim guidance*, Worcester, Natural England.

Natural England (2010) Technical Information Note TIN069: *Assessing the effects of onshore wind farms on birds*, Worcester, Natural England.

Needham, J. (1965) *Science and Civilisation of China*, Cambridge, Cambridge University Press.

NERI (1998) *Impact Assessment of an Off-shore Wind Park on Sea Ducks, Technical Report No. 227*, Denmark, National Environmental Research Institute (NERI).

NOP (2005) *Survey of Public Opinion of Wind Farms*, NOP (National Opinion Poll (UK)

NPL (2016) *Wind turbine Noise Model*, [Online]. Available at http://resource.npl.co.uk/acoustics/techguides/wtnm/ (Accessed 13 December 2016).

NRC (2007) *Environmental Impacts of Wind Energy Projects, Committee on Environmental Impacts on Wind Energy*, Washington DC, National Academies Press.

NWCC (2007) *NWCC Mitigation toolbox*, Washington, National Wind Coordinating Collaborative.

NWCC (2011) *Comprehensive Guide to Studying Wind Energy/Wildlife Interactions*, Washington, National Wind Coordinating Collaborative.

ODPM (2004a) *Planning Policy Statement 22 (PPS22): Renewable Energy*, Norwich, Office of the Deputy Prime Minister.

ODPM (2004b) *Planning for Renewable Energy – A Companion Guide to PPS22*, Norwich, Office of the Deputy Prime Minister.

Ofcom (2009) *Tall structures and their impact on broadcast and other wireless services*, https://www.ofcom.org.uk/__data/assets/pdf_file/0026/63494/tall_structures.pdf (Accessed 20 March 2017).

Ofcom (2017) *Fixed terrestrial links: Windfarms and Wireless Services: Coordination and Guidance*, [Online]. Available at https://www.ofcom.org.uk/manage-your-licence/radiocommunication-licences/fixed-terrestrial-links (Accessed 1 March 2017).

ORE-Catapult (2016) *SPARTA: The performance data exchange platform for offshore wind*, [Online], Glasgow, Catapult and London, The Crown Estate. Available at https://ore.catapult.org.uk/our-services/ (Accessed 1 March 2017).

ORE-Catapult (2017) Operations & Maintenance (O&M). Offshore Renewable Energy (ORE) Catapult, https://ore.catapult.org.uk/our-services/innovation-challenges/offshore-wind-innovation-challenges/operations-mantenance (Accessed 9 January 2017).

OVG (2010) *the Offshore Valuation – A valuation of the UK's offshore renewable energy resource*, [Online], Offshore Valuation Group and Public Interest Research Centre. Available at http://publicinterest.org.uk/offshore/downloads/offshore_valuation_exec.pdf (Accessed 1 March 2017).

PB (2011) *Update of UK Shadow Flicker Evidence Base – Final Report*, Newcastle Upon Tyne, Parsons Brinckerhoff.

Planning Guidance (2015) *Renewable and low carbon energy: Particular planning considerations for hydropower, active solar technology, solar farms and wind turbines*, Planning Guidance, Department of Communities and Local Government, available at https://www.gov.uk/guidance/renewable-and-low-carbon-energy#particular-planning-considerations-for-hydropower-active-solar-technology-solar-farms-and-wind-turbines (Accessed 19 March 2017).

Randhawa, B. S. and Rudd, R. (2009) *RF Measurement Assessment of Potential Wind Farm Interference to Fixed links and Scanning telemetry Devices, ERA Report Number 2008-0568* Leatherhead, Surrey, ERA Technology Ltd for Ofcom.

RenewableUK (2010a) *Wind Energy Generation Costs, Fact Sheet 04, June*, [Online], London, RenewableUK. Available at http://www.solarventus-energy.co.uk/Wind-Energy-Generation-Costs.pdf (Accessed 1 March 2017).

RenewableUK (2010b) *Onshore Costs/Benefit Study*, London, RenewableUK.

RenewableUK (2011a) *Planning Guidance: Small Wind – A good practice guide*, [Online], London. Available at http://www.caithnesswindfarms.co.uk/Small%20wind.pdf (Accessed 1 March 2017).

RenewableUK (2011b) *Planning Guidance: Small Wind – Appendices*, London, RenewableUK. Available at Available at https://c.ymcdn.com/sites/renewableuk.site-ym.com/resource/resmgr/publications/guides/small_wind_planning_a.pdf (Accessed 20 March 2017).

RenewableUK (2014) *Small Wind Turbine Standard*, London, RenewableUK. Available at http://smallwindcertification.org/wp-content/uploads/2014/08/RUK-Small-Wind-Turbine-Standard3.pdf (Accessed 20 March 2017)

RenewableUK (2015) *Guidance & Supporting Procedures on the Application of Wind Turbine Safety Rules*, London, RenewableUK. RenewableUK Available at http://c.ymcdn.com/sites/www.renewableuk.com/resource/collection/ae19eca8-5b2b-4ab5-96c7-ecf3f0462f75/Wind-turbine-safety-rules-guidance.pdf?hhSearchTerms=%22small+and+wind+and+turbine+and+standard%22 (Accessed 20 March 2017)

RenewableUK (2016a) *Offshore Renewables Aviation Guidance (ORAG); Good Practice Guidelines for Offshore Renewable Energy Developments*, London, RenewableUK. Available at http://c.ymcdn.com/sites/renewableuk.site-ym.com/resource/resmgr/publications/reports/orag_aviationguidance.pdf (Accessed 20 March 2017).

RenewableUK (2016b) *New records set in best ever year for British wind energy generation*, London, RenewableUK. Available at http://c.ymcdn.com/sites/renewableuk.site-ym.com/resource/resmgr/publications/reports/orag_aviationguidance.pdf (Accessed 20 March 2017).

RenewableUK (2017) UK wind speed database [Online]. Available at http://www.bwea.org/noabl/ (Accessed 1 March 2017).

Reynolds, J. (1970) *Windmills and Watermills*, London, Hugh Evelyn Ltd Publishers.

Risø (2009) *European wind resources over open sea*, [Online]. Available at http://www.windatlas. dk/europe/oceanmap.html (Accessed 9 November 2011).

Rodrigues, L. et al. (2014) *Guidelines for consideration of bats in wind farm projects*, Revision 2014, EUROBATS Publication Series No. 6. Available at www.eurobats.org/sites/default/files/documents/publications/publication_series/pubseries_no6_english.pdf (Accessed 20 July 2017).

Rogers, A. L., Rogers, J. W. and Manwell, J. F. (2005) 'Comparison of the Performance of Four Measure-Correlate-Predict Algorithms', *Journal of Wind Engineering and Industrial Aerodynamics*, vol. 93, no. 3, pp. 243–264.

Rogers, A. L., Rogers, J. W., Manwell, J. F. (2006a) 'Uncertainties in Results of Measure-Correlate-Predict Analysis', *European Wind Energy Conference*, February/March.

Rogers, A. L., Manwell, J. F., Wright, S. (2006b) *Wind turbine Acoustic Noise – A White Paper*, Amherst, Renewable Energy Research Laboratory, University of Massachusetts.

Van Rooij, R. and Timmer, N. (2004) *Design of Airfoil for Wind turbine Blades*, The Netherlands, Delft University of Technology.

RSPB (2015) *UK Onshore Bird Sensitivity Map*, [Online]. Available at http://www.rspb.org.uk/Images/Onshore_wind_sensitivity_2015_tcm9-432921.pdf (Accessed 1 March).

RSPB (2016a) *Lodge wind turbine*, [Online]. Available at http://www.rspb.org.uk/reserves-and-events/find-a-reserve/reserves-a-z/reserves-by-name/t/thelodge/windturbine/index.aspx (Accessed 1 March 2017).

RSPB (2016b) *The RSPB's 2050 energy vision: Meeting the UK's climate targets with nature*, [Online], Sandy, Bedfordshire, Royal Society for the Protection of Birds. Available at www.rspb.org.uk/images/energy_vision_summary_report_tcm9-419580.pdf (Accessed 1 March 2017).

SDC (2005) *Wind Power in the UK: A guide to key issues surrounding wind power development in the UK* [Online]. Available at https://research-repository.st-andrews.ac.uk/handle/10023/2306 (Accessed 1 March 2017).

Smith, A., Stehly, T. and Musail, W. (2015) *2014-2015 Offshore Wind Technologies Market Report, Technical Report NREL/TP-5000-64283* Colorado, National Renewable Energy Laboratory.

South, P. and Rangi, R. (1972) *Wind tunnel investigation of a 14 feet diameter vertical axis windmill, Report Number LTR-LA-105*, Ontario, Canada, National Aeronautical Establishment, National Research Council.

Stiedal, H (2014) *Innovate to Generate – Making Offshore Wind a Truly Global Energy Source* [Online], London, Carbon Trust. Available at https://www.carbontrust.com/media/571800/carbon-trust-innovate-to-generate-18-09-14.pdf (Accessed 1 March 2017).

Sutherland, H. J., Berg, D. E. and Ashwill, T. D. (2012) *A Retrospective of VAWT Technology*, Albuquerque, New Mexico, Sandia National Laboratories.

Tangler, J. L. and Somers, D. M. (1995) *NREL Airfoil Families for HAWTs*, Colorado, National Renewable Energy Laboratory.

Taylor, D. A. (1998) 'Using buildings to harvest wind energy' in *Building Research and Information*, E & FN Spon.

TNEI (2007) *Onshore Wind Energy Planning Conditions Guidance Note, A report for the Renewables Advisory Board and BERR*, Newcastle, TNEI Services Ltd.

Troen, I. and Lundtang Petersen, E. (1989) *European Wind Atlas*, Roskilde, Risø National Laboratory.

Udell, D., Infield, D. and Watson, S. (2010) 'Low-cost mounting arrangements for building-integrated wind turbines', *Wind Energy*, vol. 13, no. 7, pp. 657–669.

US DOE (2015) *Wind Vision: A New Era for Wind Power in the United States*, Washington, US Department of Energy.

USFWS (2010) *Wind turbine Guidelines Advisory Committee – Recommended Guidelines*, Washington, US Fish and Wildlife Service.

USFWS (2012) *Land-Based Wind Energy Guidelines*, [Online], Washington, US Fish and Wildlife Service. Available at https://www.fws.gov/ecological-services/es-library/pdfs/WEG_final.pdf (Accessed 1 March 2017).

VertAx (2017) *VertAx Wind Ltd* [Online]. Available at http://vertaxwind.com/ (Accessed 2 March 2017).

VPM (2017) *VAWTPower Management, Inc.* [Online]. Available at http://www.vawtpower.blogspot.co.uk/ (Accessed 2 March 2017).

WEC (2016) *World Energy Resources: Wind 2016*, London, World Energy Council.

WEF (2017) *Renewable Infrastructure Investment Handbook: A Guide for Investors*, [Online], Cologny, Geneva, Switzerland, World Economic Forum. Available at www3.weforum.org/docs/WEF_Renewable_Infrastructure_Investment_Handbook.pdf (Accessed 2 March 2017).

Westmill Wind Farm Co-operative (2017) *Westmill Wind Farm Co-operative* [Online]. Available at http://www.westmill.coop/ (Accessed 2 March 2017).

WHO (2014) *Burden of disease from Household Air Pollution for 2012*, Geneva, Switzerland, World Health Organization.

WHO (2016) *Air Pollution. A global assessment of exposure and burden of disease*, Geneva, Switzerland, World Health Organization.

Willow, C. and Valpy, B. (2011) *Offshore Wind – Forecasts of future costs*, Cricklade, Swindon, Wiltshire, BVG Associates for RenewableUK.

Wolfram (2011) The *Wolframs Demonstrations Project demonstration of 2D Vector Addition*, [Online]. Available at http://demonstrations.wolfram.com/2DVectorAddition/ (Accessed 2 March 2017).

YouGov (2010) *Public Attitudes to Wind Farm YouGov Survey*, London, YouGov plc for Scottish Renewables.

Zervos, A. and Kjaer, C. (2009) *Pure Power – Wind energy targets for 2020 and 2030*. Brussels, European Wind Energy Association.

Chapter 9

Wave energy

By Les Duckers and Ned Minns

9.1 Introduction

The possibility of extracting energy from ocean waves has intrigued people for centuries. However, although there are concepts over 200 years old, it was only in the latter half of the twentieth century that viable schemes began to emerge. In general, these modern wave energy conversion schemes have few environmental drawbacks, and the prospects that some of them may make a significant energy contribution are promising. In fact, in areas of the world where the wave climate is energetic and where conventional energy sources are expensive, such as remote islands, some of these schemes are already competitive. By 2020 it is expected that a number of commercial schemes will be in operation around the world, and as experience is gained costs are expected to fall.

The total wave energy resource is large, however estimates of its magnitude vary widely as the oceans have not been fully monitored over a long enough time period to establish a reasonable value. Moreover, even if a reasonable value could be determined there are difficult questions about how much of the resource is economically viable given the limitations of location and technology.

Nonetheless, it is clear that wave energy could make a major contribution to meeting world energy needs in a sustainable way. The World Energy Council has estimated the exploitable worldwide resource to be 2000 TWh per year if all potential improvements to existing devices are realized (WEC, 2010). Therefore for some countries – the UK is one example – wave energy offers a very large potential resource to be tapped. Thorpe estimates that between 15 and 20% of current UK electrical demand could be met from marine energy sources (Thorpe, 1999).

Technological developments could enable wave energy to fulfil this promise. A number of shore-mounted, near-shore and offshore prototypes are already planned or in operation. Refinements of these prototype designs could open up the possibility of harvesting vast quantities of energy from the oceans, particularly from floating offshore wave farms, consisting of tens or hundreds of devices.

The UKERC/ETI Marine Energy Technology Roadmap [ETI and UKERC, 2014) describes the steps which, once implemented, would facilitate exploitation of the vast European ocean energy resource and enable the realization of 3.6 GW of installed capacity by 2020, and close to 188 GW by 2050 (these figures encompassing both the wave devices of this chapter and the tidal technologies of Chapter 7).

History

Following the UK 'energy crisis' of 1973 and Stephen Salter's landmark paper (Salter, 1974) a large number of device concepts were invented, mathematically modelled and experimentally tested, with support from commercial sponsors and the former UK Department of Energy. Unfortunately, insufficient time and money was allocated to bring the various concepts and associated technologies to maturity, and in 1982, the Department of Energy scaled down the UK wave energy programme (Ross, 1995).

Some of the research teams involved were, however, able to sustain a minimal effort on wave energy projects. In 1989 a 75 kW prototype oscillating water column (OWC) wave energy converter was installed on Islay in Scotland. This had been fully funded by the Department of Energy following a recommendation that small-scale devices should be investigated as a source of energy for islands and remote communities where diesel normally provided the main energy source (ETSU, 1985).

Meanwhile, during this period of reduced funding in the UK, a number of other countries, notably Norway and Japan, increased their research and development wave energy programmes. With hydroelectric schemes supplying virtually all of its electricity, Norway had little immediate domestic need for wave energy, but it was keen to develop an export market for wave energy technology. In contrast, Japan did require more energy sources, but its wave climate is very modest.

Japan has conducted a substantial wave energy research programme, with many teams working on a variety of projects (see Section 9.5 for details of the Whale, BBDB and Pendulor). The late wave energy pioneer Yoshio Masuda, who began work in Japan in the 1940s, is generally considered to be the inventor of the oscillating water column (see below) and is often described as the 'father' of wave energy. He was the inspiration behind the Kaimei, the first large floating wave power station. Further details can be found in Miyazaki (1991); Miyazaki and Hotta (1991); Kondo (1993) and ISOPE (2002).

In the 1990s there was a revival of awareness of the potential of wave energy amongst politicians and others in a number of countries. In particular, a European Union initiative was launched (Caratti et al., 1993), which provided funding for a small number of projects and led to the formation of a European Wave Energy Thematic Network. Between 1992 and 2002, Thorpe carried out several surveys for the UK Department of Trade and Industry (DTI) and, by looking at the main types of device, he estimated the electricity generation costs to be around 5p per kWh (in 2001 prices), based on an annual practical wave resource of 30 TWh (Thorpe, 1992; 1998; 2001).

Over the years 2001–2017 there has been substantial UK investment in wave energy technology, both from commercial investors and from the government. Two marine energy centres have been established, in Scotland and Cornwall, and these centres offer facilities for developers to test their pilot schemes. There has been considerable growth in interest in wave energy, especially in Scotland and Ireland, where the most significant wave energy resources in the British Isles exist.

International interest in wave energy has also grown in the last 10 years with developments in India, China, the United States, Japan, Korea, South Africa, Australia, and the EU amongst others. An extensive list of wave devices and developments can be found in Chapter 14 by Thorpe in the 2010 World Energy Council survey (WEC, 2010). Country profiles can be found in Appendix E of IEA-_RETD (2012).

9.2 Introductory case studies

The case studies presented in this section provide some insight into the nature of waves and of schemes designed to harness wave energy.

The TAPCHAN device is particularly valuable in that it incorporates an element of storage, and the oscillating water column (OWC) concept is important because it has been deployed in a number of countries and represents the most common form of wave energy converter deployed to date, probably because of its simplicity and robustness.

Shoreline prototypes are generally of the oscillating water column type. They are often regarded as easier to construct and maintain than offshore devices, but, in practice, each shore-mounted device may have to be purpose-built to exactly fit the specified location, whereas offshore devices can potentially be built in production facilities in large numbers. This should eventually lead to high-volume, high-quality, low-cost fabrication and assembled structures could be towed out in calm conditions for deployment in wave farms. By using an area of ocean a few kilometres or more offshore as a wave farm, it should be possible to deploy a large array of wave energy converters and hence capture large quantities of energy, which could then be transmitted back to shore via subsea electrical cables. Offshore devices can harvest greater amounts of energy, as the waves in deep water have a greater power density than those in the shallower water near to land.

TAPCHAN

In 1985 a 350 kW prototype TAPCHAN wave energy converter, built by the company Norwave, commenced operation on a small Norwegian island some 40 km north-west of Bergen.

The name 'TAPCHAN' comes from the 'TAPered CHANnel' design of the scheme (Figure 9.1). The first, and so far only, example had a channel with a 40 metre wide horn-shaped collector. Waves entering the collector fed into the wide end of the tapered, upward sloping channel, where they then propagated towards the narrow end with increasing wave height. The channel walls on the prototype were 10 m high (from 7 m below sea level to 3 m above) and 170 m long. Because the waves were forced into an ever-narrowing channel, their height became amplified until the crests spilled over the walls into the reservoir at a level of 3 m above the mean sea level. The kinetic energy in the waves was thus converted into potential energy, and this was subsequently converted into electricity by allowing the water in the reservoir to return to the sea via a low-head Kaplan turbine system (see Figure 9.22 and Chapter 5 for details of the Kaplan turbine). This powered a 350 kW generator that delivered electricity into the Norwegian grid.

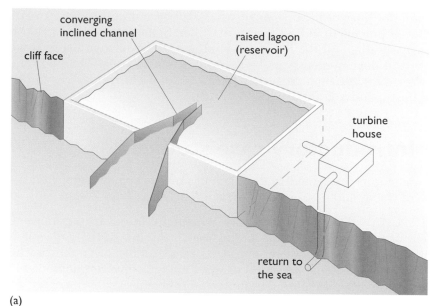

Figure 9.1 (a) The tapered channel (TAPCHAN) wave energy conversion device. (b) Aerial photograph of the Norwegian TAPCHAN: water is entering the reservoir in the centre of the image, where the channel device can just be seen between the cliff walls

The TAPCHAN concept is simple. With very few moving parts, its maintenance costs should be low and its reliability high. The storage reservoir also helps to smooth the electrical output. We shall see in Section 9.3 that ocean waves have a random nature and so most wave energy converters produce a fluctuating power output. In contrast, a TAPCHAN device 'collects' waves in the reservoir, and so the output from the Kaplan turbine is dependent on the relatively steady difference in water levels between the reservoir and the sea. A TAPCHAN device therefore has an integral storage capacity which is generally not found in other wave energy converters.

In the 1990s Norwave considered methods for reducing the cost of construction of future TAPCHANs. Among those methods is a scheme for wave prediction, to allow the Kaplan turbine to run at a greater output for some short time before the arrival of a number of large waves. This reduces the level of water in the reservoir and so makes room for those large waves. This technique may permit the designers to build schemes with smaller reservoirs and hence reduce the construction costs. A second cost-reduction method that has been proposed is to fabricate a shorter channel, and this was tried out on the existing prototype at Bergen by reducing the length of the existing channel. Unfortunately, there were some technical difficulties with the dynamiting of the concrete channel, and the ensuing commercial problems have meant that this prototype is no longer in operation (Petroncini and Yemm, 2000).

Perhaps more than other wave energy systems, the TAPCHAN approach can only be used in very particular places: to be effective it requires a good wave climate (i.e. high average wave energy, with persistent waves); deep water close to shore; a small tidal range (less than 1.0 m), otherwise the low-head hydro system cannot function properly for 24 hours a day (this therefore excludes most of the UK south of the Shetland Isles); and a convenient and cheap means of constructing the reservoir, which usually requires a natural feature of the coastline.

The Islay shoreline oscillating water columns

In the mid-1980s Queen's University, Belfast, (QUB) worked on the development of a shoreline oscillating water column (OWC) device. After surveying several Scottish islands, the island of Islay was chosen for their first OWC scheme, which was installed in a natural gully in 1989. It supplied the local grid with electricity on an intermittent basis from a 75 kW generator from 1991 until it was decommissioned in 1999.

The approach was to develop a device which could be built cheaply on islands using locally available technology and plant. The main part of the OWC consisted of a closed wedge-shaped chamber, made from modular pre-fabricated concrete components, which was placed above the natural gulley. This chamber sloped down to below the sea surface and wave motion drove the water column thus trapped within the lower part of the chamber up and down like a huge piston. A cylindrical tube connected to the atmosphere allowed air to be thus expelled and drawn into the chamber. On its way to and from the atmosphere the air drove a Wells turbine (see Box 9.3), which was directly coupled to an electrical generator.

With experience gained from this natural gully project the team from QUB then collaborated with Wavegen of Inverness to develop what they referred to as a 'designer gully' OWC to overcome some of the limitations of the first Islay project. The principal modifications applied to the second Islay OWC, called LIMPET (Land Installed Marine Powered Energy Transformer), were in the construction method and the shape of the oscillating column.

The designer gully was excavated behind a natural rock wall, which was only removed at the end of the installation (Figure 9.2). As regards the water column, the chamber in the first Islay OWC had a horizontal floor at right angles to the back wall, causing turbulence and consequent loss of

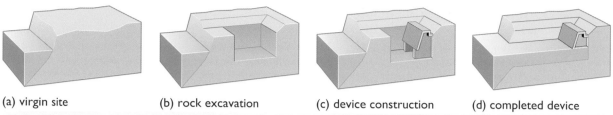

(a) virgin site   (b) rock excavation   (c) device construction   (d) completed device

Figure 9.2 Construction sequence used by Queen's University and Wavegen for the 'designer gully' LIMPET OWC

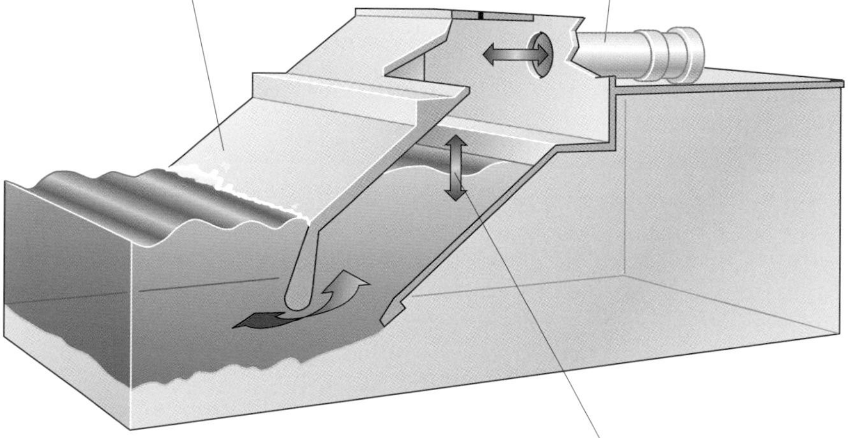

The Wells turbine rotates in the same direction regardless of the air flow, thus generating irrespective of upward or downward movement of the water column.

Reinforced-concrete capture chamber set into the excavated rock face.

Air is compressed and decompressed by the oscillating water column (OWC). This causes air to be forced through the Wells turbine and then drawn back.

Figure 9.3 Outline of the LIMPET device on Islay

Figure 9.4 Photograph of the LIMPET OWC device

energy within the chamber itself so, in order to improve the flow of water into and out of the oscillating chamber, the main part of the chamber of the designer gully OWC was built at a slope so as to efficiently change the water motion from horizontal to vertical and vice versa (Figure 9.3).

Construction of LIMPET was completed in September 2000, and Figure 9.4 shows the finished structure, consisting of a rectangular sloping chamber which ducts the airflow through two contra-rotating Wells turbines. Each turbine is coupled to a 250 kW induction generator, giving the device a 500 kW maximum power output. Details of the first years of operation of LIMPET can be found in Boake et al. (2002). LIMPET has now accumulated many thousands of hours of operation, in more than ten years, and is used by Wavegen as a test bed for air turbines (Heath, 2012).

9.3 **Physical principles of wave energy**

Ocean waves are generated by wind passing over long stretches of water (known as 'fetches'). The precise mechanisms involved in the interaction between the wind and the surface of the sea are complex and not yet completely understood, but three main processes appear to be involved.

(1) Initially, air flowing over the sea exerts a tangential stress on the water surface, resulting in the formation and growth of waves.

(2) Turbulent air flow close to the water surface creates rapidly varying shear stresses and pressure fluctuations. Where these oscillations are in phase with existing waves, further wave development occurs.

(3) Finally, when waves have reached a certain size, the wind can exert a stronger force on the upwind face of the wave, causing additional wave growth.

Because the wind is originally derived from solar energy we may consider the energy in ocean waves to be a stored, moderately high-density form of solar energy. Solar power levels, which are typically of the order of 100 W m^{-2} (mean value), can be eventually transformed into waves with power levels of over 100 kW per metre of crest length. Note that these power levels are reported in different units, one *per square metre*, and the other *per metre of crest length*. This is because the nature of waves makes it impossible to refer to power per unit area: we have to consider that the wave action takes place throughout the depth of water, and so we consider the power passing through a *1 metre wide slice of water*. We cannot therefore directly compare solar and wave power densities, but can state that the solar to wind to wave sequence gradually concentrates the power density.

A simple, 'regular' wave can be characterized by its wavelength, λ; height, H; and period, T (see Box 9.1). Waves of greater height contain more energy per metre of crest length than small waves. It is usual to quantify the power of waves rather than their energy content.

BOX 9.1 Wave characteristics and wave power

The shape of a typical wave is described as **sinusoidal** (that is, it has the form of a mathematical sine function). The difference in height between peaks and troughs is known as the **height**, H, and the distance between successive peaks (or troughs) of the wave is known as the **wavelength**, λ.

Suppose that the peaks and troughs of the wave move across the surface of the sea with a velocity, v. The time in seconds taken for successive peaks (or troughs) to pass a given fixed point is known as the **period**, T. The **frequency**, f, of the wave describes the number of peak-to-peak (or trough-to-trough) oscillations of the wave surface per second, as seen by a fixed observer, and is the reciprocal of the period. That is, $f = 1/T$.

If a wave is travelling at velocity v past a given fixed point, it will travel a distance equal to its wavelength λ in a time equal to the wave period T. So the velocity v is equal to the wavelength λ divided by the period T, i.e.:

$$v = \lambda/T$$

The power, P, of an idealized ocean wave is approximately equal to the square of the height, H (metres), multiplied by the wave period, T (seconds). The expression is:

$$P = \frac{\rho g^2 H^2 T}{32\pi} \text{ W m}^{-1}$$

or (approximately) in kW m^{-1}, $P \approx H^2 T$ kW m^{-1}

where ρ is the density of water and g is the acceleration due to gravity.

Deep water waves

In terms of wave propagation, water is considered to be 'deep' when the water depth is greater than about half of the wavelength λ. The velocity of a deep-water ocean wave can be shown to be proportional to the period as follows:

$$v = gT/2\pi$$

This leads to the useful approximation that the velocity in metres per second is about 1.5 times the wave period in seconds.

An interesting consequence of this result is that in the deep ocean the long waves travel faster than the shorter waves. This is referred to as 'dispersion', a unique feature of deep water waves which can lead to dangerous and hard to predict combinations of wave crests. Long waves can catch up with preceding shorter waves creating large combined waves where they meet.

If both the above relationships hold, we can find the deep water wavelength, λ, for any given wave period:

$$\lambda = gT^2/2\pi$$

Intermediate depth waves

As the water becomes shallower, the properties of the waves become increasingly dominated by water depth. When waves reach shallow water, their properties are completely governed by the water depth, but in intermediate depths (i.e. between $d = \lambda/2$ and $d = \lambda/4$) the properties of the waves will be influenced by both the water depth d and wave period T.

Shallow water waves

As waves approach the shore, the seabed starts to have an effect on their velocity, and it can be shown that if the water depth d is less than a quarter of the wavelength, the velocity is given by:

$$v = \sqrt{g(d)}$$

In other words, the velocity under these conditions is equal to roughly three times the square root of the water depth d – it no longer depends on the wave period.

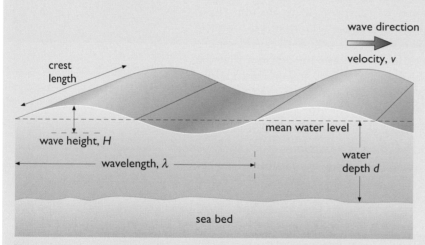
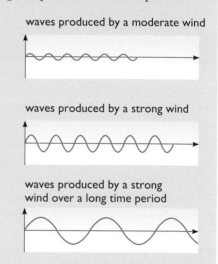

Figure 9.5 Characteristics of an idealized wave

Waves located within, or close to, the areas where they are generated are sometimes referred to as a 'wind sea'. When winds change in strength and direction, the resulting wave patterns can become quite complex. Waves can travel out of these areas with minimal loss of energy to produce 'swell waves' at great distances from their area of origin. As waves travel, there is a systematic tendency for periods and wavelengths to increase, and so swell seas are typically made up of relatively long period waves. The height and steepness of the waves generated by any wind field depends upon three factors: the wind speed; its duration; and the fetch, i.e. the distance over which wind energy is transferred into the ocean to form waves. When a steady wind has blown for a sufficient time over a long enough fetch, the waves are referred to as constituting a fully developed sea.

The UK is well situated to make use of wave energy because it lies at the end of a long fetch (the Atlantic Ocean) with the prevailing wind blowing towards it. The UK's western approaches have therefore been of most interest to wave energy engineers and developers. Although often stormy, the fetches in the North Sea are relatively short and so the wave power densities are generally smaller than those of the Atlantic Ocean.

Typical sea state

A typical sea state is actually composed of many individual components, each of which is like the idealized wave described in Box 9.1. Each wave has its own properties, i.e. its own period, height and direction. It is the combination of these waves that we observe when we view the surface of the sea, and the total power in each metre of wave front of this irregular sea is of course the sum of the powers of all the components. It is obviously impossible to measure all the heights and periods independently, so an averaging process is used to estimate the total power, as follows:

(1) By deploying a wave-rider buoy it is possible to record the variation in surface level during some chosen period of time. (Satellites are also used to detect wave heights)

(2) The average water height will always be zero, since the average value also defines the zero value, but we can obtain a meaningful figure by calculating the significant wave height, H_s. This is defined as $4 \times$ the *root mean square* of the water elevation – i.e. the instantaneous elevations are first squared, making all of the values positive, then the mean over a number of waves is calculated, then the significant wave height is calculated as four times the square root of the mean. The **significant wave height** is approximately equal to the average of the highest one-third of the waves (which generally corresponds to the estimation of height made by eye, since the smaller waves tend not to be noticed)

(3) The **zero-up-crossing period** T_z (or in abbreviated form, the zero-crossing period) is defined as the average time – counted over ten crossings or more to get a reasonable average – between upward movements of the surface through the mean level. (Note that including the downward movements would give the half-period.)

The **energy period**, T_e, is more useful to us as it characterizes the energy in the waves, whereas T_z represents all of the content of the waves including very small components which have negligible energy. The energy period is slightly longer than the zero-up-crossing period: $T_e \sim 1.12 T_z$, but this can vary considerably by site

(4) For a typical irregular sea, it can then be shown (the derivation lies outside the scope of this book) that the average total power in one metre of wave crest can be approximated by $P = 0.5 H_s^2 T_e$ where P is in units of kW per metre length of wave crest. The approximation $P = 0.5 H_s^2 T_e$ is commonly used.

Figure 9.6(a) illustrates a typical wave record and shows the significant wave height and zero-crossing period. Figure 9.6(b) shows two further wave records for the same location, recorded on different days.

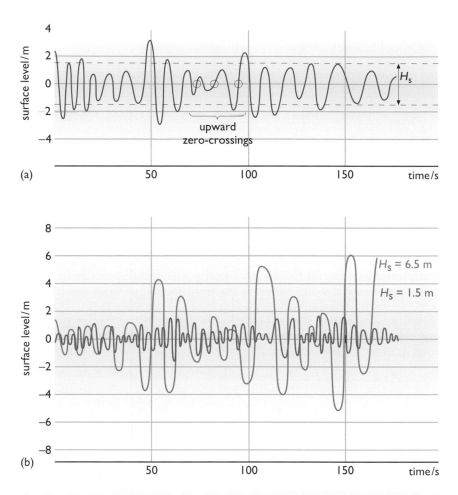

Figure 9.6 (a) A typical wave record. In this example the significant wave height H_s = 3 m (from −1.5m to +1.5m). Successive upward movements of the surface are indicated with small circles. In this case there are 15 crossings in 150 seconds, so T_z = 10 seconds. Then T_e = 11.2 seconds. From this, P = 0.5 (3^2 × 11.2) kW m^{-1} = 50.4 kW m^{-1} (b) Two further wave records are shown here for the same location but represent recordings taken on different days

Variations in the wave power at a given location

Sea level recordings made at different times or dates will of course differ, leading to different values of H_s and T_e. Suppose that each recording represents a time period of one-thousandth of a year or 8.76 hours. If we record 1000 sea states at our chosen location over a whole year, characterizing each of them by their values of H_s and T_e, we can build up a statistical picture of the distribution of wave conditions at our chosen location. This picture, or scatter diagram, gives the relative occurrences in parts per 1000 of the contributions of H_s and T_e. The example of a scatter

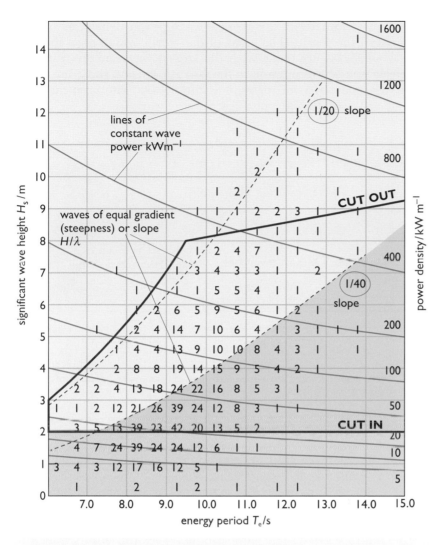

Figure 9.7 Scatter diagram of significant wave height (H_s) against energy period (T_e) for 58°N 19°W in the middle of the north Atlantic. The numbers on the graph denote the average number of occurrences of each combination of H_s and T_e in each set of one thousand 8.76 hour-long measurements made over one year. The most frequent occurrences are at $H_s \sim 2$ m, $T_e \sim 9$ s. Wave devices cannot operate in all conditions, and three areas are identified for a hypothetical device which will be in survival mode in the pink area and normal operation in the green area. Blue is the region where the incident wave energy is too low to convert effectively.

diagram shown in Figure 9.7 is for the north Atlantic and shows that the waves at this location have a high average power density. In water 100 m deep at South Uist (Hebrides, Scotland), for example, the annual average might typically be around 70 kW m^{-1} (or 613 000 kWh m^{-1} per year), whereas closer to shore where the depth is 40 m, the corresponding figure might be around 20–30 kW m^{-1}. These figures indicate that the north Atlantic is indeed a valuable wave energy resource.

Figure 9.8 shows estimates of the average wave power density at various locations around the world. The areas of the world which are subjected to regular wind fluxes are those with the largest wave energy resource. South westerly winds are common in the Atlantic Ocean, and often travel substantial distances, transferring energy into the water to form the large waves which arrive off the European coastline.

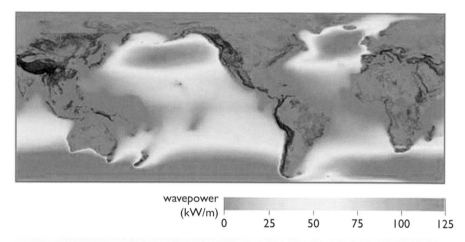

Figure 9.8 Global annual average wave power in kilowatts per metre (kWm⁻¹) of crest length (source: OAS & IEA 2017)

Wave pattern and direction

The direction of waves travelling in deep water is obviously dictated by the direction of the wind that generated them. Waves can travel vast distances across open water without much loss of energy. At any given location we can therefore expect to observe waves arriving from different sources, and hence different directions. For example, we might see waves approaching us from the south-west which were produced by the winds over the mid-Atlantic, but at the same time find that some waves have been generated by storm conditions to the south and east of our position. It is easy to imagine that the resulting wave pattern will be complex, and indeed such patterns are commonly observed. (Figure 9.9a).

A representation of the annual average power as a function of direction at a given location can be given by a 'directional rose' (Figure 9.9b).

What happens beneath the surface?

The surface profile of the ocean is the obvious evidence for the existence of waves, but we also need to understand the sub-surface nature of waves if we are to design schemes to capture energy from them (Figure 9.10).

Waves are composed of orbiting particles of water. Near to the surface, these orbits are the same size as the wave height, but the orbits decrease in size as we go deeper below the surface. The size of orbits decreases exponentially with depth.

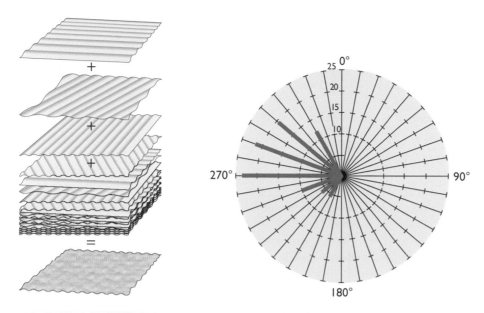

Figure 9.9 (a) The development of a complex sea – wave components produced by different weather conditions in different distant geographical areas arrive and combine in the local area (source: adapted from Claeson, 1987) (b) A directional rose for waves. The length of the line in each sector represents the average annual power in that sector. In this case most of the waves are coming from the west (source: Thorpe, 1999)

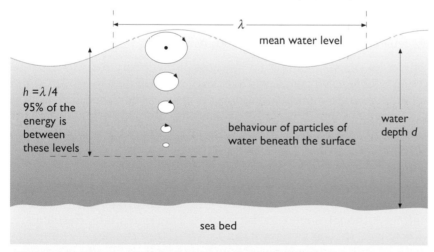

Figure 9.10 Behaviour of particles of water beneath the surface

To capture the maximum energy from a wave we could try to construct a device that was deep enough to intercept all of the orbiting parts of that wave. But this would be impractical and uneconomic, since the lowest orbits actually contain very little energy. In deciding how deep a structure to extract wave energy should be, it is useful to know that 95% of the wave energy is contained in the layer between the surface and a depth h equal to

Table 9.1 North Atlantic offshore wave conditions

	Period /s	Height /m	Power density /kW m⁻¹	Velocity /m s⁻¹	Wavelength /m
Storm	14	14	1700	23	320
Average	9	3.5	60	15	150
Calm	5.5	0.5	1	9	50

a quarter of the wavelength λ (i.e. $h = \lambda/4$). From Table 9.1 we can see that a typical north Atlantic wavelength might be 150 m, which tells us that a device would have to be 38 m deep to intercept 95% of the incident wave energy. Such a device would be too expensive, and the optimum depth for most practical devices will be between 4 and 10 m.

Moving into shallow water

There are a few areas in the world where the shoreline is formed by a steep cliff which drops into reasonably deep water. These are the areas most suitable for shore-mounted wave energy converters because the incident waves have a high power density. However, for most of the coastlines around the world the near-shore water is quite shallow. Due to the frictional coupling between the water particles at the greatest depths with the seabed, deep water waves gradually give up their energy as they move into shallower water and eventually run up the shore to the beach. The frictional effect becomes significant when the water depth is less than a quarter of a wavelength, and the power loss can be tens of watts per metre of crest length for every metre travelled inshore.

This power loss is very important, because it obviously reduces the total wave energy resource. Typically, waves with a power density of 50 kW m⁻¹ in deep water might contain 20 kW m⁻¹ or less when they are closer to shore in shallow water (depending on the distance travelled in shallow water and the roughness of the seabed). However, storm waves are also attenuated in the same way and, with careful siting of machines (below the surface, for example), can be less likely to destroy shoreline devices.

A further mechanism for energy loss as waves run up the beach is the formation of steeper, breaking waves, which are turbulent and energy-dissipating. Although such waves ('breakers') are potentially desirable for leisure activities such as swimming and surfing, they can be very damaging to structures such as wave energy converters, and so should be avoided when choosing suitable locations.

Refraction

As ocean waves approach the shore they will usually be entering shallower water and, as we saw in Box 9.1, the velocity of the waves then becomes governed by the water depth. Shallower water means lower wave velocity.

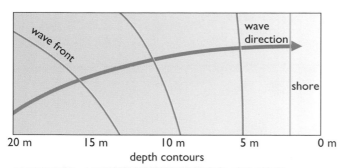

Figure 9.11 As waves travel from deep water into shallow water their velocity reduces and the resulting refraction generally causes waves to approach a beach at right-angles to the shore

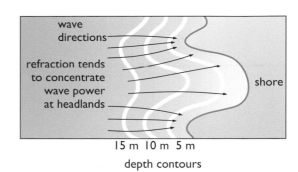

Figure 9.12 Concentration effects of refraction around a shoreline with headlands

This in turn leads to **refraction** (change of wave direction due to change of wave velocity).

Imagine that a wave crest approaches shallow water at an angle, so that one end of the crest reaches the shallow water first. This part then moves more slowly than the rest, changing its direction. The remainder of the crest progressively adopts this new direction as its velocity is also reduced on entering the shallow water. The effect of refraction, caused by the reducing depth and hence velocity, is gradually to change the direction of the crest to be roughly parallel with the shore (Figure 9.11).

Knowledge of the depth contours around a coast allows the identification of areas where waves will be concentrated, and are thus the most cost-effective for wave energy developments. Consider a shoreline with headlands (Figure 9.12) – notice how the varying water depth, as shown by the contours (white lines), causes refraction to occur. This concentrates the waves onto the headlands, and leaves the other areas with reduced wave power density.

9.4 **Wave energy resources**

Wave power resources (the annual average kW per metre of crest length) around the world were shown in Figure 9.8. As already mentioned, the World Energy Council has estimated a total worldwide resource of 2000 TWh per year (Thorpe, 1999), although this estimate might in fact prove to be rather conservative as others have suggested higher figures.

Regarding the UK, Thorpe (1999) estimated that the total annual average wave energy along the north and west side of the United Kingdom (i.e. from the south-west approaches to Shetland) ranges from 100 to 140 TWh per year at the near shore to about 600 to 700 TWh per year in deep water. Figure 9.13 shows the seasonal variation in significant wave height, a measure of the available resource, in British waters (BERR, 2008). The variation is quite marked, with a greater resource being available, on average, during the winter months, coinciding with the highest demand for energy. The proportion of this resource that could actually be harnessed to produce electrical power depends, of course, on various practical and technical constraints.

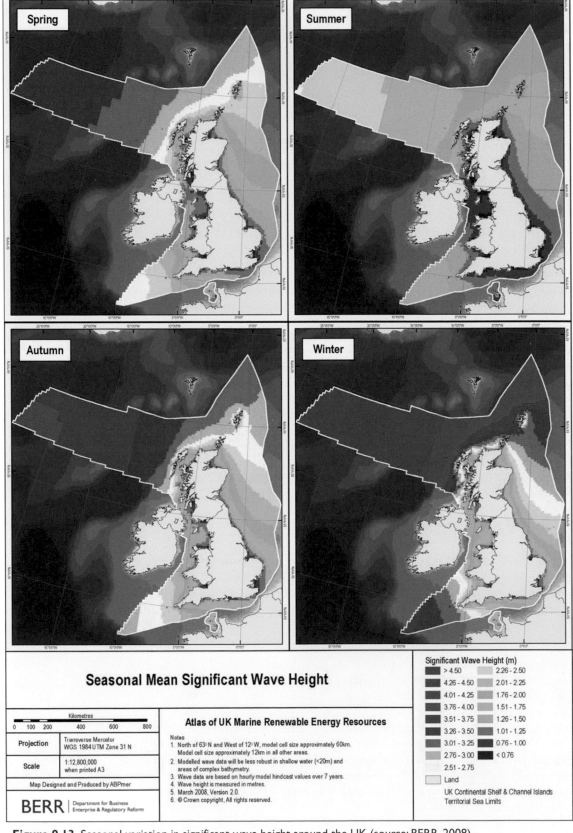

Figure 9.13 Seasonal variation in significant wave height around the UK (source: BERR, 2008)

Thorpe (1992) estimated that the technical resource (i.e. the resource technically available regardless of cost, see Chapter 11) is between 7 GW and 10 GW annual average power, which is equivalent to 61 to 87 TWh per year, depending on the water depth. In comparison, total UK annual electricity production in 2010 was approximately 360 TWh. Table 9.2 gives a breakdown of the resource at different water depths.

Table 9.2 The natural and technical wave energy resource for the north and west side of the UK

Water depth/m	Average natural resource		Average technical resource	
	/GW	/TWh per year	/GW	/TWh per year
100	80	700	10	87
40	45	394	10	87
20	36	315	7	61
Shoreline	30	262	0.2[1]	1.75

[1] The technical shoreline resource is very dependent on details of the local shoreline structure, for example the nature and shape of the rock formations and of gullies and beaches.
(source: Thorpe 1992, 2001)

9.5 Wave energy technology

In order to capture energy from sea waves it is necessary to intercept the waves with a structure that will react in an appropriate manner to the forces applied to it by the waves. In the case of a shore-mounted device, like TAPCHAN and most OWC devices, the structure is firmly fixed to the seabed, and the waves make water move in a useful way. For other types of device some part of the structure may be fixed, perhaps anchored to the seabed, but another part may be a float which moves in response to the waves by pulling against the anchor. In this case the relative motion between the anchor and the float provides the opportunity to extract energy.

Very loosely tethered floating structures can also be employed, but a stable frame of reference must still be established so that the 'active' part of the device moves relative to the main structure. This can be achieved by taking advantage of inertia, or by making the main structure so large that it spans several wave crests and hence remains reasonably stable in most sea states.

A body in the sea subject to waves can respond via six types of movement. These are sway, roll and yaw, which are not generally harnessed in wave energy conversion technology, and heave, surge and pitch (Figure 9.14), which are harnessed to varying degrees in most wave energy converters. these three modes of movement can be defined as follows:

- pitch – waves cause the device, or part of it to rotate about its axis
- heave – waves cause the device to rise and fall vertically
- surge – waves cause the device to move horizontally backwards and forwards.

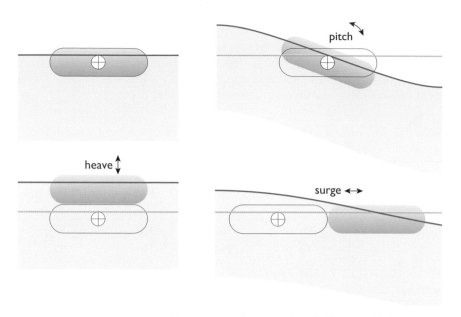

Figure 9.14 The pitch, heave and surge responses of a floating object to incident waves

It is easy to imagine a heaving device, rising and falling with the waves, but such devices have too high a natural frequency to be particularly effective (see the subsection on multi-resonant OWCs for more details). Surge devices tend to have low natural frequency and so can respond to a wide range of wave frequencies. Theoretically, surging motions are twice as energetic as heaving ones, thus making it preferable to harness the surge component of waves (see for example Salter and Lin (1995)).

Economics demands that a device should survive at sea for at least five years between major services. During that time some of its components will have to execute 15 to 30 million cycles, placing severe constraints on material selection and strain levels. A structure designed to operate at a particular design wave power density will also have to endure storms with power densities ten to thirty times higher than the design value.

The physical size of the structure of a wave energy converter is a critical factor in determining its performance. The appropriate size can be estimated by considering the volume of water involved in the upper particle orbits in a wave. In most circumstances a wave energy converter will have to have a swept volume which is similar to this volume of water in order to capture all of the energy contained in the wave. A variety of wave energy concepts are discussed below. The precise physical size and shape of each device will be governed by its mode of operation, but as a rough guide the swept volume must be of the order of several tens of cubic metres per metre of device width. A device with a swept volume much smaller than this would have a limitation on the total energy that it could capture from each typical wave cycle: although it might still be capable of capturing most of the energy from small waves it would be restricted in its response to larger waves, thereby reducing its overall efficiency.

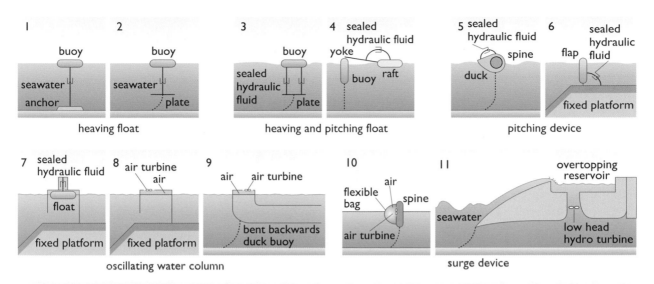

Figure 9.15 Schematic representation of various types of wave energy converter (source: based on Falnes and Løvseth, 1991)

There are many different configurations of wave energy converter, and a number of ways of classifying them have been proposed. One approach is to classify by mode of operation (Figure 9.15).

Another approach is to consider the device location (Figure 9.16) – here the three general classifications are:

- fixed to the seabed, generally in shallow water (e.g. Breakwater OWC)
- tethered in intermediate depths (eg Oyster)
- floating offshore in deep water (eg AWS-III).

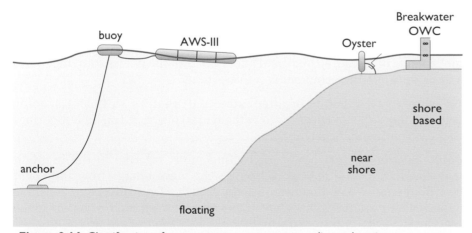

Figure 9.16 Classification of wave energy converters according to location

Alternatively the geometry and orientation of the wave energy converter may be used (Figure 9.17). Here the options are:

- terminators
- attenuators
- point absorbers.

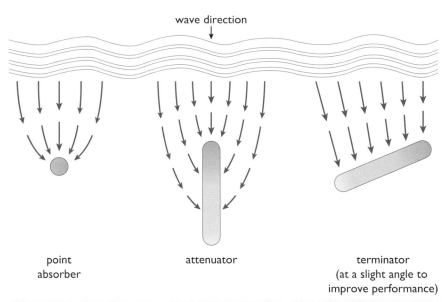

wave direction

point
absorber

attenuator

terminator
(at a slight angle to
improve performance)

Figure 9.17 Classification of wave energy converters according to size and orientation

Terminator devices have their principal axis parallel (or almost parallel, since a slight angle helps to increase the stability of the whole body) to the incident wave front, and they physically intercept the waves. Depending on the depth, a 300 m long device in a wave climate of 60 kW per metre thus intercepts up to 300 × 60 kW = 18 000 kW. Typically 1/3 of this might be converted into electricity (i.e. 6 000 kW). The terminator might be moored to a leading buoy and anchor, which allows the terminator to pivot as the wave direction changes.

Attenuators have their principal axis perpendicular to the wave front, so that wave energy is gradually drawn in towards the device as the waves move along it. A leading buoy and anchor permit the device to swing round and orientate to the principal wave direction.

An array of terminators or attenuators will require sufficient spacing between individual devices to allow them to swing round on their moorings without fouling their neighbours.

Point absorbers must have small dimensions relative to the incident wavelength ($d<<\!/\lambda4$) and work by drawing wave energy from the water beyond their physical dimensions (Falnes, 2011). Typically, being sensitive to waves from all directions, they do not require freedom to orientate to the incoming waves, and so can usually be tightly moored. In principle, they could be extremely slim vertical cylinders which execute large vertical excursions in response to incident waves, but in practice the hardware involved tends to mean that they are a few metres in diameter and absorb energy from perhaps twice their own width. Tethered buoy systems, for example, act as point absorbers.

Point absorber theory (Budal and Falnes, 1975), which is derived from radio antenna theory, suggests that if the body is assumed to be small with respect to the wavelength, the power it may absorb is related, not to the size of the body, but to the wavelength. Therefore, point absorber theory offers the wave energy converter designer the opportunity to create a small device capable of absorbing large amounts of energy from outside its physical bounds. Theoretically the maximum power, P_{max}, absorbed by a perfect point absorber is determined by the mode (heaving or surging) in which it oscillates, and is given by:

$$P_{max} = \varepsilon\left(\frac{\lambda}{2\pi}\right)P_i$$

Where P_i is the incident wave power per metre crest length, and ε is 1 for a heaving device and 2 for a surging device. It is important to note that, from this equation, a body oscillating in surge is theoretically capable of absorbing twice the power of that of a body oscillating in heave, and that for both this could be several times the power incident upon the width of the device. The theory also assumes linear, plane, regular waves with no mixing of period or direction.

the principle

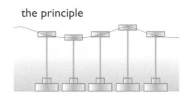

BOX 9.2 Point absorbers in practice

In theory a perfect wave energy point absorber responding to a heaving motion could capture the energy from a wave front with a length equal to $\lambda/2\pi$ metres. For example, a wave with a period of 6 seconds would have a wavelength of between 56 m and 72 m, depending on whether the water is deep or shallow, and so a perfect (1 m wide) point absorber would absorb the energy from a width of between about 9 m and 12 m. In reality, however, the capture width is much less due to the limitations of the vertical amplitude of the motion of the point absorber and the ratio of diameter to wavelength. The irregular, multidirectional waves that occur at sea also cause a reduction in efficiency.

The mathematical explanation of this is outlined by Nielsen and Plum (2000) who go on to report experimental results. These devices are said to have a **capture width ratio**, i.e. the ratio of apparent diameter to physical diameter, greater than 1. Figure 9.18 shows an example of a float system working off a heaving motion. The concept of **latching**, or holding the float under water for a second or so before allowing it to follow the wave, has been developed to maximize the energy capture by permitting a large amplitude of motion of the float, which is needed for optimal performance (Falnes and Lillebekken, 2003; Falnes, 2011).

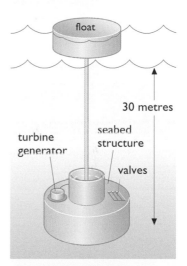

Figure 9.18 Tethered buoy wave energy converter

Because full-scale devices are generally very large, as a result of needing to be on a similar scale to the wavelengths that they capture, most developments of wave energy technology involve the testing of scale models. Indoor, purpose-built tanks, lakes and reservoirs have been used by research teams. In 1977 a team at the University of Edinburgh constructed a wide wave tank in which it was possible to create repeatable multi-directional mixed-sea conditions, and so test the effects of varying specific design parameters of wave energy converters.

Various tanks, such as the one shown in Figure 9.19, use the same wave-making technology, and new ones include current generators to model the effect of tidal flow on waves.

Figure 9.19 Plymouth COAST Lab Ocean Basin, 35 m x 15.5 m for wave current testing

two-directional axial air flow

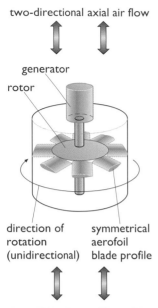

generator

rotor

direction of   symmetrical
rotation   aerofoil
(unidirectional) blade profile

Figure 9.20 The Wells turbine

Power take off

All wave energy converters require a way to turn their primary motions into usable power, usually electricity. The system that does this is called the power take-off system. There are a wide range of methods to convert these primary motions into linear or rotary motions for a generator.

Many of the devices tested to date are of the oscillating water column (OWC) type. In these devices, an air chamber pierces the surface of the water and the contained air is forced out of and then into the chamber by the approaching wave crests and troughs. On its passage from and to the chamber, the air passes through an air turbine (generally of the Wells type) which drives a generator and so produces electricity. The Wells turbine (Figure 9.20) is an axial-flow device which rotates in the same direction irrespective of the direction of airflow, and has aerodynamic characteristics particularly suitable for wave applications (Box 9.3). A very attractive feature of OWCs is that the cross-sectional area of the air turbine can be much smaller than the area of the moving water surface. This reduction in available area for airflow acts like gearing to increase the air velocity through the turbine such that it can operate at high speed.

BOX 9.3 The Wells turbine

The Wells air turbine, invented by Professor Alan Wells, is self-rectifying – that is to say, it can accept airflow in either axial direction. To achieve this, the aerofoil-shaped blade profile must be symmetric about the plane of rotation, untwisted and with zero pitch, i.e. the chord line must be in line with the plane of rotation. The vector diagrams in Figure 9.21 show how this occurs. Note that the diagrams are very much like those used in Chapter 7 to explain the properties of wind aerofoils. As the blade moves forward, the angle of attack, which is the angle between the relative airflow velocity and the blade velocity, is small and this produces a large lift force (L). The forward component of L provides the thrust which drives the blade forwards.

The Wells turbine operates in much the same way as would a horizontal-axis wind turbine with symmetrical, untwisted blades and with zero pitch angle. Consider the nearest blade with air flowing in an upward direction (Figure 9.21(a)). If we now work in the frame of reference of the blade – we do this by making the blade appear to be stationary to us (even though it is moving) by considering the blade velocity vector to be in the opposite direction to the blade's actual direction of movement – we get Figure 9.21(b). Note that because the blade chord is in line with the plane of rotation, the angle of attack α is the same as the relative wind angle ϕ referred to in Section 7.4 of Chapter 7.

If we now resolve these vectors we get Figure 9.21(c). From this diagram we can see that there will be a net forward force on the blade acting in the plane of rotation if $(L \sin \alpha) - (D \cos \alpha)$ is greater than zero. The reaction components are of little interest but the rotor bearings must be capable of carrying these forces with minimal friction losses. If the net forward thrust is greater than zero, then the blade will be driven forwards and can usefully extract energy from the airflow. The shape of the blade is extremely important here since it will dictate the values of the lift and drag coefficients C_L and C_D and hence the magnitude of the forward thrust.

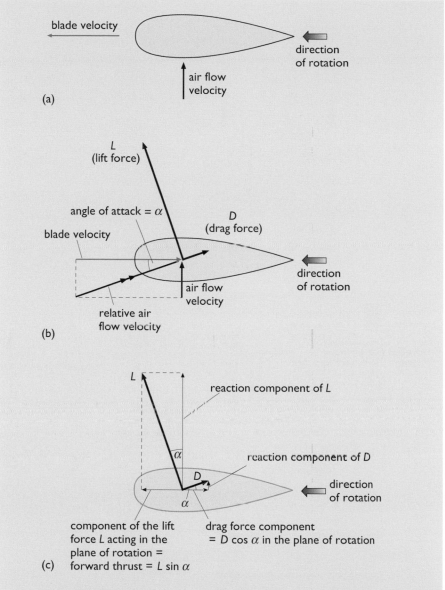

Figure 9.21 The Wells turbine (a) airflow and blade velocity; (b) relative air velocity and lift and drag forces; (c) forces in the plane of rotation

There is a linear relationship between airflow and pressure drop for the Wells turbine rotating at a constant speed. This means that the Wells turbine has a constant impedance to airflow. Careful choice of the design parameters ensures that this impedance matches the requirements of the wave climate at the chosen location. Impedance matching like this maximizes the transfer of energy from the wave to the generator. The Wells turbine is ideally suited to wave energy applications because of its constant impedance, whereas the impedance of a conventional air turbine varies with airflow. A conventional air turbine may have a superior peak efficiency, but a Wells turbine will perform well over a range of air flows giving it a better wave cycle efficiency. A further beneficial characteristic of the Wells turbine is that, at the sizes typically employed, it can rotate at high speed (1500–3000 rpm) so that the electrical generator can be attached directly to the shaft of the turbine, obviating the need for a gear box to raise the generator speed.

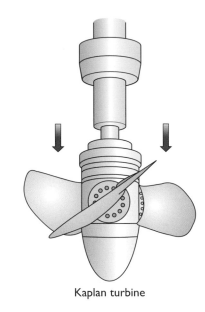

Kaplan turbine

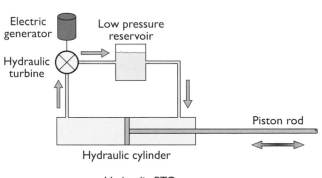

Hydraulic PTO

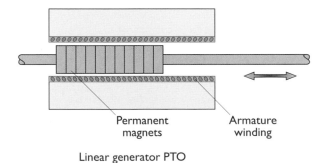

Linear generator PTO

Figure 9.22 Selected power take-off options used in wave energy devices

Other common power take-off systems include hydraulic, linear electrical generators, impulse air turbines and low-head hydro turbines.

The latter are generally Kaplan turbines and look very much like ship propellers (see Chapter 6).

Hydraulic systems use a reciprocating motion to pressurize hydraulic oil, which can be passed through a turbine to generate the rotary motion required for a generator. Hydraulic systems can offer excellent controllability, but this comes at the cost of an expensive and complex control system.

Linear generators are actuated in a similar way to hydraulic PTOs, but permanent magnets are mounted on a bar which reciprocates along a tube surrounded by coils. Linear generators are also very controllable, but require very tight tolerances for efficient generation, meaning that they must be well protected, especially from bending.

Some wave energy converters simply generate a head of water using a pump, either for use with a hydro turbine or to make potable water through reverse osmosis.

A few devices use a large mass which can slide or rotate based on the motion of the device's structure. These motions can be coupled to a rotational generator or directly used in a linear generator.

Fixed devices

Fixed seabed and shore-mounted devices are usually terminators, and these have been amongst the most common types of wave energy converter to have been tested as prototypes at sea. Having a fixed frame of reference, being closer to a grid, and with good access for maintenance purposes, they have obvious advantages over floating devices.

However, such devices have the disadvantage that they generally operate in shallow water and hence at lower wave power levels. In shallow water the waves become steeper and more damaging. Another drawback of shore-mounted devices is that of geographical location – only a limited number of sites are suitable for deployment: to optimize output, they need to be positioned in an area of small tidal range, otherwise their performance may be adversely affected. It is also worth noting that mass production techniques are unlikely to be totally applicable to shore-mounted schemes, as site-specific requirements will demand a tailored design for each device, thus adding to the production costs.

OWC-based devices

Fixed oscillating water column devices have been investigated in a number of locations with slight variations being made to the basic concept – this subsection considers bespoke OWC installations, then OWCs combined into other civil engineering projects, before finally considering more advanced OWC designs. For an excellent review of oscillating water columns see Heath (2012).

In addition to the Islay OWC described earlier, an OWC was installed on the island of Pico, part of the Azores, in the north Atlantic. The 500 kW capacity unit is in a slightly sheltered bay which helps to protect it from the excesses of the largest storms. The scheme consists of a 12 metre by 12 metre concrete structure built *in situ* on the rocky sea floor close to the shoreline (it spans a natural harbour), and is equipped with a horizontal-axis Wells turbine-generator capable of producing an average annual output of 124 kW, but rated for a maximum instantaneous value of 525 kW. The Wells air turbine is designed to operate between 750 and 1500 rpm, has eight blades of constant chord and a diameter of 2.3 m (Falcão, 2000). A second turbine with pitch-controlled blades, which permit the blades to be set at the optimum pitch angle for air flow in one direction and then set in the opposite sense when the flow reverses, is to be operated as a contrast to the fixed-pitch one (Le Crom et al., 2009, 2010).

Wave energy activity is developing in China – much of the Chinese work is linked to Japan, either in concept or by the exchange of ideas and staff. Some of the work concentrates on OWC powered navigation buoys (developments in floating devices are discussed below), some on theoretical modelling. A shoreline mounted 3 kW OWC, in a natural gully on Dawanshan Island in the Pearl River, has been successful enough to warrant upgrading with a 20 kW turbine. Dawanshan Island is being developed as a test site. For developments in China see Zhang (2014).

Harbour and breakwater schemes

When OWCs are incorporated into breakwaters, the double benefits of the breakwater – the provision of a harbour and the generation of electricity – mean that the function of electrical production does not have to justify the total capital cost, and so the generated cost of electricity may be more economically attractive.

In July 2011 a 16 chamber, OWC wave energy scheme with a total installed capacity of 296 kW was inaugurated in a breakwater in Mutriku (Figure 9.23) in the Basque Country by the Basque utility Ente Vasco de la Energia (Renewable Energy Focus, 2011).

Figure 9.23 Aerial view of the Mutriku OWC breakwater. (source: Torre-Enciso et al. 2009)

Wavegen produced the power take-off (PTO) units based on their experience of the LIMPET. Each of the 16 Wells turbines are rated at 18.5 kW and have two five-bladed runners. The plant completed its first five years of operation in July 2016, during which time it produced 1.3 GWh of electricity.

There was significant opposition to the breakwater from ecologists on the grounds of the environmental impact of the concrete structure on sediment transport and the noise of these turbines. There was a good deal of noise from the OWC chambers during construction, notably during storms, however this was reduced to a very moderate and acceptable level by using noise attenuators on the turbines and enclosing the turbines in a turbine house. Environmental monitoring of sediment transport and noise emissions continue, but the general perception is that it has been a successful project.

Japanese developers have incorporated an OWC into a 20 m section of an extension to the harbour wall at Sakata, on the north-west coast of Japan. The unit was installed in 1989 and contains a Wells turbine rated at 60 kW.

In India, trials of a multi-resonant OWC device, installed in a breakwater and employing a Wells turbine of 2 m in diameter, that drives a 150 kW generator (Figure 9.24), commenced off the Trivandrum coast in 1991 (Ravindran et al., 1995).

The device was estimated to be capable of delivering an average of 75 kW during the monsoon period from April to November, and 25 kW from December to March. A power conversion system incorporating two units has been installed so that in the higher power period both units can be operational. Experiments have been conducted to produce desalinated water from the plant, and to test an impulse turbine (Jayashankara et al., 2009).

Since the annual average wave power density along the Indian coast is only between 5 and 10 kW per metre, it is perhaps surprising to see such research and development activity. The Trivandrum OWC did not deliver the anticipated output into the local grid and was decommissioned in 2011.

A Chinese team has also described a breakwater OWC design (Liu et al., 2009).

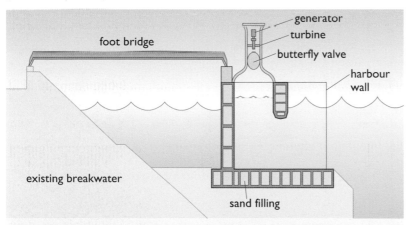

Figure 9.24 Cross-section through the Indian breakwater OWC

The resonant OWC

Most wave energy devices are resonant systems – they have a natural resonance period or time over which they repeat their motion. This is just like the motion of a child on a swing who moves with a period of a few seconds. The period is dictated by the length of the swing's ropes – longer ropes give a longer period.

In the case of OWCs the natural period is governed by the length of water from the water or wave surface outside to the water surface inside the device.

This can be termed the coupling length (l) and the arrangement behaves like a 'U' tube containing water. The period of oscillation (T) is given by:

$$T = \pi \sqrt{\frac{2l}{g}}$$

For example if $l = 25$ m then T is about 7 s. To put this into a wave energy context: if an OWC has a coupling length of 25 m then it will resonate with a period of 7 s, and it will be very responsive to incoming waves which have an energy period of 7 s. If the incoming waves have greater or smaller periods the response of the OWC will be reduced. This is true of most wave energy devices – the resonance they exhibit makes it difficult to extract

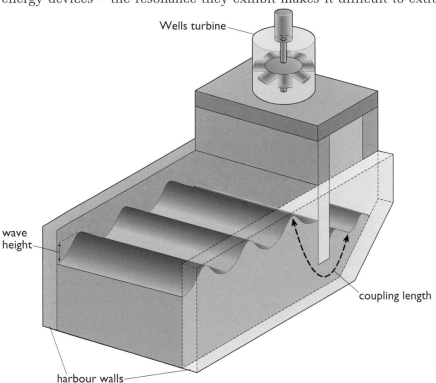

energy from a wide range of wave conditions or wave periods.

Figure 9.25 A resonant oscillating water column

Non-OWC fixed devices

Other, non-OWC fixed device prototypes have also been tested in Japan and Scotland. A number have used a mechanical linkage between a moving component, such as a hinged flap, and the fixed part of the device. An example of this approach is the **Pendulor** (Figure 9.26).

The Pendulor is a paddle, hinged at the top, which is fitted one-quarter of a wavelength from the back wall of a caisson. This is at the first anti-node (i.e. the point of maximum amplitude of a series of waves) and so the paddle is subjected to the maximum possible wave movement (note that the paddle can be located at the anti-node for only one particular wavelength). In the regions of Japan where Pendulor devices have been tested, the seas generally have wavelengths close to the design wavelength for much of the year.

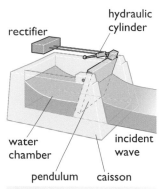

Figure 9.26 The Japanese Pendulor device

Figure 9.27 The 300 kW Waveroller deployment in 2012 (source: AW-Energy Ltd)

A push–pull hydraulic system converts the mechanical energy from the movement of the Pendulor paddle into electrical energy. Two prototypes with nominal power outputs of 5 kW have been operational on Hokkaido, Japan since the early 1980s. The Korean wave energy conversion study group (KORDI) started working on a Floating Pendulor Study in 2010 under a collaborative agreement with a Japanese research group with the intention of accelerating the development of this concept. Information on the modelling of arrays of pendulor can be found in Gunawardene et al. (2016).

In Portugal, a working prototype of the Waveroller hinged device was trialled for 26 months between 2012 and 2014. This device is designed to respond to surge forces. The first large scale Waveroller wave power device (Figure 9.27) was installed near Peniche in Portugal. Testing is currently underway on a full size PTO for deployment on the first commercial demonstrator.

Other developers have tried the concept with more or less success, notably Aquamarine's Oyster which was deployed at the European Marine Energy Centre (EMEC) in Orkney, Scotland starting with a 315 kW prototype in 2009 and a larger 800 kW machine in 2011 (Aquamarine Power, 2011a).

Floating devices

Floating systems, with the main body of the structure floating on the surface but moored to the seabed, have attracted attention over the years and are the subject of activity in many current projects. Floating wave energy conversion devices should be able to harvest more energy than fixed, on-shore devices, since the wave power density is greater offshore than in shallow water. There are also fewer restrictions to the deployment of large arrays of such devices.

In this section the devices are broken into the categories described in Figure 9.17:

- Terminator devices such as the Duck, and AWS III; floating OWCs such as the Whale; and floating overtopping converters such as Wavedragon.
- Point absorbers such as Seabased, and OPT's power buoy.
- Attenuator devices such as Pelamis and Bombora.

BOX 9.4 Support from EMEC and Wave Hub

The European Marine Energy Centre (EMEC) mentioned in Chapter 6, Box 6.6, helps to solve many of the problems that have dogged the construction of prototype wave-power generators in the past. For example, storm-proof moorings and armoured cables facilitate the simple and cheap installation and testing of new devices. Also EMEC can accommodate up to four machines at a time, so designers can directly compare devices under identical conditions, giving them a chance to spot small design 'tweaks' that will improve the efficiency of energy generation.

Another facility providing an infrastructure for the demonstration and proving of arrays of wave energy generation devices over a sustained period of time is the Wave Hub, a grid-connected offshore facility 16 kilometres off the north coast of Cornwall. It holds a 25 year lease on 8 sq km of seabed and can accommodate four separate wave energy schemes each with a capacity of 4–5 MW.

Terminator devices

OWC-based devices

A number of floating OWC-based devices have been developed – the Kaimei, a converted barge fitted with a number of floating OWC devices and the first sea-going Wells turbine, was first tested in Japan in 1977 – as mentioned previously, this was the first large floating wave power station.

As mentioned earlier, China has an expanding interest in wave energy – numerous OWC-powered navigation buoys have been deployed and the BBDB concept, originally proposed by Masuda, has been tested at model scale (Masuda et al., 2000). Experiments on the BBDB concept (illustrated in Figure 9.28) have been conducted at the Guangzhou Institute of Energy Conversion (Masuda et al., 2000; Xianguang et al., 2000).

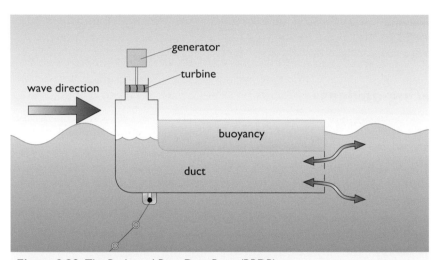

Figure 9.28 The Backward Bent Duct Buoy (BBDB)

Ocean Energy Ltd has further developed the Backward Bent Duct Buoy and a quarter scale hull has achieved over 20 000 hours of operation in live sea trials at the wave energy test site at Spiddal in Galway in Ireland (Figure 9.29). There it was subjected to a wide range of wave conditions including a severe storm when wind speeds reached 30–35 metres per second and a wave height of 8.2 m was recorded. During the testing the airflows and power output from the tests scaled up predictably and the hull behaviour was also consistent and reliable.

The Ocean Energy buoy was used to trial a novel air turbine in 2011, but there has not been any development since then.

Figure 9.29 Ocean Energy buoy with Wells turbine (courtesy of Ocean Energy)

Other variations on this theme include the Swan DK3, which is based on the L-shaped Backward Bent Duct Buoy (Meyer and Nielsen, 2000) and the Whale, a floating forward facing OWC-based device (Washio et al., 2001).

An alternative design of floating OWC has been developed by Oceanlinx. It is designed for use in harbours or rocky outcrops where the water at the coastline is deep. As wave fronts approach they are amplified up to three times at their focal point by the parabola-shaped collector, before entering a 10 metre by 8 metre OWC structure. Oceanlinx's most recent OWC deployment involved its Mk3 floating device. The unit was a one-third scale demonstration version of a full scale 2.5 MW device. It was installed offshore from the eastern breakwater of Port Kembla Harbour from February to May 2010. The unit supplied electrical power directly into the local grid and had over 5000 operating hours experience of generating electricity (Oceanlinx, 2011). It broke free of its moorings during a severe storm in May 2010 and ran up onto the beach. Oceanlinx's next 1 MW greenWAVE demonstrator suffered an incident during deployment in March 2014 and ended up stranded in the wrong place (Tidal Energy Today, 2015). Despite these setbacks, the company still intends to continue developing the concept.

Figure 9.30 Artist's impression of an Oceanlinx greenWAVE unit

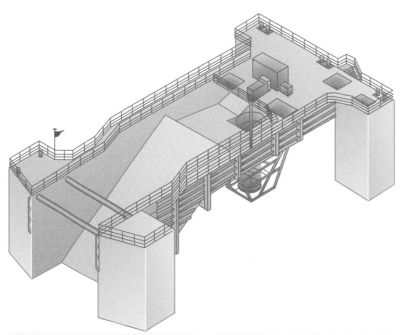

Figure 9.31 The Floating Wave Power Vessel (FWPV)

Overtopping devices

A pilot version version of the Floating Wave Power Vessel, based on ship construction was anchored in 50–80 m depth of water 500 m from the Shetland coast in 2002. This device was developed by the Swedish company Sea Power International as part of a Scottish government-backed scheme. It was designed to have a maximum power output of 1.5 MW, producing about 5.2 GWh per year (Lagström, 2000). The device functions by capturing the water from waves that run up its sloping front face. The captured water is returned to the sea via a standard Kaplan hydroelectric turbine (see Chapter 5). In many respects this device may be compared to a floating version of the TAPCHAN.

A Danish version of this floating TAPCHAN concept is the Wave Dragon. Again waves run up a tapered channel and a head of water collects in a reservoir – the water then returns to the sea via a set of simplified Kaplan hydroelectric turbines. A quarter-scale version was launched at Nissum Bredning in northern Jutland in March 2003, and this delivered its first power to the grid in June 2003. More turbines were installed at sea in September 2003, demonstrating that maintenance at sea is possible (Sorensen et al., 2003).

Duck

The Edinburgh Duck concept (Figure 9.32), conceived by Professor Stephen Salter at Edinburgh University in the 1970s, was originally envisaged as many cam-shaped bodies linked together on a long flexible floating spine which was to span several kilometres of the sea. The spine would be oriented almost parallel to the principal wave front, making the Duck largely a terminator. The Duck was designed to extract energy by pitching to match the orbital motion of the water particles, as discussed earlier. To generate power each cam-shaped body – or duck – either moves relative to the spine, producing high pressure in a hydraulic system, or drives a set of gyroscopes mounted in the noses of the 'ducks'). This matching can be nearly 'perfect' at one wave frequency and the efficiency in long waves can be improved by control of the flexure of the spine through its joints. The concept is theoretically one of the most efficient of all wave energy schemes, but it is likely to take some years to fully develop the engineering necessary to utilize this concept at full scale. The control and engineering challenges of optimizing power conversion efficiency whilst producing steady electrical energy from a 'randomly' rocking body have been studied in great detail by Salter and his team.

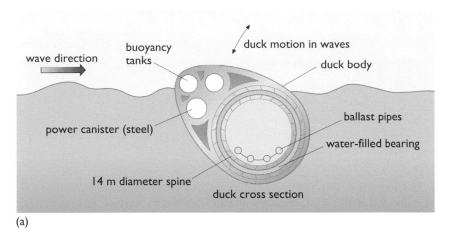

(a)

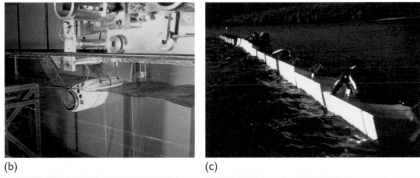

(b)   (c)

Figure 9.32 (a) The Edinburgh Duck wave energy converter; (b) Duck model being tested in a wave tank; (c) A scale model of the Duck being tested in Loch Ness, Scotland by Coventry University

AWS-III/Clam

AWS Ocean Energy has recently refined the key aspects of AWS-III technology, which initially appeared as the Circular Clam, developed at Coventry University in the UK in the 1980s.

The important point about AWS-III is its highly sensitive diaphragms which accept energy from the surge motion of the waves. As pointed out earlier in this chapter, the surge component of waves can be twice as energetic as the heave.

AWS-III consists of nine interconnected air chambers, or cells, arranged in rows of six and three on a catamaran with Wells turbines in each cell. At full scale the cells would be 16 m wide and 8 m deep, and would be deployed in deep water (100 m). Each cell is sealed against the sea by a flexible reinforced rubber diaphragm. Waves cause the movement of air between cells. Air, pushed from one cell by the incident wave, passes through at least one of the Wells turbines on its way to fill other cells. As the air system is sealed, this flow of air will be reversed when the positions of wave crest and trough on the circle change. Figure 9.33 shows an artist's impression of a small array of AWS-III devices (AWS, 2011).

Figure 9.33 Small array of AWS-III devices

Point absorber devices

There are many point absorbers in development and even production around the globe. Point absorbers tend to be smaller and cheaper than other types of wave energy converters, but also have lower power outputs with a maximum limit of about 300 kW (Falnes, 2011). Achieving this level of output requires sophisticated control systems which are still in development, so the most point absorbers to date have rated powers of 50–150 kW. Nevertheless, their cost and relative ease of deployment mean that they can be funded to commercial scale to prove they work.

Seabased's WEC is one example of a point absorber. A float on the surface is connected via a flexible cable to a linear generator sitting on the seabed and is held in place by a gravity foundation.

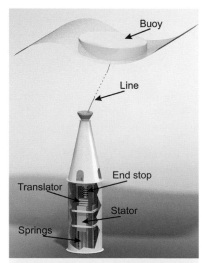

Figure 9.34 Schematic of Seabased's WEC

Seabased completed installation of 36 units as the first part of a 10 MW array in Sotenäs, near Smögen on the Swedish west coast in 2015.

Carnegie is another good example of an active developer with a point absorber.

Figure 9.35 Seabased WECs during deployment showing the linear generator and foundations on deck and two versions of the float in the water

general view

The wave energy converter termed 'PowerBuoy', developed by Ocean Power Technology (OPT), consists of a modular buoy-based system which drives generators using mechanical force developed by the vertical movement of the device. Each module is relatively small, permitting low-cost regular maintenance, leading to expected lifetimes of at least 30 years.

An early version of the PowerBuoy was deployed in December 2009 approximately 1.2 km off the coast of Oahu, Hawaii in a water depth of 30 m. To date, this device has operated and produced power from over 3 million power take-off cycles (one take-off cycle is the term for one operational cycle of one significant wave cycle) and 4400 hours of operation. This project is part of OPT's on-going programme with the US Navy to develop and test the PowerBuoy technology and has contributed to the development of OPT's systems (OPT, 2011).

OPT has, in recent years, been targeting a niche market of offshore applications requiring an uninterruptable power supply (UPS) for many years. Applications include unmanned scientific and naval instruments deployed at sea for long periods. OPT currently offers two versions of their PowerBuoy rated at 3 kW and 15 kW. These can accommodate batteries to ensure continuous power even through calm periods.

Inertial devices

It is possible to increase the power captured by means of controlling the inertial mass, see Flocard and Finnigan (2012). The Pitching and Surging FROG, developed by Professor French and his colleagues at Lancaster University, is a reaction wave energy converter which achieves energy-absorbing behaviour by the movement of internal inertial mass

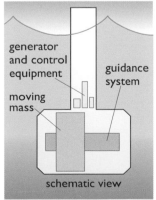

Figure 9.36 The Pitching and Surging FROG wave energy converter

(McCabe et al., 2003). Hence, it is a compact structure which does not require a large spine to provide a stable frame of reference (Figure 9.36).

Attenuator devices

There are no attenuators currently in development, however one of the most advanced wave energy converters deployed to date was an attenuator by Pelamis Wave Power.

Pelamis

The Pelamis, or 'sea snake', can trace some of its ancestry to the Edinburgh Duck; it consists of a number of floating cylindrical tubes, connected to each other by active joints. The tubes are arranged at a slight angle off the down-wave direction and so act as attenuator devices. The wave-induced heaving and swaying of the tubes is resisted by hydraulic rams that pump high-pressure oil through hydraulic motors via smoothing accumulators, and the hydraulic motors drive electrical generators to produce electricity.

The ability to withstand high wave power densities in storms has been a key goal of the designers; the Pelamis is capable of inherent load shedding, which means that the spine is not subjected to the full structural loadings that would otherwise be imposed on it during a storm. As a long thin structure with very little reserve buoyancy, it sits down the waves rather than across them – see Figure 9.37 – and so passes through the potentially damaging storm waves. It also becomes detuned in long storm waves, where the waves are much longer than the device.

Figure 9.37 The Pelamis P2

The prototype 750 kW devices were 150 m long, 3.5 m in diameter, and composed of five modular sections. To reduce risk, wherever possible use was made of existing technology that has been proven offshore. Several deployments of small arrays were achieved by Pelamis, but the company went into administration in 2014 unable to service the large debts built up during the development of its business.

Bombora

A near-shore underwater concept is shown in Figure 9.38. Here double sets of membrane covered air chambers run out from the shore to act as attenuators. Bombora has designed a 60 MW wave farm at Peniche in Portugal and expects delivered energy to cost 15 to 30 p/kWh depending on assumptions about discount rate, operating and maintenance costs, and performance (Bombora, 2017).

Figure 9.38 Bombora's flexible membrane seabed mounted device

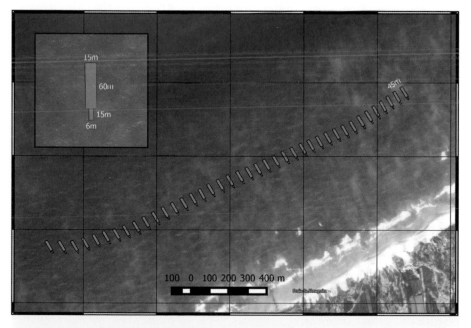

Figure 9.39 Potential 60 MW Bombora farm layout at Peniche, Portugal

Technology co-location

A way of reducing costs is to share infrastructure between different technologies. This could be by co-locating devices to share the cost of moorings, deployment, operation and maintenance (O&M) or export cables. Or it could be by sharing a support structure such as Floating Power Plant's (FPP) machine. Their P37 demonstrator is shown in Figure 9.40 and was rated at 33 kW from wind and 50 kW from wave.

Figure 9.40 Floating Power Plant's P37 demonstrator near Lolland in Denmark

FPP's commercial device, the P80, is rated at 5–8 MW for wind and 2–3.6 MW wave power, depending on the location and turbines used.

Good wind and wave resources are often co-located, so combining or co-locating offshore wind turbines makes sense. FPP's concept uses hinged floats with hydraulic PTOs on a platform which also hosts wind turbines. As well as the clear benefits to this system in terms of shared deployment, O&M, electricity export, and effective use of the available resources there are less obvious benefits such as that the wind turbines stabilise the platform and increase the efficiency of the wave energy converters. The wave energy extraction means that there is an area behind the platform with reduced wave height which allows O&M activities in a wider range of sea-states. This is very valuable as the availability of O&M weather windows has a significant effect on project finances. This is evidenced by the fact that O&M vessels for the nearby Vindeby wind farm in Denmark would go to the P37 and await suitable conditions for staff to board the conventional wind turbines.

Another interesting combination is Albatern's squid, which has been co-located with fish farms, with the objective of removing the need for shore power.

OPT has deployed its machines in remote locations to power US Navy sensing platforms where solar power provides an inadequate source of power.

9.6 Arrays

Figure 9.41 Diagram of array

Modelling the performance of individual wave energy devices is beyond the scope of this book. Some simple OWCs and buoys can be treated as mass-spring-damper systems and then the response to waves can be modelled using the approach found in McCormick (1981).

As mentioned, though, the large potential of waves lies offshore in water depths of 50 m or more and we can expect arrays of devices to be deployed to harness the wave energy. Modelling the performance of arrays of wave energy converters requires that we consider the influence of devices on their neighbours and of rows of devices on successive rows. The first row is exposed to the full incident wave power and captures a proportion of that power leaving a reduced power density to proceed to the second row and so on. This means that successive rows are relatively less productive. The accurate modelling of arrays becomes imperative for the prediction of wave energy farm productivity. See Folley (2016) for details of modelling of individual devices and arrays.

9.7 Economics

Reducing the capital, operation and maintenance costs is the key to the successful economic implementation of wave energy stations. The capital cost per kW of establishing a wave-energy run power station could be twice that of a conventional station running on fossil fuels; and the capacity factor is likely to be much lower than a conventional station due to the variability of the wave climate. Therefore, wave energy costs can only be competitive if the capital costs are reduced and running costs are significantly below those for a conventional station. Naturally the 'fuel' or wave energy costs are zero, leaving the operation and maintenance costs as the determining factor. Schemes will therefore have to be reliable in their energy conversion and robust enough to survive the wave climate for many years. This means schemes designed for long lifetimes and with small numbers of moving parts (to minimize failures). The oscillating water columns and TAPCHAN schemes are good examples of what is required.

On the other hand the wave energy pioneer Stephen Salter has long argued against 'simplicity' and in favour, for instance, of sophisticated control and power conversion systems to maximize the useful energy that can be captured from the large and expensive structures that had traditionally been required to intercept the waves (Salter et al., 2002).

The UK Committee on Climate Change (CCC, 2011) has calculated the cost of electricity from a possible future (2030) 50 MW array of shoreline wave energy converters, with a capital cost of £2200 per kW and life expectancy of 40 years. Using a 10% discount rate and expressed in £(2010):

■ for a low capacity factor (15%) the cost of electricity is 29.1p kWh^{-1}
■ for a high capacity factor (22%) the cost of electricity is 19.9p kWh^{-1}.

As with the hydro and tidal plants discussed in Chapters 5 and 6 the discount rate used is an important factor, and is discussed further in Chapter 10 and Appendix B.

This illustrates the importance of achieving a high capacity factor — making as much use of the device as possible. The cost figures reported here are high and reflect the fact that the fixed devices cannot usually benefit from mass production as they would be purpose built for a specific location, and also that such devices would generally operate in shallow water where there is a much reduced wave energy climate.

The total capital investment required for wave energy schemes is dependent on overall average efficiencies and on location. Many of the devices detailed in this chapter have average efficiencies of around 30%. Frequency response characteristics and limitations of swept volume and survival when operating in very energetic seas are responsible for the generally low overall efficiency. At the time of writing, the capital cost is typically around £3000–£4500 per installed kW, although the cost of particular schemes may vary markedly from this.

Large schemes are technically demanding because of the high structural loads imposed by the north Atlantic wave climate. As time has passed there have been improvements in design, performance and construction techniques, together with rationalization of some of the problems and a move to smaller schemes. Smaller schemes are technically simpler and less financially risky, and hence the capital costs and insurance costs are reduced, with commensurate reduction in produced energy costs.

Wave energy technology has moved into the commercial world and several developers are already deploying prototypes and, in some cases, executing schemes generating electricity or desalinating sea water at favourable prices. Coupled with incentives for avoided carbon dioxide emissions, the economic prospects for commercial wave energy exploitation appear to be good.

As wave energy is considered to be environmentally benign (see below), if the technology can be successfully developed it is likely to become an attractive commercial and political proposition, and this should result in an extensive installation programme with wave farms deployed in many locations. The cost of such installations will certainly fall from current levels due to refinement to designs, making them more efficient, and to lower production costs when they are (where possible) mass produced.

9.8 Environmental impact

Wave energy converters may be among the most environmentally benign of energy technologies for the following reasons:

- They have little potential for chemical pollution. At most, they may contain some lubricating or hydraulic oil, which will be carefully sealed from the environment
- They have little visual impact except where shore-mounted
- Noise generation is likely to be low – generally lower than the noise of crashing waves (there might be low-frequency noise effects on cetaceans, but this has yet to be confirmed)
- They should present a small (though not insignificant) hazard to shipping
- They should present no difficulties to migrating fish
- Floating schemes, since they are incapable of extracting more than a small fraction of the energy of storms, will not significantly influence the coastal environment. Of course, a scheme such as a new breakwater incorporating a wave energy device will provide coastal protection, and may result in changes to the coastline. Concrete structures will need to be removed at the end of their operating life
- Some stakeholders, such as fishermen and surfers, are concerned that wave energy schemes could detract from their activities or livelihoods, but it is usually possible to find a compromise that is acceptable to all parties
- Near-shore wave energy schemes will release (from, for example, construction and material transport) an estimated 11 g of CO_2, 0.03 g of SO_2 and 0.05 g of NO_x for each kWh of electricity generated (Thorpe, 1999), making them very attractive in comparison to the conventional UK electrical generating mix of coal, gas and nuclear plants (see Chapter 11). Thus wave energy can make a significant contribution in meeting climate change and acid rain targets.

9.9 Integration

The electrical output from a wave energy scheme can be used directly but it is much more likely that the electricity will be fed into a grid. The electrical issues associated with such schemes include variability of supply of electricity due to the nature of waves, phasing, power factor correction and transmission losses (Freris and Infield, 2008).

Wave energy for isolated communities

If the grid is small and serves a small, remote community, great care must be taken in integrating the electrical output from a wave energy scheme: the output from a wave energy scheme will vary with time (except in the case of units such as the TAPCHAN) and may cause voltage or frequency fluctuations. Energy storage or other methods of smoothing may thus be necessary.

Many small communities currently depend on diesel generators for their electricity. A diesel generator is best run at a constant output close to its design capacity – say 50 kW. Therefore, if a diesel unit is the sole source of electricity, the load from the grid should always be matched to 50 kW. Clearly, the consumers will cause the load to vary as they switch appliances on and off, so to allow for this, a 'dump' load can be incorporated into the system. The diesel output is directed to this dump load when the load on the grid falls – for instance at night when the demand is low, energy can be used for space heaters, freezers, water heaters or to drive washing machines, etc.

Similarly, the incorporation of a varying electrical output from a wave energy scheme into the grid can partially be accommodated by the use of such dump loads located in houses, schools, factories etc. to stabilize the grid (Figure 9.42).

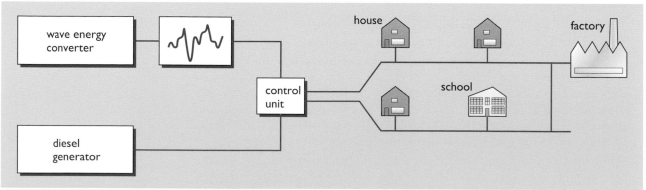

Figure 9.42 An integrated wave energy/diesel system

By careful overall design of an integrated scheme, a remote community could enjoy significant gains in electricity supply from a wave energy scheme. If this produces most of the energy the reduction in diesel oil consumption would be substantial, and since it is costly to transport diesel oil to remote locations, the cost savings could be large. Ideally the diesel generator would be held in reserve and only brought into use when the wave activity is too low to meet demand. An energy storage system such as batteries would be very useful in smoothing the supply and minimizing the call on the diesel generator.

Wave energy for large electricity grids

When the electrical outputs of several wave energy units are added together, the total output will be generally smoother than for a single unit. If we extend this to an array of several hundred floating devices, then the summed output will be smoother still. In addition, any fluctuations in output will be less important if the electricity is to be delivered to large national systems like those of the UK, where in most locations the grid is 'strong' enough to absorb contributions from a fluctuating source. Figure 9.43 illustrates a typical scheme.

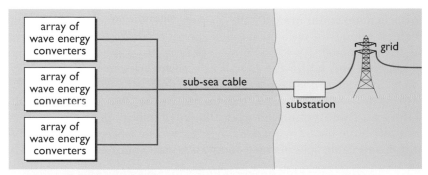

Figure 9.43 Electrical connections for an array of wave energy devices

Finally, although we have dwelt upon short-term fluctuations of seconds or minutes, the wave resource also varies on a day-to-day and season-by-season basis. For those countries in the north-east Atlantic such as the UK, Ireland, Spain and Norway, the seasonal variation shown in Figure 9.44 demonstrates that wave power output reaches a maximum in the bad weather of winter when the electrical demand is greatest (see for example, Rugbjerg et al., 2000 and Petroncini and Yemm, 2000).

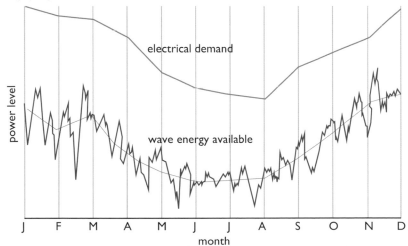

Figure 9.44 Seasonal availability of wave energy and electrical demand for the UK

9.10 Summary

The concept of extracting energy from ocean waves has intrigued people for centuries. The detailed understanding of waves, and of the energy that they contain, is complex, as is the description of a typical sea state which is composed of many waves, with different frequencies moving in different directions at different velocities. However, by developing wave theory even in a simple form we can build up a working knowledge which is suitable for designing wave energy converters.

There are some important outcomes from even a basic appreciation of waves – firstly that the **world wave energy resource is extremely large** (and even now has not yet been fully assessed). If only a small proportion of this could be harnessed it would make a major contribution to meeting the world's energy requirements.

The **waves in deep water are very energy rich**, but of course the conditions are difficult to operate in and so **wave energy converters for this environment must be highly robust**. As waves move towards the shore they lose some of their energy, and while the operating conditions are less strenuous, different technological challenges are raised in designing economic devices for capturing the wave energy. All told, there are over 1000 patents on wave energy devices, resulting in **well over 100 wave energy projects** – in this chapter a few have been reviewed as being representative of some of the more significant categories of wave energy converter shown in Figure 9.15.

In the UK, while the wave climate is conducive to wave power developments, the political climate has not always been so favourable. In the 1970s and 1980s the UK government's brief to wave energy teams was to design 2 GW schemes. In hindsight we can appreciate that this was remarkably ambitious – analogous to expecting someone to design a Boeing 747 in the early days of aviation without going through the evolution of the biplane, single seater monoplane, jet engine, etc. Attitudes are changing very quickly, however, prompted by the need to address global climate change, by the issue of long-term resource security of fossil fuel supplies and by the increasingly competitive economics of wave energy. Supporting this effort are the EMEC and Wave Hub test sites which have been established in Cornwall and the Orkney respectively.

UK teams conducted much of the early work in the development of wave energy systems, but as this chapter records many other countries are now very active. **Commercial involvement has had an important impact** and we can now see private enterprise schemes and concepts emerging from countries such as Norway, Australia, Sweden, Denmark and the USA, as well as the UK. Across the globe, several companies have already commissioned their first wave stations, and more are obtaining permission to exploit sea areas and deploy sea trials.

It should be remembered that the generation of electricity is not the only option for the delivered energy. **Desalination, coastal protection, water pumping, mariculture, mineral recovery from sea water and hydrogen generation are among the benefits being developed**.

The cost of energy from the current generation of wave energy converters is high, but wave energy developers are confident that with time, experience and technological improvements costs will reduce, rendering **wave energy an environmentally attractive and sustainable industry**. Indeed some commentators have made the point that no other energy technology (coal, oil, nuclear, wind) has ever started commercial production from such a low unit cost as was initially expected of wave energy. If wave energy costs follow the example of wind power then rapid cost reductions can be expected (Aquamarine Power, 2011b).

The development of wave energy technologies has been a long process, but the economics of current designs are potentially attractive. Proving the long-term survival and cost effectiveness of designs as technologies mature should make the prospect of wave energy stations on a large scale a real possibility – we may well see such converters deployed in large numbers to harvest the considerable wave energy that is present at some locations.

References

AquaEnergy Group (2010) *AquaEnergy Group* [Online]. Available at http://aquaenergygroup.com (Accessed 18 November 2011).

Aquamarine Power (2011a) 'The Danish wind industry 1980–2010', *International Journal of the Society for Underwater technology,* vol. 30, no. 1, pp. 27–31.

Aquamarine Power (2011b) *Aqua Marine Power* [Online]. Available at http://www.aquamarinepower.com/ (accessed 18 November 2011).

AWS (2011) *AWS Ocean Energy* [Online]. Available at http://www.awsocean.com (Accessed 18 November 2011).

BERR (2008) *Atlas of UK Marine Renewable Energy Resources* [Online]. Available at http://www.renewables-atlas.info/ (Accessed 18 November 2011).

Boake, C. B., Whittaker, T. J. T., Folley, M. and Ellen, H. (2002) 'Overview and initial operational experience of the LIMPET wave energy plant', *The Proceedings of the Twelfth (2002) International Offshore and Polar Engineering Conference*, Kitakyushu, Japan, 26–31 May. The International Society of Offshore and Polar Engineers [Online]. Available at http://citeseerx.ist.psu.edu/viewdoc/download?doi=10.1.1.473.2780&rep=rep1&type=pdf (Accessed 22 March 2017).

Bombora (2017) *Bombora Wave Power* [Online]. Available at http://www.bomborawavepower.com.au/ (Accessed 24 April 2017).

Boström, C., Rahm, M., Svensson, O., Strömstedt, E., Savin, A., Waters, R. and Leijon, M. (2011) 'Temperature Measurements in a Linear Generator and Marine Substation for Wave Power', *Journal of Offshore Mechanics and Artic Engineering*, vol. 134, no. 2 [Online]. DOI:10.1115/1.4004629 (Accessed 28 March 2017).

Budal, K. and Falnes, J. (1975) 'Power generation from ocean waves using a resonant oscillating system', *Marine Science Communications*, vol. 1, pp. 269–288.

Caratti, G., Lewis, A. and Howett, D. (eds) (1993) Wave energy R&D: Proceedings of a workshop held at Cork 1–2 October 1992. Brussels, Commission of the European Communities

Carbon Trust (2010) *Carbon Trust* [online]. Available at http://www.carbontrust.co.uk (Accessed 18 November 2011).

CCC (2011) *The Renewable Energy Review: May 2011* [Online], London, Committee on Climate Change. Available at https://www.theccc.org.uk/archive/aws/Renewables%20Review/The%20renewable%20energy%20review_Printout.pdf (Accessed 22 March 2017).

Claeson, L (1987) '*Energi från havets vågor*', (in Swedish), Sweden, Energiforskningsnämnden, report no. 21.

ETI and UKERC (2014) *Marine Energy Technology Roadmap 2014* [Online], Loughborough/London, Energy Technologies Institute and the UK Energy Research Centre. Available at https://s3-eu-west-1.amazonaws.com/assets.eti.co.uk/legacyUploads/2014/04/Marine-Roadmap-FULL-SIZE-DIGITAL-SPREADS-.pdf (Accessed 17 March 2017)

ETSU (1985) *Wave Energy: The Department of Energy's R&D Programme 1974–1983*, Oxfordshire, Energy Technology Support Unit, Report R26.

EU-OEA (2010) *Oceans of energy: European Ocean Energy Roadmap 2010–2050* [Online], Brussels, European Ocean Energy Association. Available at https://www.icoe-conference.com/publication/oceans_of_energy_european_ocean_energy_roadmap_2010_2050/ (Accessed 22 March 2017).

Everett, B., Boyle, G. A., Peake S. and Ramage, J. (eds) (2012) *Energy Systems and Sustainability: Power for a Sustainable Future* (2nd edn), Oxford, Oxford University Press/Milton Keynes, The Open University.

Falcão, A. F. de O. (2000) 'The shoreline OWC wave power plant at the Azores', paper presented at *Fourth European Wave Energy Conference*, Aalborg, Denmark, 4-6 December.

Falnes, J. (2011) 'Heaving buoys, point absorbers and arrays', Philosophical Transactions of the Royal Society A 2012, vol. 370, no. 1959, pp. 246–277 [Online]. DOI: 10.1098/rsta.2011.0249 (Accessed 23 March 2017).

Falnes, J. and Lillebekken, P. M. (2003) 'Budal's latching controlled buoy type wave power plant', *Fifth European Wave Energy Conference*. Cork, Ireland. 17–20 September. Hydraulics & Maritime Research Centre [Online]. Available at https://brage.bibsys.no/xmlui/bitstream/handle/11250/246687/549640_FULLTEXT01.pdf?sequence=1&isAllowed=y (Accessed 22 March 2017).

Falnes, J. and Løvseth, J. (1991) 'Ocean wave energy', *Energy Policy*, vol. 19, no. 8, pp. 768–775.[Online]. DOI: 10.1016/0301-4215(91)90046-Q (Accessed 22 March 2017).

Flocard, F. and Finnigan, T. D. (2012) 'Increasing power capture of a wave energy device by inertia adjustment', Applied Ocean Research, vol. 34, pp. 126–134 [Online]. DOI: http://dx.doi.org/10.1016/j.apor.2011.09.003 (Accessed 31 March 2017).

Folley, M. (2016) *Numerical Modelling of Wave Energy Converters: State of the Art Techniques for Single Devices and Arrays*, Cambridge, Massachusetts, Academic Press.

Freris, L. and Infield, D. D. (2008) *Renewable Energy in Power Systems*, Chichester, Wiley.

Gill, A. T. and Rocheleau, R. E. (2009) *Advances in Hawaii's Ocean Energy RD&D*, paper presented at Eighth European Wave and Tidal Energy Conference. Uppsala, Sweden, 7–10 September.

Gunawardane, S. P., Kankanamge, C. J. and Watabe, T. (2016) 'Study on the Performance of the 'Pendulor' Wave Energy Converter in an Array Configuration', Energies, vol. 9, no. 4 [Online]. DOI: 10.3390/en9040282 (Accessed 31 March 2017).

Heath, T. (2012) 'A Review of Oscillating Water Columns', *Phil. Trans, R, Soc. A*, vol. 370, pp. 235–245 [Online]. DOI: 10.1098/rsta.2011.0164 (Accessed 22 August 2017).

IEA-RETD (2012) *Offshore Renewable Energy*, Oxfordshire, Earthscan/ Routledge.

ISOPE (2002) *The Proceedings of the Twelfth (2002) International Offshore and Polar Engineering Conference, Volume I: Ocean Resources and Engineering', Twelfth (2002) International Offshore and Polar Engineering*

Conference. Kitakyushu, Japan, 26-31 May. ISOPE [Online]. Available at http://www.isope.org/publications/proceedings/ISOPE/ISOPE%202002/TOC.PDF (Accessed 24 March 2017).

ITPEnergised (2017) ITPEnergised [Online]. Available at http://www.itpenergised.com (Accessed 15 September 2017).

Jayashankara V. S., Anand A. T., Geetha A. S., Santhakumar B. V., Jagadeesh Kumar A. M., Ravindran C. T., Setoguchi D. M., Takao E. K., Toyota F. S. and Nagata F. (2009) 'A twin unidirectional impulse turbine topology for OWC based wave energy plants', *Renewable Energy,* vol. 34, pp. 692–698.

Kondo, M. (1993) *Proceedings of International Symposium on Ocean Energy Development*, Hokkaido, Japan, Muroran Institute of Technology/Cold Region Port and Harbor Engineering Research Center.

Kraemer, D. R. B., Ohl, C. O. G. and McCormick, M. E. (2000) 'Comparisons of experimental and theoretical results of the motions of a McCabe wave pump', *Fourth European Wave Energy Conference.* Aalborg, Denmark, 4–6 December. Denmark, Energy Centre Denmark, pp. 211–218.

Lagström, G. (2000) 'Sea Power International – Floating Wave Power Vessel (FWPV)', *Fourth European Wave Energy Conference.* Aalborg, Denmark, 4–6 December. Denmark, Energy Centre Denmark, pp. 141–146.

Le Crom, I., Brito-Melo, F., Neumann, A. and Sarmento, A. J. N. A. (2009) 'Numerical Estimation of Incident Wave Parameters Based on the Air Pressure Measurements in Pico OWC Plant', *Eighth European Wave and Tidal Energy Conference EWTEC 2009.* Uppsala, Sweden, 7–10 September [Online]. Available at http://www.homepages.ed.ac.uk/shs/Wave%20Energy/EWTEC%202009/EWTEC%202009%20(D)/papers/132.pdf (Accessed 23 March 2017).

Le Crom, I., Brito-Melo, A., Neumann, F. and Sarmento, A. J. N. A. (2010) 'Portuguese Grid Connected OWC Power Plant: Monitoring Report', *Twentieth International Offshore and Polar Engineering Conference.* Beijing, 20–25 June. The International Society of Offshore and Polar Engineers [Online]. Available at http://www.isope.org/publications/proceedings/ISOPE/ISOPE%202010/data/papers/10TPC-1166LeCro.pdf (Accessed 23 March 2017).

Liu, Z., Shi, H., and Hyan, B-S. (2009) 'Practical design and investigation of the breakwater OWC facility in China', *Eighth European Wave and Tidal Energy Conference EWTEC 2009.* Uppsala, Sweden, 7–10 September [Online]. Available at http://www.homepages.ed.ac.uk/shs/Wave%20Energy/EWTEC%202009/EWTEC%202009%20(D)/papers/275.pdf (Accessed 23 March 2017).

Masuda, Y., Kuboki, T., Xianguang, L. and Peiya, S. (2000) 'Development of terminator type BBDB', *Fourth European Wave Energy Conference*, Aalborg, Denmark, 4–6 December. Denmark, Energy Centre Denmark, pp. 147–152.

McCabe, A. P., Bradshaw, A., Widden, M. B., Chaplin, R. V., French, M. J. and Meadowcroft, J. A. C. (2003) 'PS FROG MK 5: an offshore point absorber wave energy converter', paper presented at *Fifth European Wave Energy Conference.* Cork, Ireland, 17–20 September.

McCormick, M. E. (1981) Ocean Wave Energy Conversion, Chicester, Wiley.

Meyer, N. I. and Nielsen, K. (2000) 'The Danish wave energy programme second year status', *Fourth European Wave Energy Conference*. Aalborg, Denmark, 4–6 December. Denmark, Energy Centre Denmark, pp. 10–18.

Miyazaki, T. (1991) 'Wave energy research and development in Japan'

Oceans 91 Symposium. Honolulu, Hawaii, 1–3 October.

Miyazaki, T. and Hotta, H. (eds) (1991) 'Proceedings of Third Symposium on Ocean Wave Energy Utilization', (in Japanese), *Third Symposium on Ocean Wave Energy Utilization*. Tokyo, Japan, 22–23 January. Japan, Japan Marine Science and Technology Center.

Nielsen, K. and Plum, C. (2000) 'Point absorber – numerical and experimental results', *Fourth European Wave Energy Conference*. Aalborg, Denmark, 4–6 December. Denmark, Energy Centre Denmark, pp. 219–226.

Nolan, G., O'Cathain, M., Ringwood, J. V. and Murtagh, J. (2003) 'Modelling, simulation and validation of the power take-off system for a hinge barge wave energy converter', paper presented at the *Fifth European Wave Energy Conference*. Cork, Ireland, 17–20 September.

Oceanlinx (2011) *Oceanlinx* [Online]. Available at http://www.oceanlinx. com/ (Accessed 18 November 2011).

OES (2017) *An International Vision for Ocean Energy 2017* [Online]. Portugal, Ocean Energy Systems Technology Collaboration Programme. Available at https://www.ocean-energy-systems.org/library/vision-and-strategy/document/oes-vision-for-international-deployment-of-ocean-energy-2017-/ (Accessed 23 March 2017).

OPT (2010) *Ocean Power Technologies* [Online]. Available at http://www. oceanpowertechnologies.com/ (Accessed 18 November 2011).

OPT (2011) *Ocean Power Technologies* [Online]. Available at http://www. oceanpowertechnologies.com/projects.htm (Accessed 18 November 2011).

Petroncini, S. and Yemm, R. W. (2000) 'Introducing wave energy into the renewable energy marketplace', *Fourth European Wave Energy Conference*. Aalborg, Denmark, 4–6 December. Denmark, Energy Centre Denmark, pp. 33–41.

Ravindran, M., Jayashankar, V., Jalihal, P. and Pathak, A. G. (1995) 'Indian wave energy programme: progress and future plans', paper presented at the *Second European Wave Power Conference*. Lisbon, Portugal, 8–10 November.

Renewable Energy Focus (2011) *Renewable Energy Focus* [Online]. Available at http://www.Renewableenergyfocus.com (Accessed 4 September 2011).

Ross, D. (1995) *Power from the Waves*, Oxford, Oxford University Press.

Rugbjerg, M., Nielsen, K., Christensen, J. H. and Jacobsen, V. (2000) 'Wave energy in the Danish part of the North Sea', *Fourth European Wave Energy Conference*. Aalborg, Denmark, 4–6 December. Denmark, Energy Centre Denmark, pp. 64–71.

Salter, S. H. (1974) 'Wave power', *Nature*, vol. 249, pp. 720–724.

Salter, S. H., Taylor, J. R. M. and Caldwell, N. J. (2002) 'Power conversion mechanisms for wave energy', *Proceedings of the Institution of Mechanical*

Engineers Mech. Engineers Part M: Engineering for the Maritime Environment, vol. 216, no. 1. [Online]. DOI: 10.1243/147509002320382103 (Accessed 23 March 2017).

Salter, S. and Lin, C. P. (1995) 'The sloped IPS wave energy converter', paper presented at the *Second European Wave Power Conference*. Lisbon, Portugal, 8–10 November.

Sorensen, H. C., Christensen, L. and Hansen, L. K. (2003) 'Development of Wave Dragon from scale 1:50 to prototype', paper presented at the *Fifth European Wave Energy Conference*. Cork, Ireland, 17–20 September.

Thorpe, T. W. (1992) *A Review of Wave Energy*, Oxfordshire, Energy Technology Support Unit, Report R-72.

Thorpe, T. W. (1998) *Overview of Wave Energy Technologies*, London, Marine Foresight Panel, AEAT-3615.

Thorpe, T. W. (1999) *A Brief Review of Wave Energy* [Online]. Available at http://lamtengchoy.com/main/uploads/others/1298524568_8914.pdf (Accessed 24 April 2017).

Thorpe, T. W. (2001) 'The UK market for marine renewables', paper presented at the *All-Energy Futures Conference*. Aberdeen, February.

Tidal Energy Today (2015) 'Oceanlinx stranded wave energy device removal stalls', *Tidal Energy Today*, 14 April 2015 [Online]. Available at http://tidalenergytoday.com/ (Accessed 24 April 2017).

Torre-Enciso, Y., Ortubia, I., de Aguileta, L. and Marqués, J. (2009) 'Mutriku Wave Power Plant: From the Thinking out to the Reality', *Eighth European Wave and Tidal Energy Conference EWTEC 2009*. Uppsala, Sweden, 7–10 September [Online]. Available at https://tethys.pnnl.gov/sites/default/files/publications/Torre-Enciso_et_al_2009.pdf (Accessed 23 March 2017).

Wacher, A. and Neilsen, K. (2010) 'Mathematical and Numerical Modeling of the AquaBuOY Wave Energy Converter', *Mathematics-in-Industry Case Studies Journal*, vol. 2, pp. 16–33.

Washio, Y., Osawa, H. and Ogata, T. (2001) 'The open sea tests of the offshore floating type wave power device "Mighty Whale" – characteristics of wave energy absorption and power generation', *Ocean 2001 Symposium*. Honolulu, Hawaii, 5–8 November. IEEE [Online]. DOI: 10.1109/OCEANS.2001.968786

WEC (2010) *Survey Of Energy Resources*, London, World Energy Council.

Xianguang, L., Peiya, S., Wei, W. and Niandong, J. (2000) 'The experimental study of BBDB model with multipoint mooring', *Fourth European Wave Energy Conference*, Aalborg, Denmark, 4–6 December. Denmark, Energy Centre Denmark, pp. 173–185.

Yemm, R., Pizer, D., Retzler, C. and Henderson, R. (2012) 'Pelamis: Experience from concept to connection', *Philosophical Transactions of the Royal Society A*, vol. 370, no. 1959, pp. 365–380.

Zhang, Y., Lin, Z., Liu, Q. (2014) 'Marine renewable energy in China: Current status and perspectives', *Water Science and Engineering*, vol. 7, no. 3, pp. 288–305 [Online]. DOI: 10.3882/j.issn.1674-2370.2014.03.005 (Accessed 23 March 2017).

Chapter 10

Deep geothermal energy

By James P. Warren

10.1 Introduction

Geothermal energy – heat from the Earth – is one of the less well recognized forms of renewable energy. This chapter explains the nature of deep geothermal energy, why it is treated as renewable energy, its usage over the past hundred years and its probable future.

Several energy technologies are currently labelled 'geothermal' and it is important to distinguish between them:

- Deep geothermal energy usually draws heat from hot rocks or aquifers that can be a kilometre or more below the Earth's surface.
- Ground source heat pump technology, described in Chapter 2, particularly that used for domestic heating, is perhaps best described as 'shallow geothermal energy' and only uses flows of heat between the air and the top layer of the ground down to a depth of about 40 metres.
- Aquifer Thermal Energy Storage, also described in Chapter 2, is another 'shallow geothermal' heat pump technology and uses heat flows to and from underground aquifers, typically 100 metres below the surface.
- Heat pumps may also be used in conjunction with deep geothermal wells to raise the temperature of water up to a usable level for heating buildings. As such they do have a place in this chapter.

Deep geothermal energy is one of the few renewable energy sources that is constantly available. Although it used to be thought the preserve of regions prone to volcanic activity, that has now changed; new technologies have made this energy source more widely available and it is realistic to think that geothermal energy could meet a significant fraction of the world's energy demand in the near and longer term future. This chapter documents its current usage and outlines the steps that are being taken to make this forecast a reality.

Geothermal energy – the mining of geothermal heat

In the continuing search to find cost-effective forms of energy that neither contribute to global warming nor threaten national security, geothermal energy has become a significant player. It is one form of 'renewable' energy that is independent of the Sun, having its ultimate source within the Earth. It is a comparatively diffuse resource; the amount of heat flowing through the Earth's surface, 10 000 EJ y^{-1}, is tiny in comparison with the massive 5 400 000 EJ y^{-1} solar heating of the Earth which also drives the atmospheric and hydrological cycles. Fortunately, there are many places where the Earth's heat flow is sufficiently concentrated to have generated natural resources in the form of steam and hot water (180–250 °C), available in rocks within feasible drilling distance of the ground surface and suitable for electricity generation. These are the so-called 'high-enthalpy' resources (see Box 10.1).

BOX 10.1 Enthalpy

Enthalpy is a thermodynamic term defined as the heat content of a substance per unit mass, and is a function of pressure and volume as well as temperature. For any given substance the enthalpy can be thought of as a kind of energy store that provides, or accepts, energy in the form of heat. This idea is reflected in the original derivation of the term – it comes from the Greek words for 'heat inside'. Most importantly, enthalpy includes the latent heat of vaporization of that substance. Thus the conversion of water into steam at its boiling point involves a large increase in enthalpy, even though the actual temperature increase may be small. Enthalpy is usually denoted by (H) in chemistry nomenclature, and the standard units for enthalpy are joule per kilogram; however, other variants exist and are used.

Geothermal resources are usually classified as 'high enthalpy' (water and steam at temperatures above about 180–200 °C), 'medium enthalpy' (about 100–180 °C) and 'low enthalpy' (<100 °C). The term 'enthalpy' is used because temperature alone is not sufficient to define the useful energy content of a steam/water mixture. A mass of steam at a given temperature and pressure can provide much more energy than the same mass of water under the same conditions. The distinction is important to geothermal practitioners so the term has entered into general use.

For the purposes of this chapter, however, it is usually sufficient to think of temperature and enthalpy as going hand in hand.

The techniques for exploiting the resources are very simple in principle, and are analogous to the well-established techniques for extracting oil and gas. One or more boreholes are drilled into the reservoir, the hot fluid flows or is pumped to the surface and it is then used in conventional steam turbines or heating equipment. Geothermal wells are drilled to depths of 700 to 3000 m.

Obviously, electricity is a more valuable and versatile end-product than hot water, so most attention tends to be focused on those resources capable of supporting power generation, i.e. hot enough to make electricity generation economic. By 2015 world electrical power-generating capacity from geothermal resources had reached 13.3 gigawatts electrical (13.3 GW_e), a small but significant contribution to energy needs in some areas (Matek, 2016). A further 26 countries intend to have geothermal generation plants in service by 2020 (see Table 10.1), with total world capacity forecast to increase to 18.5 GW_e.

About a further 75 gigawatts of thermal capacity (75 GW_t) is also being harnessed in non-electrical 'direct use' applications, principally for space heating, agriculture, aquaculture and a variety of industrial processes (Bertani, 2015). Many of these applications occur outside high-enthalpy regions where geological conditions are nevertheless suitable to allow warm water (less than 100 °C) to be pumped to the surface. In many cases, the heat from the water can be used directly, and in other cases the temperature may be increased by the use of heat pumps. These are 'low-enthalpy' resources; in 2015, these installations supplied a total of 73 000 GWh, equivalent to over 6 million tonnes of oil.

In almost all these situations heat is being removed faster than it is replaced, so the concept of 'heat mining' is appropriate. Although geothermal resources are non-renewable on the scale of human lifetimes and, strictly, fall outside the remit of this book, they are included because they share many features with true renewable resources. For example, geothermal energy is a natural energy flow rather than a store of energy like fossil or nuclear fuels. At one time it was thought that many high-enthalpy resources were indeed renewable, in the sense that they could be exploited indefinitely, but experience of declining temperatures in some steam producing fields and simple calculations of heat supply and demand show that – locally – heat is being mined on a non-sustainable basis (see Box 10.2). Nevertheless, and especially as techniques become available for extracting energy from rocks that do not support natural water flows (the so-called Hot Dry Rock concept, now more correctly termed Enhanced – or Engineered – Geothermal Systems (EGS): see Section 10.4), the volumes of rock that are potentially exploitable are so large in comparison to an individual reservoir that – to use an agricultural analogy – the resource could be 'cropped' on a rotational basis. Once a particular zone has been depleted, further boreholes could be drilled to a deeper layer, or a few kilometres further away, and production continued. After a few such operations, each with a lifetime of 20–30 years, the original zone would have regenerated to economically exploitable levels.

Table 10.1 Geothermal electricity generation: installed capacity and energy produced in 2010 and 2015 and forecast capacity in 2020

Country	Energy Produced in 2010 (GWh)	Energy Produced in 2015 (GWh)	Installed in 2015	Forecast for 2020-2025
United States	16 603	16 600	3450	5600
Phillippines	10 311	94 646	1870	2500
Indonesia	9600	9600	1340	3500
Mexico	7047	6071	1017	1400
Italy	5520	5660	916	1000
Iceland	4597	5245	665	1300
New Zealand	4055	7000	1005	1350
Japan	3064	2687	519	570
El Salvador	1442	1442	204	300
Kenya	1430	2848	594	1500
Costa Rica	1131	1511	207	260
Turkey	490	3127	397	600
Papau-New Guinea	450	432	50	70
Russia	441	441	82	190
Nicaragua	310	492	159	200
Guatemala	289	237	52	140
Portugal	175	196	29	60
China	150	150	27	100
France	95	115	16	40
Germany	50	35	27	60
Ethiopia	10	10	7.3	50
Austria	3.8	2.2	1.2	6
Thailand	2	1.2	0.3	1
Australia	0.5	0.5	1.1	20
Romania	0	0.4	0.1	5

Source: Bertani, 2015

BOX 10.2 Deep geothermal extraction and recharge rates

The following simple calculations show the order of magnitude discrepancy between commercial geothermal extraction rates and thermal recharge by the Earth's natural heat flow. They demonstrate that it is *not the heat flow but the heat store that is being exploited*, even in high-enthalpy areas.

(a) The currently exploited area of Tuscany, northern Italy, totals about 2500 km^2. This is a generous estimate that ignores the fact that the active fields are only a small subset of this. The average heat flow is about 200 mW m^{-2} so the total heat flow through this surface is:

$$2500 \times 10^6 \text{ (m}^2\text{)} \times 200 \times 10^{-3} \text{ (W m}^{-2}\text{)} = 500 \text{ MW}_t$$

The region currently supports generating capacity of >700 MW$_e$; at a mean generating efficiency of 15%, this would require 4600 MW$_t$. Moreover, steam supplies have been proved to support an electricity-generating capacity of 1500 MW$_e$, which would require an input of approximately 10 000 MW$_t$. So the ratio of forecast commercial production rate to thermal recharge is at least:

$$\frac{10\ 000 \text{ MW}_t}{500 \text{ MW}_t} \text{ or } 20\text{:}1.$$

(b) In the Imperial Valley of California, USA, a commercial lease of 4 km^2 is expected to support generating capacity of some 40 MW$_e$ (which will require at least 250 MW$_t$). Assuming an average heat flow of 200 mW m^{-2}, the thermal recharge is less than 1 MW$_t$, giving a ratio of:

$$\frac{250 \text{ MW}_t}{1 \text{ MW}_t} \text{ or } 250\text{:}1.$$

(c) The well field in Krafla, Iceland, covers an area of some 50 km^2. It currently generates some 60 MW$_e$, implying heat extraction of some 400 MW$_t$. To recharge the field on a sustainable basis, the heat flow would have to be 8 W m^{-2}. Again, this is much higher than could be expected even in this very active area. Of course, there is recharge by the Earth's natural heat flow but a thermally depleted reservoir may take many tens or even hundreds of years to regenerate. However, in regions of large geothermal resources this is generally a minor problem.

Note: MW$_t$ denotes megawatts of thermal energy; MW$_e$ denotes megawatts of electrical energy.

The source of heat

Heat flows out of the Earth because of the massive temperature difference between the surface and the interior: the temperature at the centre is around 7000 °C. So why is the Earth hot? There are two reasons: first, when the Earth formed around 4600 million years ago the interior was heated rapidly as the kinetic and gravitational energy of accreting material was converted into heat. If this were all, however, the Earth would have cooled within 100 million years. Of much greater importance today is the second mechanism: the Earth contains tiny quantities of long-lived radioactive isotopes, principally thorium 232, uranium 238 and potassium 40, all of which liberate heat as they decay. These radiogenic elements are concentrated in the upper crustal rocks. Cumulative heat production from these radioactive isotopes (approximately 500 EJ y^{-1} today) accounts for about half the surface heat flow, though the exponential decay laws for radioactivity imply that heat production was about five times greater soon after the Earth formed. Heat is transferred through the main body of the Earth principally by convection, involving motion of material mainly by creep processes in hot deformable solids. This is a very efficient heat transport process resulting in rather small variations of temperature across the depth of the convecting layer. Closer to the surface, across the outer 100 km or so of the Earth, the material is too rigid to convect because it is colder, so heat is transported by conduction and there are much larger increases of temperature with depth (i.e. larger 'thermal gradients': see Section 10.2). This rigid outer boundary layer, or shell, is broken into a number of fragments, the **lithospheric plates**, which move around the surface at speeds of a few centimetres a year in concert with the convective motions beneath (Figure 10.1). Only this last point is of direct relevance to geothermal exploitation; our ability to drill into the Earth is restricted to the upper few kilometres, so we have to look for mechanisms and locations where the Earth's interior heat is brought within our reach.

From our point of view, it is mainly at the boundaries between plates, particularly where they are in relative extension or compression, that heat flow reaches a maximum. Here, the heat energy flowing through the surface averages around 300 milliwatts per square metre (300 mW m^{-2}) as compared with a global mean of 60 mW m^{-2}.

However, along plate margins heat flow can be even more concentrated locally because rock material reaches the surface in a molten form, resulting in volcanic activity that is often spectacular. Storage of molten, or partially molten, rock at about 1000 °C just a few kilometres beneath the surface strongly augments the heat flow around even dormant volcanoes. These heat flows result in high thermal gradients, really the hallmark of high-enthalpy areas, which are further enhanced in the upper regions by induced convection of hot water. Over geological periods of time, this high heat flow has resulted in large quantities of heat being stored in the rocks at shallow depth, and it is these resources that are mined by geothermal exploitation and commonly used for electricity generation. In areas of lower heat flow, where convection of molten rock or water is reduced or absent, temperatures in the shallow rocks remain much lower, so any resources are likely to be suitable only for direct use applications. So we see that high-enthalpy resources, including all those currently exploited for geothermal electric power (see Figure 10.1), are associated with volcanically active plate margins or localized hot spots like the Hawaiian islands. Boiling mud pools, geysers and volcanic vents with hot steam are characteristic features of such geothermal areas.

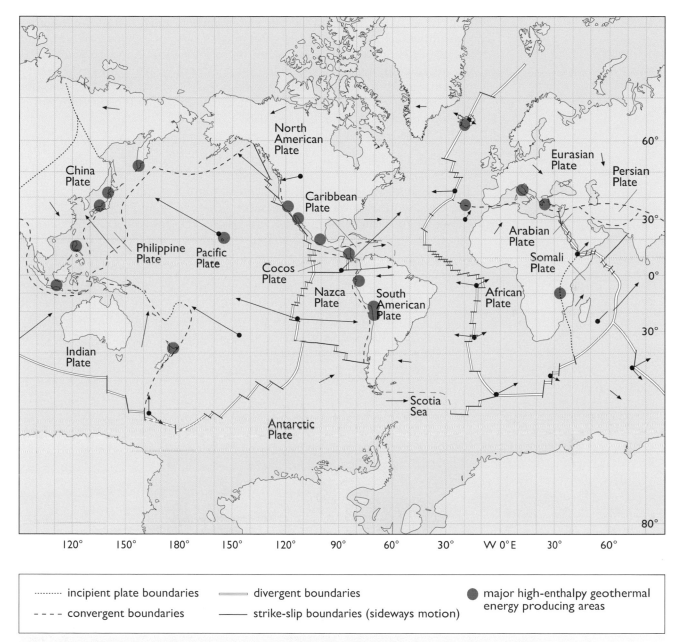

Figure 10.1 Map of the Earth's lithospheric plates indicating the relative speeds of motions by the lengths of the arrows (generally 1–10 cm per annum). Large dots indicate major high-enthalpy geothermal energy-producing areas

Historical perspective

The historical exploitation of geothermal resources dates back to Greek and Roman times, with early efforts made to harness hot water for medicinal, domestic and leisure applications. Roman spa towns in Britain generally sought to exploit natural warm water springs with crude but reliable plumbing technology. The early Polynesian settlers in New Zealand, who lived undisturbed by European influence for several hundred years until the eighteenth century, depended on geothermal steam for cooking and warmth, and hot water for bathing, washing and healing. Indeed, the healing properties of geothermal waters are renowned throughout the modern world and have important medical benefits.

By the nineteenth century, progress in engineering techniques made it possible to observe the thermal properties of underground rocks and fluids, and to exploit these with rudimentary drills. In Tuscany, where the indigenous geothermal fluids were exploited as a source of boron from the eighteenth century onwards, natural thermal energy was used in place of wood for concentrating and processing the solutions. An ingenious steam collection device, the *lagone coperto* ('covered pool'), sparked a rapid growth in the Italian chemical industry, resulting in flourishing international trade. The generation of electrical power started in 1904, fostered by Prince Piero Ginori Conti, and 1913 saw the arrival of the first 250 kW_e power plant at Larderello, marking the start of new industrial activity. Today the Larderello power station complex has a capacity approaching 800 MW_e and a rebuilding programme is in progress that will take the capacity to some 1000 MW_e (Figure 10.2).

(a)   (b)

Figure 10.2 (a) The old geothermal power station at Larderello (b) The newer array system showing the smaller sized geothermal power units, at Larderello

The Wairakei field in New Zealand was the second to be developed for commercial power generation, though not until the early 1950s. It was followed closely by the Geysers field in northern California where electricity was first generated in 1960. With an installed capacity that peaked at 2800 MW_e in the early 1990s, the Geysers field is still the most extensively developed in the world, though it will be overtaken soon by the Philippines. The Geysers field, however, illustrated the dangers of uncoordinated exploitation; the field was exploited independently by several different companies. In particular, little of the extracted water was reinjected into the reservoir after use. The result was a decline in steam pressure and a reduction in output capacity by several hundred megawatts. It was this decline that accounted for the very low net growth in world capacity during the second half of the 1990s. Fortunately, the problem was recognized in time, and reinjection is now practised widely; steam pressures and volumes are now recovering.

With the notable exceptions of Italy, the most volcanically active country in mainland Europe, and Iceland, which lies on the volcanic ridge of the central Atlantic, the chief geothermal nations are clustered around the Pacific rim. Japan, the Philippines, Indonesia and Mexico have shared in recent technological developments; the installations in El Salvador and Nicaragua are strategically vital to the economies of those nations, and several other countries, notably Costa Rica, Ecuador and Chile, also produce geothermal electricity.

Meanwhile, schemes making direct use of geothermal heat for district heating and agricultural purposes have advanced, with the major producers being China, the USA, Sweden, Turkey, Japan, Norway and Iceland. France developed substantial heating systems in the 1970s and 1980s, but there have been significant developments in eastern Europe. Many of these countries have now moved away from coal as a districting heating fuel and have moved to gas as a heating fuel, but there are notable pockets of interest in geothermal systems. These have benefited from new expertise in high-grade materials for pipelines, pumps and heat exchangers that can withstand the corrosive effects of **brines**.

10.2 The physics of deep geothermal resources

Primary ingredients

Geothermal resources of most types must have three important geophysical characteristics, as shown in Figure 10.3: an aquifer containing water that can be accessed by drilling; a cap rock to retain the geothermal fluid; and a heat source.

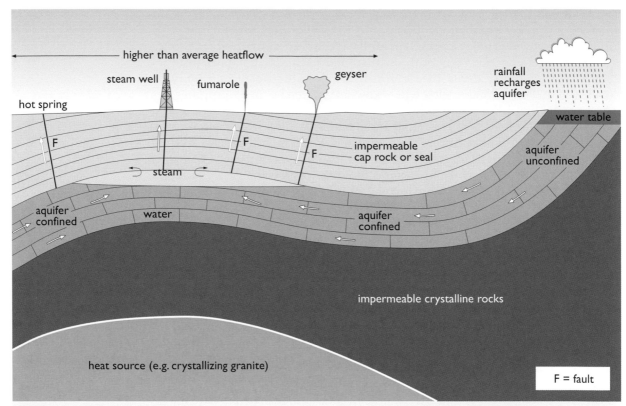

Figure 10.3 Simplified schematic cross-section to show the three essential characteristics of a geothermal site: an aquifer (e.g. fractured limestone with solution cavities); an impermeable cap rock to seal the aquifer (e.g. clays or shales); and a heat source (e.g. crystallizing granite). Steam and hot water escape naturally through faults (F) in the cap rock, forming fumaroles (steam only), geysers (hot water and steam), or hot springs (hot water only). The aquifer is unconfined where it is open to the surface in the recharge area, where rainfall infiltrates to keep the aquifer full, as indicated by the water table just below the surface. The aquifer is confined where it is beneath the cap rock. Impermeable crystalline rocks prevent downward loss of water from the aquifer

First, what is an aquifer? **Natural aquifers** are porous rocks that can store water and through which water will flow. **Porosity** refers to the cavities present in the rock, whereas the ability to transmit water is known as **permeability**. A geothermal aquifer must be able to sustain a flow of geothermal fluid, so even highly porous rocks will only be suitable as geothermal aquifers if the pores are interconnected. In Figure 10.4, rocks (a) and (c) are porous and likely to be highly permeable, whilst (b) and (d) have low porosity and permeability. Example (e), however, has low permeability despite its high porosity whereas cavities developed in (f) by dissolution of more soluble components give high porosity and permeability. Permeability due to fracturing ('fracture permeability'), as in (g), is particularly important in many geothermal fields.

A good measure of the permeability of a rock is its hydraulic conductivity (K_w). Darcy's Law states that the speed (v) of a fluid moving through a porous medium is proportional to the hydraulic pressure gradient causing the flow:

$$v = K_w \frac{H}{L} \tag{1}$$

(a)
high porosity
– rounded grains,
uniform size
(good sorting)

(b)
low porosity
– rounded grains,
many sizes
(poor sorting)

(c)
medium porosity
– angular grains,
uniform size
(good sorting)

(d)
very low porosity
– angular grains,
many sizes
(poor sorting)

(e)
vesicular porosity
– may not be
interconnected,
e.g. basalt

(f)
solution porosity
– mild solution
along crystal
boundaries
e.g. limestone

(g)
porosity along
fractures or
bedding planes

Figure 10.4 The relationship between grain size, shape and porosity in sedimentary rocks, especially sandstones (a–d); vesicular porosity in crystallized lava flows due to gas bubbles (e); and solution porosity resulting from rock dissolution, especially where acid groundwaters attack limestone (f). Porosity also develops in rocks along original planes of weakness, especially bedding planes and fractures (joints and faults) (g)

Here, H is the effective head of water driving the flow, and is measured in metres of water. The pressure gradient, or hydraulic gradient (H/l) is the change in this head per metre of distance L along the flow direction.

The volume of water (Q) flowing in unit time through a cross-sectional area A m^2 is v times A. So Darcy's Law may also be written:

$$Q = AK_w \frac{H}{L} \qquad (2)$$

and K_w (the hydraulic conductivity) may be interpreted as the volume flowing through one square metre in unit time under unit hydraulic gradient. Some values of hydraulic conductivity for different rocks are given in Table 10.2.

Notice that the highest values of K_w occur in coarse-grained unconsolidated rocks, such as the ash layers, which are particularly common in volcanic areas, but that values are also quite high in some limestones and sandstones. These are aquifer rocks, with high permeability. It should be remembered also that fracture permeability is often important in geothermal aquifers (see Figure 10.4(g)), and is central to the Hot Dry Rock/Enhanced Geothermal System (HDR/EGS) concept. This is discussed after geothermal waters later in this section.

In a confined aquifer, such as in Figure 10.3, the fluid pressure beneath the extraction point is high because there is a **cap rock**, a relatively impermeable rock, or seal, to prevent fluid escaping upwards. A cap rock is essential if a

Table 10.2 Typical porosities and hydraulic conductivities

Material	Porosity/%	Hydraulic conductivity/m day^{-1}
Unconsolidated sediments		
Clay	45–60	$<10^{-2}$
Silt	40–50	10^{-2}–1
Sand, volcanic ash	30–40	1–500
Gravel	25–35	500–10 000
Consolidated sedimentary rocks		
Mudrock	5–15	10^{-8}–10^{-6}
Sandstone[1]	5–30	10^{-4}–10
Limestone[1]	0.1–30	10^{-5}–10
Crystalline rocks		
Solidified lava[1]	0.001–1	0.0003–3
Granite[2]	0.0001–1	0.003–0.03
Slate	0.001–1	10^{-8}–10^{-5}

[1] The larger values of porosity and hydraulic conductivity apply to heavily fractured rocks and, for limestones, may also reflect the presence of solution cavities (see Figure 10.4(f)).

[2] Granite is a coarsely crystalline rock that has cooled down slowly from a melt at depth in the Earth. Such rocks are generally non-porous and impermeable, but contain many natural fractures and acquire limited permeability.

steam field is to develop. Mudrocks, clays and unfractured lavas are ideal. The importance of cap rocks was demonstrated in the early 1980s during exploration for geothermal resources in a very obvious place, the flanks of the volcano Vesuvius. Only small amounts of low-pressure fluid were discovered because the volcanic ashes that form its flanks are apparently quite permeable throughout. Given time, alteration of the uppermost deposits or overlying sediments by hot water and steam can create clays or deposit salts in pore spaces, so producing a seal over the aquifer, and in this way many geothermal fields eventually develop their own cap rocks. For this reason, however, the youngest volcanic areas, like Vesuvius, are not necessarily the most productive from a geothermal viewpoint.

The third prerequisite for exploitable geothermal resources is the presence of a heat source. In high-enthalpy regions, abundant volcanic heat is available, but in low-enthalpy areas the heat source is less obvious. In such regions there are two main types of resource: (a) those located in deep sedimentary basins where aquifers carry water to depths where it becomes warm enough to exploit, and (b) those located in 'hot dry rocks' where natural heat production is high but an artificial aquifer must be created by enhancing the rock fractures in order that the geothermal resource may be exploited. Let us now look at each type of resource in more detail.

Volcano-related heat sources and fluids

The heat supply for a high-enthalpy field is usually derived from a cooling and solidifying body of magma (partially molten rock), which need not necessarily be centred directly beneath the geothermal field (Figure 10.5). It may seem surprising that much of the magma rising beneath a volcano is not erupted but instead reaches only a level of neutral buoyancy at which its density is the same as that of the surrounding rocks. Two factors conspire to halt the rising magma: first the pressure of overlying rocks reduces as the magma ascends; this promotes the separation of liquid magma from its dissolved gases, which are lost, increasing the density of the remaining magma; second, shallower rocks are inherently less dense than rocks at greater depth, usually because they are less compressed. So, whereas volcanic eruptions are driven by exceptionally high gas pressures, many magmas form 'intrusions', coming to rest and crystallizing beneath the surface at a depth of 1-5km.

In the 1980s, experiments were undertaken in the USA with the ultimate aim of drilling directly into or very close to magma bodies, where temperatures may be up to 1800 °C, and to harness geothermal power by cycling water through their outer margins. In preparation, the US Magma Energy Program

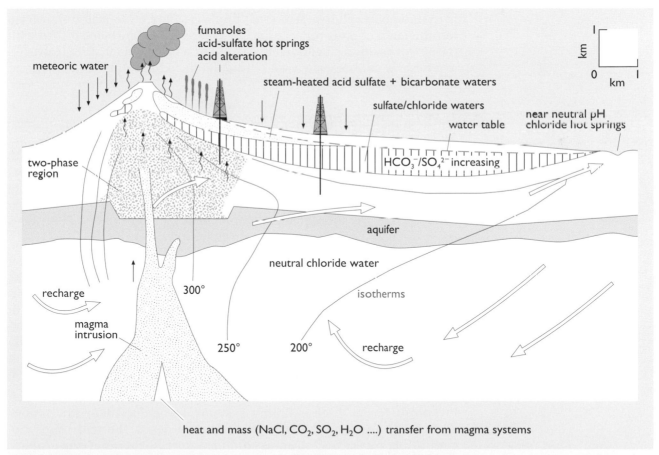

Figure 10.5 Conceptual model of a typical volcanic geothermal system in which meteoric (rain) waters percolate deep into the volcanic superstructure where they are heated by a body of magma that loses dissolved gases as it rises, and forms an intrusion (see text). The hot aqueous fluid rises and may reach the point at which water boils to form steam, producing a two-phase (steam + water) zone. The hydraulic gradient causes the geothermal fluid to migrate through any permeable rocks in the volcano flank. Here the fluids may be accessed by drilling (see drill rig symbols); the chemistry of the fluid changes during migration due to mixing with CO_2-saturated rain water

succeeded in drilling into the molten Kilauea lava lake (Hawaii) and ran successful energy extraction experiments. In any drilling operation, the drill bit is cooled and lubricated by circulating fluid (referred to as mud even though, usually, it is mainly water), which also lifts cuttings to the surface. In this case, as was expected, the circulating water solidified a thin shell of lava around the drill bit. The resulting tube of solid but thermally fractured rock acted as a heat exchanger, with heat being transferred to the drill hole by convecting magma. While a useful test, this is still a long step from drilling into a live magma chamber with high-pressure dissolved gases, and no further work has been undertaken in recent years.

A close encounter with magma occurred during the development of the Krafla field in northern Iceland when, in 1977, rising magma reached the depth of a borehole at 1138 m and three tonnes of magma was erupted through the hole in 20 minutes! Quite by chance, the development history of this field was dogged by a series of eruptions in 1975 and 1984, the first at Krafla for over 250 years, but progress improved once the eruptions ceased and the field now supports 60 MW_e of power generation. Krafla is also the site of the Icelandic Deep Drilling Project (IDDP), the aim of which is to find and exploit supercritical fluids (>350 °C) that would transform the efficiency of power generation. Various test holes are being drilled with the aim of having a single well that could provide about 50 MW_e. However, engineering on this scale is very difficult due to the nature of the rock and the site depths encountered.

Several of the world's most advanced geothermal sites (for example, in northern Italy and the western USA) are located in extinct volcanic areas. Fortunately for geothermal exploitation, because rocks are such good insulators, magmatic intrusions may take millions of years to cool to ambient conditions. Such intrusions, therefore, continue to act as a focus for 'hot fluid', or hydrothermal convective cycles in permeable strata as in Figure 10.5. The nature of the resource then depends on the local conditions of pressure and temperature in the aquifer, and this determines the extraction technology and the profitability of the site.

The range of pressures and temperatures of interest to current geothermal developments lies typically between 100 and 300 °C (though conditions in parts of some fields exceed 400 °C), and in pressure range up to about 20 megapascals (20 MPa). This is the pressure that might be encountered in an aquifer at a depth of about 2 km (see Box 10.3). As noted above, however, there is increasing research interest in the possible exploitation of really high temperature geothermal fluids, which would offer the prospect of much greater conversion efficiencies. These high temperatures are often referred to as **supercritical**.

In simple terms, high-enthalpy systems are subdivided into vapour dominated and liquid dominated, depending on the main pressure controlling phase (i.e. steam or liquid water) in the reservoir. Vapour dominated systems are the best and most productive geothermal resources, largely because the steam is dry (free of liquid water) and is of very high enthalpy. Where reservoir rocks are at pressures below hydrostatic, which promotes steam formation (perhaps 3–3.5 MPa at depths down to 2 km), there must be some barrier to direct vertical groundwater infiltration. The Larderello field in Italy (Figure 10.2) is of this type.

BOX 10.3 Pressure and depth

As noted in Chapter 6, Box 6.5, hydrostatic pressure is when water increases by about 1 atmosphere for an increase of 10 m in depth in a column of liquid. It follows that a water-filled geothermal aquifer 1 km thick will produce a pressure increase of 100 atmospheres (100 bar). (A pressure of 1 bar is approximately equal to 1 atmosphere.)

The SI unit for pressure is the pascal or, more appropriately for the high pressures in geothermal systems, the megapascal (MPa). One MPa is approximately 10 atmospheres, and the 20 MPa mentioned in the text is thus 200 atmospheres.

The pressure also affects the boiling point of water. While water boils at 100 °C at atmospheric pressure, it has to be raised to 180 °C to boil at 10 atmospheres pressure and to over 260 °C at 50 atmospheres pressure. Thus, even though the water in a deep aquifer may be at a very high temperature, it is likely to be in a liquid state. However, closer to the surface, where the pressure is lower, it may be present as steam.

Many geothermal aquifers also contain a steam zone and, since steam has a much lower density than water, the **vapourstatic** increase in pressure with depth (i.e. the pressure due to the weight of the column of steam) is much smaller than the hydrostatic increase of a column of liquid water.

In contrast, liquid-dominated systems are at higher than hydrostatic pressures, exceeding 10 MPa at depths below 1 km. Production of electricity from liquid-dominated systems benefits from the higher fluid pressures at depth, and water can 'flash' into steam en route to the surface. That is to say, the pressure-temperature curve in liquid dominated systems is below the boiling curve for water at all depths, but when the thermal aquifer is punctured by boreholes, the reduced pressure in the well means that the rising water crosses the boiling point curve on its way to the surface. However, the steam is often wet and of lower enthalpy, which adds to the technical problems for electricity production. The famous Wairakei field in New Zealand is liquid-dominated but, typically for such systems, has developed a two-phase zone as pressures have fallen during exploitation. Fortunately, the groundwater zone has a relatively low permeability, which suppresses the tendency for natural venting of steam over most of the Wairakei area.

The heat source in sedimentary basins

An important key to understanding many geothermal resources is the heat conduction equation:

$$q = K_T \frac{\Delta T}{z} \tag{3}$$

This is analogous to Darcy's Law, but here q is the one-dimensional vertical **heat flow** in watts per square metre (W m^{-2}). ΔT is the temperature difference across a vertical height z, and $\Delta T/z$ is thus the **thermal gradient**. The constant K_T relating these quantities is the **thermal conductivity** of the rock (in W m^{-1} °C^{-1}) and is equal to the heat flow per second through an area of 1 m^2 when the thermal gradient is 1 °C per metre along the flow direction.

Values of K_T for most rock types are quite similar, in the range 2.5–3.5 W m^{-1} °C^{-1} for sandstones, limestones and most crystalline rocks. However, mudrocks (clays and shales) are the exceptions, with lower values of 1–2 W m^{-1} °C^{-1}. These are also among the most impermeable rocks (Table 10.2), so mudrocks contribute two of the essential characteristics for geothermal resources: they act as impermeable cap rocks and as an insulating blanket, enhancing the geothermal gradient above aquifers in regions of otherwise normal heat flow.

So, even under conditions of average heat flow (60 mW m^{-2}), it is possible to obtain temperatures of 60 °C within the top 2 km of the Earth's crust. Box 10.4 demonstrates how the differing insulating properties of rocks influence the way the temperature varies with depth. To maintain the same vertical heat flow, low-conductivity rocks require a steeper temperature gradient than a relatively good conductor, and are accordingly important in augmenting temperatures at depth.

BOX 10.4 Thermal gradient and heat flow

Consider the situation where there is a steady upward flow of heat through the top few kilometres of the Earth's crust. We can use Equation 3 to relate this flow to the temperature at any depth if we know the thermal conductivity of the rock.

If, for instance, the temperature is found to be 58 °C at a depth of 2 km (2000 metres) and the surface temperature is 10 °C, the temperature gradient is:

$(58 - 10)/2000 = 0.024$ °C m^{-1}

and if the thermal conductivity of the rock is 2.5 W m^{-1} °C^{-1}, the heat flow rate is:

$2.5 \times 0.024 = 0.060$ W m^{-2}

or 60 mW m^{-2}.

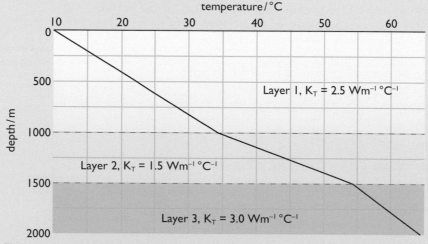

Figure 10.6 Variation of temperature with depth across three zones of differing thermal conductivity, K_T

Suppose, however, that this same 60 mW flows up through several layers with different thermal conductivities. Equation 3 tells us that the thermal gradient must be different in each layer, with the temperature changing most rapidly through the layer with the lowest conductivity, as in Figure 10.6.

This has led to exploration programmes aimed at locating natural waters in areas of thick sedimentary rock sequences containing mudrocks and permeable limestones or sandstones. For example, the Paris area is at the centre of a 200 km wide depression in the crystalline **basement rocks**.

Exploration for hydrocarbon resources in the 1960s and 1970s found very little oil or gas, but was extremely successful in locating hot water between 55 °C and 70 °C at depths of 1–2 km. While low-enthalpy water resources are unsuitable for power generation (no high-pressure steam can be produced, and temperatures are too low to permit an acceptable generation efficiency), they can be of considerable benefit in meeting demands for low-grade heat (space heating, etc.). However, to be economic they must be located close to a heat load. The Paris area is ideal in this respect. Similar resources occur in some of the sedimentary basin areas of the UK, such as beneath the Yorkshire–Lincolnshire coast and in Hampshire, where the only commercial UK geothermal scheme operates in Southampton, but most are remote from suitable heat loads.

Finally, there are two extensions of the criteria discussed above which make some sedimentary basin resources more attractive:

(1) There are some basins where the background heat flow is above average, and these have led to large-scale non-electrical applications of geothermal energy in a number of countries (Table 9.2). The geological reasons for the association of high heat flow with sedimentary basins are not altogether surprising: stretching processes within the Earth's outer plate layer induce thinning that can radically raise the heat flow as well as creating a surface depression on which sedimentation occurs. Beneath the south Hungarian Plain, for example, geothermal gradients as high as 0.15 °C m^{-1} have been recorded and 120 °C water occurs at 1 km depth.

(2) In other areas, larger sedimentary thicknesses may occur. For example, high-pressure fluids at temperatures of 160–200 °C occur at 3–5 km depth in the Gulf of Mexico, southern Texas and Louisiana. Because of chemical processes occurring as a result of the depth and temperature of burial and the efficient sealing of the aquifers by impermeable rocks, pressures greatly exceed hydrostatic and 100 MPa has been recorded in local pockets of fluid. The fluids are highly saline brines with trapped gas, especially methane. These so-called **geopressured brines** are a potentially important geothermal resource for power generation, a resource that has remained untapped to date but on which there is intermittent government funding for research in the USA. The great advantage of geopressured resources is that they offer three kinds of energy: geothermal heat, 'hydraulic' energy (in water at high pressure), and the chemical energy in the large quantities of methane that are found dissolved in the fluid.

Geothermal waters

Most of the foregoing has been concerned with the source of geothermal heat. Exploitation of the heat, however, requires that the geothermal water be brought to the surface, and that brings with it a different set of problems. Water that has been in contact with rock for long periods (and geothermal waters can be thousands or even millions of years old) contains dissolved minerals. Hot water tends to be more reactive than cold water, so geothermal waters can often contain around 1% of dissolved solids. Typically, these will be carbonates, sulfates or chlorides, and dissolved silica becomes significant where waters have been in contact with rocks above 200 °C. For this reason, geothermal fluids are often called 'brines'. Dissolved gases are also common, especially at higher temperatures. Techniques are available to deal with all of these, but it is essential that they be taken into consideration at the design stage of the plant. With correct design, these contaminants can all be handled and disposed of without either operational or environmental difficulty. If they are ignored, however, or the plant designed before the water has been properly characterized, the entire system can fail within a matter of months. The various techniques that can be used are beyond the scope of this chapter, but some examples are quoted in Section 10.3.

'Hot dry rocks' or engineered geothermal systems (EGS)

Our attention now turns from sedimentary strata to the underlying crystalline 'basement'. Most rocks, especially crystalline and basement rocks, do not contain sufficient water to provide a viable geothermal resource, although far greater amounts of heat are stored in such rocks than are available in aquifers. It was thought initially that deep basement rocks would indeed be dry, and so the term **hot dry rock** (HDR) was coined around 1970 to describe the heat stored in impermeable (or poorly permeable) rock strata and the process of trying to extract that heat. More recently, it has been recognized that few if any rocks are actually dry, and there is now a general acceptance of the term 'enhanced' – or 'engineered' – 'geothermal systems' (EGS) to categorize the various projects aimed at extracting this heat.

When the permeability is too poor to allow the necessary flow of fluid, what is required is the creation of an artificial heat exchanger zone within suitably hot rocks. Because rocks are (by normal standards) poor conductors of heat, very large heat transfer surfaces (of the order of square kilometres) are required if heat is to be extracted at useful rates. This can be achieved by enhancing the natural fracture system that occurs in all such crystalline rocks. Water can then be circulated through the enhanced zone so that heat may be extracted, ideally to generate steam and, hence, electrical power. Although the technology to create suitable arrangements for reproducible heat recovery has not yet been perfected, in theory at least EGS technology could be applied over a significant proportion of the Earth's surface.

Because drilling is expensive, with costs rising exponentially with depth, only the upper 6–7 km of the Earth's crust is generally used in calculating geothermal energy potential (though some hydrocarbon and research drilling has gone as deep as 15 km). Given current technical and economic constraints on drilling depths, a minimum geothermal gradient of around 0.025 °C m^{-1} is required if development is to be economic. With a typical thermal conductivity of 3 W m^{-1} °C^{-1}, this requires (from Equation 3) a heat flow of 75 mW m^{-2}, only a little above the Earth's average. In practice, however, to minimize expenditure it is customary to look for rocks with much higher heat flows (as at experimental sites in the USA, Japan and France). Granite bodies are ideal targets, because such rocks occupy large volumes of the upper crust and they crystallized from magmas that had naturally high concentrations of the chemical elements with long-lived radioactive isotopes – uranium, thorium and potassium. Here we reach a situation in which heat flow through the Earth's surface is augmented (perhaps by a factor of 2) by heat production within certain shallow crystalline rocks. If, in addition, a layer of poorly conducting sedimentary rocks overlies the granite, its 'blanketing' effect will increase the temperature gradient and make higher temperatures available at shallower depths.

10.3 Technologies for exploiting high-enthalpy steam fields

The first stage in prospecting for geothermal resources in volcanic areas involves a range of geological studies aimed at locating rocks that have been chemically altered by hot geothermal brines, and finding surface thermal manifestations, such as hot springs or mud pools. Investigations of fluid chemistry and, increasingly, the release of gases through fractured rocks allow assessment of the composition and resource potential of trapped fluids. These studies provide the first clues to the likely presence and location of exploitable resources. However, geophysical prospecting techniques, particularly resistivity surveying and other electrical methods designed to detect zones with electrically conducting fluids (i.e. brines), are probably the most effective for precise location of buried geothermal resources. Once a suitable geothermal aquifer has been located, exploration and production wells are drilled using special techniques to cope with the much higher temperatures and, in some cases, harder rock conditions than in oil and water wells. Since fluid pressures in the aquifer range up to about 10 MPa, the driller must ensure that the mud is dense enough to counteract these pressures and avoid '**blow out**', where an uncontrollable column of gas is discharged. The well is lined ('cased') with steel tubing that is cemented in place, leaving an open section or a perforated steel casing at production depths. As each string of casing has to be inserted through its predecessors as the well depth increases, the well diameter decreases with depth from perhaps 50 cm near the surface to 15 cm at production depths. A wellhead with valve gear is welded to the steel casing at ground level. This allows the well to be connected to a power plant via the network of insulated pipes that are a familiar sight in geothermal areas.

Technologies for electrical power generation depend critically on the nature of the resource – not just the fluid temperature and pressure but also its salinity and content of other gases, all of which affect plant efficiency and design. The size of any power station is determined by the economics of scale; conventional coal- or oil-fired stations are typically a few hundred megawatts per unit. A typical geothermal unit, by contrast, is usually 30–50 MW$_e$. This is because the amount of steam delivered by one well is usually sufficient to generate only a few MW$_e$, and wells are linked across the field and back to the station by pipeline. Above a certain capacity, the cost of the pipelines is such that it is cheaper to develop a separate station in another part of the field.

Given the fact that most of the costs of the electricity derived from geothermal resources are accounted for by the need to pay back the capital investment, with day-to-day operating costs being relatively minor (and insensitive to output variation), there is a great incentive to maximize the efficiency with which the relatively low-grade heat (by power generation standards) is converted into useful energy. Today there are several hundred installations operating worldwide and these include four main types, described below.

Dry steam power plant

As the name implies, this type of system (Figure 10.7(a)) is ideal for vapour-dominated resources where steam production is not contaminated with liquid. The system configuration is essentially the same as a conventional fossil fueled steam plant shown in Chapter 2 (see Fig. 2.1) except that the dry steam comes out of the ground. The reservoir produces superheated steam, typically at 180–225 °C and 4–8 MPa, reaching the surface at several hundred kilometres per hour and, if vented to the atmosphere, sounding like a jet engine at close proximity. Passing through the turbine, the steam expands, causing the blades and shaft to rotate and hence generating power.

In the simplest form of power plant, a 'back-pressure' unit, the low-pressure exhaust steam is vented directly to the atmosphere. Although such units are simple, they are also very inefficient; their main use is as temporary transportable units during the development of a new field. Once the steam supply is ensured, normal practice is then to install 'condensing' plant as shown in Figure 10.7(a).

BOX 10.5 Cooling geothermal power plants

As described earlier in Chapter 2, the Carnot efficiency of a power plant is critically dependent on the temperatures between which it operates. In a geothermal plant the inlet temperature is determined by that of the available steam from the ground. A good efficiency thus requires that the 'cold' side of the plant, the condenser, is at as low a temperature as possible. A conventional fossil fuelled power station is likely to be sited close to a river or the sea with a good supply of cooling water. This can be circulated directly through the condenser. Alternatively a 'cooling tower' can be used (see Figure 10.7(c). This contains a heat exchanger at the bottom which is sprayed with river water. The heat from the condenser evaporates this water and a plume of warm moist air rises inside the tower creating a cooling draught of air. However, this form of cooling uses large amounts of water.

The locations of geothermal plants may limit their access to good supplies of cooling water. One option is simply 'dry cooling', in which the condenser is air-cooled with a large fan, in the same manner as a car radiator. A better option is to evaporate some of the low-temperature brine exhausted from the power plant by feeding it into a cooling tower from a 'hot well' situated at the bottom (see Figure 10.7(a)). However, this may eventually lead to the water levels in any underground aquifer becoming depleted.

At temperatures typical of geothermal fluids efficiencies are low. Higher temperatures of 300–350 °C and higher pressures can produce higher efficiencies, but, despite the use of high-temperature superheated steam, rarely exceed 20%. Nevertheless, whereas a 1960s plant required almost 15 kg steam per saleable kWh in optimum conditions, modern dry steam plant with higher temperature steam and better turbine designs can achieve 6.5 kg per kWh, so a 55 MW_e plant requires 100 kg s^{-1} of steam.

Plant efficiency, and therefore profitability, is strongly affected by the presence in the geothermal fluid of so-called 'non-condensable' gases, such as carbon dioxide and hydrogen sulfide. When the turbine exhaust gases are cooled, achieving a suction effect on the turbine as the water condenses into liquid, gases that do not similarly condense cause higher residual pressures at the back end of the turbine. Even a small percentage of such gases reduces suction efficiency and so impacts on the economics

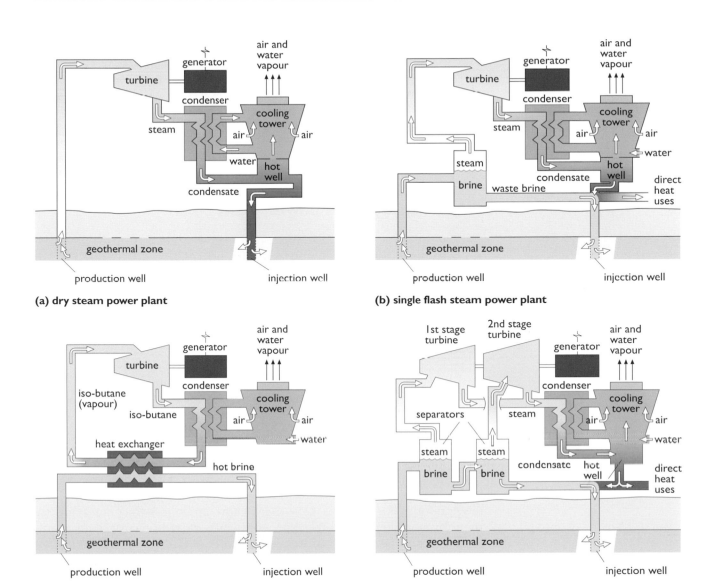

(a) dry steam power plant

(b) single flash steam power plant

(c) binary cycle / organic rankine cycle power plant

(d) double flash power plant

Figure 10.7 Simplified flow diagrams (a–d) showing the four main types of geothermal electrical energy production

of the system; for this reason, many geothermal plants are fitted with gas ejectors. However, the ejectors themselves require either a steam supply or electric power from the turbine-generator and, consequently, reduce output. Non-condensable gases have an additional economic impact: it is no longer acceptable in most places to vent them into the atmosphere, so they must either be trapped chemically or reinjected with the waste water to avoid pollution, and both options entail additional costs.

In general, dry steam plant is the simplest and most commercially attractive. For that reason, dry steam fields were exploited early and have become disproportionately well known. In fact, only the USA and Italy have extensive dry steam resources, though Indonesia, Japan and Mexico also have a few such fields. Elsewhere, and even in these countries, liquid-dominated fields are far more common. While in some areas it is common practice to reinject the spent fluid, very little was reinjected in the largest field, the Geysers in the USA, until falling fluid pressures led to a recognition among the various private operators that the field was being over-exploited. There is now an agreed cooperative reinjection policy to make the resource more sustainable; some 70% of the mass of produced steam is now reinjected, and field production has been stabilized. Not all of this fluid comes from the condensed steam from the turbines, because much is evaporated from the cooling towers. Instead, in a new and imaginative environmental development, treated sewage water is piped 48 km from a local community – thereby solving an unrelated local disposal problem as well as helping to maintain the geothermal reservoir.

Single flash steam power plant

As noted in Box 10.3, the boiling point of water increases with pressure. Thus if geothermal fluid is drawn from a deep aquifer where it is present as liquid water, it may boil on the way to the surface as the pressure is reduced. The process of steam production associated with pressure reduction is called **flashing**. The steam may arrive at the surface as 'wet' steam containing droplets of water and the steam must first be separated from this water in the plant (see Figure 10.7(b)). The separator also has the important function of protecting the steam turbine from a massive influx of water should conditions in the well change. However, it is usually better to avoid flashing in the well because this can lead to a rapid build-up of scale deposits as minerals dissolved in the fluid come out of solution, leading to the plugging of the well. For this reason, the well is often kept under pressure to maintain the fluid as liquid water and the actual flashing to steam takes place within the power plant. To deal with hot high-pressure water requires complex equipment designed to reduce the pressure in a controllable manner and induce flashing. Again, a conventional condensing steam turbine is at the heart of the plant, but lower steam pressures and temperatures (0.5–0.6 MPa, 155–165 °C) are common, so the plant typically requires more steam per kWh than would be required in a dry steam plant, say around 8 kg per kWh. Moreover, the bulk of the fluid produced, often up to 80%, may remain as unflashed hot brine which is then reinjected unless there are local direct use heating applications available. In general, therefore, reinjection wells must be available for fluid disposal both at single flash plants and at plants incorporating the newer types of technology described below. Increasingly, the provision of reinjection is becoming a standard requirement of many licensing schemes.

Binary cycle / Organic Rankine Cycle power plant

This type of power plant (Figure 10.7(c)) uses a secondary working fluid with a lower boiling point than water, such as pentane or butane, which is vaporized and used to drive the turbine. It is more commonly known as an Organic Rankine Cycle (ORC) plant (introduced earlier in Chapter 2). Its main advantage is that lower-temperature resources can be developed where single flash systems have proved unsatisfactory. Moreover, chemically impure geothermal fluids can be exploited, especially if they are kept under pressure so that no flashing ever takes place. The geothermal brine is pumped at reservoir pressure through a heat exchange unit and is then reinjected; the surface loop is closed and no emissions to the environment need occur. Ideally, the thermal energy supplied is adequate to raise the temperature of the secondary fluid to well above its boiling point. For geothermal fluid temperatures below about 170 °C, higher generating efficiencies are possible than in low-temperature flash steam plants. A disadvantage is that keeping the geothermal fluid under pressure and repressurizing the secondary fluid can consume some 30% of the overall power output of the system because large pumps are required. Large volumes of geothermal fluid are also involved; for example, the Mammoth geothermal plant in California uses around 700 kg s^{-1} to produce 29 MW$_e$. Nearly 250 binary cycle units are in operation today. The units are often small (5–10 MW$_e$ is typical). When larger systems are required, multiple sets are installed.

Double flash power plant

Recently, several attempts have been made to develop improved flashing techniques, particularly to avoid the high capital costs and parasitic power losses (e.g. circulating pumps for the secondary fluid) of binary plants. Double flash (Figure 10.7(d)) is ideal where geothermal fluids contain low levels of impurities and so the scaling and non-condensable gas problems that affect profitability are at a minimum. Quite simply, unflashed liquid remaining after the initial high-pressure flashing flows to a low-pressure tank where another pressure drop provides additional steam. This steam is mixed with the exhaust from the high-pressure turbine to drive a second turbine (or a second stage of the same turbine), ideally raising power output by 20–25% for only a 5% increase in plant cost. Even so, extremely large fluid volumes are required. The East Mesa plant in southern California, for example, commissioned in 1988, uses brine at 1000 kg s^{-1} from 16 wells to generate 37 MW$_e$; i.e. around five times as much fluid as for similar dry steam plant (though temperatures would be much higher in the latter case).

Future developments

As the geothermal industry continues to expand, there will be a need to develop technologies that can produce geothermal power from a variety of resources that are less ideal than dry steam. Increasing use is being made of geothermal fluids that are at either lower temperature than (but similar pressure to) those in dry steam fields, or at the same or higher temperature and much higher pressure. These are essentially liquid-dominated resources, albeit of high enthalpy, and they exist in large volumes. Inevitably, variants on the binary and double flash systems will continue to be developed; they are at the leading edge of current research. More recently, greater use has been made of the produced fluids by operating combined or hybrid cycles, using an ORC to extract further work from the main turbine exhaust or the separated water.

A number of other approaches are being developed to increase the efficiency of generation from lower temperature fluids. For example, the Kalina cycle (after Dr Alexander Kalina) is just one variety of the ORC that uses an ammonia-water mixture as the working fluid, the composition of which varies throughout the cycle. The net generating efficiency is expected to be up to 40% higher than for a conventional ORC system under the same conditions; at an inlet temperature of 130 °C the net efficiency is estimated at over 13% (58% of the theoretical Carnot efficiency). The early demonstration units (one in Iceland, one in Japan and two in Germany) are all essentially prototypes and have challenges linked to the corrosion, scaling and contamination stemming from the geothermal fluid mixture (Global Cement, 2012). However the operators remain optimistic and these systems continue to be tested in non-geothermal applications.

10.4 Technologies for direct use of geothermal energy

Some of the countries that are exploiting geothermal resources for non-electrical purposes have chosen to develop these direct use applications in areas flanking the main steam fields. Japan, New Zealand, Iceland and Italy are obvious examples, where wet steam or warm water at a range of temperatures is readily available for industrial, domestic and leisure applications. In this section, however, we leave these aside and concentrate principally on the low-temperature resources found in regions remote from plate boundaries, typically in sedimentary basins, several of which have been developed across central Europe. Drilling techniques resemble those discussed earlier, but the process is generally less hazardous since the geothermal fluid is found under much lower pressure and temperature conditions than in hot steam fields, and pumps are often required to bring the fluid to the surface at adequate flow rates. However, the hot water is usually too saline and corrosive to be allowed directly into heating systems, so once again corrosion-resistant heat exchangers are widely used. It should be pointed out that, given the cost of deep boreholes, a large market is required if sales of heat are to break even. Ideal loads for the systems include: vast greenhouse complexes with both overhead and underground pipes, or domestic systems with combinations of underfloor and radiator heaters. The dense multi-storey apartment blocks of the Paris suburbs are ideal markets.

The French led the development of these low-enthalpy resources in Europe. Over the past 40 years no fewer than 55 geothermally fed group heating schemes were installed in the Paris Basin, with several more in south-western France. At the design stage, a twin production and reinjection borehole system would be planned on the basis of supplying 3–5 MW_t of heat energy (25–50 l s^{-1} of water at 60–70 °C) over a lifetime of 30–50 years. The spacing of the wells must be designed to maintain high fluid pressures by reinjection while avoiding the advance of a 'cold front' (i.e. fluid at reinjection temperatures) towards the production well until capital costs are paid back, and this means that flow conditions in the aquifer need detailed study. A typical layout for a twin production well scheme, with a schematic of the heat transfer technology, is shown in Figure 10.8. Note in this example the interesting application of heat pumps to enhance the system efficiency by reducing the reinjection temperature. Heat pumps work on the same principle as refrigerators, but here produce a concentrated high temperature output. Of course, they consume electrical energy but, in the example shown in Figure 10.8, they enable the number of heated apartments to be doubled.

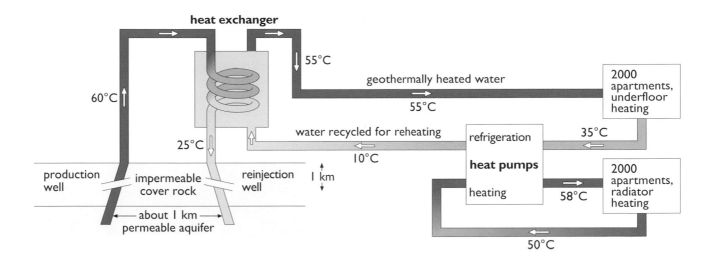

Figure 10.8 An example of a district heating scheme installed in the Paris region. This shows how the geothermal heat is exchanged to a secondary freshwater circuit. The circuit is used first to heat 2000 apartments by underfloor heating; the residual energy is then boosted using heat pumps to provide radiator heating in a further 2000 apartments. Note that the main function of the heat pumps is to lower the reinjection temperature and so extract more heat from the geothermal fluid rather than to raise the production temperature

Although the French group heating schemes were generally a great practical success, a few suffered technical problems – mainly corrosion and scaling in the wells – and were abandoned. More seriously, their economic benefits were only marginal at times of low oil prices, increasing availability of natural gas and high interest rates and several were abandoned for financial reasons during 1989–92. Nevertheless, 34 remain in operation. At the time of writing, with rising oil and gas prices (and low interest rates) such operations look increasingly attractive again. Two new schemes had been completed by January 2011 and were scheduled to begin operation during that year – it is expected that about two new schemes per year will be developed over the next decade. Currently, they produce an annual saving of over 200 000 tonnes of oil (or equivalent in other fossil fuels) in an area which, 40 years ago, had no obvious geothermal potential. The same concept applies to the analogous UK scheme in Southampton.

In Germany, reunification had a significant effect on the way in which geothermal developments occurred, as the better (though still low temperature) resources tend to be concentrated in the eastern part of the country. Although a few schemes existed before reunification, with the freeing of capital following unification, combined with concerns about CO_2 emissions from fossil fuels, the geothermal market really began to take off. Several large-scale district heating schemes are already in operation and even more are under active development. By the end of 2009 there were 162 group heating schemes making direct use of geothermal energy in Germany, with an installed thermal power of some 250 MW_t. Several small ORC power plants are also in operation or under development in the south and south-west. There are, in addition, nearly 180 000 ground source heat pumps (see below) with a total capacity of 1860 MW_t. Substantial numbers of new schemes, both large and small, are under development.

Enhanced (or engineered) geothermal systems

All conventional geothermal systems (except, perhaps, ground source heat pumps (GSHPs)) rely on the presence of water circulating through the rock to extract heat and bring it to the surface. However, even in a good aquifer more than 90% of the heat is contained in the rock rather than in the water. Moreover, the vast majority of rocks are poorly permeable at best and the occurrence of an exploitable geothermal reservoir is a rarity. On the other hand, heat exists everywhere, and the amount of energy stored within accessible drilling depths (say, down to 7000 m, where temperatures of >200 °C would be widespread) is colossal. Cooling one cubic kilometre of rock (which is about the scale of a geothermal reservoir) by 1 °C will provide the energy equivalent of 70 000 tonnes of coal.

To put this in context, a report prepared by the Massachusetts Institute of Technology for the US Department of Energy in 2006 (MIT, 2006) calculated the heat in place at various depths in the continental USA (excluding Alaska, Hawaii and Yellowstone National Park) (Figure 10.9). The estimate was 13 million exajoules (13×10^{24} joules); by comparison, the total annual primary energy consumption in the USA is about 100 exajoules. Even though only a small fraction of this resource is likely to be developable, the potential of such an energy resource cannot be ignored.

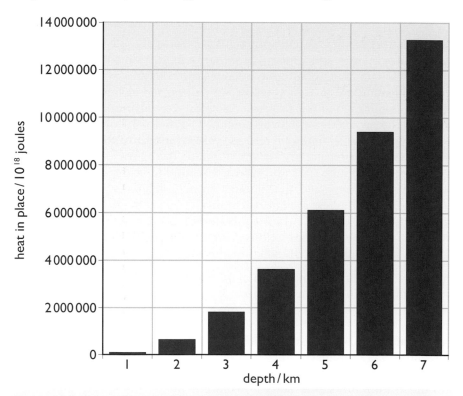

Figure 10.9 The total heat in place at various depths beneath the continental USA, excluding Alaska, Hawaii and Yellowstone (source: adapted from MIT, 2006)

This situation is not confined to the USA. All over the world, temperatures around 200 °C are accessible under a high percentage of the land mass. If this store of heat could be exploited, it would give almost every country the opportunity to generate electricity from an indigenous and (for all practical purposes) renewable resource. It is this prospect that has motivated a number of countries to spend over US$350 million over the past 40 years to find a way to exploit the resource.

The familiar concept of twin production and injection boreholes provides the basis of system designs, but here drilled into relatively hard crystalline rock and terminating several hundred metres apart. In principle, water can then be pumped down one hole, flow through the rock picking up heat, and return to the surface via the second borehole. The fundamental problem is that, as mentioned in Section 10.2, rocks are very poor conductors of heat and, to extract energy at a rate sufficient to pay back the high cost of the boreholes needed to reach these depths, very large heat transfer surfaces are needed – of the order of several square kilometres! There is now a consensus that the only practical way of achieving this figure is to work with nature, exploiting the fact that most deep rocks contain extensive networks of natural fractures. In principle, a suitable heat exchange surface can be created by opening these pre-existing fractures.

The focus has, therefore, been on learning how to stimulate and manage the fracture networks to support a useful and controllable flow of water between boreholes. The stimulation is done by using a variant of an oil industry technique known as hydro-fracturing, which consists of pumping water down the borehole at increasing pressure until fractures in the rock are opened. The progressive development of the opening fractures is followed by listening to and locating the sound of rock surfaces moving over one another. This is known as microseismic monitoring.

If the second borehole has already been drilled, it may be necessary to repeat the stimulation in that hole in order to link up with the first zone. Alternatively, the second hole may be drilled after the first stimulation to intersect the stimulated zone. A closed circuit water circulation through the fracture system is thereby generated. The trick has been in learning precisely how to control the injection conditions in each hole to ensure that water can flow through the system with minimum resistance. At the same time, the stimulated zones around each hole must link up in such a way that water losses are minimized, because water losses mean wasted pumping costs (and in many regions water itself is a valuable commodity). Water is circulated down the injection well, through the reservoir and up the production well to a heat exchanger and turbo-generator where the thermal energy is converted to electricity. (Lower-temperature district heating schemes are also under consideration, but the high capital costs of such an operation require an extremely large local market for the heat produced.)

Pioneering work took place at Fenton Hill in New Mexico in the 1970s and 1980s where the Los Alamos National Laboratory (LANL) developed two systems at temperatures of 200 °C and >300 °C. The Fenton Hill project proved the principle in 1979, when a 60 kW$_e$ ORC generator operated for a month on the produced water. The operating parameters were very far from those that would be needed in a commercial system, however, and a number of teams in various countries (USA, Japan, UK, France, Germany and, most recently, Switzerland) have worked cooperatively in the intervening years to understand and refine the techniques required. Notable among these activities was the full-scale (but comparatively shallow – 2000 m – and deliberately low-temperature) rock mechanics experiment at Rosemanowes Quarry in Cornwall, England, which began in 1975 and laid many of the foundations for our current understanding of the behaviour of natural fracture systems and how they might be managed.

At the beginning, it was assumed that basement rocks at depth would be devoid of fluids, so the technology was termed Hot Dry Rock (HDR). Over the years, however, it has become clear that 'Dry' is a misnomer; very few basement rocks have proved to be completely dry. Water has been found in fractured basement even at the deepest levels in the exploration boreholes in the Kola Peninsula, Russia (15 km) and in the Black Forest, Germany (>8 km). More recently, the term 'Enhanced' (or 'Engineered') 'Geothermal Systems' (EGS) has come to replace the original name, though the phrase 'hot dry rock' is too catchy to disappear easily. The overall technology has been broadly defined as 'any system in which reinjection is necessary to maintain production at commercially useful levels'. This redefinition also emphasizes the continuity which exists in the spectrum of reservoir permeabilities and geothermal technologies. A more detailed definition is given in the MIT report:

> The U.S. Department of Energy has broadly defined Enhanced (or engineered) Geothermal Systems (EGS) as engineered reservoirs that have been created to extract economical amounts of heat from low permeability and/or porosity geothermal resources. For this assessment, we have adapted this definition to include all geothermal resources that are currently not in commercial production and require stimulation or enhancement. EGS would exclude high-grade hydrothermal but include conduction dominated, low-permeability resources in sedimentary and basement formations, as well as geopressured, magma, and low-grade, unproductive hydrothermal resources. In addition, we have added coproduced hot water from oil and gas production as an unconventional EGS resource type that could be developed in the short term and possibly provide a first step to more classical EGS exploitation.
>
> (MIT, 2006, p. 1.10)

The countries principally involved in the research worked closely together throughout the 1980s and 1990s; this type of research makes great demands on both money and expertise and is aimed at what should prove to be a generally applicable technology, so it is an ideal subject for international collaboration. Following this logic, during the 1980s the various teams in the UK, France and Germany, with the support of the European Commission, agreed to pool their resources to develop a single experimental site at Soultz-sous-Forêts in the Upper Rhine Valley. The aim was to build on the results derived from the work in Cornwall, but at a site where temperatures at depth were expected to be higher. As it turned out, temperatures at depth were similar to those expected in SW England, but the work led eventually to demonstration of the practicability of the EGS concept, in this case a three-well system that has now started delivering power to the French national grid (see Box 10.6).

BOX 10.6 The EGS site at Soultz-sous-Forêts

The Soultz site, like Fenton Hill, benefits from the blanketing effect of 1000 m of sedimentary rock above the crystalline basement. The geothermal gradients through the sediments average 0.08–0.1 °C m^{-1}, falling to 0.028–0.05 °C m^{-1} in the crystalline basement beneath. Teams from France, Germany, the UK, Italy and Switzerland worked together on the site, eventually drilling four deep boreholes (>3800 m, 170 °C to 5000 m, 200 °C) as well as several shallower boreholes that are used for the microseismic monitoring system (Figure 10.10). Geologically, the site is located in the Upper Rhine Graben, where E–W tensional forces have stretched the crust and caused the granitic basement to subside. The basement is heavily fractured and even supports a small amount of natural fluid flow.

During more than 20 years of operation, the project at Soultz, coordinated and part-funded by the European Commission, became recognized as the world leader in developing EGS technology. It has been shown that with careful control of the pressure and density of the stimulation fluid the stimulated zone can be persuaded to develop laterally so that it can be accessed by a conventional arrangement of two or more boreholes deviated in opposite directions. In late 1997, after several years of testing and hydraulic stimulation of the fracture system at 3.5 km depth, a 4-month circulation test was carried out. 25 kg s^{-1} of water was circulated on a continuous basis between wells GPK1 and GPK2. The system operated in a closed loop, with the heat produced (ca. 10 MW$_t$) being dumped via a heat exchanger. The overall loss rate was zero and no make-up water was required. In a two-well system like this, such a result was possible only because a down-hole pump was used in the production well, altering the sub-surface pressure field to ensure that losses from the injection well could be balanced by input from the natural *in situ* fluids. Tracers added to the injected fluid proved that circulation was occurring; production also balanced by input from the natural *in situ* fluids.

Site map

Since 1987:

- EPS1 fully cored → exploration well
- GPK1 → injection well
- GPK3 → injection well
- GPK2 and GPK4 → production wells

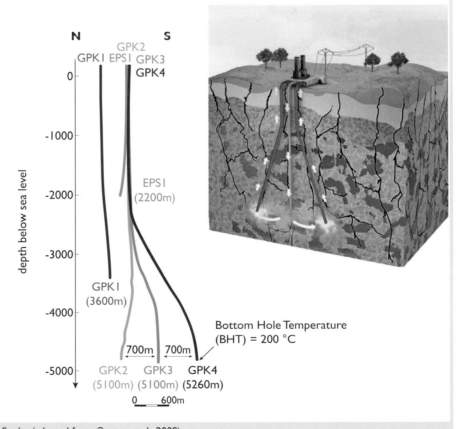

Figure 10.10 The arrangement of wells at Soultz (adapted from Genter et al., 2009)

Production declined very rapidly when reinjection was stopped, with both observations demonstrating that this is a true EGS system under the above definition. Of equal importance was the finding that the overall system impedance was less than 0.2 MPa l^{-1} s^{-1}, closer to the targets than any previous project (target parameters for EGS developments are discussed later and are summarized in Table 10.4).

Figure 10.11 The EGS site at Soultz-sous-Forêts

Following this success, it was decided to continue to the pilot plant/proof-of-concept stage. To improve the flow distribution in the reservoir, a three-well system was designed, with two production wells flanking a single injector. GPK2 was deepened, and GPK3 and GPK4 drilled, to about 5000 m, where the bottom hole temperatures were 200 °C. The wells were deviated to give bottom hole separations of 700 m. After stimulation of the fracture zones between the wells, a 1.5 MW$_e$ ORC turbo-generator was installed. The first power was produced in June 2008, and continuous production started in January 2011, albeit at a low initial flow rate (<35 l s^{-1}) as a precaution against induced seismicity (see Box 10.7).

The first power plant has been plagued by corrosion issues related to water salinity. The project aim is to finish commissioning a second experimental power plant which is hoped to double the current output over the next few years (Held et al., 2014).

As mentioned earlier, the increasing emphasis in recent years on the need to reduce CO_2 emissions and dependence on fossil fuels has caused a number of countries to increase their efforts to develop geothermal resources. Notable among these have been Germany and Australia. Germany has been involved in EGS research since the earliest days of the work at Los Alamos but, following experience gained from its work at Soultz, has begun to apply the techniques at home. Indeed, the first commercial EGS system has been completed at Landau, also in the Upper Rhine Valley about 50 km north of Soultz, to provide both heat and power to the town. It is a two-well system, with a depth of about 2700 m, producing water at 70 l s^{-1} and 160 °C. The Landau power plant was inaugurated in November 2007 and since then has been running continuously, except for some maintenance work. It is able to deliver about 3 MW_e and about 4 MW_t.

Perhaps the most surprising entrant in the geothermal stakes over the past decade has been Kenya.

Olkaria has added additional capacity using a single flash cycle system geothermal electricity generation station, located in East Africa on the Great Rift Valley in Kenya. Recall that the Great Rift Valley is a massive continental fault system that runs 6000 km from Mozambique to Jordan and has the potential to supply many generation sites. The Olkaria station is currently supplying about 49% of all the electricity for Kenya, with an estimated potential of 10 000 MW_e available in the entire country (Davies, 2015) over the long term. This has been a remarkable step change in a very short period of 2–3 years.

Olkaria, near Lake Naivasha, has been generating small amounts of electricity for 20 years, but now, with an investment of around £1 billion, they will be able to produce more electricity than the country's entire current capacity of 1600 MW (Heap, 2013) going into their fourth and fifth phases of increased generation levels. The Rift Valley produces superheated steam (230 °C) across eight rigs at approximately 3 km depth. The generation system there has been phased and the latest expansion project consisted of four identical 70 MW units, each with a design steam pressure and temperature of 6 bar and 158.7 °C (Patel, 2015). Although some may be critical of the single flash systems, they are proving simple to implement and robust to the overall circumstances.

Figure 10.12 Olkaria phase 4 expansion of geothermal station in Kenya

Figure 10.13 Part of the ol-Karia geothermal power generation complex is seen from a vantage point on the floor of the Kenyan Rift Valley, near the shores of Lake Naivasha some 120 kilometers north-east of capital, Nairobi (source: Tony Karumba/AFP/Getty Images)

Rojas (2015) notes the use of wellheads at Olakaria. The 'wellheads take the shape of a normal geothermal power plant, but in a smaller version. While a geothermal power plant is run by steam piped from tens of wells, a wellhead utilises steam from just a single well.' Once the main plant is completed the wellheads are removed and moved to different stations so that steam from the wells can then feed into the main plant. In the meantime large gains have been made with 11 wellheads producing 56 MW from heads which would normally lay idle. Kenya is the first country to use wellheads in this way and plans to continue rolling the system out over time.

Further wells along the Rift Valley are anticipated, with up to 100 eventually, and with this growth comes higher levels of electricity penetration to households and small businesses, giving a much needed injection into the local economy. Previously, periods of blackouts and unreliable electricity curbed development. The project is helping build local knowledge, capacity and investment confidence and as a result geothermal prospecting is happening in nearby Ethiopia, Rwanda and Tanzania (Heap, 2013). With nearly 70% of Africans not on the mains grid, geothermal power stations offer great promise. When coupled with solar PV systems, batteries and LEDs, for lighting and other uses such as phone charging, the technologies can create large changes in society. Examples include: raising education levels (better lighting and longer study periods); lowering health exposure (reduction of indoor stoves, paraffin lamps, and other fossil fuelled household items); and increasing general prosperity in small businesses (Heap, 2013).

10.5 Environmental implications

The main environmental concerns associated with 'conventional' geothermal energy development are focused on those to do with site preparation, such as noise pollution during the drilling of wells, and the disposal of drilling fluids, which requires large sediment-settling lagoons. Noise is also an important factor in high-enthalpy geothermal areas during well-testing operations when steam is allowed to escape, but once a field comes into production noise levels rarely exceed those of other forms of power plant. Accidents during site development are rare, though a notable exception in 1991 was the failure of a well originally drilled in 1981 at the Zumil geothermal station on the flanks of Santiaguito volcano in Guatemala. Hundreds of tonnes of rock, mud and steam were blown into the atmosphere when the well 'blew its top', apparently because of gravitational slippage of the ground beneath the site.

Longer term effects of high-enthalpy geothermal production include ground subsidence, gaseous pollution and induced seismicity. In dry steam fields, where the reservoir pressures are relatively low and the rocks are self-supporting (as at the Geysers and Larderello), subsidence is rare. In liquid-dominated systems, however, significant reduction of the higher pressures, for example due to inadequate fluid reinjection, can induce subsidence, usually on the millimetre to centimetre scale (although maximum localized subsidence of 3 m has occurred at Wairakei as a result of early exploitation without reinjection). Reductions in reservoir pressure can also have an adverse effect on the natural manifestations (geysers, hot springs) that are a common accompaniment of high-enthalpy fields and often important to the local tourist industry. Such concerns have severely restricted the development of geothermal power generation in Japan.

Geothermal 'pollutants' are chiefly confined to the non-condensable gases: principally CO_2, with lesser amounts of hydrogen sulfide (H_2S) or sulfur dioxide (SO_2), hydrogen (H_2), methane (CH_4) and nitrogen (N_2). In the produced water there is also dissolved silica, heavy metals, sodium and potassium chlorides and sometimes carbonates, depending on the nature of the water–rock interaction at reservoir depths. Today these are almost always reinjected and this also removes the problem of dealing with waste water. Traditionally, geothermal fields have received a bad press on account of their association with the 'rotten eggs' smell of H_2S. However, this and other gaseous products of old leaking plants have now been reduced so that the environmental impact of thermal production is at a minimum. Modern plants are fitted with elaborate chemical systems to trap and destroy H_2S. Interestingly, the level of atmospheric H_2S over the Geysers field is now lower than that emitted naturally from hot springs and geysers before geothermal developments began. Nevertheless, the image of polluting geothermal systems has slowed developments at several new sites. For example, environmental legislation covering the Miravalles plant, located on the periphery of a rainforest conservation area in northern Costa Rica, delayed completion of the plant for four years. A project on Mount Apo on Mindanao Island in the Philippines was turned down by the World Bank and the Asian Development Bank on social and environmental grounds. Objectors claimed that 111 hectares of forest would be threatened, 28 rivers polluted and a national park destroyed.

The position over emissions of CO_2, an important greenhouse gas, is rather more complicated. Geothermal reservoirs often contain significant quantities of CO_2, so emissions from those power plant will also be higher in CO_2 than might otherwise have been expected. On the other hand, exploitation of the field often reduces natural emissions. Leaving aside that possible benefit, however, a survey carried out by the International Geothermal Association (IGA, 2002) shows a wide variation in CO_2 emissions from existing plants, ranging from 4 g per kWh to 740 g per kWh (though the latter figure is extreme, from a field that is naturally high in CO_2 and undoubtedly leaking large quantities long before geothermal development began). The weighted average is 122 g per kWh. Typical CO_2 emission rates from fossil-fired power stations range from about 400 g per kWh for the most up-to-date natural gas fired combined cycle plant to about 900 g per kWh for the best coal-fired stations (see Chapter 11).

Induced seismicity

The question of whether there is induced seismicity around conventional geothermal sites has been much debated, and in the case of high-enthalpy systems it must be recognized that most steam fields are located in regions already prone to natural earthquakes because of their proximity to plate boundaries. There is evidence that fluid injection lubricates fractures and increases pressures, creating small earthquakes (microseismicity), especially when reinjection is not at the same depth as the producing aquifer (mainly for reasons of fluid disposal). However, in cases where reinjection is designed to maintain reservoir pressures, seismicity is not greatly increased by geothermal production.

In conventional low-enthalpy systems, where reinjection merely maintains the natural level of reservoir pressure, induced seismicity is rare or absent.

In EGS systems, on the other hand, injection occurs at higher pressures and induced seismicity is common, at least in the project development stages (see Box 10.7). Although the vast majority of induced events are small and detectable only instrumentally, a few are large enough to be felt at the surface. This is giving rise to public concern that could severely inhibit further development, and a multinational research programme has been mounted to understand and minimize these effects.

BOX 10.7 Induced seismicity in EGS projects

Although induced seismicity has rarely been a problem in conventional geothermal developments, the question has assumed far greater significance in the context of EGS projects. The stress fields in hard crystalline rocks, which are the typical targets for EGS, are invariably anisotropic, i.e. the three principal stresses (that is, the stress field resolved along three orthogonal axes) are of different magnitudes. This means that any fracture that is not precisely aligned with the stress field will have a tendency to slip. The EGS technique of stimulating the natural fracture system exploits this property; increasing the fluid pressure within the fracture reduces the forces that are keeping it closed and allows the two surfaces to slip past one another. As fractures will always be rough, the misalignment as the fracture closes will tend to keep it propped open, and this is the basis of the stimulation. This slippage of the fracture, usually of the order of millimetres, is in fact a micro-earthquake, and locating the noise generated by these events – called 'microseismicity' – is the principal tool used to follow the progress of the stimulation. A typical stimulation operation will generate tens of thousands of micro-events, virtually all of them far below the threshold of human perception.

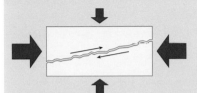

Figure 10.14 Schematic diagram of a fracture subjected to a non-uniform stress field

However, just occasionally an event will occur that is large enough to be detected at the surface, sometimes during stimulation but more commonly after the wells have been shut in. Typically, each of the stimulation procedures at Soultz resulted in up to five 'felt' events, all with local magnitudes (symbol M_L) less than 3. Events of this size are very unlikely to cause material damage – they are similar to the effect of a heavy lorry passing nearby – but they can alarm the local population.

Although this may be a matter of a perceived rather than an actual problem, public perception is becoming a critical factor in EGS planning in Europe and the USA (where reinjection into the Geysers – itself an example of EGS in an already seismically active area – has resulted in enhanced seismic activity). In Landau, a succession of small seismic events – although in an area already known for low level but persistent seismic activity – has prompted the operators to reduce flow rates (and therefore output) until the causes of the increased seismicity are better understood.

Even more seriously, a proposed EGS project in Basel, Switzerland (Figure 10.15), intended to provide heat and power to the city, was forced to close when an hydraulic stimulation operation in the first borehole triggered several small events up to M_L 3.4. Although no one was hurt and material damage was minimal, pressure of public opinion forced the abandonment of the project.

Those organizations and companies who are aiming to develop EGS systems have recognized the seriousness of this problem. It may well be true that induced seismicity is very unlikely ever to pose a real threat to life or property (except perhaps in those areas with a history of significant natural seismicity), but negative public opinion could prove to be a real showstopper. For that reason, there is a multinational research programme underway that is aiming to understand the reasons for the occurrence of these sporadic 'felt' events, and so devise methods to minimize or eliminate them.

Figure 10.15 The EGS rig at Basel

10.6 Economics and world potential

On an international scale, geothermal energy is one of the most significant 'renewable' energy resources. Its strength in this respect is that it can provide firm, predictable power on a 24-hour per day basis. Table 10.3 shows the performance of some typical high-enthalpy power plants; note the high capacity and availability factors. This high resource availability distinguishes geothermal energy from many other renewables, and results in significantly greater amounts of energy being supplied for a given installed capacity.

Table 10.3 Performance of typical geothermal power plant

	Italian 60 MW	Italian 20 MW	Japanese 50 MW
Year	1999	1999	1997-98
Installed capacity/MW$_e$	60	20	50
Maximum load/MW$_e$	55	17	48.3
Annual produced electricity/MWh	462 845	142 248	361 651
Hours of operation of plant	8748	8483	8112
Capacity factor/%	96.1	95.5	85.5
Availability factor/%	99.9	96.8	92.6

Source: IGA, 2001

There was quite spectacular growth in geothermal installed capacity of approximately 14% per year following the oil embargoes of the early 1970s, at a time when conventional generating capacity grew at between 0 and 3% per year. Stabilization of oil prices brought the growth rate down to about 8% per year by the early 1990s and cheap natural gas together with liberalization of electricity markets further reduced rates to 3% during the 1990s. This trend has now reversed, driven in part by concerns about climate change and also by rising fossil fuel prices. The underlying trend is again close to 10% per year, though it has been masked in recent years by the downturn in production of the Geysers during the period of over-exploitation.

It is difficult to discuss the economics of geothermal development except in the most general terms, because the details are so location-specific. While running costs are relatively minor, geothermal projects are capital intensive and the main element of annual costs is amortization of capital. The biggest single item is drilling costs, which rise exponentially with depth. Consequently, in common with other mining operations, costs are very dependent on the quality of the resource (depth, temperature, flow rate, etc.), and hence vary greatly from country to country and from place to place. Particularly in a new area, there is also a high initial risk that the borehole will be unsuccessful in locating an exploitable resource. Moreover, the economics are also strongly dependent on country-specific fiscal and regulatory issues like interest rates, subsidies, feed-in-tariffs, carbon credits, the cost of competing fuels, etc. On the other hand, successful geothermal projects benefit from the high availability of the resource and the consequent avoided cost of back-up plants.

The best illustration of the economics of geothermal plant, therefore, is the willingness of private industry to become involved and the rate at which new plants are being commissioned. In New Zealand, for instance, where the electricity industry was privatized during the 1990s and different fuels have to compete on equal terms with hydroelectricity, a new privately owned 55 MW$_e$ plant was commissioned in 1996, a second in 1999 and several more in the period 2003–10. Generating equipment in some of the older fields that have now been in use for nearly 50 years is also being upgraded. A recent study by the Ministry of Economic Development (Figure 10.16) showed that geothermal is currently the most competitive new-build option (Harvey et al., 2010).

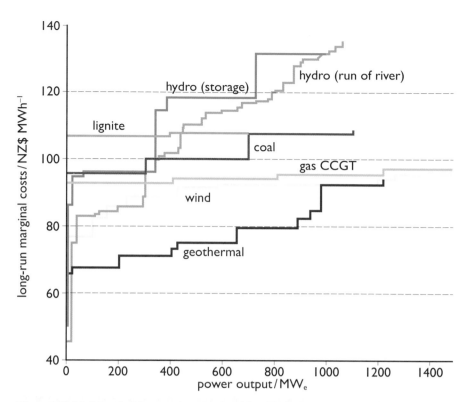

Figure 10.16 New build generation costs in New Zealand
Note: In October 2017 1 NZ$ = £0.54.

Sharing of experience and research and development costs among the different operators will be a vital factor in achieving targets. The obvious economic advantages of high-enthalpy resources in providing a good return on capital have stimulated loan investments in geothermal developments by international agencies, such as the World Bank, especially in Central and South American countries. But perhaps the greatest economic gain to society in general lies in the 112 million barrels of oil a year that is already being saved (or over 200 million barrels if the electricity would otherwise have been generated by fossil fuels at 30% efficiency).

Although geothermal resources make a significant contribution in some high-enthalpy areas (e.g. >30% of energy usage in Kenya and Iceland, 24% in El Salvador and 16% in New Zealand), the total amount of geothermal electricity produced in 2016 (some 67 TWh) accounted for only about 0.27% of global electricity consumption (Figure 10.17). Yet the long-term potential is much higher, especially in volcanically active countries, and may be realized as the technology improves. As suggested earlier, as EGS techniques improve the potential will rise dramatically, even in countries that lack high-enthalpy resources. In principle, successful development of EGS could allow most countries, even those in 'normal' areas, to generate 10% or more of their power needs from geothermal resources.

In support of this suggestion, in mid-2011 the International Energy Agency published its 'Technology Roadmap for Geothermal Heat and Power' (IEA, 2011). This roadmap foresees that by 2050 geothermal electricity generation could reach 1400 TWh per year, i.e. around 3.5% of global electricity production. It also estimates that geothermal heat could contribute 5.8 EJ annually by 2050. It qualifies the forecasts, however, by noting that

'For geothermal energy for heat and power to claim its share of the coming energy revolution, concerted action is required by scientists, industry, governments, financing institutions and the public.'

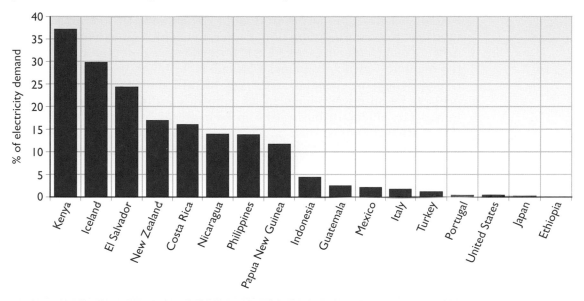

Figure 10.17 Percentage of countries' electricity demand generated from geothermal energy in 2015 (Source: IEA (2016) and Bertani (2015))

The future outlook for geothermal power systems is generally very positive with high growth rates currently, but with some issues and barriers which should be noted. The trajectory at the moment is for the existing level of 12.6 GW to be raised to 21.5 GW by 2020 or shortly thereafter. More optimistic growth rates envisage a five-fold increase in installed capacity, expected for 2030, which could result in additional global capacity of another 40 GW above 2020 targets (Delony, 2016). Production in the US, Indonesia and Mexico is expected to grow, with further activities in Chile, Kenya and parts of the Caribbean making new significant contributions to the group of 24 countries currently using geothermal. With 90 countries potentially showing interest in geothermal systems there are many more possibilities available and if all projects conceived were implemented then the profile of the technology would be raised significantly. Some countries are using set targets to drive their installations forward; India has set a short-term target of 1000 MW for 2017 with 10 000 MW goal by 2030. (Nelson, 2017). China hopes to triple their geothermal production by 2021. Growth in Indonesia is being supported by a policy of fixed price feed in tariffs to spur on development (Nelson, 2017). In Indonesia, reducing the time for geothermal permitting is also being addressed, thereby speeding up the entire process to completion. Other countries are also adopting similar permitting processes.

Barriers to growth include policy uncertainty, licensing delays, high upfront costs of drilling and the management of risk linked to exploration (Delony, 2016). Thus investors need more innovative financing mechanisms and ways to manage the risk within each project. Different countries and regions face different barriers. For example, in Indonesia processes are slowed by using very large overseas developers and at times the tendering processes required present obstacles. However, these large complex systems, once operational, can result in massive gains overall and may be worth the short term delays. Within the US, with a sluggish market, the geothermal industry and its lobby are striving hard for further legislation to provide better tax credits for geothermal, which help spur on the industry in sites like Salton Sea (Imperial Valley, California) and Salmon-Challis National Forest (Idaho). Issues in the US are linked to the long lead times until production, exploration which fails to produce, and drilling on federal lands. Mexico is trying to raise potential by changing the regulatory structure to make private investment opportunities more likely in the power generation market. In 2016, more than 44 new power projects began development in 23 countries adding another 1563 MW in capacity and contributing to a spike in global growth (Nelson, 2017). If the rate is sustained then power production could be above 23 GW in 2021, compared to the 2017 level of 12–13 GW (Nelson, 2017).

The Eastern Caribbean holds various volcanic systems that are a natural spot for geothermal development but the work is slowed by the remoteness of the area, the diverse political environment and the small load centres needed (Delony, 2016). Test sites in Monserrat, Nevis and Saint Kitts are already underway or will begin exploration studies. In Kenya, the Rift Valley holds promise of up to 1500 MW in the near term based on the current 390 MW of installed capacity. KEGC (Kenya Electricity Generating Company) – the main supplier of electricity in the country – has tended to partner with Indian and Chinese investing groups to finance various projects. The geothermal industry will achieve further growth with increased international cooperation, new investments and technological breakthroughs coupled with incentives that give parity with solar and wind (Nelson, 2017). The economics of lower-grade geothermal resources are more marginal and depend on local political and economic conditions, such as the availability and price of fossil fuels, the willingness of governments to invest in new energy concepts, the degree of environmental awareness and the related tax incentives to promote 'clean' energy commercially. The awareness of the problem of greenhouse gas emissions and climate change in recent years has nevertheless provoked many governments into support of what until recently has been a neglected 'renewable' resource.

It is worth pointing out that the economics of space heating operations are also dependent on making maximum use of the geothermal resource (which is capital intensive but has low running costs). If the geothermal system is sized to meet the maximum demand of the heat load, it will lie idle for much of the year; typically, the shape of a domestic heating load duration curve throughout the year is such that some 80% of the total energy needs can be met by a system with a power output of only 40% of the peak demand. Consequently, the geothermal component of the system should be (and, in successful schemes, is) designed to meet less than half of peak demand, with the shortfall being made up by an auxiliary fossil fuel boiler.

Looking to the economic future

The economics of future EGS developments are speculative at the time of writing, and will remain so until the technology is fully demonstrated. The best estimates, derived from recent progress in reservoir development and reductions in drilling cost, and for sites with a mean temperature gradient of 35–40 $°C$ km^{-1}, are that electricity might be produced for about

£0.1–0.17 per kWh in the early pilot plant (2010 prices), reducing to half these figures for a multi-module commercial system. There is an important caveat, however: these estimates come from cost models, not financial analysis, and the distinction is important.

Financial analysis can be applied to an existing operation or to a technology that is proven; all the steps in the process are known, and the costs of each step can be calculated. This information can be used to derive the break-even cost of the product in an unambiguous way. **Cost modelling**, on the other hand, is usually applied to an unproven technology – often one that is still being developed. It examines the possible costs of each step in terms of assumptions about the performance of the step itself and each preceding step. It says only that '...*if the technology performs in this way, then the cost will be...*'. The result is only as good as the initial assumptions. Cost modelling is a useful tool for setting the targets that various elements of the technology must achieve, or for establishing which aspects of the research offer the best opportunities for improvement, but it does not predict prices. This often gives rise to misconceptions, and ones that EGS research has suffered from, particularly in the UK. Used correctly, however, such analyses can be very useful.

Cost analyses of this type have been carried out for two-well systems by all the research teams involved in EGS, and resulted in general agreement on the target parameters to be achieved for a two-well EGS reservoir, aiming to produce 200 °C water for electricity generation over a 20-year reservoir life (Table 10.4).

Table 10.4 Target parameters for a two-well EGS system

Flow rate/kg s^{-1}	75–100
Effective heat exchange area/m^2	>2 × 10^6
Accessible rock volume/m^3	>2 × 10^8
Impedance[1]/MPa l^{-1} s^{-1}	0.1
Water losses/%	<10

[1] Strictly, impedance is not a constant, but varies with flow rate (because of pressure variation). The specified figure is the resistance to flow at the target flow rate

Until recently, none of the projects had come close to achieving simultaneously the required flow, impedance and loss rate. In 1997–8, however, the Soultz project demonstrated a closed loop circulation at 25 l s^{-1} for 4 months with zero losses and impedance close to the necessary target. It is on the basis of extrapolating the findings from this and subsequent experiments that the previously mentioned cost estimates are derived. It is still worth repeating, however, that these figures come from cost modelling, not financial analysis. Although we can make reasonable estimates of the capital costs of EGS schemes, we still do not know for certain how to construct a good reservoir, so we can only base our estimates on what its performance and properties ought to be. Predictions of the cost of power from EGS are arguably premature until further technological developments have provided better reservoir performance data. The existing developments at Landau and Soultz, and the imminent systems in Australia and Germany will go some way towards clarifying the issues.

10.7 Geothermal potential in the United Kingdom

Sedimentary basin aquifers

As in many other countries, it was the oil crisis of the mid-1970s that spurred geothermal resource evaluation in the UK. By 1984, new maps of heat flow (Figure 10.18(a)) and of promising geothermal resource sites (Figure 10.18(b)) had been produced. Three radiothermal granite zones stand out with the highest heat flow values, but significant heat flow anomalies also occur over the five sedimentary basins identified, partly because these are regions of natural hot water upflow. Many shallow boreholes were drilled during this period to measure heat flow, as were four deep exploration wells that are shown in Figure 10.18 (b) and Table 10.5. In each case the main aquifer is the permeable Lower Triassic Sherwood Sandstone (named after its most notable outcrops in the East Midlands).

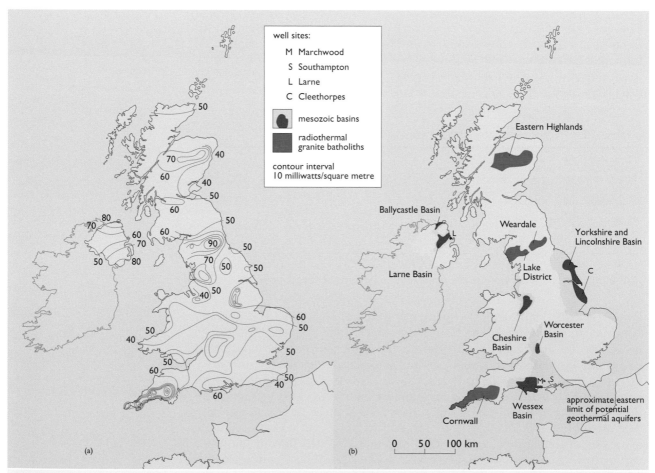

Figure 10.18 (a) Heat flow map of the UK based on all available measurements to 1984 compiled and published by the British Geological Survey. (b) Distributions of radiothermal crystalline rocks (granites) and major sedimentary basins likely to contain significant geothermal aquifers in the UK

Table 10.5 Characteristics of UK deep exploration wells

Location	Completion	Well depth /m	Bottom hole temperature /°C	Main aquifer depth /m	Temperature of aquifer/°C
Marchwood	Feb 1980	2609	88	1672–1686	70
Larne	July 1981	2873	91	960–1247	40
Southampton	Nov 1981	1823	77	1725–1749	76
Cleethorpes	June 1984	2092	69	1093–1490	44–55

Source: Downing and Gray, 1986

The shallower intersections with the aquifers at Larne and Cleethorpes are at rather low temperatures for geothermal exploitation but have reasonably high fluid flow rates because of the large aquifer thickness. Unfortunately, the other two wells, which intersect the aquifer at a better temperature, produce rather low flow rates because of the restricted vertical height of good aquifer rock. The yield is reduced not just because the sedimentary sequence is thinner in the Southampton area, but also because much of the Sherwood Sandstone proved to be more highly cemented and therefore less permeable. The resources are nevertheless substantial (Table 10.6).

Table 10.6 Potential UK geothermal energy resources at different temperatures

Basin	Potential resource at 40–60 °C /EJ	Potential resource at > 60 °C /EJ
East Yorks and Lincs	26.2	0.2
Wessex	2.8	1.8
Worcester	3.0	–
Cheshire	8.9	1.5
Northern Ireland	6.7	1.3
Total	**47.6**	**4.8**

Source: Downing and Gray, 1986

Assumptions behind these estimates are that the 40–60 °C resource would be exploited with the use of heat pumps, producing a reject temperature of 10 °C, whereas use of the >60 °C resource would not involve heat pumps, giving a 30 °C reject temperature. The latter resources could be doubled with heat pumps. For comparison, current UK electrical energy demand is around 350 TWh per year 1.3 EJ per year, or about 30 GW$_e$ as equivalent continuous power), so we are considering here quite large resources of renewable energy, but as heat rather than electricity.

So why are geothermal aquifers not being exploited much more widely? The problem is not just one of marginal economics and geological uncertainty, but rather the mismatch between resource availability and heat load, itself a function of population density. More than half the UK resources are located in East Yorkshire and Lincolnshire, essentially rural areas lacking the concentrated populations of the Paris basin. While electricity can readily be transported over long distances from source to market, hot water is more of a problem. Transmission distances for the latter under UK conditions are likely to be restricted to just a few kilometres so the resources are likely to remain undeveloped. It is interesting to note, however, that the distance limitation arises from cost rather than heat loss; some Icelandic pipelines exceed 50 km in length with a temperature drop of only about 1 °C.

The consequence of this is that Southampton, where the geothermal borehole contributes some 2 MW$_t$ to the city's group heating system, remains the only direct use of a deep geothermal resource truly put into operation in the UK. However, the project was abandoned leaving only the Southampton borehole and a geothermal 'heat only' energy scheme. This system forms part of the city centre district heating system, drawing warm (76 °C) water from the Wessex Basin Hot Sedimentary Aquifer at 1800 metres depth (DECC, 2013).

Engineered geothermal systems

Of the three principal granite zones – in the eastern Highlands, northern England and south-west England – the latter is characterized by the highest heat flow, as shown in Figure 10.18(a). However, large areas of the more northerly granite masses are covered by low thermal conductivity sedimentary rocks and so temperatures will be higher at depth than if the granite bodies came to the surface (Figure 10.6). Nevertheless, substantial areas of Cornwall and Devon are projected in Figure 10.19 as having temperatures above 200 °C at 6 km depth. One estimate for the EGS resource base in south-west England suggests that 300-500 MWt (about 10 PJ y^{-1}) could be developed in Cornwall over the next 20–30 years with much more to follow later.

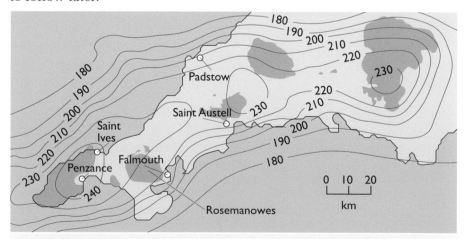

Figure 10.19 Projected temperature (°C) at 6 km depth beneath south-west England. Granite bodies that crop out at the surface are shaded

From 1976–92, Cornwall was the site of a major EGS research project that laid the foundations for the subsequent multinational European project at Soultz. Following Soultz's demonstration of the feasibility of generating power from deep fractured granite, plans are in hand for further EGS projects to begin where economically feasible. The first will start with a 5 km exploratory borehole on the northern edge of the Carnmenellis granite (labelled Rosemanowes, the site of the earlier project, in Figure 10.19), with the aim of constructing a multi-well system to generate 10 MW_e of power. The second will be on the St Austell granite and is designed to produce 4 MW_e and several MW_t of heat, primarily to supply the Eden Project (an eco-centred visitor attraction).

10.8 Legal and regulatory issues

As the exploitation of geothermal resources has moved in each country from being a technical novelty to a commercial reality, and one with serious financial implications for the developers, so it has been necessary for each country to incorporate geothermal resources into its regulatory system.

Central to this is the question of resource ownership. Not only must developers (or their banks) be confident that they have a right to the resource that they will be extracting, but there must also be provision: (a) to protect the developer's rights, and (b) to protect the rights of third parties (such as neighbouring developers or property owners). Just as every country has its own legal code, so every country with a geothermal industry has defined these rights and responsibilities according to its own background legal framework and conditions.

In the majority of cases, if not all, ownership of the resource has been reserved to the national or regional authority, which then has the power to grant licences and exact royalties under specified conditions. Typically, these conditions will impose obligations to explore and develop the resource within a defined area and within a specified period, and will include provision to protect shallow water sources and to limit the impact on neighbouring areas. For reasons of both environmental protection and reservoir management, it is now becoming common to require reinjection rather than disposal to surface. Reinjection, too, is controlled; in Italy, for example, it is not permitted to reinject into any aquifer other than the source.

Unfortunately, the UK remains one of the few countries not to have established a legal basis for geothermal development. UK mining law is particularly complicated, with different types of mineral resources being subject to different ownership, etc., so there is no easy precedent to apply to geothermal resources. In the absence of an established demand, there has been no incentive to draw up new legislation. That means that there is currently no clear definition of ownership or rights, which complicates the financing of projects. Now that interest is beginning to develop, that situation will need to change. Appropriate discussions have already begun at ministerial and parliamentary level.

10.9 **Summary**

Geothermal energy is heat which is 'gathered' from earth. It can be classed into four basic forms:

- **Deep geothermal energy**, typically from **more than 1 km below** the Earth's surface, which draws heat from hot rocks or aquifers.

- **Ground source heat pump technology** is technically **shallow (~ 40 metres)** geothermal energy and utilizes the flows of heat between the air and the top layer of the ground.

- **Aquifer thermal energy storage** is also shallow (~ **100 metres**), acting as a geothermal heat pump technology, and uses heat flows to and from underground aquifers.

- **Heat pumps** can be used with deep geothermal wells to raise the temperature of water up to a usable level for heating buildings.

About **half of all geothermal energy produced is deep geothermal**, used to produce electricity in about 25 countries. These countries are located in many instances along the lithospheric ridges or in areas with volcanic activity that provide hot spots of steam, hot water, springs, geysers and fumaroles. The other half of geothermal production can be attributed for use in non-electrical 'direct use' applications, such as space heating, agriculture, aquaculture and other industrial processes.

Deep geothermal energy is the mining of heat from high enthalpy sources – where the **Earth's heat flow is sufficiently concentrated** to have generated natural resources **in the form of steam and hot water (180–250 °C)**. The rocks must be within feasible drilling distance (about 700–3000 metres) of the ground surface.

The heat is due to radiogenic rocks, which contain radioactive elements that decay and generate heat in the surrounding crust. Geothermal resources are non-renewable on the scale of human lifetimes yet are still considered renewable since they are energy flows, not energy stores like carbon or nuclear-based fuel. **Depletion of boreholes** (or wells) over time does occur and **geothermal systems require good management** of the aquifers and geographic fields to ensure overuse does not occur. A critical part of management is understanding the typical heat exchange rates for the field in question along with good, long term monitoring of output temperatures and pressures **to ensure stability**.

Deep geothermal energy needs three things for success: **an aquifer of sufficiently hot water** that can be drilled into; **a cap rock** to retain the geothermal fluid; and **a heat source**. Aquifers contain porous rocks that can store water and through which water will flow. The hydraulic permeability helps determine the suitability of a potential geothermal site.

The most **important rock properties** within the aquifer include **grain size, shape and porosity**. How these interact with each other also determines site potential along with any fractures or weaknesses in the layers of rock. **The fluid pressure gradients** and volume of water flowing in the rock are also key considerations and systems are measured in hydraulic conductivities (cubic meters per day). Fluid pressures can reach up to 10 MPa. Cap rocks are simply impermeable rock layers which ensure the aquifer (and the heat) do not escape. Different rock types have different rates of heat conduction, but in general rocks are poor conductors.

In **high-enthalpy regions**, the heat source is likely to be near volcanic areas (extinct or active), whereas in **low-enthalpy areas** the heat source is either a deep sedimentary basin where aquifers carry water to depths where it becomes warm enough to exploit, or hot rocks. Hot dry rocks have high natural heat production and an artificial aquifer must be created by enhancing the rock fractures to exploit the source. The term **'enhanced' – or 'engineered' – 'geothermal systems' (EGS)** is applied to systems using hot dry rocks **with supplementary water or systems, which require stimulation or enhancement** to be more profitable. All rocks contain some moisture despite the term 'dry rocks'.

Hot water which has been in contact with rocks tends to contain high amounts of dissolved solids and dissolved gasses and thus can be quite reactive. **These contaminated waters are called brines** and need treatment before being used in geothermal plants. One way to avoid using brines directly is to use the geothermal water to heat 'clean' water or some other secondary fluid for use in the plant, although this incurs losses.

Drilling for geothermal fields is costly and **increases greatly with depth**. Geothermal field strata tend to be much harder than oil fields, and some testing via bore holes is undertaken to ensure there is a sufficient geothermal gradient present and that fluid pressure is high. Scientific methods such as **resistivity surveying** – searching for the electrically conducting brine – are used to locate aquifers. Wells are drilled and constructed in a similar manner to those used for oil extraction.

Four types of power plants are used to convert the steam into electricity via turbine systems. **Dry steam power plants** are employed for 'clean' steam inputs at 180–225 °C and 4-8 MPa, and are the simplest and most economically viable. Single flash steam power plants use pressure reduction, called **flashing**, before the steam enters the turbine. These systems tend to run at slightly lower temperatures and pressures than dry steam power plants. **Binary cycle**, and **Organic Rankine Cycle (ORC)** power plants employ a secondary fluid (like butane) which is vaporized and drives the turbine. ORC plants require a heat exchanger and a closed sub-system for the secondary fluid – and are more costly and complex. However, the binary cycle can run with lower input temperatures and may outperform some low-temperature single flash systems. **Double flash power plants**, put simply, use multiple separators to run two turbines in series and require large fluid volumes to operate with high efficiencies. The second stage turbine raises the cost by ~ 5% and yields ~ 20% more power.

In all of these plants **the condensate tends to be reinjected** back into ground within the geothermal zone along with a range of **cooling systems (air cooled towers)** and copious insulated pipework to ensure minimal losses occur when redirecting input steam flows.

Growth and interest in geothermal energy in countries within East Africa is growing as the Great Rift Valley is recognized as an important geothermal zone. For **developing countries geothermal investment and production can substantially raise the quality of life** by providing many other benefits in addition to making electricity. Due to poor grids, and low connectivity, **geothermal can play a major factor in economic development**.

The potential global growth of geothermal systems is significant. **Existing levels could double by 2020** and some predictions see a five-fold increase by 2030, which would result in some 60 GW global capacity. This spread from a small group of countries to a much wider base would also potentially lower costs for further development. Some developing countries like Kenya or El Salvador depend significantly on geothermal energy as a source of clean electricity.

References

Bertani, R. (2015) 'Geothermal Power Generation in the World 2010-2014 Update Report', *Proceedings World Geothermal Congress 2015*. Melbourne, Australia, 19–25 April 2015. Stanford University [Online]. Available at https://pangea.stanford.edu/ERE/db/WGC/papers/WGC/2015/01001.pdf (Accessed 10 October 2016).

Davies, M. (2015) 'Geothermal power in Kenya', *BBC News*, 18 September [Online]. Available at http://www.bbc.co.uk/news/business-34290828 (Accessed 8 March 2017).

DECC (2013) *Deep Geothermal Study Review Final Report*, London, Department of Energy and Climate Change [Online]. Available at https://www.gov.uk/government/uploads/system/uploads/attachment_data/file/251943/Deep_Geothermal_Review_Study_Final_Report_Final.pdf (Accessed 8 March 2017).

Delony J. (2016) '2016 Outlook: Future of Geothermal Industry Becoming Clearer', *Renewable Energy World*, 27 January [Online]. Available at http://www.renewableenergyworld.com/articles/2016/01/2016-outlook-future-of-geothermal-industry-becoming-clearer.html (Accessed 8 March 2017).

Downing, R. A. and Gray, D. A. (eds) (1986) *Geothermal Energy: the Potential in the United Kingdom*. London, HMSO.

Genter, A., Fritsch, D., Cuenot, N., Baumgärtner, J. and Graff, J. -J. (2009) 'Overview of the current activities of the European EGS Soultz project: from exploration to electricity production', *Proceedings, Thirty-Fourth Workshop on Geothermal Reservoir Engineering*. Stanford University, Stanford, California, 9–11 February. International Geothermal Association [Online]. Available at https://www.geothermal-energy.org/pdf/IGAstandard/SGW/2009/genter.pdf (Accessed 13 March 2017).

Global Cement (2012) *Kalina Cycle power systems in waste heat recovery applications* [Online]. Available at http://www.globalcement.com/magazine/articles/721-kalina-cycle-power-systems-in-waste-heat-recovery-applications (Accessed 8 March 2017).

Harvey, C. C., White, B. R., Lawless, J. V. and Dunstall, M. G. (2010) '2005–2010 New Zealand Country Update', *Proceedings World Geothermal Congress 2010*. Bali, Indonesia, 25–30 April. International Geothermal Association [Online]. Available at https://www.geothermal-energy.org/pdf/IGAstandard/WGC/2010/0008.pdf (Accessed 13 March 2017).

Heap, T. (2013) 'Energy revolution promises to transform East Africa', BBC News, 25 February [Online]. Available at http://www.bbc.co.uk/news/world-africa-21549380 (Accessed 8 March 2017).

Held, S., Genter, A., Kohl, T., Kölbel, T., Sausse, J., & Schoenball, M. (2014) 'Economic evaluation of geothermal resevoir performance through modeling the complexity of the operating EGS in Soultz-sous-Forêts', *Geothermics*, vol. 51, pp. 270-280.

IEA (2011) *Technology Roadmap for Geothermal Heat and Power* [Online], Paris, OECD/ IEA. Available at http://www.iea.org/publications/ freepublications/publication/Geothermal_roadmap.pdf (Accessed 10 March 2017).

IEA (2016) *Key world energy statistics* [Online], Paris, IEA. Available at https://www.iea.org/publications/freepublications/publication/ KeyWorld2016.pdf (Accessed 8 March 2017).

IGA (2001) 'Performance indicators for geothermal power plant', *IGA News*, July–September.

IGA (2002) 'Geothermal power generating plant: CO_2 emission survey', *IGA News*, July–September.

Lund J. W. and Boyd T. L. (2015) 'Direct Utilization of Geothermal Energy 2015 Worldwide Review', *Proceedings World Geothermal Congress 2015*. Melbourne, Australia, 19–25 April. Stanford University [Online]. Available at https://pangea.stanford.edu/ERE/db/WGC/papers/WGC/2015/01000.pdf (Accessed 10 October 2016).

Matek (2016) 2016 *Annual US & Global Geothermal Power Production Report*, Washington, D.C., Geothermal Energy Association [Online]. Available at http://geo-energy.org/reports/2016/2016%20Annual%20 US%20Global%20Geothermal%20Power%20Production.pdf (Accessed 15 March 2017).

MIT (2006) 'The future of geothermal energy: impact of EGS on the United States in the 21st century', *Report for US DoE*, Cambridge, Massachusetts, Massachusetts Institute of Technology.

Nelson A. (2017) '2017 Outlook: Geothermal is trending upwards, January/ February', *Renewable Energy World*, 3 February [Online]. Available at http://www.renewableenergyworld.com/articles/print/volume-20/issue-1/ features/geothermal/2017-outlook-geothermal-is-trending-upwards.html (Accessed 10 March 2017).

Patel. S (2015) 'Olkaria Geothermal Expansion Project, Rift Valley Province, Kenya', *Power*, 12 January [Online]. Available at http://www.powermag. com/olkaria-geothermal-expansion-project-rift-valley-province-kenya/ (Accessed 8 March 2017).

Rojas, F. (2015) 'Geothermal wellhead power plants are proving to be a faster way to develop new energy projects in Kenya', *ThinkGeoEnergy*, 21 October [Online]. Available at http://www.thinkgeoenergy.com/wellhead-geothermal-plants-are-a-key-pillar-of-kenyas-renewable-success/ (Accessed 8 March 2017).

World Energy Council (2016) *World Energy Resources 2016* [Online]. Available at https://www.worldenergy.org/wp-content/uploads/2016/10/ World-Energy-Resources_FullReport_2016.pdf (Accessed 8 March 2017).

Acknowledgements

The author would like to acknowledge the contributions to this chapter in previous editions from John Garnish and the late Geoff Brown.

Chapter 11

Integrating renewable energy

by Bob Everett, Godfrey Boyle, Jonathan Scurlock and David Elliott

11.1 Introduction

Renewable energy already makes a large contribution to world primary energy needs: in 2014 it amounted to about 14% (see Figure 1.2 in Chapter 1). In 2015 it contributed an estimated 19% of world final (or delivered) energy (REN21, 2017). Between 2010 and 2016 world fossil fuel consumption increased slowly, while the total consumption of nuclear electricity fell slightly. However, as described in Chapter 1, various renewable energy technologies have seen high growth rates. Between 2010 and 2016, world wind power output increased almost three-fold and solar PV output increased almost ten-fold (BP, 2017). Expansion in the manufacture of renewable technologies has also produced dramatic cost reductions (see for example Figure 4.7 in Chapter 4).

These renewable energy growth rates seem likely to continue, spurred on by concerns about climate change, and the local air pollution consequences of coal-fired electricity generation in China and India.

The previous chapters reviewed each major renewable energy source in turn, giving some consideration to the contributions that each could make to future needs. Modern industrial societies demand energy in extremely large quantities and in many different forms. In order to meet these requirements, a vast world-wide network of energy supply and distribution systems has been built up. To what extent will these existing energy networks need to be modified and supplemented if renewables are to continue to make an increasing contribution? Can renewables deliver our energy, not only in significant *amounts* and at an acceptable *price*, but also in the right *form*, at the right *time*, and in the right *place*? What are the factors that currently make it difficult for renewable energy sources to compete with conventional sources? What can society do to assist their adoption and use?

These are some of the questions addressed in this chapter. Many of the topics discussed in earlier chapters will be revisited, and an attempt made to draw some of the threads together. Initially we examine the situation in the UK before looking further afield.

Firstly, Section 11.2 looks briefly at existing energy systems and how they currently supply our various demands. A prime reason for using renewable energy is to reduce global greenhouse gas emissions. Section 11.3 discusses the relative emissions of conventional energy and different renewable energy sources.

Next, Section 11.4 discusses the available 'resource' or 'potential' for each renewable energy technology, and how practical and economic constraints impose limits on the contributions that each might make.

Section 11.5 looks at the current financial cost of renewable energy (particularly renewable electricity), concentrating on the UK.

Sections 11.6 and 11.7 look at renewable energy technologies in more detail, investigating the extent to which they can deliver energy in suitable forms, in the right place and at the right time. This includes the problem of integrating electricity from renewables into national electricity grids. Finally, Section 11.8 considers some possible system solutions, including international expansion of electricity grids to enable continent-wide electricity transmission, and the potential of hydrogen as an energy carrier and transport fuel.

11.2 **The existing UK energy system**

Figure 1.5 in Chapter 1 has illustrated the UK's primary and delivered (or final) energy use. Delivered energy takes five main forms:

- *liquid fuels*: about 45% in 2015 – almost entirely oil and its derivatives – petroleum (gasoline), diesel, kerosene, etc., although with an increasing percentage of ethanol and biodiesel (about 1.6% of liquid fuel use by energy content) (BEIS, 2016a)

- *gaseous fuels*: about 30% in 2015 – mostly methane ('natural gas'), plus a small amount of 'bottled' liquefied petroleum gas (LPG – propane and butane)

- *solid fuels*: in the recent past this has been almost entirely coal and coke. In 2015, these contributed less than 3%; however, with a resurgence of biomass as a fuel there are increasing contributions from fuel wood and straw, contributing about an extra 1%

- *electricity*: about 19% in 2015 – mostly from fossil-fuelled or nuclear power stations, with nearly a quarter of this coming from hydroelectricity and other renewable sources

- *heat*: less than 2% in 2015 – although only deployed to a limited extent in the UK, this involves distributing heat energy to industry and buildings directly in the form of hot water or steam, in district heating (DH) (also known as community heating) schemes. The heat can be provided by centralized boilers (in some cases fuelled by refuse incineration) or by waste heat from power stations in combined heat and power (CHP) schemes. It can also include heat from deep geothermal sources or large scale heat pumps.

These forms of energy flow into the various distribution networks and are delivered to final consumers in the main energy-using sectors of the economy, normally categorized as *domestic, services, industry* and *transport*.

Within each of these sectors, this delivered energy is used to provide the *energy services* that we actually need.

Firstly there are services based around *heat*, i.e. warm homes, cooked food and heat for industrial processes. Heat is required in many forms – from tepid water to superheated steam – for washing, cooking, space heating and industrial processing. It can be provided either by burning fuels close to where the heat is needed, by piping heat in from a more distant CHP or DH plant, or by using electricity, either in resistance heaters or in more sophisticated devices like microwave ovens. However, a word of caution is necessary here. A warm home does not necessarily require a large supply of heat energy. The desired level of energy service can also be achieved through improved insulation and similar energy-conserving measures, such as those in the *Passivhaus* retrofit described in Chapter 3.

Secondly there are energy services based on *motive power*. A significant proportion of the energy for this is in the form of electricity for electric motors driving everything from CD players through to lifts and heavy industrial machinery. Motive power is also needed, obviously, for transport (cars, trucks, buses, trains, ships and aircraft, etc.) In most cases, the form of delivered energy currently used to supply it is oil-derived, although electricity is widely used for electric railways and in the rising technology of electric cars. Again a word of caution is necessary. Much of the end energy service for transportation can be described as *mobility* yet this does not necessarily require the physical transportation of people. In many cases the use of telecommunications can provide an alternative energy service.

Finally, there are *electricity-based energy services*. Electricity can be used to provide heat, motive power and lighting, each of which can also be provided by other forms of delivered energy. However, electricity is essential for those systems that simply cannot function with any other energy form – not just computers and communications systems but also more specialized electro-chemical processes, such as the manufacture of aluminium or chlorine.

Distribution

It is worth dwelling a little on the changing nature of our energy distribution systems. Most of the UK's energy demand occurs within the relatively small areas of the major cities. Many of these are built in sheltered inland valleys to deliberately avoid the worst excesses of wind, wave and tidal energy. Many large industrial towns grew up around their fuel supplies: coal or wood. Some faded to obscurity when their fuel ran out, or were replaced by cheaper alternatives. Others survived by importing fossil fuels.

At the end of the nineteenth century, the UK was a major producer and exporter of coal. Most European cities ran mainly on coal, brought by rail or by sea. It was burned when and where it was needed, or was converted into electricity or 'town gas' in local plants. The precise forms in which we make use of electricity (for example as a 230 volt AC supply) and gas (as a low-pressure piped supply) had largely been fixed by the beginning of the twentieth century. Yet, at that time, oil as we know it currently scarcely existed as a commercial commodity.

Today coal, oil, natural gas and electricity are distributed not just country wide but internationally. Coal, once the major fuel in the UK, is now mainly just a fuel for power stations. Some is still locally mined, but most is imported by ship. Although coal generated over 20% of the UK's electricity in 2015, the UK government has pledged to close all remaining coal power stations by 2025.

The UK was a major importer of oil until the 1970s when discoveries in the North Sea made it not merely self-sufficient but a net exporter. There is an extensive network of pipelines conveying crude oil from the North Sea fields and also an internal network conveying oil around the country. The UK's North Sea oil production peaked in 1999 and is now declining: the country became a net oil importer in 2006.

Until the 1960s, the UK gas supplies were 'town gas' made from coal, which consisted of a mixture of carbon monoxide, methane and hydrogen. Then the large discoveries of natural gas (almost pure methane) in the North Sea revolutionized the industry. Hundreds of town gas production plants (gasworks) were closed down and a completely new infrastructure of long distance distribution pipes was built to bring in the North Sea gas. This had a higher energy density than town gas and meant that all gas boilers and cookers had to be fitted with new burners. However, it also meant that the energy-carrying capacity of the existing pipework was doubled. This was useful since there was an immediate rise in sales of gas-fired central heating systems. Natural gas has been *the* fuel of the last quarter of the 20th century. It has a lower proportion of carbon than coal and can be burned at high efficiency in combined cycle gas turbine (CCGT) power stations (see Chapter 2). A whole new generation of these plants was constructed in the UK in the 1990s, leading to a further large increase in gas demand.

Looking beyond the UK, there is an impressive Europe-wide system of pipelines for gas distribution, stretching from gas fields in Algeria in the south to Norway in the north, from Siberia and the Caspian Sea in the east to the Republic of Ireland in the west. A major pipeline such as that shown in Figure 11.1 can cost £1 million or more per kilometre to lay.

Although it is less visible this pipe network annually carries three times as much energy as the electricity grid in the UK.

Figure 11.1 A high-pressure gas pipeline being laid. Such pipes can carry several gigawatts of power, i.e. a similar amount to that carried by a 400kV electricity power line

These pipelines are also used not just to transport gas, but to store it, by varying the gas pressure. This is known as 'line packing'. This mode of storage has replaced the telescopic low pressure gas holders that were a highly visible feature of most UK towns. In the UK, gas is also stored under pressure in sealed underground salt caverns.

The UK's North Sea gas production peaked in 2000 and, like oil production, has been declining, with the country becoming a net gas importer after 2005. This situation is unlikely to be changed significantly by recent UK shale gas discoveries. Some natural gas is imported via pipeline from the Norwegian and Dutch areas of the North Sea, but other imports arrive by ship as liquefied natural gas (LNG) from countries such as Algeria in North Africa and Qatar in the Persian Gulf. This is liquefied by cooling to $-162\,°C$ and can then be transported by insulated tanker and stored in large insulated storage tanks (see Figure 11.2).

(a)

(b)

Figure 11.2 (a) The LNG tanker, Arctic Princess, docking at the Grain LNG terminal in Kent; (b) insulated LNG tanks at the Grain terminal. Each tank in the foreground holds 190 000 m³ of LNG stored at $-162\,°C$

Despite these large investments in infrastructure it is unclear whether or not natural gas will have a long-term future as a fuel in Europe as fossil fuels are phased out.

Turning to electricity, currently it is mostly generated in large plants, usually of 200 MW or more capacity, however the contribution from smaller generators is growing. Although much of the UK's electricity demand is in the south, it has been more convenient in the past to site coal-fired power plants in the mining areas of Wales or the north of England. This has also avoided past problems of sulfur dioxide emissions reaching urban areas. Currently, most of these plants are reliant on foreign coal imported by sea. The UK's nuclear power stations are mostly sited on the coast for ease of access to cooling water. Gas-fired power stations are more likely to be sited closer to their loads, and supplied with large quantities of natural gas through the high pressure pipelines described above. Since impurities such as sulfur are largely removed from the gas at source, these power stations are responsible for fewer pollution problems than their coal-fired counterparts.

The electrical output of these power stations is distributed to consumers via the National Grid, a network of high-voltage cables that covers the UK, with links to neighbouring countries (Figure 11.3). Since they are likely to be major carriers of renewably generated electricity in the future, we will look at the national grids of the UK and the Republic of Ireland in more detail later.

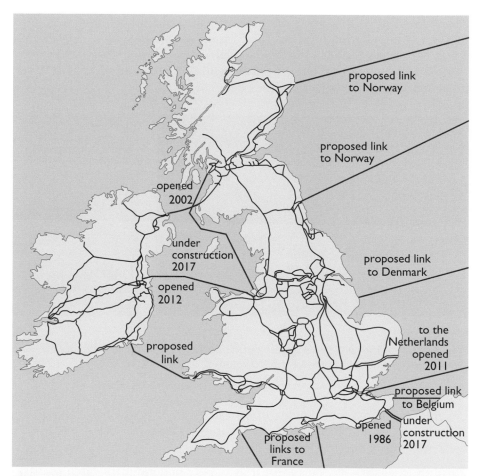

Figure 11.3 The existing UK and Irish high voltage electricity grids, with links (existing and proposed) to Norway, Denmark, the Netherlands, Belgium and France

Finally there are heat distribution networks. Although district heating (DH) is currently little used in the UK, in Denmark it supplies 60% of homes. Distribution pipes such as that shown in Figure 11.4 carry heat from DH and CHP plants, such as that shown at Avedøre in Denmark in Chapter 5. The Copenhagen DH network spans a distance of 50 km from east to west.

11.3 How can renewable energy decrease greenhouse gas emissions?

Figure 1.4 in Chapter 1 has shown the rising global emissions of CO_2 from fossil fuel combustion. Box 2.1 in Chapter 2 described the process of CO_2 production from the combustion of a fuel such as methane. The chapter also described how coal, a high carbon fuel, has a higher CO_2 emission factor than oil or methane – i.e. when burned it produces more CO_2 per unit of heat emitted.

How much CO_2 will be saved if a renewable energy source is substituted for a fossil fuel one? The answer depends critically on which fossil fuel is being replaced and whether the energy is being used for electricity generation or for heat or transport use.

Electricity generation in 2014 accounted for about 40% of global CO_2 emissions and consumed two-thirds of world coal production (IEA, 2016). This makes electricity-generating renewables of particular interest.

Electricity-generating renewables

The production of electricity from fossil fuels involves a considerable production of waste heat. The figure for the CO_2 emitted per kilowatt-hour of electricity depends both on the emission factor of the fuel and on the generation efficiency. Table 11.1 shows average values for UK generation in 2015. In round numbers, the production of 1 kWh of electricity from coal produces about 0.9 kg of CO_2 and about 0.4 kg from gas.

Figure 11.4 District heating pipes being laid in Denmark. These have a single casing that contains both the flow and return water pipes and some insulation

Table 11.1 CO_2 emission factors for UK electricity in 2015 and an approximate external cost of associated carbon emissions

Fuel	Emission for electricity /kg CO_2 kWh^{-1}	Assumed factor emission for fuel[1] /kg CO_2 kWh^{-1}	Implied average generation efficiency[2] /%	External cost at an assumed carbon price of £50 per tonne CO_2
Coal	0.92	0.32	35	4.6
Gas	0.38	0.20	54	1.9
Average - UK generation mix	0.33			1.7

1 Based on net calorific or lower heating value of fuel
2 i.e. electricity emission factor ÷ fuel emission factor Sources: BEIS, 2016b and 2016c

The UK generation mix includes gas, coal, nuclear power and a number of other sources. Figure 11.6 shows the breakdown for 2015. Although about a fifth was from coal, nearly a half came from low-carbon sources such as nuclear and renewables. The overall average emission factor has been declining rapidly and in 2015 was 0.33 kg CO_2 kWh^{-1}.

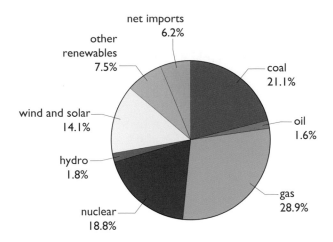

Figure 11.5 UK electricity generation fuel mix 2015 (source: BEIS, 2016a)

Carbon dioxide emissions at the power station are, of course, not the full story. There are also CO_2 and methane emissions involved in the extraction and supply of their fuel. Where fuels have been imported, some of these emissions may take place in different countries and even on the high seas. An overall greenhouse gas emission factor including these factors can be expressed in units of CO_2e, i.e. the equivalent amount of CO_2 that would give the same amount of global warming as that resulting from the actual mix of greenhouse gases emitted. Including these extra factors would increase the values in Table 11.1 by 5–10%. Figure 11.6 shows the ranges of estimates of overall greenhouse gas emissions for European power plants per kilowatt-hour of electricity generated. The wide ranges for conventional fossil-fuelled plants reflects the different generation efficiencies. The lowest emission values are likely to be from CHP plants. The range for hydroelectricity reflects the uncertainty over methane emissions from reservoirs.

Figure 11.6 Ranges of greenhouse gas emissions for a variety of electricity-generating technologies in Europe: the 2015 average performance figures for UK plant are for direct CO_2 emissions alone (sources: EEA, 2011; Fritsche and Rausch, 2009; BEIS, 2016b)

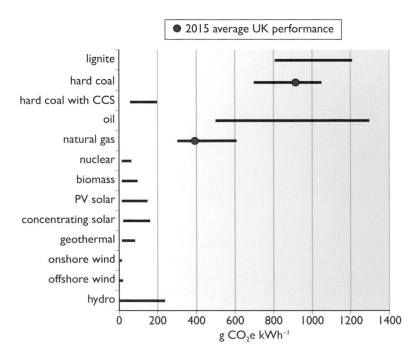

All electricity-generating technologies involve some emission of CO_2 and other greenhouse gases over their full life cycle. This is likely to be true even for future fossil-fuelled plants using **carbon capture and storage (CCS)**, where most of the CO_2 (possibly 90%) is separated and sent for sequestration in deep underground aquifers – a topic that will be described further in Section 11.8. However the typical emission levels for renewables are a fraction of those for conventional fossil-fuelled generation.

It is sometimes argued that the CO_2 savings from renewable energy should be compared with coal-fired generation, since this is a form of generation that needs to be removed as rapidly as possible. For a country such as China, where nearly 80% of electricity is generated using coal, this is a reasonable argument. However in most UK and European assessments the savings are likely to be compared with gas, which is the fossil fuel of choice for electricity generators.

Any economic comparison of rival technologies should include 'external costs', i.e. the cost of the damage done by the pollution emitted to the environment. Much effort has been put into trying to calculate a **carbon price**, a figure for the global damage done by the emission of a tonne of CO_2 (or any other greenhouse gas). It is likely that this lies somewhere between £30 t CO_2^{-1} and £60 t CO_2^{-1} (see Everett et al., 2012). Table 11.1 includes values for the external costs of the electricity using a sample price of £50 t CO_2^{-1}. Given that in 2015 large UK industrial consumers paid about 5.6 p kWh^{-1} for their electricity the inclusion of this external cost represents a significant increase.

Finally, greenhouse gases are not the only pollutants emitted. As described in Chapter 5, Table 5.8, coal-fired electricity generation may involve the emission of large amounts of sulfur dioxide (SO_2), a major contributor to acid rain, and particulates, which are a major cause of the current smogs in China. Modern coal-fired plants can be equipped with flue gas desulfurization (FGD) equipment to cut SO_2 emissions and precipitators and filters to reduce particulate emissions. All forms of combustion (including that of biomass) can potentially produce oxides of nitrogen (NO_x), which can also contribute to acid rain, although careful burner design can minimize their production.

Thus overall, renewable energy technologies for electricity generation have a role to play in reducing global CO_2 emissions, acid rain emissions and harmful particulates.

Heat and transport fuels

Fuels may also be used directly to generate heat or in transport applications. As explained in Chapter 2, different fuels have different carbon dioxide emission factors. The figures given in Table 2.1 are the 'direct' CO_2 emission factors for the CO_2 produced at the point of combustion. However, there are also extra emissions of CO_2 and other greenhouse gases involved in the extraction, transport and refining of these fuels. These are known as 'energy indirect'. For some fuels, such as transport biofuels, these indirect emissions can be quite significant.

Table 11.2 gives values for their emission factors per kWh of heat produced.

Table 11.2 CO_2 and greenhouse gas emission factors for different heating and transport fuels based on higher heating value (HHV)

Fuel	CO_2 emission factor /kg CO_2 kWh^{-1}	Direct and indirect greenhouse gas emission factor /kg CO_2e kWh^{-1}
Domestic coal	0.31	0.39
Natural gas	0.18	0.21
Light heating oil / diesel fuel (100% mineral)	0.25	0.30
Petrol (100% mineral)	0.24	0.29
Bioethanol	-	0.07
Biodiesel	-	0.12
Wood Pellets	-	0.05

Source: BEIS, 2016c

Where heat is generated from fossil fuels, the conversion efficiencies are much higher than for electricity generation, typically 85% for modern gas or oil-fired condensing boilers. A kilowatt-hour of useful heat from a gas-fired boiler thus involves greenhouse gas emissions of about 0.25 kg CO_2e, while that from an oil-fired boiler is about 0.36 kg CO_2e.

Ideally, sustainably grown biomass should not involve significant greenhouse gas emissions. However, fossil fuel energy use and greenhouse gas emissions are normally involved in fertilizer production, planting, harvesting, processing and transport. Thus the overall greenhouse gas emission factor for wood pellets is assumed to be about 0.05 kg CO_2e per kWh. Heating a home by burning wood in a well-designed stove with an efficiency of 70% is still likely to involve the emission of about 0.07 kg CO_2e per kWh in greenhouse gases.

The picture for transport fuels is more complex. Conventional fossil fuel-based petrol (gasoline) and road diesel fuel (DERV) produce CO_2 when burned, but further greenhouse gas emissions (about 20%) are involved, particularly in refining. Overall, it is estimated that burning a litre of 100% fossil fuel-based petrol involves the (direct and indirect) emission of nearly 2.8 kg CO_2e of greenhouse gases. For a litre of DERV the figure is about 3.2 kg CO_2e.

UK petrol is currently sold blended with up to 5% ethanol, which may come from a range of sources as described in Chapter 5. Ideally this should be a zero-carbon fuel, but in practice there are again emissions involved in production. The life-cycle greenhouse gas emissions from combustion of a litre of bioethanol may only be 25% of that of conventional petrol. Similarly, the combustion of a litre of biodiesel, produced from rape seed oil, is likely to involve 30–50% of the emissions of conventional diesel fuel. These figures are obviously highly dependent on the precise production method used for the biofuel.

11.4 **How much renewable energy is available?**

Let us now consider what contribution could be made by renewable energy, concentrating on the UK. It is easy to make sweeping statements about the very large natural energy flows available. These form the 'total, theoretical or available resource'. Defining how much energy a particular source or a specific technology might actually supply in practice requires looking at the various constraints – technical, social and economic – on its use. These whittle down the 'resource' or 'potential' towards something more realistic. A possible ranking of such constraints is given in Box 11.1. The 'economic potential' at the end is likely to be far smaller than any notion of a total available resource (see Figure 11.7).

BOX 11.1 **Resource size terminology**

Fossil fuel *resources* are stores of energy, potentially available for use once extracted from the ground. These are large but finite and can only be consumed once. Fossil fuel *reserves* are more limited; these consist of those resources that are known *and* currently considered economic to extract within a whole range of practical, legal and environmental constraints.

In contrast, renewable energy resources are potential *flows* of energy, analogous to the *rate of consumption* of fossil fuels. It is customary to specify their potential *annual contributions*, which for sustainable sources should be available repeatably year after year. It is important to bear in mind this distinction when comparing the potential contributions from fossil and renewable sources.

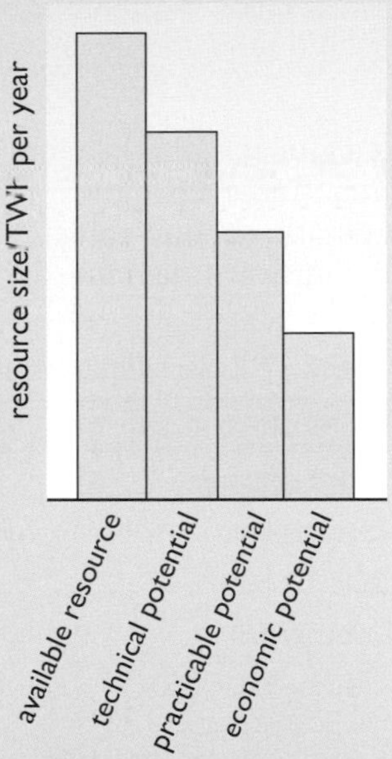

There are many definitions of the annual resource or potential of specific forms of renewable energy, differing mainly in the extent to which the definitions take into account the practical and economic limitations on its use. The following terms have all featured in recent UK energy literature.

The **available, theoretical or total resource** – the total annual energy delivered by the source; for example the total energy carried by ocean waves or the wind, or the total incident solar energy.

The **technical potential** (also referred to as the **accessible resource**) – the maximum annual energy that could be extracted from the accessible part of the available resource using *current mature technology*. Although this could change with technological advances, it may ultimately be

Figure 11.7 A notional ranking of renewable energy resource sizes – the 'technical', 'practicable' or 'economic' potentials may be considerably less than the 'available' resource.

limited by basic laws of physics (determining for instance the properties of wind turbines or the efficiency of heat engines). It is also limited by basic accessibility constraints due to:

- practical difficulties such as the presence of roads, buildings and lakes

- institutional restrictions and the need to avoid such areas as National Parks, Sites of Special Scientific Interest (SSSIs), Areas of Outstanding Natural Beauty (AONBs), etc.

The **practicable potential** (or *practical resource*) – the technical potential, reduced by taking into account:

- constraints on using or distributing the energy – such as transportation problems, access to the electricity grid or problems of a variable supply

- further limitations on land or technology use due to public acceptability. It may be difficult to quantify these since they may only become apparent when planning permission is sought and environmental objections are expressed.

The **economic potential** – the amount of the technical potential that is economically viable. Any judgement about this requires the specification of an acceptable energy cost and a discount rate that sets the cost of borrowing money for investment. (See Section 11.5 and Appendix B).

Figure 11.7 gives a notional ranking of the different sizes of the potentials.

In 2016 the UK's electricity demand was about 300 TWh. Taking, for example, onshore wind, the theoretical resource is estimated at about 4000 TWh y^{-1}, i.e. over ten times the electricity demand. The technical potential is estimated to be about 660 TWh y^{-1} and the practical resource about 80 TWh y^{-1} (CCC, 2011a). So far only about a quarter of this resource has been developed; in 2016 onshore wind generated 21 TWh (BEIS, 2017).

These kinds of estimates, together with assumptions about economic conditions, underlie the various energy scenarios described in Chapter 12.

At a global scale the technical potential for renewable energy is extremely large. Table 11.2 gives ranges of estimates for some different electricity-generating renewables. Actual global electricity generation in 2016 from all sources, renewable and non-renewable, was about 25 000 TWh.

Table 11.3 Global technical potential for some electricity-generating renewables

Technology	Technical potential / TWh y^{-1}
Hydroelectricity	~14 000
Wind power	24 000–160 000
Ocean energy	2000–90 000
Geothermal	33 000–310 000

Source: IPCC, 2011

As described in Chapter 1, global primary energy demand in 2014 was about 575 EJ. Estimates of the technical potential for direct *solar* energy range between 1575 and 50 000 EJ y^{-1} (IPCC, 2011). Estimates for the global technical potential for *bioenergy* range to over 1500 EJ y^{-1}, (see Chapter 5, Table 5.11). As described in Chapter 5, there are potential conflicts between land use for growing food and for energy crops, which may limit the future contribution of bioenergy to below 200 EJ y^{-1}.

Note: 1 EJ = 278 TWh

Renewable potential for the UK

Detailed assessments of the future UK potential for different renewable energy technologies have been made over the past 20 years, concentrating particularly on electricity-generating sources. The rapid decarbonization of electricity supply is considered a major route to cutting UK greenhouse gas emissions by 80% by 2050. As pointed out in Box 11.1, the technical resource for onshore wind alone is about twice the 2016 UK electricity demand. However, for the immediate future what is required are estimates of the practical and economic resources. Table 11.4 shows estimates of the UK practical resource for some electricity-generating renewables with some comments on their limitations.

Table 11.4 Estimates of UK practical resource for some electricity-generating renewables

Technology	Practical resource / TWh y^{-1}	Limitations
Solar PV	140	Variability. Costs, though these have been falling rapidly
Hydro	8	Few available sites for reservoir hydro and limited resource for run of-river plants
Tidal range (barrage)	~40	Costs and environmental acceptability. A Severn Barrage would make up half of the resource
Tidal stream	18–200	Costs. Still at an early stage of development
Onshore wind	80	Variability. Public acceptability
Offshore wind	>400	Variability. Costs are now becoming competitive with fossil-fuelled generation.
Wave	~40	Variability. Costs. Still at an early stage of development
Deep geothermal	35	A longer term possibility still not demonstrated in the UK

Source: CCC, 2011a

In comparison, the UK potential for domestically-sourced biomass may be more limited. Again, there are potential conflicts between the need for land to grow food, wood for construction purposes, biofuels for transport and biomass for heat and electricity generation. In its 2011 report (CCC, 2011b), the UK Committee on Climate Change suggested a 2050 bioenergy potential of 400–750 PJ (about 5–9% of UK 2015 primary energy consumption). They recommended that the use of bioenergy for non-CHP electricity generation should not be promoted, given the many alternatives listed above. More recent studies, described in Chapter 5, have suggested that bioenergy may supply 10–12% of future UK energy needs.

11.5 Is renewable energy available at an acceptable financial cost?

As pointed out in the previous chapters, the costs of electricity from technologies such as PV and wind have been falling. Some renewable energy technologies are undoubtedly cheaper than others; and their output has to be compared with the costs of rival options. For electricity-generating renewables this includes other low-carbon technologies such as nuclear power and fossil-fuelled generation using carbon capture and storage (CCS). The following section examines the estimated cost for some UK electricity-generating technologies, but is of wider applicability. The methods used to calculate the 'levelized cost of electricity (LCOE)' over the lifetime of a plant are described in Appendix B and have been used in examples in previous chapters.

Electricity-generating renewables in the UK

As noted above, in 2015 large industrial electricity consumers in the UK paid about 5.6 p kWh^{-1} for their electricity. It has been estimated that electricity from a new base load (i.e. almost continuously operating), gas-fuelled, high efficiency H class CCGT plant, to be built in 2020, would cost 6.6 p kWh^{-1}. This assumes a 7.8% discount rate on the capital repayments and includes a 1.9 p kWh^{-1} carbon price (as in Table 11.1).

Figure 11.8 shows estimated levelized costs of a range of electricity generating renewables. These are compared with the costs of electricity from this gas CCGT plant and two low-carbon alternatives, a nuclear power plant and a coal-fuelled integrated gasification combined cycle (IGCC) plant with carbon capture and storage (CCS).

As mentioned earlier, any fair comparison of fossil-fuelled and low carbon technologies should take into account a carbon price. The values used in the calculations for Figure 11.8 are on a sliding scale – rising in future years – but are equivalent to an average rate of about £50 per tonne of CO_2 emitted. The costs of coal generation with CCS still include a small carbon price because it is assumed that only 90% of the CO_2 can be captured. It also includes extra costs for the transport and disposal of the captured CO_2, probably in saline aquifers deep underground, most likely beneath the North Sea.

As described earlier, many of the other technologies involve low levels of emissions of greenhouse gases, but their carbon costs are not included in Figure 11.8.

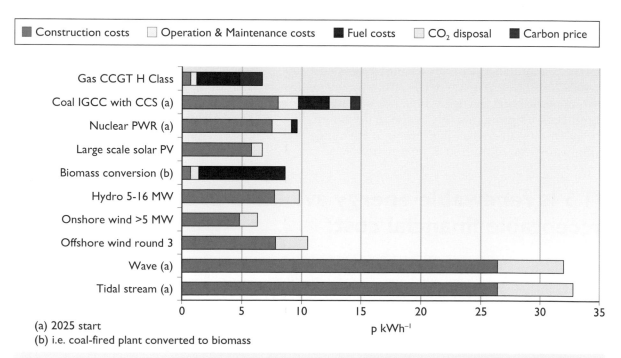

(a) 2025 start
(b) i.e. coal-fired plant converted to biomass

Figure 11.8 Cost comparison of conventional and low-carbon electricity-generation technologies assuming a 2020 start, except where noted. The assessment uses different 'hurdle rate' discount rates for each technology depending on their perceived risk. Costs are in 2014 pence per kWh (source: BEIS, 2016d)

Prices, investment risk and hurdle discount rates

It is obvious from Figure 11.8 that some technologies are currently far more expensive than others, yet they are supported because they may become cheaper in the future. The biomass example is an existing conventional steam power plant converted to be fuelled by wood. Like the CCGT plant this has relatively low capital costs.

The other renewable energy technologies shown are all capital-intensive. These technologies are all assumed to have reasonably long lifetimes of 15 years or more. Their fuel costs are zero and the bulk of the electricity costs consist of the repayments on the initial capital outlay. The electricity price is thus heavily influenced by the capital cost per kW of installed capacity and the discount rate applied to the finance (as explained in Appendix B, this is the 'real', inflation corrected, interest rate charged). The importance of discount rate has been discussed in relation to hydro schemes in Chapter 6, Box 6.9, and in relation to a proposed Severn Barrage in Chapter 7. In practical projects the discount rate is likely to be determined by the perceived investment risk for private investors and the possibility of poor performance or even complete failure.

The analysis for Figure 11.8 uses 'hurdle rates': the estimated minimum rate of financial return at which investors might commit capital to a project (NERA Consulting, 2015). Some technologies, such as onshore wind, PV, and hydro have been treated as 'mature technologies'. They are considered 'low risk' investments and finance may be available at a discount rate of under 8%. Although large-scale PV has been more expensive than conventional gas CCGT generation in the past, by 2020 it is expected to be highly competitive as production methods continue to improve and the scale of production continues to increase.

Offshore wind has been considered a slightly more risky investment, with finance likely to be available at a discount rate of about 9%. Again capital costs are high (typically twice those per kW of capacity of onshore wind), but are falling with increasing volume production and installation experience.

There are then 'new' technologies such as tidal stream and wave power that are still only at the prototype stage. Future capital and installation costs can only be roughly estimated and will depend on future production volumes. A 'high risk' discount rate of over 10% has been assumed for investment for these, giving a high electricity cost.

The key lessons from the past deployment of onshore (and offshore) wind and PV are:

- Volume production and deployment produces reductions in capital and installation costs.
- Demonstrated successful deployment reduces the perceived financial risks and makes capital investment available at lower discount rates.

The result has been reduced levelized electricity costs that are now becoming increasingly competitive with conventional fossil-fuelled electricity generation.

Tidal barrages and lagoons, described in Chapter 7, have not been included in the assessment shown in Figure 11.8. Although these use conventional hydro turbine technology, they are, at the moment, 'first of a kind' projects.

In an earlier assessment they were considered to be a risky investment with a discount rate of 15%, resulting in an electricity price of over 50p kWh^{-1} (Mott MacDonald, 2011). If this perceived risk could be reduced, then a discount rate of 10% would reduce the electricity price to about 30 p kWh^{-1}. It has been argued that capital-intensive projects with significant prospects of reducing national CO_2 emissions should be funded with (government) loans at a very low 'social' discount rate of 3.5%. This would dramatically reduce the electricity price to only about 11 p kWh^{-1}. Such low interest finance would, of course, also reduce the costs of rival low-carbon technologies, particularly nuclear power (see CCC, 2011a).

Heat-generating renewables in the UK

As pointed out in Chapter 1, over a third of the end-use of fuel in the UK is for low-temperature heating. Domestic space heating alone accounts for about 17% of UK primary energy consumption and domestic water heating a further 5% (BEIS, 2016e). As described in Chapter 3, insulation and similar energy-conserving measures can result in dramatic reductions in space heating demand, even in existing buildings.

At present, natural gas is the prime UK heating fuel. However, it is not clear what role it will play in the future as fossil fuel use is reduced. If large amounts of electricity from low-carbon sources can be made available (particularly in winter) at an acceptable price from renewable energy and nuclear power, then heat pumps may be a preferred alternative. Biomass may also be a useful heating fuel, particularly in rural areas beyond the gas grid, and solar water heating is another option. Table 11.4 shows some levelized cost estimates for alternative domestic heating options, taking into account the capital costs of the heating system. The higher figures of the ranges are for well-insulated houses with low energy demands.

Table 11.5 Levelized costs of heat from different sources for domestic use

Heating type	Cost / p kWh^{-1}
Natural gas boiler	6.8–10.5
Air-source heat pump	10.9–16.5
Ground-source heat pump	12.8–18.6
Biomass boiler	15.3–23.7
Solar thermal water heating	26.6

Note: Based on 2011 costs, 15 year equipment life and 8% discount rate
Source: CCC, 2011a

Balancing renewables and energy efficiency

Considering the balance between investment in improved energy efficiency and investment in low-carbon energy supply is very important. Conservation and efficiency measures are usually more cost-effective than many low carbon supply options, and they narrow the gap between energy demand and the available supply of renewable energy.

The technology comparison shown in Figure 11.8 above is in terms of energy price, including a notional carbon price. Another way of making a comparison is to look at the different costs of saving a tonne of CO_2 using different technologies. These can be set out as a **marginal abatement cost curve (MACC)**. Figure 11.9 shows an example of the potential savings for the UK domestic or residential sector in 2020 (CCC, 2008). It assumes that many basic measures such as improved loft insulation have already been taken up by then. The technologies are shown ranked from the lowest cost on the left to the most expensive on the right.

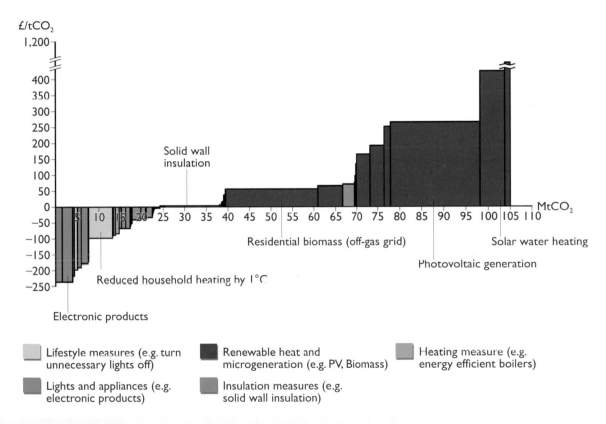

Figure 11.9 Marginal abatement cost curve (MACC) for different technologies in the UK domestic sector for 2020. UK total CO_2 emissions for 2015 were just over almost 400 Mt (source: CCC, 2008)

There is a wide range of options. Many efficiency options have an overall negative cost. Replacing refrigerators with more efficient models reduces electricity demand. Insulating solid walls in millions of homes can reduce the national heat demand. Both of these are likely to have a lower cost than renewable energy supply options such as photovoltaic (PV) electricity or solar water heating. This might be taken as a reason to delay the deployment of renewables until a full programme of energy-efficiency retrofits has been carried out. In practice, achieving the necessary rate of reduction in national CO_2 emissions means that all options need to be pursued simultaneously.

World-wide renewable electricity prices

The UK is a country with good wind, wave and tidal resources. However, it cannot be described as particularly sunny, nor does it have a well-developed geothermal resource.

Figure 11.10 shows the changes in levelized costs of some renewable electricity generating costs on a world-wide basis between the end of 2009 and 2016, expressed in US cents per kWh. At the time of writing (early 2017), US\$1 = £0.80, so a UK large industrial consumer electricity price of 5.6 p kWh^{-1} ≈ 7 US cents per kWh.

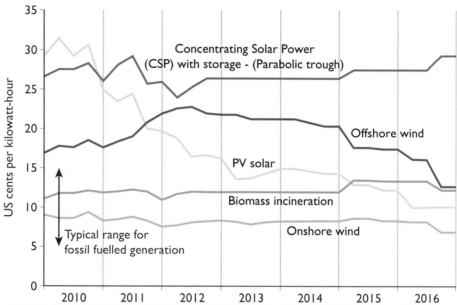

Figure 11.10 World-wide levelized costs for selected renewable energy sources between the end of 2009 to end of 2016, together with estimated fossil fuel generation costs (sources: UNEP/Bloomberg New Energy Finance, 2017; IRENA/IEA, 2017)

This shows that, in the right locations, biomass sourced electricity and onshore wind have been competitive with fossil-fuelled generation since 2009. Also, cost reductions in PV and offshore wind have now brought them into the fossil-fuelled cost range.

However, wind and PV have disadvantages of variability. Biomass is capable of producing 'firm' continuous generation, but so is Concentrating Solar Power (CSP) with molten salt storage (see Chapter 3). The costs for this are still US¢ 25–30 per kWh, but this could fall with further development.

Many countries (including the UK) are now operating auctions for new renewable energy generation, mostly large projects greater than 100 MW capacity. In 2016, these produced extraordinarily low bid prices. For example, under 3 US cents kWh^{-1} for PV projects in the Middle Eastern states of Dubai and Abu Dhabi, and for one in a very sunny part of Chile in South America. Even in Germany and the Netherlands solar PV bids have been in the range 8–9 US cents kWh^{-1}. A new 600 MW offshore wind farm in Denmark has also received an auction bid price of 5.5 US cents kWh^{-1} (although this does not include grid connection costs to the land). These low bid prices reflect not only the ability to produce large renewable energy projects at a low capital cost, but also the availability of finance (in sums of hundreds of millions of pounds) from investors who see these as 'low-risk' projects.

II.6 Are renewable energy supplies available *where* we want them?

So the potential contribution of renewables to UK (and global) energy needs could be very large. Let us take our first practical constraint – can renewables supply our energy *where* we want it?

It would be helpful if renewable supplies were located near the points of maximum energy demand in the major cities. Some can be as solar thermal or photovoltaic panels can be fitted to the roofs of buildings.

Energy crops are likely to be available in rural areas, but perhaps not those where high value food crop production is seen as more important. Fuels such as wood and forestry wastes need to be gathered and then transported from where they are grown to local or national centres of demand. This may involve some pre-processing such as the conversion of wood or straw into pellets with a higher energy density. As has been described in Chapter 5, biomass in the form of wood has only half the energy density of coal, and only about a third of that of fuel oil, so transporting it may be more expensive. Careful thought needs to be given to the forms of transport used to minimize the amounts of energy and CO_2 emissions involved.

Municipal solid wastes (MSW), forestry and agricultural wastes are widely used as fuels. In combined heat and power (CHP) plants they are used to make both electricity and heat. In district or community heating plants, they are just used for heat production alone.

In all these cases, the heat can be distributed through city-wide heating networks, as is commonplace in Denmark. Such networks can also be used to distribute geothermal heat where it is accessible, such as in Southampton in the UK.

Although wind power can be used at a local scale, its large-scale use in the UK (and most of north-west Europe) involves the use of large turbines situated at prime onshore and offshore sites with high wind speeds. Similarly prime sites for wave, large hydro and tidal energy are likely to be relatively remote from the main centres of demand. Integrating large-scale renewable electricity supplies into the existing distribution systems needs careful planning, investment and probably some further changes to market structures.

The electricity grid in the British Isles

In the UK and Ireland, new electricity-generating renewables feed into the existing National Grid structures that have been built up over the past 80 years (see Figure 11.3). In England and Wales, the National Grid was developed to distribute electricity from the large coal-fired stations sited close to the main mines in Yorkshire and Wales to the main load centres in the major cities, which have become increasingly concentrated in the south of England. Extra grid links have been added from coastal nuclear power stations. There are also undersea links to France and the Netherlands. A direct link from England to Ireland was also completed in 2012.

The electricity system of Scotland has been relatively self-contained, with the demand being met by a mix of gas, nuclear and hydroelectric plants. It is connected to England by two grid links on land. Environmental objections to any expansion of these links has led to the laying of an offshore cable (shown in Figure 11.3) through the Irish Sea between Cheshire and south-west Scotland. The Scottish grid is also linked directly to that of Northern Ireland. This, in turn, is linked to that of the Republic of Ireland where Dublin, on the east coast, forms a major load, in part supplied by large power stations sited on the west coast running on imported coal (which may be replaced by new power plants running on imported liquefied natural gas).

Wave, wind and tidal power

Figure 11.11 summarizes some of the information from earlier chapters on the locations of wind, wave and tidal energy sources in the British Isles. As can be seen, these electricity-generating sources are widely distributed. Both the UK and the Irish Republic have very significant potential for renewable energy supplies. Comparison of this map with Figure 11.3 raises the question of how the existing national grids will need to evolve in order to match the new energy sources to the loads.

The best *wave* power resources are in Scotland, the Atlantic coast of Ireland and the south-west of England. Current prototype development is concentrated in Scotland.

The best *onshore wind* resources are in Scotland, Wales, Cornwall, the north and west of England and the west of Ireland. Projects in all of these areas are currently under rapid development.

The prime areas for *offshore wind* development are the shallow waters of the North Sea and Irish Sea. As shown in Figure 11.11, it would only require an area of sea 30 km by 40 km to supply 10% of the UK's electricity needs. To date, offshore wind farm development has been close inshore. However, future UK large-scale development is likely to take place further out to sea. As has been described in Chapter 8, several 'Round 3' zones have been specified for this and the development of floating wind turbines may allow their siting in a wider range of offshore locations.

The potential for *tidal barrages* is concentrated on a few large estuaries, particularly the Severn and the Solway Firth, though 'lagoon' type structures could be built in more open sea areas. The potential for *tidal stream* devices is in similar estuary locations, but there are many possibilities around prominent headlands, notably the Pentland Firth between Orkney and the Scottish mainland, and the 'Race of Alderney' between Alderney in the Channel Islands and Cap la Hague on the French coast.

As Figure 11.3 shows, the UK and Irish grids mostly do not run in a concentrated 'point to point' manner. They are built as a 'mesh' and some lines are apparently duplicated. This is deliberate to allow the system to continue operating despite the failure of any one given line. This gives flexibility of operation and potentially allows the insertion of extra sources into the existing system. The National Grid is also a key to dealing with the time variability of some renewable energy supplies, as discussed below.

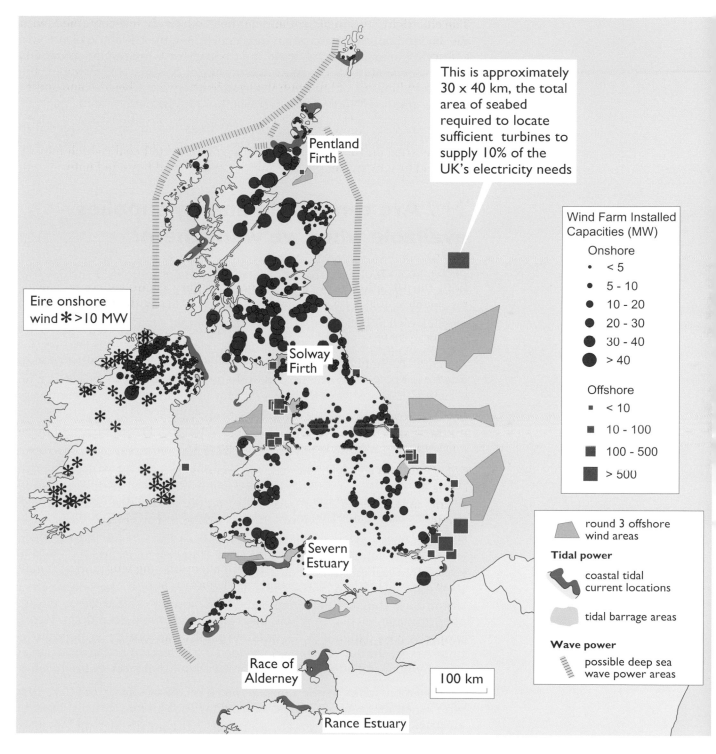

This is approximately 30 x 40 km, the total area of seabed required to locate sufficient turbines to supply 10% of the UK's electricity needs

Pentland Firth

Eire onshore wind ✳ >10 MW

Solway Firth

Severn Estuary

Race of Alderney

Rance Estuary

100 km

Wind Farm Installed Capacities (MW)

Onshore
- • < 5
- • 5 - 10
- • 10 - 20
- • 20 - 30
- • 30 - 40
- • > 40

Offshore
- ■ < 10
- ■ 10 - 100
- ■ 100 - 500
- ■ > 500

round 3 offshore wind areas

Tidal power

coastal tidal current locations

tidal barrage areas

Wave power

possible deep sea wave power areas

A pan-European view

There are other pressures on the development of 'national' grids. The European Commission would like to develop the pan-European electricity market. From this trading perspective the existing link from France to England and the 500 MW link from Scotland to Northern Ireland have been

Figure 11.11 Geographical distributions of onshore and offshore wind farms, and possible future areas for development of offshore wind farms, wave and tidal energy projects in the British Isles (sources: BEIS, 2016b; SEAI, 2007; Offshore Valuation Group, 2010; BERR, 2008; Energia, 2017)

identified as 'bottlenecks', limiting international trade in electricity. Where electricity from renewable sources is concerned, the UK and Ireland have plentiful resources of wind, wave and tidal power, which could potentially be complemented by the hydro resources of Norway and Sweden, the biomass resources of central Europe and the solar resources of southern Europe and even North Africa (see Section 11.8 below).

The creation of good international grid links is considered essential to prevent what have been called 'congested grid corridors' in the flow of electricity from renewable sources.

11.7 Are renewable energy supplies available *when* we want them?

Our demand for energy is not constant. It varies widely over the day, the week and the year. In the UK, we need more energy for heating buildings in winter than in summer: the UK consumes over twice as much natural gas in a typical December as it does in a summer month.

Generally, in the 'developed' countries, there are few constraints on our demand for energy. Our electricity supply systems are organized in such a way that power is virtually certain to be available whenever we turn on a switch. Gas is always there waiting for us to turn on the cooker. Every major highway has petrol stations at regular intervals ready to serve us when we drive in.

Complex infrastructures have been put into place to enable supply to meet demand. A greater need for, say, gas will set off a whole series of pumps distributing supplies from storage facilities through a nationwide set of mains. Any failure of the infrastructure to deliver energy on demand – be it gas, electricity, heating oil or petrol – causes a consumer outcry.

Biomass fuels have many of the advantages of fossil fuels. Most of them can be easily stored and used on demand. However, most other renewable sources are variable in their output, although they may, to a certain extent, be predictable. Wind, wave and solar energy are dependent on weather conditions. They are not completely 'firm' supplies – that is, we cannot always guarantee that they will be available when we need them. Hydro power is also dependent on weather conditions, but its availability is improved by the built-in storage provided by the reservoirs that form part of most hydro installations. Tidal power is variable, but entirely predictable.

This section looks at the consequences of these characteristics, first with renewable sources of heat and then with the more difficult topic of electricity.

Renewables as heat suppliers

Heat is a relatively easy topic to consider. Individual solar water heaters can be mounted on the roofs of houses or other buildings. They effectively reduce the demand for other forms of water heating and do so mainly in the summer.

As pointed out above, a wide range of biomass fuels including household and agricultural wastes, together with fossil fuel sources, can be used to produce heat for district or community heating networks. Such systems also allow the input of heat from solar sources, heat pumps, geothermal aquifers, or even surplus wind power. Most heat stores associated with community heating schemes, such as that shown in Figure 11.12, hold only sufficient supplies for a day or so. Using them frees the system designers from concerns of how the heat demand of the local buildings varies over the day. 'Inter-seasonal stores', which store heat from summer through to winter, are being developed for use with large solar water heating schemes, as described in Chapter 3. However, these do have to be physically very large to minimize the ratio of surface area to volume.

Municipal solid waste is not a fuel that is desirable to store for long periods. Whatever the means used to turn it into heat and electricity – mass burning, gasification or pyrolysis – it is best to do so steadily year-round. But other biomass fuels can be stored and used to meet winter space heating demands.

Experience in many countries has shown that it is economically viable to insulate existing buildings to levels where their space heating demands are dramatically reduced. They can then be heated easily with relatively low-temperature district heating systems and the problem of supplying winter demand is not so difficult.

Figure 11.12 The Pimlico Accumulator in central London holds enough hot water to supply over 3000 homes for a day

Integrating electricity from renewables

At present (2017), integrating electricity from renewable sources remains a matter of some concern. Meeting the UK government's current aim for over 30% of electricity supplies to come from renewable sources by 2020 requires the connection of up to 50 GW of renewable electricity generation capacity.

To understand the problems involved in coping with variable inputs of such sizes it is worth describing the existing system in some detail. As will be shown, to a large extent the practical resource limitations of *where* and *when* overlap.

How the current UK electricity system works

Electricity is a much more flexible and valuable commodity than heat, but it is difficult to store. Essentially, it has to be generated immediately to suit the demand. In the UK, as in many other countries, electricity demand varies enormously over the year, and hour by hour throughout the day.

Figure 11.13 shows sample national daily demand patterns for summer and winter.

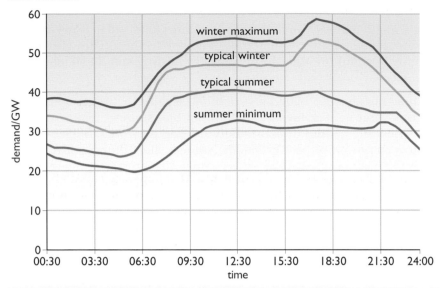

Figure 11.13 UK summer and winter daily electricity demand profiles in 2010/2011 (source: National Grid, 2011)

At night, demand is relatively low (in summer below 25 GW), but it picks up very rapidly early in the morning and flattens out during the day. In winter months there is a pronounced peak (sometimes reaching over 55 GW) in the early evening.

We can look at electricity demand on such a large aggregated scale because of the existence of the National Grid and regional distribution networks. Although individual consumers use electricity in an intermittent manner, turning on a light here and a heater there, they each usually do so at slightly differing times. This averages out to a smoothly varying electricity demand that the grid system can cope with relatively easily. This effect is known as **diversity of demand**. There are exceptions when consumer behaviour becomes synchronized, which will be discussed later.

The history of the UK National Grid's development has been described in some detail in Everett et al., 2012. When its construction was initiated in the 1930s, electricity was generated locally in hundreds of small separate power stations with low average generation efficiencies. Initially the grid linked together the most efficient power stations within regions of the UK, allowing mutual backup and coverage for peak demand. These all had to operate using alternating current (AC) at a common frequency, chosen to be 50 cycles per second (50 Hz). Hundreds of small, inefficient, local coal-fired stations were closed down.

After experiments in 1937, it was found that it was safe to connect together all the regional systems of England, Scotland and Wales into a single national network. The output of individual stations was controlled centrally in order to optimize the overall system performance. It has largely been run as such since 1938.

At that time, many of the power stations and local distribution networks were privately owned. Although the UK industry was nationalized in 1947 and then privatized in 1989, this has not affected the national nature of the distribution system.

The grid has been repeatedly strengthened and reinforced since 1934. The main links now operate at 275 kV and 400 kV with lines capable of carrying 2 GW or more. The overall philosophy has been one of large generating plants, justified on the basis of economies of scale, centralized control, and distribution outwards and downwards in voltage to the consumer (i.e. from left to right in Figure 11.14). At the lowest voltage, the system consists of buried cables in the streets capable of supplying both 400 V, 3-phase AC and 230 V, single-phase AC (the norm for domestic consumers).

This is a picture of 'centralized generation', yet increasingly there is a trend towards **embedded generation** – smaller generators situated more locally to the loads. Solar PV panels or small CHP units may be installed in houses, schools or local leisure centres, connected to the grid at the 400/230 V level. Individual onshore wind turbines of around 1 MW output and industrial-scale CHP units are likely to be connected at the 11 000 V level. Larger offshore wind farms may have high-voltage links running ashore at up to 150 000 V. Future Round 3 offshore wind farms may be connected into international grid link cables.

In practice, the grid provides enormous flexibility of operation. The demands of Oxford can be met from a power station in Yorkshire or a storage plant in Wales. Across Europe, the electricity systems operate on the common frequency of 50 Hz. Because they are linked across national borders, power plants are able to share loads or export surpluses of electricity.

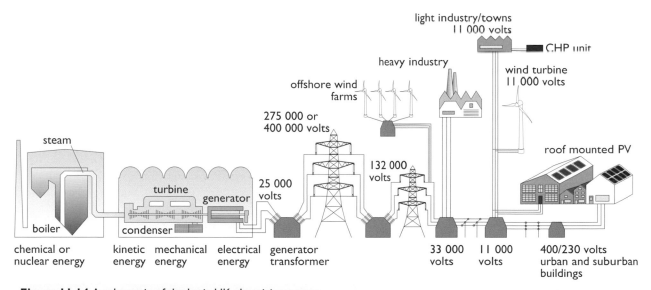

Figure 11.14 A schematic of the basic UK electricity system

Matching supply and demand

The high-voltage electricity grid in England, Wales and Scotland is centrally controlled by a **transmission system operator**, the National Grid Company. The total demand pattern is met by a mixture of generating plants, owned by different companies, which compete in a centrally controlled **power pool**, a competitive market in electricity. At any given time a computer model run by the system operator estimates what the national demand will be in the following few hours and invites bids for the supply of electricity. Power station owners reply (or rather their computer programs do), the cheapest offers are accepted and the system is adjusted to bring the appropriate stations online by remote control. This task is known as **scheduling** and power plants whose output can be varied by remote control are described as **dispatchable**. The stations with the lowest running costs supply the **base load** (a term that usually means long-term, continuous generation). The lowest cost sources are likely to be renewable generators, such as wind and solar power, which have zero fuel costs and nuclear, which has low fuel costs. Above them there is competition between coal-, biomass-, and gas-fuelled power stations. Although hydro power has zero fuel costs its high flexibility means that it is normally (in the UK at least) kept for supplying peak demands. At any time the competitive price is chosen to be sufficient to bring enough stations online to meet demand. The prices at times of peak demand can be very high indeed.

The bidding process only applies to generators above a certain size. Small generators below 1 MW (including most small renewable electricity generators) are not normally constrained as to when they run, but are only paid a fixed rate for their electricity rather than a 'system price'.

The ability of different plants to provide for a changing load depends on their scale. A large coal-fired or nuclear power station may take 24 hours to reach full output from cold. A normal CCGT station may be able to produce some power within an hour but could take several hours to reach full output. Following a changing demand can be achieved by running stations at part load, but this reduces their overall efficiency (and increases their CO_2 emissions per kilowatt hour of electricity generated). It is best if they are run continuously at close to full output. Nuclear power stations are particularly unsuitable for 'load following'.

However, demand is not conveniently constant. There can be rapid changes in demand for several reasons:

- The daily increase in human activity in offices and factories can increase demand in the morning by 12 GW in the space of two hours.

- Simultaneously showing a popular TV programme over the whole country can produce a synchronization of behaviour of consumers. It can send a large proportion of the population rushing to use their electric kettles at the same time during commercial breaks, leading to a **demand pickup** of over 1 GW in a matter of minutes.

- The sudden breakdown of a large power station or failure of a major transmission line may mean that generating capacity of 600 MW or more could suddenly disappear, again requiring a response in a matter of minutes. The construction of the two 1600 MW nuclear reactors at Hinkley Point in the UK creates its own large grid backup problems.

The flexibility of hydroelectricity can go a long way to meeting these rapid changes in demand. In countries with limited hydro resources the traditional way of dealing with demand changes was to maintain 'spinning reserve' – large coal-fuelled power stations generating electricity at part load, but with sufficient steam reserves to cope with any sudden increase in demand. This role has been continued in the power stations such as Drax in Yorkshire, which have been converted to burn wood. However, running stations in this manner reduces their efficiency and wastes fuel. So, over the years a range of other technologies has been brought into use to cope with this problem (Box 11.2).

BOX 11.2 Matching electricity supply to short-term demand fluctuations

Electricity has to be generated on demand and the voltage and frequency of the AC supply have to be held within relatively tight limits. A range of technologies is used to meet rapid and possibly unexpected increases in demand.

Pumped storage plants

These have been described in Chapter 6. At times of low demand, surplus electricity is used to pump water into high level reservoirs. At times of sudden peak demand, they can use the stored potential energy of the water to generate electricity with as little as 10 seconds' notice. The UK currently has four such plants, two in Wales and two in Scotland. Their combined peak output power is about 2.7 GW, approximately 5% of the UK's typical winter electricity demand. A further 400 MW scheme in Scotland has been approved. Other countries have similar schemes. Typically, such plants have an overall storage efficiency of 70–80%. Although they have a high peak power output, their water storage is usually only sufficient for a few hours operation to act as cover while alternative generation plants are started and connected.

Open cycle gas turbine and diesel peaking plants

These can be run up to full power in half an hour or less. Small 'open cycle' gas turbines (OCGT), lacking the steam stage of the CCGT, run on natural gas or light heating oil and typically supply between 10 MW and 100 MW of electricity. Figure 11.15 shows a plant that houses two 66 MW turbines. Smaller diesel generators of around 1 MW output can be brought online at full power in minutes. The disadvantage of both types is that they consume fossil fuel and are less thermally efficient than larger 'base load' power stations.

Compressed air energy storage (CAES)

In a gas turbine power plant a large amount of energy is used to compress the incoming combustion air before it is burned. A CAES plant uses off peak electricity to compress air to typically 100 atmospheres pressure and store it in an underground cavern.

At times of peak demand, this is fed to a peaking gas turbine, reducing its gas consumption by more than 60%. The first commercial plant was a 290 MW unit built in northern Germany in 1978 using storage in a salt mine. A 110 MW plant opened in Alabama in the USA in 1991 and a 476 MW plant in Texas has an anticipated operation starting date of summer 2020.

Figure 11.15 A 132 MW open cycle gas turbine (OCGT) plant in north London. It has two turbines that run on light heating oil

Rechargeable batteries

Lead acid batteries have been used by electricity utilities for peaking power and emergency backup since the late nineteenth century. They are widely used in association with wind farms to smooth rapid fluctuations in output, as well as for electricity storage for off-grid applications. Lead acid technology is limited by the number of cycles that a battery can be put through before it needs replacing.

Lithium ion batteries have now entered large scale mass production as a power source for electric vehicles and have a much longer life expectancy than lead acid batteries. They are being used in a range of applications including smoothing the output of wind farms, smoothing local peak demands in distribution systems and as emergency backup in the event of grid failure. Large battery packs of 30 MW capacity or more are now being used for grid support in Japan and a 100 MW system has been ordered for use in Australia.

Sodium sulfur batteries use a different battery chemistry with a positive electrode of molten sulfur and a negative electrode of molten sodium. The chemicals combine to produce sodium polysulfides and electricity. The battery has to be kept at 300 °C for the reaction to take place. When the battery is recharged, the elemental sulfur and sodium are regenerated. A number of MW scale plants have been built for electricity utilities in Japan.

Unlike most conventional batteries where the key active chemicals are solid, in a *flow battery*, they are liquids and can be stored in tanks separately from the battery itself. This is similar to the hydrogen fuel cell (see Box 11.3 later), which is a 'gas battery'. A 200 MW flow battery based on the element vanadium is being commissioned for grid stabilization in the Dalian Peninsula in north-east China (UETechnologies, 2016). This potentially has four hours capacity, comparable in performance to a small pumped storage plant.

Other short-term storage systems, such as those using flywheels or producing liquid air, are also under development.

Which is best?

Balancing the electricity grid of an entire country requires large amounts of power and appreciable amounts of stored energy. At present only pumped storage systems have power ratings of over 1 GW and the capacity to supply this for more than an hour or so, but such systems require suitable sites in mountainous regions. Peaking gas turbines can produce 100 MW or more and be installed almost anywhere, but consume fossil fuel in a relatively inefficient manner. CAES systems can supply 100 MW or more, but require special geology suitable for underground high-pressure air storage.

Rechargeable batteries can potentially be installed anywhere, but it is only in the last ten years that they have been available in ratings of 50 MW or more and are typically only used to supply that power for periods of an hour or two. They have the advantage of very rapid response – fractions of seconds rather than minutes. They are more likely to be used to absorb short surges and to correct control instabilities in local distribution systems. They may also act as 'starting batteries' to allow a conventional power station to recover from a grid failure and perform what is known as a **black start**.

Including renewable energy

Where do the renewables fit into all this? Different renewable electricity sources have different characteristics, particularly as regards their variability or intermittency. Strictly speaking, the word 'intermittency' implies sudden changes in output such as in a grid or power station failure. The output of large numbers of dispersed wind turbines or solar PV installations may be better described as 'variable' but possibly with a high (but predictable) rate of change of output, or **ramp rate**.

Although renewable energy statistics often quote the 'installed capacity', it is important to remember that a megawatt of wind power will not produce as much electricity over the year as a MW of 'firm' generation. As described in Chapter 1, Section 1.1, the capacity factor (or load factor) for a particular technology is used to relate annual output to installed capacity.

Let us consider the 'firm' renewables first. These may have practical capacity factors of 80% or more.

Firm renewable electricity sources

Biomass and waste plants

Generation plants using municipal solid waste, wood pellets, waste wood, biogas or landfill gas are usually relatively small, typically in the output range 100 kW to 50 MW. As such, they are likely to be connected to the system at 11 kV or 33 kV. Some, such as municipal waste and landfill gas plants run fairly continuously. Others, such as large wood-fired plants are dispatchable and run when required under the control of the transmission system operator. They have a major role to play in complementing the variable output of other renewable electricity sources.

Some smaller plants are unmanned and run under automatic control, visited by small teams for fuelling and periodic maintenance. Biomass may also be co-fired with coal in large multi-megawatt power stations or combusted in former coal-fired plants converted to run solely on wood pellets, connected to the grid at 400 kV.

Geothermal plants

Although these are not currently used in the UK, geothermal electricity-generation plants in other countries may be of 100 MW or more capacity and their grid connection and operation is the same as that of conventional coal, gas or nuclear plants. Since they have zero fuel costs they are likely to be run almost continuously, providing 'base load' electricity generation.

Hydroelectricity

Hydro power is perhaps the most desirable (and dispatchable) of all renewable electricity sources from the point of view of flexibility of supply. Water can be stored in reservoirs for months or even years, yet the generators can be wound up to full power and turned off again in minutes. Smaller 'run-of-river' plants are likely to operate almost continuously, subject to the availability of sufficient water flow. Larger plants are likely to be used to meet peak grid loads. The long-term average capacity factor for UK hydro plants is 35–40% (BEIS, 2016b). In the UK, most plants are in the range 100 kW to 100 MW and are connected at voltages of 11 kV or above. Elsewhere in Europe, such as in Norway and Sweden, hydro power plays a major role, with the output and, in some cases, the pumped storage capacity sold across international borders.

Although hydropower can be used as a solution to short-term variations in demand, there may be long-term year-to-year variations in its potential, depending on rainfall.

Variable renewables: the bigger picture

The remaining renewable electricity sources – solar, wind, wave and tidal power – can all make time-variable contributions to what is a time-variable demand. It is worth first considering the timing of renewable supply and demand on a European scale.

The Sun and the rotating Earth

The Sun rises in the east and sets in the west. Taking Greenwich Mean Time as a reference, dawn and the start of the working day thus arrives earlier in Germany than in the UK. Since the Earth rotates once every 24 hours, the Sun traverses 15° of longitude every hour. Berlin and Birmingham are approximately at the same latitude but displaced by 15° of longitude (see Figure 11.16). Thus the whole 'working day' in Berlin (with its associated electricity demand pattern) is likely to take place an hour earlier than in the UK.

The Sun obviously directly affects the timing of electricity *demand* for lighting, and also any solar electricity *generation*: the peak output of a south-facing PV panel in Berlin is likely to occur one hour earlier than one sited in Birmingham.

Solar energy is, of course, not uniformly available throughout the day or over the year. There is more solar radiation available in summer than in winter and more in southern Europe than in northern Europe (see Figures 3.7 and 3.8 in Chapter 3).

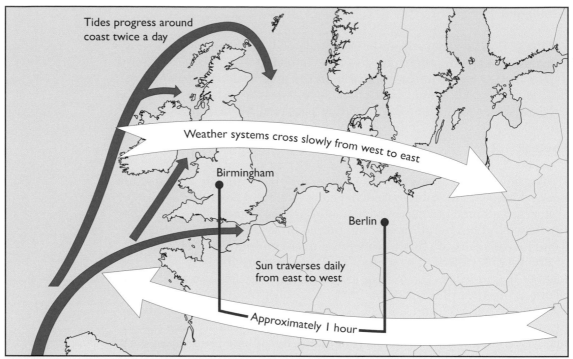

Figure 11.16 Solar, wind and tidal phenomena that influence potential electricity demand and renewable electricity generation across Europe

Wind power

As described in Chapter 8, the bulk of Europe's wind resource is in the North West. It derives from sequences of cyclones or depressions with accompanying strong winds. These weather systems move slowly from the Atlantic across Europe from west to east. They are particularly strong in autumn, winter and spring and typically about 1000 km across, taking a few days to traverse Europe. Thus on one day the wind may be blowing strongly in Ireland while Denmark is becalmed. Two days later the situation may be reversed.

The availability of stronger winds in winter means an increased wind power resource, which in part matches the increased winter electricity demand.

Tidal power

Although we normally think of tides rising and falling at one particular location, they can also be seen as progressing along the Atlantic coastline of Europe twice a day (the Mediterranean only has very small tides). The height of the tides also varies on a long two-week cycle between spring and neap tides.

Variable renewables: the details

Different variable renewables will produce electricity in different patterns. Figure 11.17 illustrates the possible electrical output of a range of them.

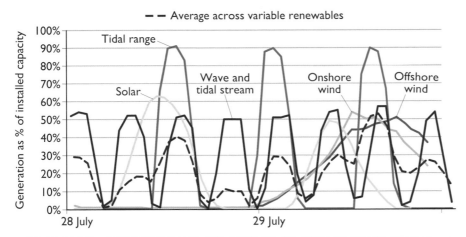

Figure 11.17 Variability of possible renewable generation technologies (over two illustrative days for an assumed 2030 mix) (source: CCC, 2011a)

The PV solar output (in yellow) appears as pulses of generation about 12 hours long once per day. That of a large tidal barrage would appear twice a day, but would slowly shift in timing through each month. That from tidal stream plants might appear four times a day, again shifting in timing throughout the month. The output from wind and wave power will depend on the precise weather conditions and may produce little or nothing for a day or more. Aggregating them together does not necessarily produce a particularly smooth result.

Solar power

In the UK, small photovoltaic systems at the kilowatt scale are connected locally at the 230 V or 400 V level. These only make up a quarter of the country's PV capacity. Larger arrays and solar farms can have outputs rated from hundreds of kilowatts to tens of megawatts. These are likely to be connected to the grid at an 11 kV level or higher. Although the UK is not sunny enough for large-scale CSP plants (described in Chapter 3), in other countries where they are deployed these generate at the multi-megawatt scale and are connected to the electricity grid at high voltage.

Naturally, PV and CSP plants that do not have thermal storage only produce electricity during the day and their output will be higher in summer than in winter. A typical capacity factor for PV in the UK is about 11%.

At the end of 2016, the UK had 11.5 GW of installed PV capacity. This does not mean that in total they will manage to produce 11.5 GW at midday on a sunny day. Figure 11.17 suggests that typical peak output might only be 50–60% of the total installed capacity. Different rooftop arrays all face in slightly different directions and some regions are likely to be clouded over.

Large amounts of solar generation may give rise to integration problems. These have already become apparent in the State of California, which has a target of getting 33% of its electricity from renewables by 2020, mostly from PV and wind. There has been a large amount of PV installation in particular since 2013, and by October 2016 California had over 14 GW of solar generating capacity (California Energy Commission, 2016a). It also had about 6 GW of wind capacity whose output also tends to peak in the middle of the day. The consequences for the supply of electricity from other generating plants (mostly gas-fuelled) are shown in the so-called 'Duck chart' (describing its shape), shown in Figure 11.18.

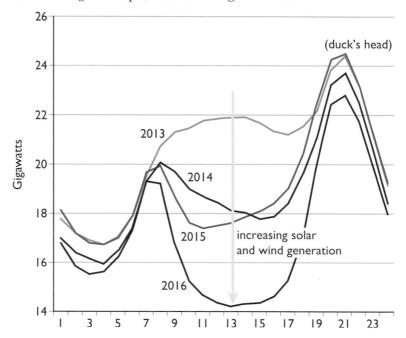

Figure 11.18 The 'Duck chart'. Demand on the California electricity grid net of wind and solar energy, i.e. that required to be supplied from conventional electricity generation sources'. Averages for 28 March–3 April for four successive years (source: California Energy Commission, 2016b)

In 2013 (the blue line in Figure 11.18), California had a similar electricity load curve to the UK (see Figure 11.13) rising steeply in the morning to a plateau from around 9.00 until 17.00, and then on to an evening peak before falling rapidly around midnight. However, by 2016 the increased PV and wind contribution meant that the remaining generating capacity in California was having to carry out two cycles of 'ramping up' and 'ramping down' each day, the rise to the evening peak being particularly steep.

This chart clearly shows the need for energy storage in the system. Seeing the coming problem, the California Public Utilities Commission asked the state's supply companies to install 1.3 GW of electricity storage by 2020. This request sparked interest in battery storage and may tip the balance for future solar development away from PV towards CSP plants with molten salt thermal storage.

Wind and wave power

As described in Chapter 8, large modern onshore grid-connected wind turbines are likely to be rated at 2–3 MW. These are likely to be connected to the grid at the 11 kV level. Larger wind farms of 50 MW or more are likely to be connected at the 33 kV level.

Offshore wind turbines are even larger and new designs could soon reach 10 MW. Offshore wind farms may have total capacities of 300 MW or more and necessitate special high voltage links to the grid onshore. The London Array in the Thames estuary has a 150 kV link to the shore, which runs from an offshore substation (Figure 11.19).

Although wave power is much less developed, it is likely that individual shoreline devices will be rated at about 1 MW and offshore 'wave farms' could be 100 MW or more, with similar grid connection problems to offshore wind farms.

The output of wind and wave power generators is, to a certain extent, predictable, using detailed weather forecasting techniques.

Typically, a 1 MW onshore wind turbine will produce 300–400 kW on *average over a year* (i.e. a capacity factor of 0.3–0.4). An offshore wind turbine will have a higher capacity factor of over 0.4 and approaching 0.5 in very windy Round 3 projects in the UK. It will produce full output on a windy day but nothing on a calm one. At modest wind speeds, its output may vary considerably from minute to minute.

Figure 11.19 An offshore substation for the London Array offshore wind farm. It is connected to the shore by a 150 kV cable

Similarly, the output of a wave power device is dependent on the random nature of the waves. On a stormy day it will run at full power, on a flat calm one it will produce nothing, and on an intermediate day, its output will be variable. A capacity factor of 30% has been assumed for the electricity cost shown in Figure 11.8.

Tidal power

As described in Chapter 7, tidal power is variable, but highly predictable.

Barrage schemes such as the proposed Severn Barrage, generating only on the ebb tide, could produce a pulse of power of up to 8 GW about six hours long every 12.4 hours (as shown in Figure 11.17). Although the peak spring output could be 8 GW, that on a neap tide a week later could only be 4 GW.

From an integration point of view, it would be better to have a large double-basin (or even triple-basin) scheme in this location with one basin kept 'high' and the other 'low' and with flexible generation and pumping between them. Although expensive, this could produce something closer to 'firm' renewable electricity.

Tidal lagoons, such as the one planned for Swansea, might operate both on the ebb tide and flow, producing a pulse of power every 6.2 hours, as will tidal stream turbines (see Figure 11.17). Their output will peak every 6.2 hours, on the incoming tide and again on the ebb tide. The output is also subject to a longer two-week cycle between spring-tide and neap-tide, the peak spring-tide output being roughly four times that of the peak neap-tide output (Carbon Trust, 2005).

The output of a single tidal turbine might consist of a rather unpromising sequence of pulses of power three to four hours long. However, as explained in Chapter 7, the output of two identical devices in different locations where the times of high tides differed by about three hours (such as Portland Bill in Dorset and the Isle of Wight on the southern English coast) could add up to a more constant supply of electricity.

The variability problem: a summary

In the past, electricity generation has been carried out using fossil- and nuclear-fuelled plants largely designed for continuous base-load operation. This tendency has been spurred on by the philosophy of market competition between generators to give the lowest costs. Dealing with the variability of demand or the unexpected loss of output from a large generation plant has been a secondary consideration. Part-loading of coal-fired plants and the techniques described in Box 11.2 have proved adequate.

However, the UK electricity grid of the future will have very large inputs (many tens of gigawatts) from renewable energy sources, with the variability characteristics outlined above. The associated problems, and possible solutions, are already apparent in countries such as Denmark and Germany where large amounts of wind power have been deployed.

Although an individual wind turbine or tidal stream turbine may have a highly variable output, if such sources are widely spaced, then just as diversity of demand adds up to a smoothly varying total demand on a national grid, so **diversity of supply** can also smooth out the local variations in output. When the wind stops blowing in Scotland, it may still be blowing in Wales. However, there is still a question of what should be done for the 'week with no wind power'.

Another integration problem is what to do on 'the *night* with *too much* wind power'. The voltage and frequency stability of the National Grid has to be kept within tightly defined limits. Traditionally this has been ensured by the flywheel inertia and stored steam pressure of enormous 600 MW coal-fired generating plants. Yet, on a summer night with low demand and plenty of wind, it is possible that all of these would have been turned off and the system would then be *wholly powered by variable renewables*. This problem has already been encountered in Denmark, where, in 2016, wind power provided over 40% of the country's electricity. Although in the past wind turbines have simply been allowed to run as and when there is adequate wind, it is now necessary to design them for **curtailment**, accepting a command from the transmission system operator to cut back on maximum output to avoid overloading the grid.

The next section describes some technical possibilities for the integration of large amounts of renewable electricity into the electricity grid. It may, however, require a switch from a philosophy of *competition* between electricity generators to one of providing *complementary* sources of generation, with the overall goal of minimizing greenhouse gas emissions.

11.8 Some partial solutions

Future large-scale reductions in greenhouse gas emissions in the UK and many other countries will require a number of tasks to be carried out that overlap in many ways:

- The decarbonization of electricity generation as described in Section 11.3
- The development of alternative transport fuels to replace fossil-fuel based petrol and diesel
- The reduction of heat demand in the building stock
- The development of replacements for oil and natural gas as heating fuels.

At present there are three main low-carbon contenders for future energy, and particularly electricity, supplies:

- coal and natural gas-fuelled systems using carbon capture and storage (CCS)
- nuclear power
- renewable energy.

At the time of writing, the large-scale generation of electricity using CCS has yet to be demonstrated, although many of the various component technologies have been.

Nuclear power is a mature technology, but has problems of public acceptability, following the nuclear accident at Fukushima in Japan in 2011. There are also problems of 'perceived financial risk' given that two large power plants being built in Europe, at Olkiluoto in Finland and Flamanville in France, are both several years behind schedule in their construction.

The large-scale deployment of renewable energy is certainly not problem-free and will require some restructuring of the existing energy (and particularly electricity) systems. Some remedies to these problems include:

- development of demand management in electricity grids
- strengthening and extension of electricity grids, both to provide short-term time diversity of supply and to tap into a wider range of supplies over the year
- development of complementary rather than competitive electricity generation
- development of hydrogen as a possible transport fuel and replacement for natural gas.

Electricity demand management

At present, in grid arrangements such as the UK's, while electricity *generation* is highly controlled there is less scope to influence electricity *demand*.

It would be convenient if electricity demand varied as little as possible over the day and night. This can be encouraged through the use of 'off-peak' electricity tariffs. Electricity for heating (or other) purposes is supplied at cheaper rates at night and the heat can then be stored until the next day.

This allows existing power stations to continue running into the night, rather than having to build extra ones to cope with higher peak demands.

In the UK, only a relatively small amount of electricity is used for heating purposes. However, this may increase if the use of electric heat pumps is developed, particularly to replace oil heating in areas beyond the reach of the gas grid.

Taking this approach further, there are many electrical loads that are not immediately needed. One example is large-scale water pumping which, given adequate storage, could be controlled to smooth electricity demand. At times of any impending shortage of renewable electricity supply, the pumps would be turned off. At present the mechanisms for such 'load shedding' are not well developed. However the concept of a **smart grid** involves a parallel flow of both electricity and price information that would allow increased flexibility in electricity demand and supply.

In 2014 the UK National Grid Company set up the Demand Side Balancing Reserve (DSBR), which allows companies to provide bids for electricity demand reductions (for which they get paid) to the transmission system operator. The companies might be large industrial plants, or they might be aggregators who gather up smaller potential load reductions from a large number of customers to bid into the system.

Electricity grid strengthening and enlargement

There are many reasons for strengthening national electricity grids and building international grid links. Existing electricity grids have grown up around the power stations of the past. Future large-scale renewable electricity sources are likely to be diverse and widely scattered. In the UK, connecting the considerable wind resource of Scotland to loads in the south has already required the strengthening of grid links in central Scotland and of those further south into England.

Although existing links can be upgraded by increasing the voltage and using multiple conductor wires, constructing new overhead transmission lines in areas of scenic beauty is likely to face stiff opposition. Typically, a 75 km long overhead 400 kV AC line capable of carrying 3 GW costs about £2 million per kilometre to construct (PB Power/IET, 2012). Placing sections of the cable underground may avoid visual objections but it is much more expensive, with prices of £10 million or more per kilometre.

The connection of offshore wind farms to land necessarily requires undersea cables, but a new 385 km long 2.2 MW grid link between Cheshire in England and Ayrshire in Scotland is being laid in the sea (see Figure 11.3), avoiding the environmental problems of a new land-based power line through regions of natural beauty.

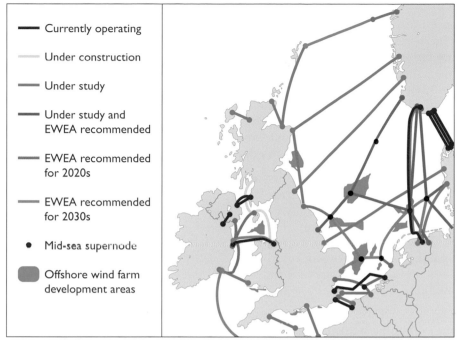

Figure 11.20 Possible future offshore and international grid links (source: updated from Offshore Valuation Group, 2010). EWEA = European Wind Energy Association.

The economic justification for the first undersea link between England and France, laid in 1986, was that peak electricity demands (and peak electricity prices) occur earlier in mainland Europe than in the UK. Such links are not cheap: the 1000 MW BritNed link between England and the Netherlands, which is 260 km long and opened in 2011, cost €600 million – over £2 million per km.

Although high voltage alternating current (AC) is normally used for land-based transmission lines, this is difficult to use for undersea cables, so the electricity is converted to high voltage direct current (HVDC) for transmission, and reconverted to AC at the opposite end. The conversion is carried out using high power semiconductor devices. Although the equipment is expensive, it is also worth using HVDC technology in new long distance land-based links over distances of more than 300 km, since the overall transmission losses can be lower than for high voltage AC.

Using grids to smooth out variability

There are many ways in which electricity grids can be used to smooth out the variability of renewable electricity supplies. For example, as described above, the movement of the tides around the coast, potentially allows the generation of electricity at different times of the day at different locations. These different outputs could be combined to approximate more continuous generation – but only given adequate grid capacity between these coastal sites and the ultimate customer electricity load.

Figure 11.16 has also shown how weather systems cross Europe from west to east, bringing with them alternating patterns of high and low wind speeds. The development of a chain of national and international grid links, from the far west of Ireland across the UK, Germany and Denmark to the Baltic Sea, could reduce the total variability of the output of wind farms in these countries. If the wind was not blowing in Ireland, it is probable that it would be blowing somewhere else along the line of grid connections. It is, however, still likely that there could be no wind for a week or more at any of the wind farms, so grid expansion cannot be regarded as a complete 'solution' to the variability problem.

The UK is not alone in considering such grid strengthening. The electricity systems of Denmark, Norway, Sweden and Finland operate as a single market, the Nordpool. Denmark is a small country and has strong grid links with Sweden and Norway. This has allowed it to develop its wind power in the knowledge that it can rely on the flexible backup of over 40 GW of Swedish and Norwegian hydropower. A 580 km long 700 MW undersea cable between Norway and the north of the Netherlands was completed in 2008.

There are likely to be considerable advantages in further north-south grid strengthening within Europe, allowing northern countries to tap into the considerable solar energy resource available in Spain and Italy, or even further into North Africa.

Complementary electricity generation

As described in the previous section, the current electricity system in the UK and many other countries is based on competition between generators. In winter, under the present market arrangements, a large coal- or gas-fuelled power plant might run continuously for months on end – as indeed it would have been designed to do. The introduction of many tens of gigawatts of wind and PV power into the UK electricity grid, and similar amounts in other European countries, changes the situation dramatically. Fossil-fuelled or large biomass plants may be required to run at part load or shut down at frequent intervals, possibly twice a day, yet still be needed to restart at a few hours' notice.

This raises two questions. The first is how 'surplus' renewable electricity might be stored or used. Box 11.3 suggests some options.

The second, and more difficult, question is what kind of electricity generation plant might be complementary in operation to wind and other variable renewable energy sources – and available to cover 'the week with no wind'. Such a plant must also be economic even though it may only be operating for 50% of the year or less. Also, ideally it should have zero CO_2 emissions.

BOX 11.3 Ways to use surplus electricity

Pumped storage

This technology has been described earlier in Box 11.2. Although it is relatively efficient at storing electricity, typically 80%, it requires large high-level reservoirs and large volumes of water. In a plant such as Cruachan in Scotland, described in Chapter 6, Section 6.5, the storage of 1 kWh requires the movement of at least 1 tonne of water between two reservoirs differing in height by 360 m. Cruachan's total storage capacity is about 800 GWh.

Heat (and cold) storage

Electricity can be used to heat water, either in domestic hot water cylinders or larger heat stores connected to district heating (DH) systems. As mentioned in Box 5.11, the specific heat capacity of water is 4.2 kJ kg^{-1} K^{-1}. Heating 1 kg of water from 10 °C to 60 °C thus requires 210 kJ. 1 kWh (3.6 MJ) of electricity thus provides enough energy to heat over 17 kg water. This low-temperature heat energy cannot be converted efficiently back into electricity because of the limitations of Carnot's Law, described in Chapter 3. However, the energy can potentially be put to good use and the required volume of storage involved is modest compared to pumped storage. If the 2500 m^3 DH store shown in Figure 11.12 was electrically heated it could absorb nearly 150 MWh of electricity.

In warmer countries, there is the option of 'ice storage' for large commercial buildings, which uses surplus electricity to produce ice that can be stored and then used to reduce the afternoon peak air-conditioning load.

Production of hydrogen and synthetic natural gas

If direct current electricity is passed between two electrodes immersed in water, hydrogen and oxygen can be collected at the electrodes. This process is known as **electrolysis**. Practical systems have efficiencies of 50–80%. Assuming an efficiency of 75%, it requires approximately 50 kWh of electricity to produce 1 kg of hydrogen. In principle this might be used to generate electricity in a fuel cell (see below) or even a CCGT with an efficiency of 50%, returning about 25 kWh of electricity, together with some low-temperature heat that could be used for district heating. The possible further conversion of hydrogen into methane (synthetic natural gas) is described briefly at the end of this section.

Liquid air energy storage (LAES)

This system, currently under development in Birmingham, UK, uses surplus electricity to liquefy air, which can then be stored in large insulated tanks for as long as required. This liquid air (or more likely liquid nitrogen, for safety reasons) can then be used to generate electricity. It is warmed to ambient temperature (using low temperature waste heat from an existing power plant or industrial process) to produce high pressure air, which can be used to drive a gas turbine.

The most likely option in the UK in the near-term future for complementary generation is the use of large biomass steam power plants such as those at Drax. These converted former coal-fired plants are well-suited to 'load following'.

Another UK candidate is the conventional gas-fuelled CCGT power station. Most existing UK CCGTs have been designed for continuous high-efficiency operation at full output and may require up to three hours to reach full power from a cold start. Additionally, the temperature cycling involved in repeated stop-start operation is likely to cause metal fatigue in the heat recovery steam boiler, leading to increased maintenance costs.

The needs for more flexible operation, however, have been addressed in new CCGT plant designs. They have a high startup 'ramp rate', reaching full power in under 40 minutes, a rapid 'ramp down' to part load in a matter of seconds, and a redesigned boiler to minimize problems of metal fatigue (Balling, 2011).

As for the economics of such plants, it can be seen from Figure 11.8 that repayments on capital costs for CCGTs and large 'biomass conversion' coal plants only make up about 10% of the final electricity cost: the single largest item for them is the fuel cost. A consequence of this is that reducing the capacity factor from the assumed base load operation to 50% only increases the electricity cost by about 15% (author's estimate based on data in BEIS, 2016d).

Another option is the use of large combined heat and power (CHP) generation plants, producing both electricity and heat, particularly those plants using steam turbine technology, which are widely used in Denmark. In the past these have been fuelled by coal or natural gas but are now increasingly fuelled by biomass. These make good use of the waste heat from electricity generation by feeding it into a local district heating network.

In summer, when heat demand is low, they maximize their electricity output. In winter they produce more heat, but at the cost of a reduced electricity output and generation efficiency. Although the total fuel input may remain constant, the relative proportions of heat and electricity produced can be changed quite rapidly. The inclusion of a large heat store in the system allows flexibility of operation, enabling the plant to vary its electricity output quite rapidly. Smaller CHP generation plants based on internal combustion engines and fuelled by natural gas or biogas are also capable of rapid changes in output and flexible operation.

In the longer term, fuel cells (described below) powered by hydrogen are also good candidates for complementary generation, since they, too, are capable of rapid changes in output.

Nuclear power stations are normally designed for virtually continuous base-load operation.

What are the likely overall system costs of integrating large amounts of variable renewables into the UK grid? A 2006 report by the UK Energy Research Centre (UKERC) concluded that at a 20% penetration of wind the system costs are 'estimated to add between £5 and £8 per MWh (0.5–0.8 p kWh^{-1}) to electricity prices' (UKERC, 2006). However, in Denmark, wind already supplies more than 40% of the nation's electricity. Might this be achieved in the UK? A 2016 update report by the UKERC sees this as possible but stresses the need for a flexible electricity system to cope (UKERC, 2017).

Is it possible to run a grid on 100% renewable energy? In 2007 researchers at the University of Kassel in Germany demonstrated what they termed a 'combined power plant', an aggregation of 36 decentralized power plants based on wind, hydropower, solar and biogas energy, running together and designed to meet 1/10 000th of the total, varying, load on the German grid (Kombikraftwerk, 2007). They see this as a model for the future operation of grids as more renewable energy sources are brought into operation.

Hydrogen – a fuel of the future?

Hydrogen has been widely advocated as a potential 'energy carrier' for the future. Its use as a fuel has a number of advantages:

- it can act as a store of renewable energy from season to season
- it can provide a transport fuel not dependent on the world's limited reserves of oil
- the only by-products of its combustion are water and a very small amount of nitrogen oxides, and even these NO_x emissions can be reduced to zero if fuel cells (see Box 11.4 below) are used.

Fossil fuel-based hydrogen

The production and use of hydrogen as a fuel is well understood. Before the arrival of natural gas, 'town gas' produced from coal in the nineteenth century consisted mainly of a mixture of hydrogen and carbon monoxide. Hydrogen is already used in large quantities as a feedstock for the chemical industry, mainly in the manufacture of fertilizers and in oil refining. Currently, it is mainly produced by steam 're-forming' of natural gas (methane), a process that necessarily also produces carbon dioxide:

$$2H_2O \quad + \quad CH_4 \quad \rightarrow \quad CO_2 \quad + \quad 4H_2$$

Steam   +   Methane $\rightarrow$ Carbon dioxide   +   Hydrogen

Modern coal and biomass gasification can also produce a similar stream of CO_2 and hydrogen.

The CO_2 produced does not necessarily have to be emitted into the atmosphere. It can be 'captured', compressed and piped for long-term storage in locations such as saline aquifers deep underground. Such large scale CO_2 disposal or **sequestration** has been carried out for over a decade by the Norwegian company, Statoil, in its Sleipner natural gas field (in this case the CO_2 is an unwanted contaminant of the natural gas).

More generally the separation of CO_2 produced by electricity generation or industrial processes, and its subsequent sequestration, is known as **carbon capture and storage (CCS)**.

A future fossil fuel-based 'hydrogen economy' could involve the extensive production of hydrogen from coal or gas, combined with CCS. The hydrogen could be stored, distributed by pipeline and used to generate electricity in efficient CCGT or fuel cell power stations, where the main combustion product would simply be water vapour. Hydrogen would largely become a substitute for the existing supply of natural gas.

Large-scale CCS generation plants and similar technologies are potential competitors to renewably generated electricity. The cost of electricity from a coal-fuelled Integrated Gasification Combined Cycle (IGCC) plant with CCS has been estimated at nearly 15p kWh^{-1} (see Figure 11.8).

This technology basically consists of a gasifier producing hydrogen which is then fed to a CCGT. On its own this technology could potentially provide 'firm' generation without any variability problems. It has been suggested that the gasifier might be run continuously (Starr, 2007), producing hydrogen that could be stored or used elsewhere, while the CCGT could operate as and when required forming the basis of a complementary generation technology to variable renewables. Problems of temperature cycling in the CCGT power plant could be reduced by keeping it warm using waste heat from the gasifier.

Renewably based hydrogen

There are also a number of ways that renewable or 'solar' hydrogen can be produced without CO_2 by-products:

- by the electrolysis of water using electricity from non-fossil sources
- by the **thermal dissociation** of water into hydrogen and oxygen using concentrating solar collectors (probably in desert areas). To do this directly would require very high temperatures, over 2000 °C, but with more complex processes using extra chemical compounds the same result may be achievable at temperatures of under 700 °C. These processes have not yet been developed on a commercial scale
- by the gasification of biomass in a similar manner to that for coal, converting the carbon content of the biomass into CO_2. If the biomass has been sustainably grown then this CO_2 should be re-absorbed in new biomass as it is grown. More hopefully, biomass gasification coupled with CCS raises the interesting possibility of producing hydrogen with *negative* overall CO_2 emissions.

Other techniques under investigation include the use of photoelectrochemical cells that produce hydrogen directly from water, and artificial chemical photosynthesis.

Using hydrogen

When burned, 1 kg of hydrogen will produce 120 MJ of heat, assuming that the resulting water is released as vapour (i.e. assuming its lower calorific value). A further 20 MJ can be obtained if the water vapour can be condensed to water. Although this is nearly three times the energy per unit *mass* of petrol or diesel fuel, hydrogen has the disadvantage of being a gas with a low energy per unit *volume* at atmospheric pressure. Hydrogen can be stored in a number of forms:

- in small quantities as a gas in pressurized containers, typically at 300–700 atmospheres. These containers obviously have a weight penalty for transport applications
- in large quantities, such as like natural gas which is stored in sealed salt caverns (ETI, 2015)
- by absorbing it into various metals, where it reacts to form a metal 'hydride': the hydrogen can be released by heating
- as a liquid, although this requires reducing its temperature to −253 °C and the use of highly insulated storage. As mentioned earlier in the chapter, natural gas (methane) is already widely shipped in liquid form, at a temperature of −162 °C. However, it is not yet clear whether or not liquid hydrogen in bulk can be shipped safely in a similar manner.

Hydrogen can also be pumped through pipelines. Here it has a disadvantage: at atmospheric pressure, its energy density is only 10 MJ m^{-3}, about a quarter of that of natural gas. This limits both the rate of throughput through the existing high pressure gas pipeline network, and the energy storage capacity of the pipelines and any underground storage facilities. Therefore hydrogen, in energy terms, will not be a direct substitute for natural gas. However, a nearer term option would be to add about 10–15% of hydrogen by volume to the existing natural gas flows, effectively reducing their overall carbon content. This would, hopefully, avoid the modification of existing burners or redesign of other end-use technologies.

A pilot 325 kW 'power to gas' electrolysis plant injecting hydrogen into the gas grid has been demonstrated in Germany. Since the electric power absorbed can be varied very quickly, this technology can be used for rapid response grid balancing (ITM, 2015).

It has been suggested that countries with plentiful supplies of renewable energy, such as Iceland, which has large untapped reserves of hydroelectricity and geothermal power, could switch to electrolytically generated hydrogen, giving up the use of fossil fuels entirely. This would include a switch to using hydrogen for road vehicles.

Starting in 2006, the European Union's HyFleet Clean Urban Transport for Europe (CUTE) project has demonstrated the use of hydrogen-fuelled buses in cities around the world. Some of these have used internal combustion engines, but eight running in London since 2011 are powered by fuel cells driving electric motors (see Figure 11.21). Several public hydrogen filling stations (see Figure 11.22) have now opened in the UK to serve fuel cell cars now being sold. The principles of the fuel cell are described in Box 11.4.

Figure 11.21 A London bus powered by a hydrogen fuel cell

Figure 11.22 A public hydrogen filling station in London. It supplies high pressure hydrogen at 300 and 700 atmospheres pressure

BOX 11.4 Fuel cells

In Chapter 2 it was explained that all heat engines are inherently limited in the efficiency with which they can convert heat into motive power, and hence into electricity if the engine is driving a generator. Normally, most of the energy in the input fuel emerges as 'waste' heat (although of course this can often be harnessed and put to good use).

The fuel cell (see Figure 11.23) enables hydrogen and oxygen fuel to be converted to electricity at a potentially higher efficiency than could be achieved by burning them in a heat engine. A fuel cell is in principle a battery in which the active elements are not solids (such as the lead and lead dioxide in a car battery) or liquids (as in the flow battery plant mentioned above), but gases. Indeed, when the fuel cell was first invented by Sir William Grove in 1839 he called it a 'gas battery'.

The principle of operation of the fuel cell is similar to electrolysis but in reverse: gases such as hydrogen and oxygen (or air) are pumped in and DC electricity is the output.

The only by-product is water and there are virtually no pollutants. There is some waste heat, but much less than in most combustion-based generation systems, and there are no moving parts. As in other types of battery, the voltage from an individual cell is low, typically about 0.7–0.8 volts, and multiple cells are connected in series to get a useful working voltage.

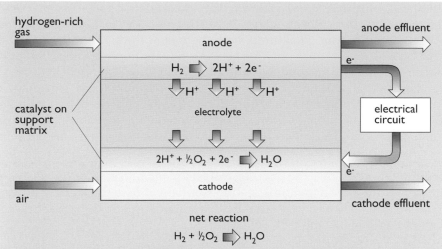

Figure 11.23 Principle of operation of a fuel cell

Despite its early invention, development work did not really start until the late 1950s. Now there is a whole range of different types including:

The Alkaline Fuel Cell (AFC). This was the first to be developed, for the US Gemini and Apollo space programmes in the 1960s. Although simple and with low manufacturing costs, AFCs are limited by the need to remove any CO_2 from the air supply to prevent contamination of the potassium hydroxide electrolyte.

The Solid Polymer Fuel Cell (SPFC). This is being developed in two forms:

■ the Proton Exchange Membrane Fuel Cell (PEMFC) – a strong candidate for transport and portable power applications, this has been demonstrated in cars and buses and is available in sizes of up 250 kW. It requires very pure hydrogen.

■ the Direct Methanol Fuel Cell – this is another candidate being developed for transport applications. Although methanol is poisonous, it is easier to handle as a fuel than hydrogen.

The Phosphoric Acid Fuel Cell (PAFC). This runs at a temperature of 180–250°C. It is one of the most developed of the fuel cell types and is available commercially, usually as a 200–450 kW unit packaged hydrogen producing steam reformer to allow it to run on natural gas. It is becoming widely used for electricity supplies for commercial buildings.

Several other fuel cell types are under development.

Typical fuel cell electricity generation efficiencies are currently in the range 35–60% (Arup, 2016). While this is better than the 25–30% that might be expected from small reciprocating engines running on natural gas or hydrogen, it is still only just competitive with the 45–60% efficiencies of new large CCGT power stations.

Fuel cell developers are aiming to produce devices with costs competitive with more conventional plants, and with higher efficiencies. For the moment, the strength of the fuel cell concept lies in its lack of noise, low pollution at the point of use, and flexibility as a 'gas battery' since, as with other batteries, its power output can be changed very rapidly, often within fractions of a second.

For future vehicle applications there is stiff competition between fuel cells and battery-electric technology. Fuel cells offer advantages of rapid refuelling compared to possible lengthy recharging times for battery-electric vehicles. Both are potentially limited by the availability of certain chemical elements: many fuel cells require 'noble metals' such as platinum as catalysts; battery-electric vehicles may be limited by global supplies of lithium for lithium-ion batteries, or lanthanum for nickel metal hydride batteries.

Most fuel cells also require very high purity hydrogen to avoid poisoning of the catalyst. It must not be contaminated with the small amounts of sulfur compounds that may be present in natural gas used for steam reforming, or with carbon monoxide.

The overall economics are also likely to depend on the relative costs of low carbon electricity for charging batteries and fuels such as hydrogen or methanol from low carbon sources for fuel cells.

The large-scale use of hydrogen would require many steps, such as those illustrated in Figure 11.24, each of which would require large capital investments. For example, surplus hydro or wind power could be electrolyzed to produce hydrogen. The hydrogen would then be stored, either as high-pressure gas or as a low-temperature liquid, before being pipelined or transported by insulated road/rail tankers to interim storage facilities. The largest insulated tanks for liquid hydrogen at present (10 000 m³) can store about 90 TJ of energy. There would also need to be provision to re-gasify the liquid hydrogen and distribute it by pipeline. Finally, there are the individual end uses:

- in homes for cooking and heating – possibly as a mixture with natural gas
- in fuel cell power stations (of all sizes) to generate electricity and useful heat
- in road vehicles using fuel cells or hydrogen-fuelled internal combustion engines.

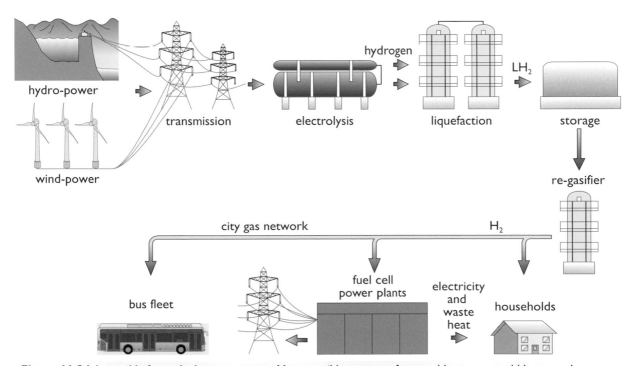

Figure 11.24 A possible future hydrogen economy. Many possible sources of renewable energy could be tapped to manufacture hydrogen, which could be shipped to consumers and used in a variety of ways

Bringing such a vision to the current energy marketplace would require placing a high value on hydrogen's carbon-free and pollution-free qualities, since hydrogen would be in competition with other ways of delivering energy services, particularly the use of electricity for surface transport and electric heat pumps for heating.

Hydrogen would also be in competition for heating purposes with natural gas, for which there is an extensive existing storage and distribution infrastructure. Indeed it is possible that it may prove more economic (and safer) to convert electrolytically generated hydrogen and carbon dioxide into methane (i.e. synthetic natural gas) and water, using the reverse of the steam reforming process described above. This could be extended to producing bio-kerosene as an aviation jet fuel. This could be interpreted as turning renewable energy into a fuel similar to fossil fuels, but the intention would be to use carbon dioxide from sustainably grown biomass sources, which uses carbon that has come from the atmosphere and will be returned to it when the methane is burned (see Sterner et al., 2010, and Wenzel, 2010).

11.9 **Summary**

Renewable energy will need to provide the **energy services** currently supplied by **fossil fuels.**

Delivered energy is that **actually received by the consumer**. This may take the form of:

- **Liquid fuels**, mainly **refined from crude oil** and used for **transport**
- **Gaseous fuels**, such as **natural gas** and used for **heating and cooking**
- **Solid fuels**, such as **coal, coke and wood** and used for **heating**
- **Electricity**, used for a range of purposes
- **Heat** distributed through **district heating systems**

Energy services are what we actually use energy for. **Energy efficiency** has an important role to play in determining **how much delivered energy** is required **to provide a given energy service**.

- **Warm homes and offices, hot water and cooked food** require the use of **heat**, which can come from the burning of fuels or the use of electricity.
- **Motive power**, driving machines or providing transportation requires **fuel-driven engines** or **electric motors**.
- **Mobility** does not necessarily mean physically transporting people. Telecommunications can provide a partial substitute.
- **Electricity-based services** include **telecommunications, computing** and **electro-chemical processes** such as the manufacture of aluminium.

Until the 1960s, **coal** was the dominant fuel in the UK, but its role is now almost completely disappearing.

Currently, **natural gas** is the main UK heating fuel. It can be imported via long-distance pipelines or by ship as **liquefied natural gas (LNG) cooled to -162°C** and can be stored in **large insulated storage tanks**. Gas is distributed throughout the UK in pipelines, which are also used to store it by varying the pressure. This is known as '**line packing**'.

Heat can also be distributed through **district heating schemes** from **combined heat and power (CHP)** plants or large **heat pump** schemes. This is widely done in Scandinavia.

Electricity is mainly **generated in large gas, coal or nuclear power plants**. It is distributed throughout the UK using **a network of high voltage power lines, the National Grid. Undersea cables** link this to the networks in other countries. **Many new links are currently being built.**

Worldwide, **oil is the dominant transport fuel**.

Renewable electricity sources have an important role in cutting global CO_2 emissions because of the **high levels of CO_2 emitted from fossil fuelled generation**. In 2014, **electricity generation** accounted for **40% of global CO_2 emissions** and **two thirds of the world's use of coal**.

A **kilowatt-hour of electricity** produced from coal in the UK involves the emission of about **0.9 kg CO_2**. One produced from **natural gas** only produces about **0.4 kg CO_2**. UK electricity is currently produced from a mix of gas, coal, nuclear and renewable energy. In 2015 the **average emission factor** per kilowatt-hour was only **0.33 kg CO_2** and this is likely to fall in the future. Electricity generation can also involve the emission of **other greenhouse gases** such as **methane**.

Overall, per kilowatt-hour of electricity generated, the **highest greenhouse gas emissions** are from **coal, lignite and oil** power plants and the **lowest** from **nuclear, renewables** and future possible plants using **carbon capture and storage (CCS)**.

Coal and oil-fired electricity generation can also produce **sulfur dioxide and particulates**, which are a major source of **local air pollution** in **China and India**.

All forms of fuel combustion, even biomass, can potentially produce polluting **nitrogen oxides (NO_x)**.

The **CO_2 emissions of fuels** can be **direct**, i.e. those emitted at the point of use, or they may include **indirect greenhouse gas emissions** involved in the extraction and transport of fuels (often from other countries).

Greenhouse gas emissions are expressed in units of **kg CO_2 equivalent**, the amount that gives the **same warming effect as one kilogram of CO_2**.

The **environmental damage** produced by emitting CO_2 into the atmosphere can be expressed as a **carbon price**. A value of **£50 per tonne of CO_2** has been assumed in this book.

Per kilowatt-hour of heat energy produced, **coal has the highest greenhouse gas emission factor, followed by oil and natural gas**. Renewable fuels such as wood have a low emission factor. The combustion of a **litre of fossil-fuelled based transport fuel** (petrol (gasoline) or diesel) involves the emission of about **3 kg of CO_2 equivalent** of greenhouse gas emissions.

Fossil fuel **resources** are expressed in terms of **finite amounts of energy that can only be consumed once. Renewable energy resources** are expressed in terms of sustainable **annual energy flows**.

- The **available resource** is the **total energy flow delivered by a particular source**, for example the total solar radiation falling on a country.
- The **technical potential** is the maximum flow that could be extracted **using current mature technology**.
- The **practical resource** is the amount limited by factors such as public acceptability and problems of distribution.
- The **economic potential** is the amount that is economically viable.

Globally, the **technical potential** for **hydroelectricity** is about **a half of the world's current electricity use**. The technical potentials for electricity from **wind power, ocean energy and geothermal energy** all **far exceed it**. The technical potential for **direct solar energy use far exceeds the current world primary energy demand**.

For the **UK**, the **technical potentials** for electricity from **solar PV and tidal stream energy** are each about a half of the country's current electricity use and the **potential for offshore wind power exceeds it**.

The **potentials for bioenergy**, both worldwide and for the UK are **limited by concerns about competing land that is needed to grow food**.

Electricity generated by a brand new combined cycle gas turbine (CCGT) plant in the UK costs an estimated **6.6 pence per kilowatt hour**, including a carbon price for the CO_2 emitted. Other **competing technologies** with electricity **prices in the range 6.3–10.5 pence** per kilowatt hour are: **nuclear, solar PV, electricity from biomass, hydro and both onshore and offshore wind**. Wave and tidal power and electricity from coal-fuelled carbon capture and storage plants are considered more expensive.

The **levelized cost of electricity (LCOE)** is the overall **average cost per kilowatt hour over the lifetime of a particular power plant**. Most renewable energy technologies have **high capital costs** and **zero fuel costs**. The **electricity cost per kilowatt hour** depends critically on the **interest rate** or **discount rate** used for the finance. This in turn depends on the **perceived project risk** to investors.

Prices for electricity from solar PV and both onshore and offshore wind have been falling. Key lessons are:

- **Volume production and deployment produces reductions in capital and installation costs**
- **Demonstrated successful deployment reduces the perceived financial risk and makes capital available at lower discount rates.**

Costs for **heat energy** at the domestic scale from **heat pumps and solar water heaters are higher than for heat from natural gas fuelled boilers**.

It is often **cheaper to save a tonne of CO_2 emissions** using **energy efficiency** than by **supplying extra renewable energy**. The relative costs can be set out in a **marginal abatement cost curve (MACC)**.

At the **world level**, renewable electricity prices have been falling. In appropriate countries **electricity from biomass, solar PV, and both onshore and offshore wind** can be **competitive with fossil fuelled generation**.

National auctions for **large renewable electricity projects** have produced **low bid prices competitive with fossil fuelled generation** in many countries.

In the **British Isles** the best sites for:

- **onshore wind** are in **Scotland, Wales, Cornwall and west and north-west Ireland**.
- **offshore wind** are in the shallow waters of the **North Sea** and the **Irish Sea**
- **tidal barrages** are in the **Severn estuary** and the **Solway Firth**
- **tidal stream** projects are around **coastal headlands** and particularly the **Pentland Firth** and the **Race of Alderney**.
- **wave power** are along the **coasts of Scotland** and the **Atlantic coasts of Ireland** and **south-west England**.

The **National Grid** is key to **distributing electricity** from these sites **to the locations of electricity demand**. A **strong electricity grid** is necessary to avoid **grid congestion** in the distribution of renewable energy.

Biomass-fuelled combined heat and power plants can be sited **close to cities** to allow both their **electricity and waste heat** to be used.

For **heat generating renewables** such as **solar thermal** and **heat pumps**, hot water storage is possible in large insulated tanks. This can easily be done on a daily basis. **Interseasonal heat storage** from summer through to winter with very large stores **is being demonstrated in Denmark**.

Gas and electricity distribution networks already have to deal with **large variations in demand** from hour to hour, day to day and throughout the year.

For example, there can be **rapid changes to electricity demand** due to:

- workers starting in offices and factories in the morning
- **demand pickup** at the end of a TV programme shown over the whole country.

The **failure of a large conventional power station** or **transmission line** may also require bringing alternative generation online at short notice.

The **existing electricity grid** has been developed to distribute electricity from **large fossil and nuclear power stations**. Increasingly, it is having to deal with **smaller distributed** and **embedded renewable electricity generators**.

Matching electricity supply and demand in the National Grid is controlled by a central **transmission system operator** who accepts **price bids for electricity from different generators** and decides which generators will run. Many smaller renewable electricity sources always run when they can and don't need to bid.

Matching short term peaks in demand can be done using:

- **pumped storage plants**
- **open cycle gas turbine and diesel peaking plants**
- **compressed air energy storage** in conjunction with combined cycle gas turbine (CCGT) power plants

■ **rechargeable batteries**. Current battery technologies available include **lead acid, lithium-ion, sodium-sulfur and vanadium flow batteries.**

Renewable sources of electricity generation can be classified as **firm** or **variable**.

Firm sources, which are always potentially available, include those fuelled by **biomass, wastes** and **geothermal energy. Hydro power plants** are likely to be continuously available, but may only be used for covering peaks in demand. **Variable sources** include **solar PV, concentrating solar power (CSP), wind and wave** and **tidal power.**

The output of **solar generation** in any particular location varies with the **local time of day.** However the rotation of the Earth means that this is 1 hour earlier in Berlin than it is in Birmingham.

The output of **wind turbines** and **wave power plants** at any location are **dependent on the local weather.** In Europe, **weather systems tend to move slowly from west to east.** Winds are **stronger in winter than in summer.**

The output of **tidal barrage** and **tidal stream** power plants at any location are dependent on the **local state of the tides.** However, the **tides progress along the European coast twice a day** and also **vary on a two-week cycle between spring and neap tides.**

Small individual house PV systems are connected to the grid at low voltage. Larger PV arrays and **concentrating solar power (CSP) plants** are likely to be **connected at a high voltage.**

Having **large amounts of PV electricity generation** in the supply mix can give rise to **a rapid rise in demand for alternative generation in the evening when the sun sets.**

Large onshore wind turbines are likely to be rated at **several megawatts power** and **connected to the grid at the 11 kV level.** Onshore and offshore **wind farms** are likely to be **connected to the grid at a higher voltage level.**

Although the **output of wind turbines is variable**, it is to a certain extent **predictable** using **detailed weather forecasting** techniques. The **capacity factor** of a large **onshore wind turbine** is about **30–40%** and for an **offshore turbine** about **40–50%.**

Tidal barrage schemes are likely to produce large pulses of electricity generation **lasting about 6 hours once every 12.4 hours.** Tidal stream turbines **produce a sequence of pulses of generation 3 to 4 hours long.** For both technologies the **magnitude varies** on the **two week spring-neap tidal cycle.**

Diversity of supply – Although individual wind turbines and tidal stream turbines may have a highly variable output, **adding together the outputs of widely spaced sources can smooth out the local variations of output. This does not solve the problem of what to do on the 'week with no wind',** which still requires some alternative source of generation.

Curtailment – it may be necessary to 'switch off' variable renewable electricity generation sources when their total supply exceeds demand.

Large scale **reductions in greenhouse gas emissions** will require the:

- **decarbonization of electricity generation using renewable energy**
- development of **alternative low carbon transport fuels**
- **reduction of heat demand** in the building stock
- development of **replacements for oil and natural gas** as heating fuels

The **large scale deployment of renewable energy** will require a **flexible electricity system** including the:

- development of **demand management in electricity grids**
- **strengthening and extension of electricity grids**. This is already taking place on a Europe-wide scale.
- development of **complementary** rather than **competitive electricity generation**. Variable renewables such as wind and PV do not generate electricity continuously. Other sources are needed to cover the periods when they are not available. Options include:
  - **large biomass-fuelled power stations**;
  - **natural gas-fuelled CCGT power plants** with a high 'ramp rate', the speed at which they can be brought online up to full power;
  - **combined heat and power (CHP) plants** with **large associated thermal heat stores**.
  - development of **methods to usefully store surplus electricity**. These include **pumped storage**, large thermal stores for **liquid air, heat for district heating systems** or **ice for air-conditioning purposes**.

Hydrogen could be used as an **energy carrier** in the future as a replacement for fossil fuels. It can be produced by:

- the **steam reforming of methane** from natural gas
- the **gasification of biomass**
- the **electrolysis of water** using electricity from renewable energy sources
- the **thermal dissociation of water** using high temperature solar heat.

It can be stored in several ways including:

- **as a gas in small pressurized containers**
- **as a liquid** at very low temperature **in large insulated storage tanks**.

Fuel cells or 'gas batteries' can produce electricity from hydrogen fuel and oxygen from the air. This is done **without combustion** and their **efficiency can be 35–60% as it is not limited by the Second Law of Thermodynamics**. Applications include:

- **Fuel cell vehicles** – cars and buses are now commercially available and public hydrogen vehicle filling stations are being opened.
- **Fuel cell electric power plants** producing both electricity and potentially useful heat.

Hydrogen could also be:

- used as a **cooking and heating fuel** for domestic applications
- combined chemically with carbon dioxide to produce **synthetic methane** and **kerosene** for many applications.

References

Arup (2016) *Five Minute Guide: Hydrogen* [Online], London, Arup. Available at http://publications.arup.com/publications/f/five_minute_guide_hydrogen (Accessed 10 April 2017).

Balling, L. (2011) *Fast cycling and rapid start-up: new generation of plants achieves impressive results* [Online], Germany, Siemens AG. Available at https://www.energy.siemens.com/hq/pool/hq/power-generation/power-plants/gas-fired-power-plants/combined-cycle-powerplants/Fast_cycling_and_rapid_start-up_US.pdf (Accessed 7 August 2017).

BEIS (2016a) *UK Energy in Brief 2016* [Online], London, Department for Business, Energy and Industrial Strategy. Available at https://www.gov.uk/government/statistics/uk-energy-in-brief-2016 (Accessed 29 March 2017).

BEIS (2016b) *Digest of UK Energy Statistics* (DUKES) [Online], London, Department for Business, Energy and Industrial Strategy. Available at https://www.gov.uk/government/collections/digest-of-uk-energy-statistics-dukes (Accessed 29 March 2017).

BEIS (2016c) *UK Government GHG Conversion Factors for Company Reporting* [Online], London, Department for Business, Energy and Industrial Strategy. Available at https://www.gov.uk/government/publications/greenhouse-gas-reporting-conversion-factors-2016 (Accessed 31 March 2017).

BEIS (2016d) *Electricity generation costs* [Online], London, Department for Business, Energy and Industrial Strategy. Available at https://www.gov.uk/government/publications/beis-electricity-generation-costs-november-2016 (Accessed 4 April 2017).

BEIS (2016e) *Energy Consumption in the UK: data tables* [Online], London, Department for Business, Energy and Industrial Strategy. Available at https://www.gov.uk/government/statistics/energy-consumption-in-the-uk (Accessed 6 April 2017).

BEIS (2017) *UK Energy in Brief 2017* [Online], London, Department for Business, Energy and Industrial Strategy. Available at https://www.gov.uk/government/statistics/uk-energy-in-brief-2017 (Accessed 3 September 2017).

BERR (2008) *Atlas of UK Marine Renewable Energy Resources* [Online], London, Department of Business, Energy and Regulatory Reform. Available at http://www.renewables-atlas.info (Accessed 8 April 2017).

BP (2017) *BP Statistical Review of World Energy 2017* [Online], London, The British Petroleum Company. Available at http://www.bp.com (Accessed 9 August 2017).

California Energy Commission (2016a) *California Energy Commission: Tracking Progress – Renewable Energy* [Online], California, California Energy Commission. Available at http://energy.ca.gov/renewables/tracking_progress/documents/renewable.pdf (Accessed 10 April 2017).

California Energy Commission (2016b) *California Energy Commission: Tracking Progress - Resource Flexibility* [Online], California, California

Energy Commission. Available at http://www.energy.ca.gov/renewables/tracking_progress/documents/resource_flexibility.pdf (Accessed 10 April 2017).

Carbon Trust (2005) *Variability of UK Marine Resources* [Online], London, Carbon Trust. Available at https://tethys.pnnl.gov/sites/default/files/publications/Carbon_Trust_2005.pdf (Accessed 16 April 2017).

CCC (2008) *Building a low-carbon economy – the UK's contribution to tackling climate change* [Online] Norwich, TSO. Available at http://archive.theccc.org.uk/archive/pdf/TSO-ClimateChange.pdf (Accessed 3 April 2017).

CCC (2011a) *The Renewable Energy Review* [Online], London, Committee on Climate Change. Available at https://www.theccc.org.uk/publication/the-renewable-energy-review/ (Accessed 1 August 2017).

CCC (2011b) *Bioenergy Review* [Online], London, Committee on Climate Change. Available at https://www.theccc.org.uk/publication/bioenergy-review/ (Accessed 1 August 2017).

EEA (2011) *LCA emissions of energy technologies for electricity production* [Online], Copenhagen, European Environment Agency. Available at http://www.eea.europa.eu/data-and-maps/figures/lca-emissions-of-energy-technologies (Accessed 29 March 2017).

Energia (2017) *Windfarm locations* [Online], Dublin, Viridian Energy Ltd. t/a Energia. Available at https://www.energia.ie/business/products-and-services/energia-renewables/windfarm-locations (Accessed 12 October 2017).

ETI (2015) *Hydrogen Storage insight report* [Online], Loughborough, Energy Technologies Institute. Available at http://www.eti.co.uk/storing-hydrogen-underground-in-salt-caverns-and-converting-it-into-a-reliable-affordable-flexible-power-source-could-help-meet-future-uk-peak-energy-demands-according-to-the-eti/ (Accessed 4 April 2017).

Everett, B., Boyle, G. A., Peake, S. and Ramage, J. (eds) (2012) *Energy Systems and Sustainability: Power for a Sustainable Future*, 2nd edn, Oxford, Oxford University Press/Milton Keynes, The Open University.

Fritsche, U. and Rausch, L. (2009) *Life Cycle Analysis of GHG and Air Pollutant Emissions from Renewable and Conventional Electricity, Heating, and Transport Fuel Options in the EU until 2030* [Online], The Netherlands, European Topic Centre. Available at http://acm.eionet.europa.eu/reports/docs/ETCACC_TP_2009_18_LCA_GHG_AE_2013-2030.pdf (Accessed 29 March 2017).

IEA (2016) *World Energy Outlook 2016*, Paris, International Energy Agency.

IEA/IRENA (2017) *Perspectives for the Energy Transition 2017* [Online], Paris, International Energy Agency/Abu Dhabi, International Renewable Energy Agency. Available at http://www.irena.org/DocumentDownloads/Publications/Perspectives_for_the_Energy_Transition_2017.pdf (Accessed 6 April 2017).

IPCC (2011) *Special report on renewable energy sources and climate change mitigation* [Online], Geneva, Intergovernmental Panel on Climate Change. Available at http://www.ipcc.ch/report/srren/ (Accessed 3 April 2017).

ITM (2015) *Thüga Group's P2G plant exceeds expectations* [Online]. Available at http://www.itm-power.com/news-item/thuga-groups-p2g-plant-exceeds-expectations (Accessed 15 April 2017).

Kombikraftwerk (2007) *Background Paper: the Combined Power Plant* [Online]. Available at http://www.kombikraftwerk.de/fileadmin/downloads/Background_Information_Combined_power_plant.pdf (Accessed 10 April 2017).

Mott MacDonald (2011) *Costs of low-carbon generation technologies* [Online], Brighton, Mott MacDonald. Available at https://www.theccc.org.uk/archive/aws/Renewables%20Review/MML%20final%20report%20for%20CCC%209%20may%202011.pdf (Accessed 4 April 2017).

National Grid (2011) *Seven Year Statement 2011*, National Grid Company, [online]. Available at http://www2.nationalgrid.com/UK/Industry-information/Future-of-Energy/Electricity-ten-year-statement/SYS-Archive/ (Accessed 14 September 2017).

NERA Consulting (2015) *Electricity Generation Costs and Hurdle Rates Lot 1: Hurdle Rates update for Generation Technologies* [Online], London, NERA Economic Consulting. Available at https://www.gov.uk/government/publications/nera-2015-hurdle-rates-update-for-generation-technologies (Accessed 5 April 2017).

Offshore Valuation Group (2010) *The Offshore Valuation: a valuation of the UK's offshore renewable energy resource* [Online], Machynlleth, Wales, Public Interest Research Centre. Available at http://www.mng.org.uk/gh/scenarios/offshore_valuation_full.pdf (Accessed 8 April 2017).

PB Power/IET (2012) *Electricity Transmission Costing Study* [Online], London, Institution of Engineering and Technology. Available at http://www.theiet.org/factfiles/transmission-report.cfm (Accessed 10 April 2017).

REN21 (2017) *Renewables 2017 Global Status Report* [Online], Paris, REN21 Secretariat. Available at http://www.ren21.net/status-of-renewables/global-status-report/ (Accessed 14 August 2017).

SEAI (2007) *Wind farms 2007* [Online], Dublin, Sustainable Energy Authority of Ireland. Available at http://www.seai.ie/Renewables/Wind_Energy/Wind_Maps/Wind_Farms_2007/ (Accessed 15 April 2017).

Starr, F. (2007) '*Flexibility of Fossil Fuel Plant in a Renewable Energy Scenario: Possible Implications for the UK*', in Boyle, G. (ed) *Renewable Electricity and the Grid: the Challenge of Variability*, Earthscan, London, pp. 121–140.

Sterner, M., Jentsch, M., Saint-Drenan, Y-M., Gerhardt, N., von Oehsen, A., Specht,. M., Baumgart, F., Feigl, B., Frick, V., Stürmer, B., Zuberbühler, U. and Waldstein, G. (2010) *Renewable (power to) methane: Storing renewables by linking power and gas grids* [Online]., Available at http://publica.fraunhofer.de/eprints/urn_nbn_de_0011-n-4088886.pdf (Accessed 4 April 2017).

UETechnologies (2016) *200MW / 800MWh Energy Storage Station to be Built with RONGKE POWER's Vanadium Flow Battery, UE Technologies* [Online]. Available at http://www.uetechnologies.com/news/71-200mw-800mwh-energy-storage-station-to-be-built-with-rongke-power-s-vanadium-flow-battery (Accessed 15 April 2017).

UKERC (2006) *The costs and impacts of Intermittency: An assessment of the evidence on the costs and impacts of intermittent generation on the British Electricity Network* [Online], London, UK Energy Research Centre. Available at http://www.ukerc.ac.uk/publications/the-costs-and-impacts-of-intermittency.html (Accessed 17 April 2017).

UKERC (2017) *The costs and impacts of intermittency – 2016 update* [Online], London, UK Energy Research Centre. Available at http://www.ukerc.ac.uk/publications/the-costs-and-impacts-of-intermittency-2016-update.html (Accessed 17 April 2017).

UNEP/Bloomberg New Energy Finance (2017) *Global Trends in Renewable Energy Investment 2017* [Online], Frankfurt, Frankfurt School - United Nations Environment Program Collaborating Centre/ New York, Bloomberg New Energy Finance. Available at http://climateobserver.org/reports/global-trends-renewable-energy-investment-2017/ (Accessed 13 April 2017).

Wenzel, H. (2010) *Breaking the Biomass Bottleneck of the Fossil Free Society* [Online]. Available at https://concito.dk/files/dokumenter/artikler/rapport-breaking_the_biomass_society_sept.2010presse_21_1791740804.pdf (Accessed 1 August 2017).

Further Reading

Elliott, D. (2016) *Balancing green power: how to deal with variable energy sources*, IOP Publishing Ltd, Bristol.

NIC (2016) '*Smart Power*', UK National Infrastructure Commission, London: [Online]. Available at http://www.gov.uk/government/uploads/system/uploads/attachment_data/file/505218/IC_Energy_Report_web.pdf (Accessed 14 September 2017).

Chapter 12

Renewable energy futures

By Stephen Peake and Bob Everett

12.1 Introduction

The world has embarked on a transition in our energy system away from fossil fuels towards renewable energy technologies. This final chapter brings together elements from the previous chapters and gathers some different views on the possible speed and scale of this transition.

Section 12.2 reviews the current (2017) situation for renewable energy and looks at how Denmark, in particular, has promoted renewable energy use since the 1980s by drawing up various *energy scenarios* to assist policy making.

Section 12.3 looks at global energy scenarios to the year 2040 from three different organizations, each with a slightly different outlook.

Section 12.4 reviews some of the future policy promises and, finally, Section 12.5 looks at investment needs between now and 2040.

12.2 The transition to renewable energy

The transition to renewable energy is being driven by a range of pressures:

- the need to provide clean, sustainable energy supplies in developing countries
- concerns about the finite reserves of fossil fuels, particularly oil
- the need to avoid local air pollution, particularly in China and India
- concerns about the safety and economics of nuclear power
- most importantly, the need to cut greenhouse gas emissions to meet climate change targets.

Access to reliable electricity and clean cooking fuels

In 2014, globally, around 1 billion people still did not have access to reliable, affordable mains electricity. An astonishing 3 billion people (over 40% of the world's population) still did not have access to modern cooking fuel (IEA and World Bank, 2017). Things are slowly (relatively) changing, with around 100 million people per year currently gaining access to mains electricity and/or modern cooking fuel.

Concerns about fossil fuel depletion

As mentioned in Chapter 1, at current production rates, world coal reserves (i.e. the amount that is *economic to extract*) might last another 150 years. However, the reserves of oil and gas are only expected to last around 50 years (BP, 2017). Concerns about world oil reserves together with the steep world oil price rises of 1973 and 1979 (see Chapter 1, Figure 1.3) have, from the 1980s onwards, given a strong impetus to develop renewable energy.

Local air pollution

Also described earlier in Chapter 1, energy use in China has increased dramatically since 1980, largely fuelled by coal. In part this has been driven by a policy of providing electricity to everyone. 98% of China's population now has access to mains electricity, however a large proportion of Chinese cities are suffering from serious long-term air pollution as a consequence. It has been estimated that 4000 people die *every day* in China from air pollution (Rohde and Muller, 2015). Most of this pollution consists of particulates and sulphur dioxide emitted by coal-fuelled power plants.

Safety of nuclear power

The safety of nuclear power has been a matter of major concern since the disaster at Fukushima in March 2011, when three adjacent nuclear reactors suffered a meltdown after being hit by an earthquake and being swamped with seawater by a tsunami. The disaster caused radioactive contamination of the surrounding land and the evacuation of 200 000 people. Six years on, in 2017, the clean-up work is still ongoing and it is likely to continue for many decades. Estimates for the total clean-up cost vary, but it could be nearly US$190 billion (Rich, 2017).

Reducing global CO_2 emissions

As described in Chapter 1, following the 21st United Nations 'Conference of the Parties' meeting (COP 21) held in Paris in late 2015, a total of 195 countries committed to curbing their greenhouse gas emissions 'consistent with holding the increase in the global average temperature to well below 2°C above pre-industrial levels and pursuing efforts to limit the temperature increase to 1.5 °C above pre-industrial levels' (UNFCCC, 2015).

Over 170 countries had adopted renewable energy targets and over 145 had policies to support renewable energy. These policies have mainly concentrated on renewable electricity generation. Over 65 countries have policies to support renewable transport fuels, focusing mainly on first generation biofuels for road transport.

To have a reasonable chance of limiting temperature increase to 2 °C (in other words around another 1 °C from the 2016 level) global energy-related CO_2 emissions must decline rapidly to only 50% of the current level by 2035, and approach zero by 2060.

12.3 **Current renewable energy policy situation**

Falling renewable energy prices

The recent rapid global growth in renewable energy production is largely the result of three decades of encouragement following the oil price increases of 1973 and 1979. It has involved *research and development* in new technologies such as wind and PV power, and *policy support* for investment in both new renewable energy technologies and existing well-developed ones, such as hydropower. As pointed out in Chapter 11, volume production and successful large-scale deployment has resulted in rapidly falling prices for large scale wind and solar PV, to the point where they are now competitive with, or even underpricing, conventional fossil-fuelled and nuclear electricity generation. In many parts of the world, the need to provide subsidies for renewable energy is retreating and it is likely that in the coming decades renewable energy will continue to grow robustly on price advantage alone, eventually becoming the dominant source of our energy supply (Martinot, 2014).

However, at present, finance in the global energy system is dominated by fossil fuels. In 2015, total global investment in the energy sector was around US$1.8 trillion (million million). This may sound a large sum but globally it is only about 2.5% of the Gross World Product (or expenditure on everything) or US$250 per person on the planet. 57% of this was accounted for by fossil fuels (see Figure 12.1). However, in recent years, a significant amount of investment has begun to flow into modern renewable energy technologies, mainly solar PV, wind and hydropower. In 2015, for the first time, more than half of global total electricity capacity additions were from renewables – overtaking new investments in fossil fuelled and nuclear powered electricity generation.

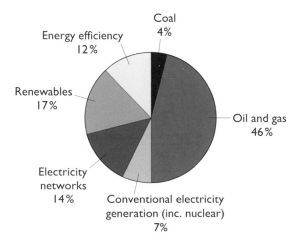

Figure 12.1 Shares of global investment in energy sector, 2015. (Source: adapted from IEA 2017)

Chapter 1 has described China's rapid energy expansion since 1980. This is no longer reliant on coal, particularly given that the high air pollution levels in the country have caused considerable political embarrassment. Since 2013 policies for the expansion of coal consumption have been reversed and, as shown in Figure 1.10, the country's coal consumption seems to *have already* peaked. This fall in part explains the flattening off and slight decline of coal's contribution to global CO_2 emissions (Figure 1.4). China is now a major global driving force in reducing the cost of renewable energy systems and leads the world in hydro, solar PV, wind, geothermal and solar water heating capacity. It is also the world's largest manufacturer of wind turbines, solar PV and solar water heating systems (REN21, 2017).

Meeting CO_2 emission targets – Europe in the lead

In many ways the European Union (EU) has led the way in setting targets for cutting greenhouse gas (and particularly CO_2) emissions. The 2009 European Union Renewable Energy Directive (sometimes known as the 20:20:20 Directive), passed in 2009 (CEC, 2009), set a target for Europe to achieve a 20% reduction in carbon emissions from its 1990 level, combined with a 20% contribution from renewable energy source to *gross final energy consumption* by 2020. As pointed out in Box 1.1 of Chapter 1, the use of this rather than primary energy gives a fairer view of the contribution of electricity generating renewables, such as PV or wind power.

Under the Paris Agreement at COP21, individual countries have made COP 21 'pledges' of **Intended Nationally Determined Contributions (INDCs)**, as described in Chapter 1. Some of these are statements of intended cuts in greenhouse gas emissions, others include specific renewable energy targets. The EU's INDC pledges a reduction in emissions of at least 40% below 1990 levels by 2030.

Can renewable energy be promoted effectively?

Denmark is a good example of a country that started on promoting a renewables energy future back in 1980, and there is clear evidence of how it has progressed since. During the 1960s, Denmark's energy use expanded rapidly in an era of cheap oil (see Figure 12.2 (a)). Its only indigenous energy production was a small amount of lignite mining. By 1972, it had become almost totally reliant on imported oil and was badly hit by the ten-fold oil price rise between 1973 and 1979, shown in Chapter 1, Figure 1.3, which created considerable political embarrassment.

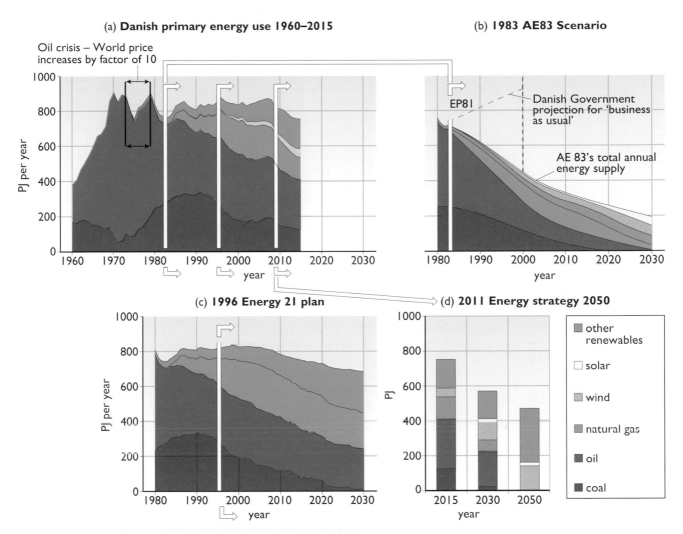

Figure 12.2 (a) Primary energy consumption in Denmark, 1960–2015; (b) Denmark's 1983 Alternative Energy scenario (AE83), and the Danish Government's 1981 'business as usual' projection (EP81); (c) 1996 Energy 21 scenario; (d) Projected energy use in 2030 and 2050 in 2009 IDA Climate Plan 2050 scenario. Notes: In Figure (a) 'renewables' includes wastes and imports. In Figure (c) 'other renewables' includes wind. Figure (d) excludes energy use in the oil and gas industries. (Sources: Norgaard and Meyer, 1989; Eurostat 1993; MEE, 1996; DEA, 2016a; DEA, 2016b; IDA, 2009)

The situation encouraged energy researchers from Danish universities to produce, in 1983, an 'Alternative Energy scenario' (AE83). It outlined a programme aimed at cutting oil and coal imports to zero by 2030, by increasing the proportion of renewables (wind, solar and biofuels) and sharply reducing total energy demand. The plan did not include nuclear power, which was very unpopular in Denmark at the time. The projection, shown in Figure 12.2 (b), contrasted with the official 1981 Danish government 'Business as usual' projection (EP81), which suggested a continued growth in energy demand (see Norgaard and Meyer, 1989).

BOX 12.1 What is an energy scenario?

An energy scenario is a fairly detailed vision of how the pattern of energy use might evolve in the future. It may involve a projection of 'business as usual (BAU)' – what is likely to happen if current policies and energy prices continue. For example, the International Energy Agency, an organization that was set up after the oil price crisis of the late 1970s and has traditionally focused its studies on the availability of oil and natural gas, has for many years produced detailed 'World Energy Outlook' scenarios.

Alternatively, a scenario may be a vision that is useful for shaping actions to be done in the present to achieve a desired outcome, such as cutting global fossil fuel emissions and/or phasing out nuclear power. Environmental organizations such as Greenpeace have commissioned a number of energy scenarios, such as their 'Energy Revolution' scenario (see Section 12.4).

A scenario is certainly *not* a prediction – as described in the main text, past energy scenarios for Denmark have all been wrong in one way or another, but have been useful in shaping the nation's energy policy.

In practice, the Danish government solved its immediate energy security problems in several ways:

- It pursued a policy of energy conservation, imposing high energy taxes and new regulations to encourage the insulation of buildings and promote the use of Combined Heat and Power (CHP). The resulting cuts in energy use for space heating were quite impressive. Between 1972 and 1985, the total area of heated building floor area increased by 30%, but the total amount of energy used to heat it decreased by 30% (Dal and Jensen, 2000).

- It switched its power stations from oil firing to coal firing. Consequently, national reliance on oil dropped from 93% in 1972 to only 43% 20 years later in 1992.

- It developed its own oil and gas resources in the North Sea. The results started to appear in the early 1980s and by 1997 Denmark had become a net energy exporter.

- It built up its renewable resources, particularly biomass and wind power, and has now set up a healthy trade in exporting large wind turbines.

Actual Danish energy consumption up to 2015 has not followed either the AE83 projection or the government EP81 projection, which confirms the point that scenarios are only pictures of what *could* happen rather than what *will* happen.

By the mid-1990s, the focus of energy policy had shifted from self-sufficiency to climate change. Denmark is a low-lying country and any future sea level rises would have serious consequences. The switch from oil to coal and natural gas meant that national CO_2 emissions remained at their 1970s levels. In 1996, the centre-left government adopted a policy of cutting these by 20% from their 1988 levels by 2005, and by 50% by 2030. This was set out in their *Energy 21* plan (MEE, 1996). Under this scenario, total primary energy use would fall only slightly and the use of coal would be replaced by increased use of natural gas and renewable energy (Figure 12.2(c)). It was suggested that, by 2030, half of Denmark's electricity could come from renewable sources. In order to implement this policy, there were continued high energy taxes and support for renewable energy under a Feed-In Tariff (FIT) scheme.

However, in 2001, a new right-wing government was elected that took the attitude that the country's environmental initiative was 'ahead of schedule'. The *Energy 21* plan was dropped, orders for three offshore wind farms cancelled and many tax incentives for energy efficiency were scrapped. Despite this, the existing policies still ensured a slow increase in renewable energy use (see Figure 12.2(a)).

Issues of CO_2 emissions and energy security could not be ignored, though. Danish oil production reached a peak in 2004, as did its natural gas production in 2006. Both have declined significantly since then, although Denmark was still a net gas exporter in 2015.

A 2009 report from the Danish Society of Engineers (IDA) concluded that switching to 100% renewable energy would be feasible (IDA, 2009). In May 2011, the Danish government produced a new 'Energy Strategy 2050' stating a long-term goal of a complete transition away from the use of fossil fuels by that date (Danish Government, 2011). This is essentially the same goal as the 1983 AE83 scenario but with the target date delayed by 20 years. It suggested:

- a continuing reduction in overall energy demand

- a reduction in the use of coal in electricity generation, to be replaced by renewable energy, for example in the Avedøre 2 CHP power plant described in Chapter 5.

- a continued expansion of the use of renewable energy, particularly solid biomass and wind power. There was a target that by 2020 renewables should contribute over 30% of final (i.e. delivered) energy use, i.e. about 200 PJ per year, with a 10% contribution to the transport sector. In the electricity sector it was suggested that 60% should come from renewables, 40% being from wind power.

A comparison of IDA projected 2030 and 2050 energy use with actual use in 2015 is shown in Figure 12.2(d). The IDA scenario is more ambitious in terms of energy efficiency and less reliant on natural gas than the now 'abandoned' *Energy Plan 21*.

Overall, there are several lessons to be drawn from Danish energy policy since the 1970s:

- Primary energy use and GDP do not go hand in hand – between 1972 and 2015, Danish GDP nearly doubled, yet primary energy consumption fell by nearly 15%.

- Reducing total energy demand may be difficult, particularly if the country has its own oil and gas reserves. Despite high taxes and progressive policies, Danish road transport fuel consumption increased by 50% and electricity demand has more than doubled between 1972 and 2015.

- The use of CHP with district heating (DH) can be successfully promoted. By 2013 over 60% of Danish homes were fed by DH and over 70% of all district heating was produced by CHP (DEA, 2015a).

- Renewable energy use can also be promoted effectively. By 2014 it supplied over half of the country's electricity, nearly 40% coming from wind power (DEA, 2015b). The renewable contribution to primary energy in 2015 was nearly 200 PJ, a far higher figure than that suggested in the AE83 scenario (although about 25% is imported as biomass).

- Denmark's high energy taxes and promotion of district heating have resulted in disagreements with the European Commission in Brussels, which has had a policy of free markets and harmonization of energy prices. The Danish government successfully argued that these policies were for the protection of the environment, which is also a prime objective of European policy.

12.4 Scenarios for global energy use

Meeting global climate change targets: three scenarios compared

What will the requirements to limit global temperature rise and the pledges made at COP 21 mean for the future rate of deployment of renewable energy?

There is a whole range of energy technologies that can be used to cut global CO_2 emissions. The main ones are:

- Energy efficiency
- Fuel switching from coal to gas
- Carbon capture and storage (CCS)
- Nuclear power
- Renewable energy

Each of these has their own relative advantages and disadvantages, and need to be assessed in terms of cost, environmental acceptability, resource limitations and possible rates of deployment. The relative balance between these different options can differ markedly between scenarios produced by different organizations.

As we have seen in Chapter 1, Figure 1.4, global CO_2 emissions from fossil fuel burning appear to have levelled out following their rapid rise over the past 30 years. The 'New Policies' scenario from the International Energy Agency (IEA) assumes that the various pledges made by countries at the 2015 COP21 meeting are carried out. Given that most of these only cover the period up to 2025, it also assumes that equivalent new policies will continue after that date. This scenario sees global CO_2 emissions rising only slightly between 2017 and 2050 (see Figure 12.3 (a)).

Two other scenarios address the need to cut CO_2 emissions drastically (Figure 12.3(b)). One has been produced by the International Energy Agency in conjunction with the International Renewable Energy Agency (IRENA), an organization specifically set up to promote renewable energy. This scenario (referred to here as the 'IEA/IRENA 66%' scenario) assumes a future emissions trajectory with an estimated 66% probability of keeping the global mean temperature rise below 2°C. The second is the 2015 Energy [R]evolution scenario from Greenpeace (simply referred to here as 'Greenpeace') with an almost identical future emissions trajectory.

These three projections for global total final energy consumption (i.e. *delivered energy* to the consumer) and the contributions from renewable energy are shown in Figure 12.3 (b).

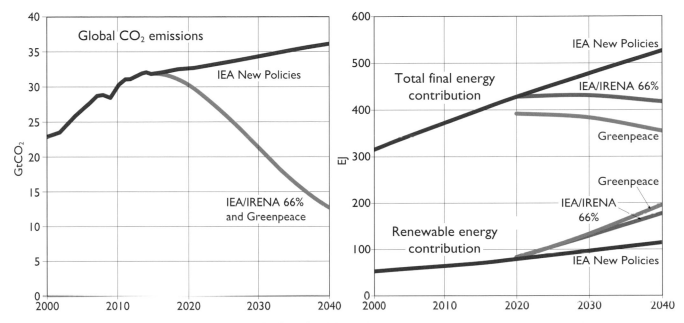

Figure 12.3 Projections of (a) global CO_2 emissions from fossil fuel combustion (b) world total final energy consumption and renewable energy contribution for IEA New Policies scenario, IEA/IRENA 66% 2°C scenario and Greenpeace Energy [R]evolution scenario (sources: IEA, 2016; IRENA, 2017; Greenpeace, 2015)

There are differences in the attitudes to the technologies between scenarios.

The IEA New Policies scenario sees world total final energy consumption continuing to increase in a linear manner with only a slight acceleration in the deployment of renewable energy. However, it does also include expanding contributions from nuclear power and carbon capture and storage.

Compared to the IEA New Policies scenario, the IEA/IRENA scenario sees:

- a higher level of energy efficiency, leading to a flattening in final energy consumption and then a fall towards 2040
- a higher contribution from renewable energy (about 50% more in energy terms by 2040).

The Greenpeace scenario:

- includes an even higher level of energy efficiency, rapidly deployed, resulting in a more rapidly falling total final energy demand
- sees the phase-out of nuclear power
- sees no future for CCS
- has a very high rate of deployment of renewable energy.

The falling levels of final energy consumption in the IEA/IRENA and Greenpeace scenarios do not mean that they are advocating a decrease in actual *energy services*. An energy efficient LED lamp can, after all, produce five times more useful light for the same amount of delivered electricity than a traditional incandescent lamp.

According to the IEA New Policies scenario, in 2014 renewables made up an estimated 18% of global total final energy consumption. The projected shares for 2040 are:

- IEA New Policies – 22%
- IEA/IRENA 66% – 43%
- Greenpeace – 55%.

These high proportions of renewable energy are achieved by a *combination* of energy efficiency reducing demand and expansion of renewable energy supply.

At the simplest level renewable energy use can be broken down into three types:

- those burned to produce *heat* such as traditional biomass and including biofuels for cooking and transport
- other heat producing renewables such as solar water heating and heat pumps
- renewable energy used to produce *electricity*, such as PV and wind, but also including electricity generated from bioenergy.

Hydroelectricity deserves special consideration because of possible environmental objections to its expansion.

The projected breakdown of renewable energy for 2040, as shown in Figure 12.4, reveals different attitudes to these basic types of renewable energy.

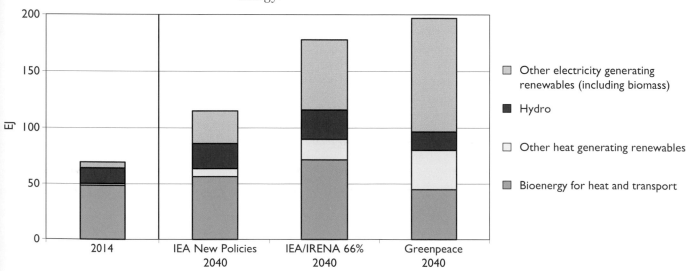

Figure 12.4 Contributions to total global total final energy demand of different types of renewable energy in the three scenarios (note: in the Greenpeace scenario 'other electricity generating renewables' includes hydrogen generated from electricity from renewable sources) (sources: IEA, 2016; IRENA, 2017; Greenpeace, 2015).

The IEA New Policies scenario sees a slow growth in bioenergy and hydro power, a higher growth rate in heat generating renewables, such as solar water heaters, and a relatively high growth rate for other electricity generating renewables, such as wind and PV. The IEA/IRENA scenario has more optimistic growth rates for all of the options. In comparison, the Greenpeace scenario assumes a *low* growth rate in bioenergy and hydro power because of environmental concerns, but compensates with an extremely high growth rate in other electricity generating renewables.

Electricity generation details

The projections for global electricity generation for the IEA/IRENA and Greenpeace scenarios are shown in Figure 12.5 (a) and (b).

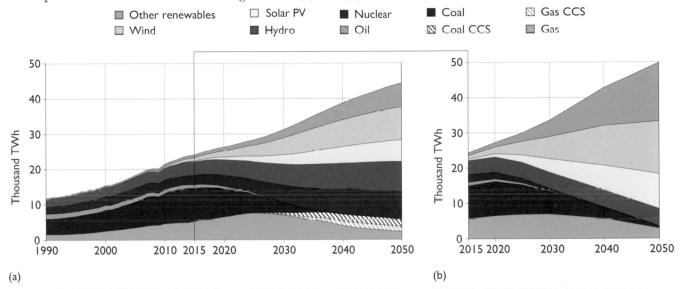

(a)   (b)

Figure 12.5 Global electricity generation by source 1990–2050: (a) IEA/IRENA 66% 2°C scenario and (b) Greenpeace Energy [R]evolution scenario (sources: IRENA, 2017; Greenpeace, 2015)

Both of these scenarios see an enormous expansion in future renewable electricity generation by 2050, particularly wind and PV, and also, in the shorter term, an expansion of gas-fuelled generation, peaking around 2025 and then falling by 2050.

The IEA/IRENA 66% scenario sees an expansion of nuclear power and also the introduction of coal and gas-fuelled carbon capture and storage around 2030.

The Greenpeace scenario has a slightly higher level of electricity generation by 2050. Some of this is used for hydrogen production for transport and other uses. It sees the phase-out of nuclear power by about 2035 and no role for CCS. Also, there is only a small expansion of hydroelectric generation.

Renewable energy growth rates

The recent high growth rates for some technologies have already been described in Chapter 1. Table 12.1 shows how the three scenarios see them continuing into the future.

Table 12.1 Annual growth rates in renewable energy contributions

	2010–2015	2014-2040		
		IEA New policies	**IRENA**	**Greenpeace**
Solar PV	42.0%	9%	13%	15%
CSP	9.7%	12%	>20%[a]	>20%
Wind	17.0%	6%	10%	11%
Geothermal	3.7%	5%	>10%[a]	14%
Hydro	2.9%	2%	2%	1%
Bioenergy[b]	2.9%[c]	1%	2%	1%

Notes: [a]estimated from figures for capacity additions for 2030 and 2050 [b]primary energy [c]growth rate 2009-2014.
(Sources: IEA, 2011; IEA, 2016; IEA/IRENA, 2017; Greenpeace, 2015)

However, it is not a matter of expanding all forms of renewable energy equally. The expansion of hydropower is limited by environmental concerns and biomass use for energy is limited by concerns over competition for land use with food production.

The key elements of the IEA/IRENA 66% and Greenpeace scenarios are summarized in Table 12.2 below.

Table 12.2 Key elements of the IEA/IRENA 66% and Greenpeace scenarios to 2040

Technology	IEA/IRENA 66%	Greenpeace
Nuclear power	yes	no
Carbon Capture and Storage	yes	no
Renewables proportion of world final energy demand in 2040	43%	55%
Expanding wind, PV, CSP and geothermal	high growth rates	high growth rates
Expanding biomass	2% annual growth rate	1% annual growth rate concentrated on generating electricity
Expanding hydro	2% annual growth rate	1% annual growth rate
Hydrogen economy	negligible	yes
Energy efficiency	moderate	very high
Increased gas used by 2025	yes	yes

Should either the IEA/IRENA 66% or the Greenpeace scenario turn out to be even remotely true, we are at the turning point of a historic energy transition. Very shortly – by 2025 – renewables could overtake coal in supplying the largest share in the world's primary energy supply (not seen since 1890), and by 2030 could overtake oil (not seen since 1948).

12.5 **Future investment needs and programmes of action**

Expanding the various renewable energy technologies will need increased investment and specific programmes of action (many already in progress). It is first worth reflecting on the likely competition from fossil fuels and future energy policies in different countries and regions of the world.

Competition from fossil fuels

Coal – given its high CO_2 emission factor (see Chapter 2) this is a fuel that needs to be phased out as rapidly as possible. Indeed it seems likely that world demand for coal has now (2017) peaked. The national policy pledges given at COP21 have been backed up by programmes of 'disvestment' (removing investment) in coal mining by large pension funds and national sovereign wealth funds. The IEA/IRENA 66% and Greenpeace scenarios see a rapid fall in global coal use, particularly for electricity generation. By 2040 its use is projected to be well under a half of its current level. At current extraction rates the world still has a 150 year supply of economically extractable coal left, and this may mean that what remains might be on the future world market at low prices. The scenarios suggest that 80% of the world's coal reserves will still be in the ground in 2050. There have been warnings to investors that money left invested in what is likely to become a future unwanted commodity will become a so-called stranded asset.

Oil – this is the world's prime transport fuel. As shown in Chapter 1 Figure 1.3, the world has experienced wildly varying oil prices, from US$20 up to almost US$120 and back to US$50 a barrel all since 1998. Worries about future prices have, in part, encouraged the switch to electricity as a future transport fuel. If the IEA/IRENA 66% or Greenpeace scenarios see the future correctly, then world consumption could fall to 60% of its current level by 2040 and 50% of the world's reserves will still be in the ground in 2050. This may mean that the remaining oil will be on the world market at low prices. This would again leave stranded assets, which could pose financial problems for those with oil fields, particularly in the Middle East.

Natural gas – all of the scenarios described above include fuel switching from coal to natural gas in electricity generation and see an increase in global gas demand between now and 2030. Given that the world reserves might only last about 50 years at current production rates, natural gas prices may well rise. This is perhaps good news for those countries with large gas reserves.

Possible regional energy policies

The outlook for some of the key nations and world regions can be summarized as follows:

- **China** – It is likely to concentrate on reducing its serious local air pollution problems while pursuing its export objectives as a major manufacturer of renewable energy technologies.

- **India** – Also has problems of severe air pollution. In 2016 almost 20% of its population did not have access to mains electricity and providing modern energy services is a key policy goal. It is also dependent on domestic coal supplies of relatively poor quality. This gives a particular impetus to developing its large solar and wind resources.

- **Middle East** – Some countries have begun to openly accept that their oil revenues will not last forever. The remaining profits from oil sales (which may not be that large if prices stay low) will need to be invested in alternative energy sources. Gas-rich states (e.g. Qatar) are at an advantage though, again, revenues from gas sales will need to be ploughed back into solar energy.

- **Africa** – In North Africa there is considerable potential for solar energy projects. Tunisia, for example, is looking to build large-scale concentrating solar projects to export electricity to southern Europe and there is a similar potential in many other North African countries. Central Africa also has an enormous untapped solar and hydro resource potential.

- **Europe** - It is likely to press on with its renewable energy targets, particularly if climate change impacts such as droughts start to seriously affect southern regions, such as Spain and Italy.

Investment needs

As shown earlier in Figure 12.1, a half of the US$1.8 trillion global energy investment in 2015 went into fossil fuels. About a third went on searching for and developing new oil and gas fields. This process is not going to cease, but it is an area where future annual energy investments are likely to *fall*. How much investment is required to limit global warming to 2°C? There is a wide range of answers from various official governmental sources and non-governmental sources, and the IEA/IRENA 66% and Greenpeace scenarios are part of this overall context (Peake and Ekins, 2016).

The IEA/IRENA 66% scenario sees overall investments in *energy supply*, including electricity generation, only increasing slowly from its 2015 level of about US$1.6 trillion through to 2050. However the proportion devoted to fossil fuels falls by a factor of four over this period and that devoted to renewables increases by a factor of about three. There is also a modest expenditure on other low-carbon technologies such as nuclear power and carbon capture and storage. The most dramatic change is the investment in *energy efficiency*, which is projected to rise from its 2015 level of about US$250 billion up to over US$2.5 trillion per year over the period 2041–2050. This includes investment in technologies such as electric vehicles.

The Greenpeace scenario does not support nuclear power or carbon capture and storage and sees a large increase in expenditure on renewable energy rising to US$1.8 trillion per year over the period 2041-2050, i.e. a sixfold increase on its 2015 level.

Programmes of action

In 2011, the United Nations (UN) Secretary-General Ban Ki-moon launched the *Sustainable Energy For All (SE4All)* initiative. This is seen by the UN as a critical part of solving some of the world's key global health, climate change and energy security issues. By 2030 this initiative aims to:

- ensure universal access to modern energy services
- double the historic global rate of improvement in energy efficiency
- increase the share of renewable energy in global final energy consumption to 36% (IEA and World Bank, 2017; IRENA, 2017).

The '100% Renewable Energy Movement'

There is a growing movement of cities, localities and municipalities around the world that have committed to what has become the '100% Renewable Energy Movement'. Examples of countries that already get almost 100% of their electricity from renewable energy include Norway in Europe and Paraguay in South America. Other localities with a near 100% contribution include the Orkney Islands in Scotland and the province of Quebec in Canada. Looking forward to the period 2025–2035, it is likely that dozens of cities around the world are likely to achieve 100% renewable electricity supply, including Jeju (Republic of Korea), Malmö (Sweden), Munich, Osnabruck (Germany), San Francisco, Rochester and San Diego (USA).

Looking further ahead to 2050, a number of cities have declared their intention to obtaining 100% of their *total* energy supply from renewable sources (not just electricity). These include Copenhagen in Denmark, Frankfurt and Hamburg in Germany, Fukushima in Japan, Sydney in Australia, and Vancouver in Canada (REN21, 2016).

A balancing act

What happens next in our energy systems is a balancing act performed at two main levels. Firstly, there is the overall balance of investment (and market sentiment) between fossil fuel systems, renewable technologies and other alternatives. Secondly, markets must then decide where and how within the 'alternatives to fossil fuels' market renewable energy investments should be directed. Renewable energy is politically perceived to compete with (a) nuclear power and (b) the application of carbon capture and storage technologies as a way of making fossil fuels acceptable. In reality these may prove to be both fairly weak contenders.

12.6 Summary

The world is now fully engaged in a transition away from fossil fuels towards renewable energy technologies. The **momentum** for change comes from a variety of pressures and drivers including human development needs, energy security concerns, air pollution and climate change. On the one hand, many people in the world still lack access to clean energy. **In 2014, globally, around 1 billion people did not have access to reliable, affordable mains electricity and 3 billion people did not have access to modern cooking fuel**.

Limiting the temperature increase to 2 °C will require that global energy-related CO_2 emissions must decline rapidly to only 50% of the current level by 2035, and approach zero by 2060. Global energy finance remains dominated by fossil fuels. **In 2015, total global investment in the energy sector was around USD$ 1.8 trillion, 57% of which was accounted for by fossil fuels.**

By the end of 2015, over 170 countries had adopted renewable energy targets and over 145 had policies to support renewable energy. These policies have mainly concentrated on renewable electricity generation. Over 65 countries have policies to support renewable transport fuels, focusing mainly on first generation biofuels for road transport.

The transition towards renewables is also happening at a more local level. **A growing movement of cities, localities and municipalities around the world have committed to the '100% Renewable Energy Movement'.**

China's coal consumption seems to have already peaked. China is both the world's largest consumer and manufacturer of many renewable energy technologies. In terms of installed capacity, it leads the world in hydro, solar PV, wind, geothermal and solar water heating capacity. It is also the world's largest manufacturer of wind turbines, solar PV and solar water heating systems.

In Europe, **Denmark is a good example of a country that started on a renewables future back in 1980**. Between 1972 and 2015, Danish GDP nearly doubled, yet primary energy consumption fell by nearly 15%. Despite high taxes and progressive policies, Danish road transport fuel consumption increased by 50% and electricity demand has more than doubled over the same period.

An **energy scenario is a projection of possible future energy use**. It is not a prediction and may be used as a tool to guide future policies. **There is a limited number of options that can be used to cut global CO_2 emissions. The main ones are:**

- **Energy efficiency**
- **Fuel switching from coal to gas**
- **Carbon capture and storage**
- **Nuclear power**
- **Renewable energy.**

Future energy scenarios differ in their enthusiasm for different combinations of these technologies.

Energy scenarios may include:

- A Business As Usual scenario – a view of the future as simply a continuation of past trends
- New Policies scenario – including policy promises and targets made by individual nation states.
- A scenario to meet a particular policy target. The IEA/IRENA 66% and Greenpeace Energy [R]evolution scenario both aim to cut CO_2 emissions drastically.

In 2014, renewables made up an estimated 18% of global total final energy consumption. By 2040 the share is likely to rise within the range 22–55% according to a range of scenarios:

■ **22% – IEA New Policies**

■ **43% – IEA/IRENA 66% Scenario**

■ **55% – Greenpeace Energy [R]evolution Scenario.**

The high proportions of renewable energy in the last two scenarios are achieved by a *combination* of energy efficiency reducing demand and an expansion of renewable energy supply.

In 2011, the United Nations launched the Sustainable Energy For All (SE4All) initiative, which aims to increase the share of renewable energy in global final energy consumption to 36%.

Renewables have an enormous potential to contribute to electricity, transport and building energy use. Many electricity generating renewables may have continuing growth rates of more than 10% per annum over the coming decades.

References

BP (2017) *BP Statistical Review of World Energy 2017* [Online], London, The British Petroleum Company. Available at https://www.bp.com/content/dam/bp/en/corporate/pdf/energy-economics/statistical-review-2017/bp-statistical-review-of-world-energy-2017-full-report.pdf (Accessed 17 August 2017).

CEC (2009) *Directive 2009/28/EC on the promotion of energy from renewable sources* [Online], Commission of the European Communities. Available at http://eur-lex.europa.eu (Accessed 14 August 2017).

Dal, P. and Jensen, H. S. (2000) *Energy Efficiency in Denmark*, Copenhagen, Danish Energy Ministry.

Danish Government (2011) *Energy Strategy 2050* [Online], Copenhagen, Danish Ministry of Climate and Energy. Available at http://dfcgreenfellows.net/Documents/EnergyStrategy2050_Summary.pdf (Accessed 23 August 2017).

DEA (2016a) *Monthly energy statistics: data tables* [Online], Copenhagen, Danish Energy Agency. Available at https://ens.dk/en/our-services/statistics-data-key-figures-and-energy-maps/annual-and-monthly-statistics (Accessed 23 August 2017).

DEA (2015b) *Energy statistics 2014* [Online], Copenhagen, Danish Energy Agency. Available at https://ens.dk/sites/ens.dk/files/Statistik/energystatistics2014.pdf (Accessed 4 August 2017).

DEA (2016a) *Monthly energy statistics: data tables* [Online], Copenhagen, Danish Energy Agency. Available at http://www.ens.dk (Accessed 25 March 2017).

DEA (2016b) *Key Figures from DEA's Preliminary Energy Statistics 2015* [Online], Copenhagen, Danish Energy Agency. Available at https://ens.dk/sites/ens.dk/files/energistyrelsen/Nyheder/2015/hovedtal_fra_energistyrelsens_forelobige_energistatistik_for_2015.pdf (Accessed 25 March 2015).

Eurostat (1993) *Energy Statistical Yearbook 1992*, Luxembourg, The Statistical Office of the European Communities.

Greenpeace (2015), *Energy [R]evolution: a sustainable world energy outlook 2015* [Online], Amsterdam, Greenpeace International. Available at http://www.greenpeace.org/international/Global/international/publications/climate/2015/Energy-Revolution-2015-Full.pdf (Accessed 9 August 2017).

IDA (2009) *The IDA Climate Plan 2050* [Online], Copenhagen, Danish Society of Engineers. Available at http://www.energyplan.eu/report-idas-climate-plan-2050-background-report-from-august-2009/ (Accessed 11 August 2017).

IEA (2011) *World Energy Outlook 2011* [Online], Paris, International Energy Agency. Available at https://www.iea.org/publications/freepublications/publication/WEO2011_WEB.pdf (Accessed 14 August 2017).

IEA (2016) *World Energy Outlook 2016*, Paris, International Energy Agency.

IEA (2017) *World Energy Investment 2016 Corregendum* [Online], Paris, IEA. Available at http://www.oecd.org/about/publishing/Corrigendum-WEI2016.pdf (Accessed 14 August 2017).

IEA and the World Bank (2017) *Sustainable Energy for All 2017: Progress toward Sustainable Energy* Paris, International Energy Agency / Washington, DC, World Bank. Available at http://www.se4all.org/sites/default/files/GTF%20Executive%20Summary%202017.pdf (Accessed 4 August 2017).

IRENA (2017), *REthinking Energy 2017: Accelerating the global energy Transformation* [Online], Abu Dhabi, International Renewable Energy Agency. Available at http://www.irena.org/DocumentDownloads/Publications/IRENA_REthinking_Energy_2017.pdf (Accessed 21 August 2017).

Martinot, E. (2014) *Reflections on Renewable Energy Past, Present and Future* [Online]. Available at http://www.martinot.info/renewables2050/2014/351 (Accessed 4 August 2017).

MEE (1996) *Energy 21*, Copenhagen, Danish Ministry of Environment and Energy.

Norgaard, J. S. and Meyer, N. I. (1989) 'Planning implications of electricity conservation: the case of Denmark' in Johansson, T. B., Bodlund, B. and Williams, R. H. (eds) *Electricity – Efficient End Use*, Lund, Lund University Press.

Peake, S. and Ekins, P. (2016) 'Exploring the financial and investment implications of the Paris Agreement', *Climate Policy*, pp. 1–16 [Online]. DOI: https://doi.org/10.1080/14693062.2016.1258633.

REN21 (2017) *Renewables 2017 Global Status Report* [Online]. Available at http://www.ren21.net/status-of-renewables/global-status-report/ (Accessed 14 August 2017).

Rich, M. (2017) 'Struggling With Japan's Nuclear Waste, Six Years After Disaster', *New York Times*, 11 March [Online]. Available at https://mobile.nytimes.com/2017/03/11/world/asia/struggling-with-japans-nuclear-waste-six-years-after-disaster.html (Accessed 14 August 2017).

Rohde, R. A. and Muller, R. A. (2015) 'Air Pollution in China: Mapping of Concentrations and Sources' *PLoS ONE*, vol. 10, no. 8 [Online]. DOI: 10.1371/journal.pone.0135749 (Accessed 9 August 2017).

UNFCCC (2015) *Historic Paris Agreement on Climate Change*, United Nations Framework Convention on Climate Change, [online] available at http://newsroom.unfccc.int/unfccc-newsroom/finale-cop21/ (accessed 14 August 2017)

Appendix A

Energy arithmetic – a quick reference

As explained in Chapter 1, this book has been written using standard scientific SI units; however many other units are widely in use. This appendix aims to provide a quick reference to ways of expressing very large and very small numbers, units and conversions between units. Section A1 describes the methods of specifying quantities in terms of powers of ten. Section A2 gives conversion factors between the main units used for energy and power throughout this book; it includes conversions to and from some commonly used US units. Section A3 relates some older or non-SI units to their SI equivalents. Formal definitions and more detailed accounts of the basis of the SI units can be found in *Quantities, Units and Symbols* (The Royal Society, 1975).

A1 Orders of magnitude

Discussions of energy consumption and production frequently involve very *large* numbers, and accounts of processes at the atomic level often need very *small* numbers. Two solutions to the problem of manipulating such numbers are described here: the use of a shorthand form of arithmetic and the use of prefixes.

Powers of ten

Two million is two times a million, and a million is ten times ten times ten times ten times ten times ten (that's six tens in all, being multiplied together). This can be written mathematically as:

$$2\ 000\ 000 = 2 \times 10 \times 10 \times 10 \times 10 \times 10 \times 10 = 2 \times 10^6.$$

The quantity 10^6 is called *ten to the power six* (or *ten to the six* for short), and the 6 is known as the **exponent**. The advantage of using this power-of-ten form is particularly obvious for very large numbers. World primary energy consumption in the year 2009, for instance, was 502 000 000 000 000 000 000 joules, and it is certainly easier to write (or say) 502×10^{18} joules than to spell out all the zeros.

The method can also be used for very small numbers, with the convention that one tenth (0.1) becomes 10^{-1}; one hundredth (0.01) becomes 10^{-2}, etc. The separation of two atoms in a typical metal, for instance, might be about 0.25 of a billionth of a metre, which is 0.000 000 000 25 m. In more compact form, it becomes 0.25×10^{-9} m.

Scientific notation is a way of writing numbers that is based on this concept, but with a more specific rule. Any number, whatever its magnitude, is written in scientific notation as a number between 1 and 10 multiplied by the appropriate power of ten. So in this form, the numbers above would be written slightly differently:

502×10^{18} becomes 5.02×10^{20}

0.25×10^{-9} becomes 2.5×10^{-10}.

In practice, styles like those on the left in each of these examples are commonly used, and a number in this form will be accepted by computers and scientific calculators – but it will be reproduced by them in either 'long decimal' or strict scientific notation, depending on the mode selected.

Prefixes

The powers of ten provide the basis for the prefixes used to indicate multiples (including sub-multiples) of units. The following table, parts of which appeared in Chapter 1 as Table 1.1, shows these in decreasing order.

Table A1 Prefixes

The sequence of the first six rows in this table may be conveniently remembered (in reverse order) with the phrase 'King Midas's Golden Touch Poisoned Everything'.

Symbol	Prefix	Multiply by	... which is
E	exa-	10^{18}	one quintillion
P	peta-	10^{15}	one quadrillion
T	tera-	10^{12}	one trillion
G	giga-	10^{9}	one billion
M	mega-	10^{6}	one million
k	kilo-	10^{3}	one thousand
h	hecto-	10^{2}	one hundred
da	deca-	10	ten
d	deci-	10^{-1}	one tenth
c	centi-	10^{-2}	one hundredth
m	milli-	10^{-3}	one thousandth
μ	micro-	10^{-6}	one millionth
n	nano-	10^{-9}	one billionth
p	pico-	10^{-12}	one trillionth

The terms 1 billion = 10^9 and 1 trillion = 10^{12} are as used in this book. However, alternative long forms where 1 billion = 10^{12} and 1 trillion = 10^{18} may be found in older English books and in other languages.

In US usage the prefix M can be used to denote 1000 and (more commonly) MM to denote 1 million. In other circumstances US data may use M to denote 'metric', i.e. 1 Mt = 1 metric tonne. In this book we use the convention shown in Table A1.

Indian usage commonly includes the terms 1 lakh = 100 000 and 1 crore = 10 million.

A2 Units and conversions

Chapter 1 discussed some of the units used in specifying quantities of energy or power in the 'real world'. The tables in this section summarize the relationships between some of these units.

Energy

Commonly used energy units include the joule, the kilowatt-hour, the tonne of oil equivalent (toe) and tonne of coal equivalent (tce), i.e. the energy content of a tonne of an 'average' sample of these fuels. Tables A2 and A3 give the conversion factors between the most frequently used multiples of these units, on the 'household' scale (A2) and on the larger scale of national or world data (A3). The oil and coal equivalents are here expressed to two significant figures only. Since different data sources may use slightly different values and there may be confusions between lower and higher heating values (see Chapter 4, Box 4.5), you are advised always to take note of the precise conversion factors used for the energy content of fuels adopted by any source you use. Note that the factors for kWh and TWh assume a 100% conversion efficiency.

Table A2 Energy conversions at the household scale

	MJ	GJ	kWh	toe	tce
1 MJ =	1	0.001	0.2778	2.4×10^{-5}	3.6×10^{-5}
1 GJ =	1000	1	277.8	0.024	0.036
1 kWh =	3.60	0.0036	1	8.6×10^{-5}	1.3×10^{-4}
1 toe =	42 000	42	11 667	1	1.5
1 tce =	28 000	28	7778	0.67	1

Table A3 Energy conversions at the national scale

	PJ	EJ	TWh	Mtoe	Mtce
1 PJ =	1	0.001	0.2778	0.024	0.036
1 EJ =	1000	1	277.8	24	36
1 TWh =	3.60	0.0036	1	0.086	0.13
1 Mtoe =	42	0.042	11.667	1	1.5
1 Mtce =	28	0.028	7.778	0.67	1

Power

The kilowatt-hour is the standard unit of *energy* used for UK gas and electricity bills. *Power* is the rate at which energy is used, consumed, transferred or transformed. Its normal units are the watt and its multiple the kilowatt. A power of one kilowatt is equal to a rate of use of energy of one kilowatt-hour per hour. A power of one watt is equal to a rate of use of one joule per second.

Table A4 shows the quantities of energy per hour and per year for different constant rates in watts. Note that it can also be used to show that, for instance, 1 kWh is 3.6 MJ and 1 TWy (terawatt-year) is 31.54 EJ or 750 Mtoe.

Table A4 Power and rate of use of energy

Rate	Joules		Kilowatt-hours per year	Tonnes of oil equivalent per year	Tonnes of coal equivalent per year
	per hour	per year			
1 W	3.6 kJ	31.54 MJ	8.76	0.75×10^{-3} toe*	1.13×10^{-3} tce*
1 kW	3.6 MJ	31.54 GJ	8760	0.75 toe	1.13 tce
1 MW	3.6 GJ	31.54 TJ	8.76×10^6	750 toe	1130 tce
1 GW	3.6 TJ	31.54 PJ	8.76×10^9	0.75 Mtoe	1.13 Mtce
1 TW	3.6 PJ	31.54 EJ	8.76×10^{12}	750 Mtoe	1130 Mtce

* i.e. the energy equivalent of 0.75 kg of oil or 1.13 kg of coal

US energy units

Energy statistics in the USA are quoted in British Thermal Units (BTU). Until the 1990s UK statistics also used the BTU and a related unit, the therm.

Table A5 gives conversion factors at the 'household scale'.

Table A5 Energy conversions at the household scale (US energy units)

	MJ	Thousand BTU	kWh	Therm	toe
1 MJ =	1	0.948	0.2778	9.48×10^{-3}	2.4×10^{-5}
1000 BTU =	1.055	1	0.293	0.01	2.5×10^{-5}
1 kWh =	3.60	3.412	1	0.034	8.6×10^{-5}
1 therm =	105.5	100	29.3	1	2.5×10^{-3}
1 toe =	42 000	40 000	11 667	400	1

The power unit of the BTU h^{-1} is commonly used, even in the UK, for the ratings of domestic boilers and air conditioning plant; 1000 BTU h^{-1} = 0.293 kW.

Table A6 gives energy conversion factors at the 'national' level.

Table A6 Energy conversions at the national scale (US energy units)

	EJ	Quadrillion BTU (Quad)	TWh	Billion therms	Mtoe
1 EJ =	1	0.948	277.8	9.48	24
1 Quad =	1.055	1	293.1	10	25
1 TWh =	3.60×10^{-3}	3.41×10^{-3}	1	0.0341	0.086
1 billion therms =	0.1055	0.1	29.3	1	2.5
1 Mtoe =	0.042	0.040	11.667	0.40	1

A3 Other quantities

Table A7 gives the SI equivalents of a few other metric units and some older units that remain in common use. The final column shows the inverse relationships. For brevity, scientific notation is used for numbers greater than 10 000 or less than 0.1.

Table A7 SI equivalents

Quantity	Unit	SI equivalent	Inverse
Mass	1 oz (ounce)	$= 2.835 \times 10^{-2}$ kg	1 kg = 35.27 oz
	1 lb (pound)	$= 0.4536$ kg	1 kg = 2.205 lb
	1 ton (= 2240 lb)	$= 1016$ kg	1 kg = 0.9842×10^{-3} ton
	1 short ton (= 2000 lb)	$= 907$ kg	1 kg = 1.102×10^{-3} short tons
	1 t (tonne)	$= 1000$ kg	1 kg = 10^{-3} t
	1 u (unified mass unit)	$= 1.660 \times 10^{-27}$ kg	1 kg = 6.024×10^{26} u
Length	1 in (inch)	$= 2.540 \times 10^{-2}$ m	1 m = 39.37 in
	1 ft (foot)	$= 0.3048$ m	1 m = 3.281 ft
	1 yd (yard)	$= 0.9144$ m	1 m = 1.094 yd
	1 mi (mile)	$= 1.609 \times 10^{3}$ m	1 m = 6.214×10^{-4} mi
Speed	1 km h^{-1} (kph)	$= 0.2778$ m s^{-1}	1 m s^{-1} = 3.600 kph
	1 mi h^{-1} (mph)	$= 0.4470$ m s^{-1}	1 m s^{-1} = 2.237 mph
	1 knot	$= 0.514$ m s^{-1}	1 m s^{-1} = 1.944 knots
Area	1 in^2	$= 6.452 \times 10^{-4}$ m^2	1 m^2 = 1550 in^2
	1 ft^2	$= 9.290 \times 10^{-2}$ m^2	1 m^2 = 10.76 ft^2
	1 yd^2	$= 0.8361$ m^2	1 m^2 = 1.196 yd^2
	1 acre	$= 4047$ m^2	1 m^2 = 2.471×10^{-4} acre
	1 ha (hectare)	$= 10^{4}$ m^2	1 m^2 = 10^{-4} ha
	1 mi^2	$= 2.590 \times 10^{6}$ m^2	1 m^2 = 3.861×10^{-7} mi^2
Volume	1 in^3	$= 1.639 \times 10^{-5}$ m^3	1 m^3 = 6.102×10^{4} in^3
	1 ft^3	$= 2.832 \times 10^{-2}$ m^3	1 m^3 = 35.31 ft^3
	1 yd^3	$= 0.7646$ m^3	1 m^3 = 1.308 yd^3
	1 litre	$= 10^{-3}$ m^3	1 m^3 = 1000 litres
	1 gallon (UK)	$= 4.546$ litres	1 m^3 = 220.0 gallons (UK)
	1 gallon (US)	$= 3.785$ litres	1 m^3 = 264.2 gallons (US)
	1 barrel	$= 159$ litres	1 m^3 = 6.3 barrels
	1 acre-ft	$= 1.233 \times 10^{3}$ m^3	1 m^3 = 0.811×10^{-3} acre-ft
	1 bushel	$= 3.637 \times 10^{-2}$ m^3	1 m^3 = 27.50 bushels

(Continued)

Table A7 SI equivalents   *(Continued)*

Quantity	Unit	SI equivalent	Inverse
Force	1 kgf (weight of 1 kg mass)	= 9.807 N	1 N = 0.102 kgf
	1 lbf (weight of 1 lb mass)	= 4.448 N	1 N = 0.2248 lbf
Pressure	1 bar ($\approx$1 atmosphere)	= 10^5 Pa (pascals)	1 Pa = 10^{-5} bar
	1 kgf m^{-2}	= 9.807 Pa	1 Pa = 0.102 kgf m^{-2}
	1 lbf in^{-2} (or psi)	= 6895 Pa	1 Pa = 1.450×10^{-4} psi
Energy	1 barrel of oil equivalent (boe)	= 5.7 GJ (lower heating value)	1 GJ = 0.175 boe
	1 cal (calorie)	= 4.2 J	1 J = 0.24 cal
	1 ft lb (foot pound)	= 1.356 J	1 J = 0.7375 ft lb
	1 eV (electron-volt)	= 1.602×10^{-19} J	1 J = 6.242×10^{18} eV
	1 MeV	= 1.602×10^{-13} J	1 J = 6.242×10^{12} MeV
Power	1 HP (horse power)	= 745.7 W	1 W = 1.341×10^{-3} HP

Reference

The Royal Society (1975) *Quantities, Units and Symbols*, London, The Royal Society.

Appendix B

Levelized costs of renewable energy

B1 Introduction

This book describes a wide range of renewable energy technologies; some, such as solar water heaters, produce only heat, while others, such as biomass combined heat and power (CHP), produce both heat and electricity. The technologies for which costs are most carefully studied are those electricity-generating ones that are in direct competition with fossil or nuclear-fuelled generating plants. This appendix aims to provide some background on how to approach the costing of electricity from such plants and explains the key terms used, although more details for each technology may be described in the individual chapters.

An ideal power plant would be available to produce electricity at any hour of the day or night throughout the year. There are, however, various constraints on many renewable energy technologies which limit the total amount of electricity that they might produce over a year.

The cost of electricity from any generating plant, whether it is fossil-fuelled, nuclear or renewably-powered has to take into account four main elements:

- fuel costs
- operation and maintenance (O&M) costs, including waste disposal
- initial capital costs
- final decommissioning costs.

In practice, for many renewable energy technologies the fuel costs are zero but there are large initial capital costs which, in some cases, can make up 75% or more of the final electricity cost. Nuclear power plants are in a similar position with low fuel costs but high capital costs as well as significant decommissioning costs. At the other extreme, generating plants fuelled by natural gas have low capital costs and high fuel costs, with the cost of the fuel making up over a half of the final electricity cost.

One accepted method of making comparisons between such different types of plant is to calculate an overall **levelized cost** for the electricity in pence per kilowatt hour over the life of the plant. This value for the electricity spreads ('levels') the costs of fuel, O&M, capital and interest repayments across a given time period, usually the life expectancy of the plant. Since inflation progressively erodes the purchasing power of money over time, such an estimate necessarily has to be expressed in money of a particular year, for example £(2015).

Levelized costs allow us to say that, for example, the cost of electricity from a land-based wind turbine is 6.1p per kWh, while that from a wood-fuelled generation plant is 8.6p per kWh (even though the plants may operate for different numbers of hours per year and have different factors in their cost breakdowns).

The account below is very brief – greater detail on costings for different UK electricity generation technologies will be found in the book and reports listed in the references.

B2 Basic contributions to costings
Factors affecting the amount of electricity generated

Rated capacity

This is the maximum power output of a plant; it is usually expressed in kilowatts (kW) or megawatts (MW). A combined heat and power (CHP) plant may have *two* ratings, one for electricity (MW_e) and one for heat (MW_t). The power output of a photovoltaic panel is usually expressed in terms of its peak power output under full bright sunlight (kWp).

Annual capacity factor

This is the ratio of the total electricity generated by a plant to the maximum that it might have produced if it operated continuously 24 hours a day, 365 days a year. There are many factors that can limit the annual capacity factor including:

(1) Plant availability – the proportion of the year that the plant is in full working order and not shut down for maintenance or repairs. Many mature renewable energy technologies have availabilities of 95% or more.

(2) Energy source availability – obviously the wind does not blow all the time, nor does the sun shine all the time. Technologies such as solar, wind, wave and tidal power are all quite tightly limited by the physical availability of their energy source. There are also considerations of matching electricity supply with demand (as described in Chapter 11). Some technologies, particularly large hydro plants, may be reserved for dealing with evening peaks in electricity demand rather than continuous base load generation. Other technologies such as biomass CHP plants may mainly operate in the winter when there is both a heat and an electricity demand.

Since the demand can have a significant effect on the total annual output, the capacity factor is sometimes also referred to as the *load factor*.

Typical capacity factors can range from about 10% for PV panels in cloudy countries like the UK through to 25%–45% for wind turbines and up to almost 100% for landfill gas.

The overall electricity produced per year can thus be expressed as:

electricity per year (kWh) = rating (kW) × capacity factor × 365 × 24 (hrs)

Factors affecting the total cost

Fuel costs

Although for many renewable energy technologies the 'fuel' is free, biomass technologies, in common with those using fossil fuels, have significant fuel costs. There are two 'costs' here:

- the cost of the purchased fuel
- the contribution of the fuel to the overall cost of the electricity generated.

The purchased fuel cost is likely to be expressed in £ per GJ (£ GJ^{-1}). For example, wood purchased at £50 per tonne with a calorific value of 15 GJ per tonne has a cost of £3.33 GJ^{-1} or 1.2 p kWh^{-1}. Other fuels, such as municipal solid waste (MSW) and landfill gas may be considered to have a *zero* cost, since their use is part of an overall disposal process.

The key factor in determining the contribution of this input energy cost to the final electricity cost is the *plant efficiency:*

$$\text{efficiency} = \frac{\text{energy output}}{\text{energy input}}.$$

Efficiency values may range from 25% for MSW plant up to 35% or more for modern large wood-fired steam plant.

The contribution to the final electricity cost is likely to be expressed in pence per kWh or £ per MWh (where 1 p kWh^{-1} = £10 MWh^{-1}).

$$\text{fuel cost per kWh generated} = \frac{\text{cost of energy input}}{\text{efficiency}}$$

Even though many renewable energy technologies have zero fuel costs, it should not be thought that the conversion efficiency is unimportant for them. A 1 kWpk PV panel with an efficiency of 20% will only require half the area of one with an efficiency of 10%.

Operation and maintenance (O&M) costs

These can be subdivided into two categories: 'fixed' costs which are not related to the plant output, and 'variable' costs, which are likely to be proportional to the plant output.

Fixed costs include site insurance, grid connection fees, regular safety inspections, etc. These may be expressed in terms of a certain percentage of the initial capital cost per year. Obviously access to the plant is a key issue, which is why O&M costs for offshore wind turbines are considerably higher than those for onshore ones.

Variable costs include, for example, fuel handling and ash disposal and the replacement of equipment damaged by 'wear and tear'. Variable O&M costs are usually quoted in terms of pence per kWh of output. For simplicity, the overall O&M costs may also be expressed in this way as well.

The actual figures vary widely between technologies, from under 1p kWh^{-1} for large PV projects up to 7 p kWh^{-1} for energy from waste plants.

Capital costs and plant life

These are often the most important factors in renewable energy plants. The capital cost is usually expressed in £ per kW of rated capacity. A large onshore wind turbine may have a capital cost of £1200 per kW of generation capacity. The cost for an offshore one may be £2200 per kW or more. This is in contrast to about £500 per kW for a competing fossil-fuelled combined cycle gas turbine (CCGT) power station.

Given these high capital costs, the expected plant life is very important. Wind turbines and PV panels may have life expectancies of over 20 years. A hydro plant or a tidal barrage might be expected to last for 60 years or more. A further factor is the construction time. A large land-based wind turbine can be erected and be operational within a year, but a large hydro plant (or a nuclear power station) may take a number of years to construct. This is time during which money is tied up without any incoming cash flow from generated electricity.

B3 Calculating costs

This section starts with a simple payback time calculation before moving on to examples of increasing complexity using levelized costing.

The case study used for these calculations is of a 1 MW (1000 kW) land-based wind turbine generator – it has a capital cost of £1200 per kW, its estimated operating and maintenance costs are £37 500 per year. It has a design life of 25 years and a capacity factor of 28%. Its construction time is very short, and there are, of course, zero fuel costs.

Payback time

For the simplest of calculations we can take the capital cost, the energy output and a competing energy price to calculate a 'payback time', the number of years taken to recover the capital outlay. This simple calculation obviously ignores O&M payments and any interest payments on the capital.

So, if we assume a competing bulk (fossil fuel) electricity price of 6p kWh^{-1}:

> Annual electricity generation = rated capacity × capacity factor × 24 × 365
> > = 1 × 28% × 24 × 365 = 2453 MWh
> Total capital cost = 1000 × £1200 = £1 200 000
> Value of annual electricity generation = 2453 × 1000 × 6p/100 = £147 200
> Payback time = capital cost/value of annual output
> > = £1 200 000/£147 200
> > = **8.2 years**

Of course during this time a further $8.2 \times £37\,500 = £307\,500$ was incurred in O&M costs which, if included, would push the payback time up further.

Simple annual levelized cost

Alternatively, a 'levelized cost' can be calculated on an annual basis. The capital can be considered to be repaid in equal annual amounts over the project lifetime (such a stream of identical payments is known as an *annuity*). Again, for simplicity the example here assumes 0% interest on the capital, so although the capital is being repaid there are no extra interest payments. The average O&M costs and fuel costs (if any) are added to these annual payments, and the average cost of the electricity is then given by:

$$\text{cost per kwh} = \frac{(\text{annual capital repayment} + \text{average annual running cost})}{\text{average annual energy output}}$$

So the calculation becomes:

Cost of capital spread over 25 years = £1 200 000/25 = £48 000 y^{-1}

Operating and maintenance costs   = £37 500 y^{-1}

Total annual costs   = £85 500 y^{-1}

Overall electricity cost = £85 500 / 2453 MWh   = £34.9 MWh^{-1}

                                                    = **3.49 p kWh^{-1}**.

Using discounted cash flow

In practice, these calculations are too simplistic because a pound earned or spent tomorrow is not worth the same as a pound today, for a number of reasons.

Firstly, there is our **time preference for money**. Given a choice, most people would rather have a pound today than a pound in the future. Put another way, we would need to be offered a pound plus some additional sum, say $x\%$, next year to forgo the use of one pound today.

Secondly, we have the ability to lend out money and charge interest. We can forgo the use of a pound today in order to have a pound plus an additional sum in the future. If the interest rate is high enough and the time long enough, this sum can be appreciable. For example, if we invested £100 at a 10% rate of interest per annum, we would expect to be able to withdraw £260 in 10 years' time. We can say that £100 has a **future value** of £260. Put another way, the **present value** of £260 in ten years' time at an interest rate of 10% is only £100.

We are 'discounting' future payments, saying that sums of money in the future can be expressed in terms of smaller sums today.

We can express this relationship mathematically as:

$$V_p = \frac{V_n}{(1 + r)^n}$$

where V_p is the present value of a sum of money V_n in n years time subject to a discount rate of r.

This concept leads to a technique of economic appraisal known as **discounted cash flow** (DCF) analysis. This can be carried out using standard computer spreadsheet functions.

Thirdly, **inflation** is another factor that needs to be included. This can be thought of as a 'disease of money', progressively eroding its value over time. In order to adjust for its effects, we should use a 'real' interest rate, rather than the purely monetary one. This is simple enough:

real interest rate = monetary interest rate − rate of inflation

Generally, the terms 'discount rate' and 'real interest rate' tend to be used interchangeably.

Now if a capital sum, say V_p, is to be repaid as an annuity (annual amounts of equal size running over the duration of a loan of n years) the present value of an annuity, A, given a discount rate r is:

$$V_p = \frac{A}{(1+r)} + \frac{A}{(1+r)^2} + \frac{A}{(1+r)^3} + ... + \frac{A}{(1+r)^n}.$$

A little bit of mathematical rearrangement gives a more convenient form, in which the annuity is expressed in terms of the capital sum, discount rate and length of loan:

$$A = V_p \times r / (1 - (1+r)^{-n}$$

The ratio A/V_p is sometimes called a **capital recovery factor**.

The mathematical function for an annuity is now a standard spreadsheet function: PMT (Payment, Month, Term). However, it is not actually necessary to use a spreadsheet for very simple calculations. Before the era of computers and spreadsheets, banks used simple pre-calculated tables to work out repayments on borrowed money. This approach is still useful today. Table B1 shows the annual repayments on £1000 of borrowed capital.

Table B1 Annuitized value of capital costs (annual repayment in £ per £1000 of capital) for various discount rates and capital repayment periods

Capital repayment period / years[1]	Discount rate / %						
	0	2	5	8	10	12	15
5	200	212	231	250	264	277	298
10	100	111	130	149	163	177	199
15	67	78	96	117	131	147	171
20	50	61	80	102	117	134	160
25	40	51	71	94	110	127	155
30	33	45	65	89	106	124	152
40	25	37	58	84	102	121	151
50	20	32	55	82	101	120	150
60	17	29	53	81	100	120	150

[1] This is not necessarily equal to the total physical lifetime of the project.

Generally, a loan is agreed to be paid back over a certain number of years. However, financial institutions are often unwilling to consider loans spread over more than 25 years because of the essential uncertainty about the future, although the full working life of a hydroelectric plant may be in excess of a hundred years. Thus it is worth remembering that the capital repayment time may be shorter than the working lifetime of a scheme.

Levelized cost of electricity from a wind turbine including discounting

Using Table B1, the previous example can be revisited. Assuming a discount rate of 8% and a project lifetime of 25 years the calculation is as follows.

First, find the annuitized value of the capital cost. From Table B1 the annual repayments on £1000 over 25 years at 8% per year = £94 per year.

For a borrowed sum of £1 200 000 they will be £1200 × £94 = £112 800.

Note that this is considerably higher than the annual repayment figure of £48 000 used in the simple annual levelized cost example. (That figure can be derived using the annuitized value of £1000 over 25 years at 0% per year = £40 per year. £1200 × £40 = £48 000.)

The annual costs now are:

Cost of capital spread over 25 years	= £112 800 y^{-1}
Operating and maintenance costs	= £37 500 y^{-1}
Total annual costs	= £150 300 y^{-1}
Overall cost per kWh = £150 300/2453 MWh = £61.3 MWh^{-1}	
	= 6.13 p kWh^{-1}

Thus, using a discount rate of 8% and a project lifetime of 25 years, we find that the original estimate of 3.49 p kWh^{-1} is an underestimate and arrive at a much higher figure of 6.13 p kWh^{-1}.

The annuitized capital costs now make up over 75% of the total.

Levelized cost of electricity from a wood-fuelled generation plant including discounting

A useful comparison is to derive the levelized electricity cost for a generation method where the calculation includes a fuel cost and the plant efficiency.

A proposed wood-fuelled 100 MW generation plant using a fluidized bed boiler and a steam turbine has a capital cost of £2500 per kW. It is expected to run for 7500 hours per year at full power (i.e. it has a capacity factor of 7500/8760 = 85.6%) and has a design lifetime of 25 years. Its overall electrical generation efficiency is 36%. Operating and maintenance (O&M) costs have been estimated at 1.6p per kWh of electricity generated. It is assumed that wood will be available at £3.33 per GJ or 1.2 p per kWh over the lifetime of the plant.

If the station unit runs for 7500 hours per year, in each year it will produce: 7500 h × 100 MW = 750 000 MWh of electricity

The total capital cost is: £2500 × 100 000 = £250 million

The plant is considered a slightly more risky investment than the wind turbine and finance is only available at a discount rate of 10%, rather than 8%.

First, the annuitized value of the capital cost is needed. From Table B1 the annuitized value of £1000 over 25 years at 10% per year = £110 per year.

Therefore, for a borrowed sum of £250 million the annual repayments will be: $250\ 000 \times £110 = £27.5$ million

Annuitized capital cost per kWh of electricity produced is:

$27.5 \times 1\ 000\ 000 \times 100 / 750\ 000\ 000 = 3.67\text{p kWh}^{-1}$

The fuel cost per kWh of electricity generated is:

1.2 p kWh^{-1}/ 36%	= 3.33 p kWh^{-1}
O&M costs	= 1.60 p kWh^{-1}
Total cost per kWh	**= 8.60 p kWh^{-1}**

Note that in this example the fuel costs account for nearly 40% of the final electricity cost.

More complex calculations

The use of a simple annuitization table is fine if the construction time is very short, and the O&M payments and output are uniform over the life of the project. For a more accurate analysis, particularly for projects such as large-scale hydro and tidal schemes, which may take many years to build, and be subject to periodic refurbishment, a full 'Net Present Value' calculation is required.

This process involves five steps:

(1) Itemizing the capital and running costs for each year of the project life.

(2) Calculating the separate Present Values of all these annual costs using an appropriate discount rate, and summing them to give a Net Present Value (NPV – this is a standard spreadsheet function).

(3) Itemizing the electrical output for each year over the project lifetime.

(4) Calculating the value of the electricity produced. The levelized cost calculation assumes an inflation corrected cost (i.e. it is the same value, P pence per kilowatt hour, for the whole life of the plant). The monetary Net Present Value of the electricity produced will then be $P \times \text{NPV}$ (kilowatt-hours produced)

(5) Calculating the unit cost of electricity in pence per kWh as:

$$P = \frac{\text{Net Present Value of costs (pence)}}{\text{Net Present Value of output (kWh)}}.$$

The references listed below contain calculations of this type.

Choice of discount rate

Many renewable energy technologies involve high capital costs, a characteristic also shared by nuclear power. As discussed in Chapter 11, Section 11.5, their economics are critically dependent on the cost of capital and choice of discount rate.

Past UK evaluations of the economics of renewable energy projects have used discount rates of 8% and even 15%, these being taken as the expectations of earnings in the private sector. As described in Chapter 10, Section 10.5, in practice the 'riskiness' of the technology has to be taken into account. Mature technologies such as PV, hydro and land-based wind may be funded with lower discount rates than newer ones such as wave or tidal stream power. Recent low bid prices for large renewable energy schemes are likely to combine low capital costs, resulting from bulk manufacturing, and low investment discount rates, a result of confidence in proven performance and deployment.

References

BEIS (2016) *Electricity generation costs* [Online], London, Department for Business, Energy and Industrial Strategy. Available at https://www.gov.uk/government/publications/beis-electricity-generation-costs-november-2016 (Accessed 28 September 2017).

Everett, B., Boyle, G. A., Peake, S. and Ramage, J. (eds) (2012) *Energy Systems and Sustainability: Power for a Sustainable Future* (2nd edn), Oxford, Oxford University Press/Milton Keynes, The Open University.

Mott MacDonald (2010) *UK Electricity Generation Costs Update* [Online], Mott MacDonald. Available at https://www.gov.uk/government/publications/uk-electricity-generation-costs-mott-macdonald-update-2010 (Accessed 28 September 2017).

Mott MacDonald (2011) *Costs of low-carbon generation technologies*, [Online], Mott MacDonald, produced for the Committee on Climate Change. Available at https://www.theccc.org.uk/archive/aws/Renewables%20Review/MML%20final%20report%20for%20CCC%209%20may%202011.pdf (Accessed 28 September 2018).

Acknowledgements

Grateful acknowledgement is made to the following sources:

Cover Image: Cameron Beccario, earth.nullschool.net

Chapter 2

Figures

Figure 2.2: Adapted from Ramage, J. (1997) Figure 5.6, Energy: A Guidebook, Second edition, Oxford University Press, © Janet Ramage 1983, 1997; Figure 2.6b: Energy Saving Trust; Figure 2.6c: © National Energy Foundation 2011; Figure 2.6d: © Robin Curtis/GeoScience Ltd.

Chapter 3

Figures

Figure 3.1: Arcon Solvarme A/S, Denmark; Figures 3.4, 3.6, 3.31, 3.32 and 3.38: Courtesy of Bob Everett; Figure 3.21: Sunmark A/S, Denmark; Figure 3.22: Courtesy of M. G. Davies; Figure 3.24: Courtesy of Derek Taylor; Figures 3.26 and 3.27: Siviour, J. B. (1977) Houses as Passive Solar Collectors, ECRM/M10710. Electricity Council Research Centre; Figures 3.28 and 3.29: Courtesy of Luwoge; Figure 3.34: Department of Energy 1987; Figure 3.35: Hockerton Housing Project; Figure 3.36: Mazra, E. (1979) *The Passive Solar Energy Book*, Rodale Press Inc; Figure 3.37: Guildhall Library; Figure 3.40: Courtesy of Biblioteca Ciudades para un futuro màs sostenible, Madrid; Figure 3.42: Koza (1983), used under a Creative Commons Attribution Licence; Figures 3.43 and 3.44: Hank Morgan/Science Photo Library; Figure 3.45: David Nunuk/Science Photo Library; Figure 3.46: Adapted from Renewables Global Future Report, www.ren21.net; Figure 3.48: www.sbp.de (Schlaich Bergermann and Partner); Figure 3.49: Courtesy of GlassPoint Solar, Inc.; Figure 3.50: Courtesy of Stephen Peake.

Chapter 4

Figures

Figure 4.1: Bibliothèque Nationale de France; Figure 4.3: Popperfoto/Getty Images; Figures 4.4 and 4.5: NASA; Figure 4.6: Adapted from BP Statistical Review of World Energy 2016; Figure 4.7: Solar PV in Africa: Costs and Markets, International Renewable Energy Agency, Copyright © IRENA 2016; Figure 4.10: Adapted from Chalmers, B. (1976) 'The Photovoltaic Generation of Electricity', Scientific American, Vol. 235, no. 4, pp. 34-43, Scientific American; Figure 4.11: Adapted from Green, M. (1982) Solar Cells, Prentice-Hall; Figure 4.12: Adapted from Chalmers, B. (1976) 'The Photovoltaic Generation of Electricity', Scientific American, Vol. 235, no. 4, pp. 34-43, Scientific American; Figure 4.14: Photo credit: Heraeus; Figure 4.17: Spectrolab; Figure 4.18: Hideki Kimura, Kouhei Sagawa via Wikimedia. This file is licenced under the Creative Commons Attribution 3.0 licence

Chapter 5
Figures

Chapter 6

Figures

Figure 6.2: Adapted from Hill, G. (1984) Tinnel and Dam: the story of the Galloway Hydros, South of Scotland Electricity Board; Figure 6.3: © 2000 Scottish Power UK plc; Figure 6.4: Courtesy of www.dee-ken-fishing. org; Figure 6.8b: © David Paterson/Alamy; Figure 6.14: Calvin Larsen/ Science Photo Library; Figure 6.18: Courtesy of Gilkes, www.gilkes.com; Figures 6.19 and 6.23: Dr Bob Everett; Figure 6.22a Adapted from www. freeflowhydro.co.uk; Figure 6.22b The National Trust Photolibrary / Alamy Stock Photo; Figure 6.27: © Thomson Reuters Corporation; Figure 6.28: Trygve Bolstad/Panos; Figure 6.29: © Worldwide Picture Library/Alamy.

Chapter 7

Figures

Figure 7.1: Media Library, EDF/Gerard Halary; Figures 7.3, 7.34, 7.35 and 7.42: Peter Fraenkel, Marine Current Turbines Limited, www. marineturbine.com; Figures 7.7, 7.10, 7.11, 7.12, 7.14, 7.17 and 7.20: Tidal Power from the Severn Estuary, Volume 1, Energy Paper 46 1981, Dept of Energy. Crown copyright material is reproduced under Class Licence number C01W0000065 with permission of the Controller of HMSO and the Queen's Printer for Scotland; Figure 7.8: United Kingdom Hydrographic Office (1992) Admiralty Tidal Stream Atlas: The English Channel NP250, © British Crown Copyright 2011. All rights reserved; Figure 7.9: Twidell, J. W. and Weir, A. J. (1986) Renewable Energy Sources, E. & F. N. Spon, Routledge; Figure 7.13: Report of the legislative Assembly of Western Australia, presented by Hon. Ian Thompson, November 1991, Department of Housing and Works, Australia; Figures 7.15 and 7.16: Reproduced courtesy of Mr L. Carson and Power Magazine, March 1978; Figure 7.18: © ETSU Department of Trade and Industry; Figures 7.19 and 7.21: The Severn Barrage Project: General Report, Energy paper 57, 1989, Department of Energy. Crown copyright material is reproduced under Class Licence Number C01W0000065 with the permission of the Controller of HMSO and the Queen's Printer for Scotland; Figure 7.22: Laughton, M. A. (1990) Renewable Energy Sources, Report 22, E. & F. N. Spon, Routledge; Figure 7.23: Frontier Economics 'Analysis of a Severn Barrage', Report for the NGO steering group, 2008, © Frontier Economics Ltd, London; Figure 7.25: Tidal Electric Inc, www.tidalelectric.com; Figure 7.27: DECC (2010) 'Severn Tidal Power Feasibility Study: Conclusions and Summary Report' Department of Energy and Climate Change; Figure 7.28 © Sustainable Marine Energy Ltd; Figure 7.31 Data from Black & Veatch for The Carbon Trust. Graph by Ocean Flow Ltd; Figure 7.32: SeaGen Uncovered, Technical Insight on MCTs, Exisiting and Next Generation Devices by Frankael, P. MCT 4th International Tidal Energy Summit London Nov 2010, Marine Current Turbines Ltd; Figure 7.36: Based on an image supplied by the Engineering Business. The Patented Stingray tidal stream is being developed by the Engineering Business, who will provide project information on their website updated at regular intervals. The first Stingray generator was installed in Shetland summer 2002 with 75% DTI funding. The Stingray demonstrator is about 20m high and weighs about 40 tonnes with a nominal 150kW output in a 4 knot current, www.engb.com; Figure 7.37: Pulse Tidal Limited; Figure 7.38: Rotech Engineering; Figure 7.39: Neptune Renewable Energy Limited; Figure 7.40: Open Hydro; Figure 7.41: www.tidalstream. co.uk; Figure 7.43: The Severn Tidal Fence Group (IT Power Ltd.); Figure 7.44: Hammerfest Strøm; Figure 7.45: Atlantis Resources Corporation and the Government of Novia Scotia; Figure 7.46: By courtesy of Voith Hydro; Figure 7.47: Verdant Power; Figure 7.48: Ocean Renewable Power Company; Figure 7.52a: Renewable Sources of Electricity in the SWEB area: Future Prospects, SWEB and the Department of Trade and Industry; Figure 7.52b: Bodlund, B. et al. (1989) 'The challenge of choices: technology option for the Swedish electricity sector', Johansson, T. B., Bodlund, B. and Williams, R. H. eds. (1989) Electricity – Efficient end use and new technologies and their planning implications, Lund University Press.

Chapter 8
Figures

Figure 8.1 © Marlec; Figure 8.2: Burroughs, W. J. et al. (1996) Weather – The Ultimate Guide to the Elements, Harper Collins; Figure 8.4: Farndon, J. (1992) How the Earth Works, Dorling Kindersley Ltd; Figure 8.8: Needham, J. with the collaboration of Ling, W. (1965) Science and Civilisation in China, Vol 4, Part II. Cambridge University Press; Figure 8.10: Derek Taylor/Altechnica; Figure 8.11: John Mead/Science Photo Library; Figures 8.12, 8.16: Courtesy of Bob Everett; Figure 8.13: Courtesy of Vestas Wind A/S; Figure 8.14: Stewart Boyle; Figures 8.15a, 8.15b, 8.38: Derek Taylor/Altechnica; Figure 8.17 © Derek Taylor, Altechnica; Figure 8.32: Time for Action: Wind Energy in Europe (1991); European Wind Energy Association; Figure 8.34: SkyscraperPage.com; Figure 8.36: Source unknown; Figure 8.37: Westmill Farm Co-operative, www.westmill.coop; Figure 8.39: Adapted from GWEC (2016) 'Global Wind Energy Outlook 2016', p. 19, Global Wind Energy Council; Figure 8.40a: © Vestas Wind Systems A/S; Figure 8.40b: Courtesy of MHI Vestas Offshore Wind A/S; Figure 8.40c: Siemens AG; Figure 8.40d: Courtesy of MHI; Figure 8.41 Courtesy of DNV GL; Figure 8.42a: Untrakdrover Covered under Creative Commons Licence https://creativecommons.org/licenses/by-sa/3.0/; Figure 8.42c: (c) DCNS; Figure 8.42d: Mitsubishi Heavy Industries; Figure 8.42f: DBD Systems LLC; Figure 8.42g: © PelaStar; Figure 8.42h: © Ideol; Figure 8.42i: Kyoto University; Figure 8.42j: By Jplourde umaine. Covered under Creative Commons Licence https://creativecommons.org/licenses/by-sa/4.0/; Figure 8.43: Illustration/Statoil; Figure 8.44: Covered under Creative Commons Licence https://creativecommons.org/licenses/by-sa/3.0/; Figure 8.45a Courtesy of London Array Limited; Figure 8.45b Courtesy of London Array Limited; Figure 8.45c Courtesy of London Array Limited; Figure 8.46: Atlas of UK Marine Renewable Energy Resources, Crown copyright material is reproduced under Class Licence Number C01W0000065 with permission of the Controller of HMSO and the Queen's Printer for Scotland.

Chapter 9

Figures

Figure 9.1b: Courtesy of Les Duckers; Figures 9.4 and 9.32b: Martin Bond/Science Photo Library; Figure 9.8: © Ocean Energy Systems; Figure 9.9a Adapted from Claeson, L. (1987) 'Energy fran havets vagor', Energiforkningsnamnden nr 21; Figure 9.13: Department for Business, Enterprise and Regulatory Reform (2008) Atlas of Marine Renewable Energy Resources, Crown copyright material is reproduced under Class Licence Number C01W0000065 with the permission of the Controller, Office of Public Sector Information (OPSI); Figure 9.15: Adapted from Falnes, J and Lovseth, J (1991) Energy policy, vol 19, no 8, October 1991, Butterworth Heinemann; Figure 9.19: © Plymouth COAST Lab; Figure 9.23: © Ente Vasco de la Energía (Basque Energy Board); Figure 9.27: © AW-Energy Ltd; Figure 9.29: OceanEnergy Ltd; Figure 9.30: Oceanlinx Limited; Figure 9.31: Adapted from: Lagstrom, G. (2000) Sea Power International - Floating Wave Power Vessel (FWPV), Fourth European Wave Energy Conference, Aalborg, Denmark, pp. 211-218; Figure 9.33: © James Abell; Figure 9.34: Hong, Y. et al (2014) Figure 5, 'Review n electrical control strategies for wave energy converting systems', Rewablc and Sustainable Energy Reviews, Vol. 31, pp. 329-342, Elsevier. Reproduced by permission; Figure 9.35: © Mona Strande/Teknisk Ukeblad; Figure 9.37: Pelamis Wave Power; Figure 9.38: © Bombora Wavepower; Figure 9.39: © Bombora Wavepower; Figure 9.40: Courtesy of Floating Power Plant A/S.

Chapter 10

Figures

Figure 10.2: © Hemis/Alamy Stock Photo; Figure 10.5: Reprinted from Earth Science Reviews, Vol 19, No 1, Henley and Ellis, 'Conceptual model of a typical geothermal system' © 1983 Elsevier Science; Figure 10.10 and 10.11: G.E.I.E. "Exploitation Minière de la Chaleur"; Figure 10.12: Geothermal Wellhead Power Plants, Olkaria/ Kenya - picture source: Lydur Skulason, Green Energy Geothermal; Figure 10.13: © TONY KARUMBA/AFP/Getty Images; Figure 10.15: REUTERS/Stefan Wermuth; Figure 10.18: IPR/31-12 British Geological Survey © NERC, All Rights Reserved.

Chapter 11

Figures

Figure 11.1: Courtesy of Transco; Figure 11.2: National Grid; Figure 11.4: Courtesy of Logstor Ror; Figure 11.5 Source: Data UK Energy Brief; Figure 11.9: Committee on Climate Change (2008) Building a low-carbon economy, Crown copyright material is reproduced under Class Licence Number C01W0000065 with the permission of the Controller, Office of Public Sector Information (OPSI); Figure 11.10 Adapted from Franfurt School FS-UNEP Collaborating Centre (2017) 'Global Trends in Renewable Energy Investment 2017', Frankfurt School-UNEP Centre/BEF; Figures 11.12 and 11.15: Bob Everett; Figure 11.13: National Gird (2010) National Grid Seven Year Statement, National Grid; Figure 11.17: Committee on Climate Change;

Figure 11.18: Adapted from 'California Energy Commission - Tracking Progress', ResourceFlexibility, http://www.energy.ca.gov/renewables/tracking_progress/documents/resource_flexibility.pdf. Source: Meredith, The Duck has Landed, Energy at Haas, U.C. Berkeley, May 2, 2016. California ISO Hourly Data, March 28-April 3, Years 2013-2016; Figure 11.19: London Array; Figure 11.20: The Offshore Valuation Group.

Every effort has been made to contact copyright holders. If any have been inadvertently overlooked the publishers will be pleased to make the necessary arrangements at the first opportunity.

Index

Page numbers in *italics* refer to entries in tables or figures.